AF685245

State-of-the-Art Functional Biomaterials in China

State-of-the-Art Functional Biomaterials in China

Editors

Huiliang Cao
Yuqin Qiao

MDPI

Basel • Beijing • Wuhan • Barcelona • Belgrade • Novi Sad • Cluj • Manchester

Editors
Huiliang Cao
East China University of Science and Technology
Shanghai
China

Yuqin Qiao
Shanghai University
Shanghai
China

Editorial Office
MDPI AG
Grosspeteranlage 5
4052 Basel, Switzerland

This is a reprint of articles from the Special Issue published online in the open access journal *Journal of Functional Biomaterials* (ISSN 2079-4983) (available at: https://www.mdpi.com/journal/jfb/special_issues/func_biomat_china).

For citation purposes, cite each article independently as indicated on the article page online and as indicated below:

Lastname, A.A.; Lastname, B.B. Article Title. *Journal Name* **Year**, *Volume Number*, Page Range.

ISBN 978-3-7258-2013-9 (Hbk)
ISBN 978-3-7258-2014-6 (PDF)
doi.org/10.3390/books978-3-7258-2014-6

© 2024 by the authors. Articles in this book are Open Access and distributed under the Creative Commons Attribution (CC BY) license. The book as a whole is distributed by MDPI under the terms and conditions of the Creative Commons Attribution-NonCommercial-NoDerivs (CC BY-NC-ND) license.

Contents

About the Editors

Huiliang Cao

Huiliang Cao, a Humboldtian, obtained his Ph.D. from the South China University of Technology. He is a full professor at East China University of Science and Technology leading the interfacial Electrochemistry and Biomaterials group. His research interests include interfacial electrochemistry and advanced materials, energy storage materials, and medical devices.

Yuqin Qiao

Yuqin Qiao, obtained her PhD degree in Biochemistry from City University of Hong Kong, and now is a full professor in school of life sciences at Shanghai University. Her current research mainly focuses on biofunctionalization of materials for bone regeneration, smart responsive materials, surface/interface engineering, and medical implants.

Preface

The development of functional biomaterials, such as osteoinductive materials, antibacterial coatings, immunomodulatory implants, etc., has taken a central position in recent decades, leading to a boom in the Chinese biomaterial community in terms of the research, development, and commercialization of various biomaterials and medical devices. The Special Issue on "State-of-the-Art Functional Biomaterials in China" is the *Journal of Functional Biomaterials'* timely attempt to inspect these essential advances and future directions.

By collecting 13 research and review articles, this Special Issue covers strategies to treat device-associated infections (DAIs), multifunctional bioactive materials for tissue engineering, and fabrication techniques. DAI is a common complication that boosts global antimicrobial resistance, threatening public healthcare systems and, thus, urging the community to design strategies and synthesize instructive materials for their prevention and treatment. Moreover, DAI pathogenesis is a site-specific biological process having an unpredictable onset and possibly involving multiple resistant pathogenic strains. As such, it has inspired studies on the roles of various cell lines and protein adsorptions, such as macrophages, neutrophils, T cells, tissue cells, fibrinogen, etc. Moreover, antibacterial activity is desired in various biomedical fields, including orthopedic surgery, oral implantology, cardiovascular medicine, etc., raising concerns about its synergistic problem with medical devices' specific functions, yielding innovations in multifunctional bioactive materials. Since mammalian and bacterial cells share many adhesion and metabolism mechanisms, antimicrobial materials are likely toxic to tissue integration and undermine the therapeutic efficacy of medical devices. Therefore, biomedical devices' integration of antibacterial activity is facing challenges in clinical translations. These hybrids and other multifunctional materials orchestrate multiple mammalian cell lines that rely on finely controlled material structures, properties, and progress in fabrication techniques. Surface modification techniques, such as ion implantation, plasma electrolyte oxidation, and soft lithography, are well recognized as efficient pathways to endow conventional biomaterials with bioactivity and multifunctionality. In addition to surface modification, nanotechnologies yielding nanoparticles or metal–organic frameworks via hydrothermal, ultrasonic-assisted, electrochemical, and mechanochemical methods are also being explored in order to develop biomaterials for disinfection, immunomodulation, and drug delivery. In summary, the "State-of-the-Art Functional Biomaterials in China" Special Issue collects diverse contributions from leading experts in their respective fields associated with functional biomaterials. It bridges basic research–clinical translation collaborations and provides unique and essential insights to develop better functional biomaterials with promising clinical translations. This reprint is a valuable resource for academic researchers, industry actors, regulators, and clinicians associated with functional biomaterials' innovation and translations.

We greatly appreciate all the contributors who devoted their valuable time and effort to writing these excellent articles and making this Special Issue and reprint a reality.

Huiliang Cao and Yuqin Qiao
Editors

Journal of
*Functional
Biomaterials*

Editorial

State-of-the-Art Functional Biomaterials in China

Yuqin Qiao [1,*] and Huiliang Cao [2,3,4,*]

1 School of Life Sciences, Shanghai University, Shanghai 200444, China
2 Interfacial Electrochemistry and Biomaterials, School of Materials Science and Engineering, East China University of Science and Technology, Shanghai 200237, China
3 Engineering Research Center for Biomedical Materials of Ministry of Education, East China University of Science and Technology, Shanghai 200237, China
4 Key Laboratory for Ultrafine Materials of Ministry of Education, East China University of Science and Technology, Shanghai 200237, China
* Correspondence: qiaoyq@shu.edu.cn (Y.Q.); huiliang.cao@ecust.edu.cn (H.C.)

1. Introduction

In recent years, rapid advancements in multidisciplinary fields (materials, biology, chemical physics, etc.) as well as emerging techniques in biomedical sciences and materials processing have led to a substantial evolution of biomaterials. Distinct from traditional biomedical materials, which are designed to adapt to their microenvironments for tissue integration, functional biomaterials are expected to dynamically and actively instruct cell responses to promote tissue regeneration and tackle biomedical engineering challenges [1]. In China, functional biomaterial-based therapeutics has been identified as one of the main strategies to meet public health problems nationally. Thus, besides pursuing the innovation of functional biomaterials, accelerating their journey from the bench to clinical practice is a top priority. Academic researchers, clinical professionals, and industry partners have been encouraged to collaborate to achieve this goal and improve clinical outcomes.

This Special Issue, "The State-of-the-Art Functional Biomaterials in China", presents 13 research and review articles collecting the latest research findings and developments. The contributed manuscripts mainly focus on providing solutions to the challenges and limitations faced by current implantable biomedical devices, which can be divided into three categories: strategies to treat implant-associated infections, multifunctional bioactive materials for tissue engineering, and fabrication techniques.

Citation: Qiao, Y.; Cao, H. State-of-the-Art Functional Biomaterials in China. *J. Funct. Biomater.* **2024**, *15*, 23. https://doi.org/10.3390/jfb15010023

Received: 14 December 2023
Accepted: 11 January 2024
Published: 15 January 2024

Copyright: © 2024 by the authors. Licensee MDPI, Basel, Switzerland. This article is an open access article distributed under the terms and conditions of the Creative Commons Attribution (CC BY) license (https://creativecommons.org/licenses/by/4.0/).

2. Overview of Published Articles

Bacterial infection is a common complication associated with various biomaterial applications and has conventionally been treated by the administration of antibiotics [2]. However, the overuse and misuse of antibiotics have led to increasing antibiotic resistance, which poses a serious public health threat around the world [3]. Thus, particular attention is being given to alternative antibacterial agents or biomaterials-based delivery systems that can improve the uptake of antibiotics at specific targeted sites. Yang and co-workers (contribution 1) report the strong antimicrobial effects of silver-doped Zr implant abutments against oral microorganisms, including *P. gingivalis* and *S. mutants*. From a clinical point of view, this functionalized abutment demonstrates good biocompatibility with human gingival fibroblasts because of the stability and low leaching property of silver nanoparticles. Ye et al. (contribution 2) describe how ZnO nanoparticles maximize their antimicrobial activities by improving the bacterial phagocytosis efficiency of polymorphonuclear neutrophils. These synergized effects promote the host's resistance to pathogen invasion and alleviate subcutaneous immune cell infiltration in vivo. Besides inorganic antimicrobial agents, this delivery system offers a way to increase the effectiveness of antibiotics and is considered another promising strategy to overcome antimicrobial resistance. Li et al. (contribution 3) designed a multi-stimuli-responsive multilayer coating with controlled release of chlorhexidine acetate (CHA) to address bacterial infections that occur when medical devices (such as urinary catheters or peripheral venous catheters) are connected to open wounds in humans. The coating's multi-responsive abilities, high bactericidal capability (>99% against *E. coli* and *S. aureus*) and reusability (six times), as well as its ease of adoption onto different substrates, are the key advantages it offers in real clinical scenarios. The implantation site is a complex and dynamic microenvironment that involves a sequence of molecular, cellular, and physiological events. Generally, host immune cells (macrophages, neutrophils, T cells, etc.) are well recognized to have a major impact on the outcome of bacterial infection [4]. However, the responses of bacteria to other mammalian cells remain largely unknown. Inspired by the theory of "the race for the surface", Chen and co-workers (contribution 4) found that tantalum surfaces exhibited a lower infectious state than titanium in an implant-related tibia osteomyelitis model, despite not revealing any antibacterial activity against *S. aureus* in vitro. Bacteria preferentially adhere to the surfaces of mammalian cells rather than those of materials, which may be detached from dead-cell-containing surfaces to reduce the production of biofilms and the occurrence of implant infection. Chen et al.'s study brings a new perspective to the manufacture of implantable materials with antimicrobial activities. In another research study, Zhi and co-workers (contribution 5) demonstrate that calcium- and protein-conditioned titanium may act against bacterial colonization (*S. mutants* and *P. gingivalis*) by releasing calcium ions and creating a basic local pH microenvironment to regulate the conformation of fibrinogen. Their study solidly confirms the indirect antibacterial strategy originally proposed by Cao et al. and provides a promising approach to tackle peri-implant infections [5].

In the field of tissue engineering, antibacterial activity has become an essential requirement for diverse biomaterials [6]. Peng et al. (contribution 6) review the preparation methods of hydrogels attached to substrate surfaces, assess the advantages and limitations of each method when introducing antibacterial substances, as well as evaluate three major antibacterial strategies applied in hydrogels, namely bacterial repellency and inhibition, contact surface killing of bacteria, and release of antibacterial agents. They also discuss the current challenges for researchers in this field, such as chemical stability and sterilization, which may have long-term effects on the human body. In another review article, Zhang and co-workers (contribution 7) discuss recent progress in metal–organic frameworks (MOFs) as antibacterial agents and illustrate their possible molecular mechanisms, such as physical interaction, component release, chemical dynamic therapy (CDT), photodynamic therapy (PDT), photothermal therapy (PTT), sonodynamic therapy (SDT), and synergistic therapy. The authors also raise concerns about these frameworks' future clinical use. For example, given that MOFs are commonly composed of metal centers and organic ligands, more studies are still needed on the effects of MOF degradation products on human metabolism. Notably, not only can biomaterials be used as antibacterial agents, but they also allow for a targeted and controlled release of antibacterial agents to increase efficiency and safety. Yang and co-workers (contribution 8) summarize recent advances in bacteria-responsive drug delivery systems (DDSs) used for combating bacterial infections, categorizing them by trigger mode, including physical-stimuli-responsive, virulence-factor-responsive, host-immune-response responsive, and their combinations. Although DDSs may suffer from limited specificity and accuracy, the pattern of on-demand drug release and the implementation of multiple functions for tissue repair/regeneration make them one of the most important aspects of biomedical devices. Clinically, the available approaches and strategies to combat bacterial infections are also highly dependent on tissue microenvironment and on the types of biomaterials employed. Cao and co-workers (contribution 9) give a brief overview of antibacterial designs for implantable medical devices from the perspectives of unpredictable onset and site-specific incidence, possibly involving multiple and resistant pathogenic strains. In their paper, Cao et al. intend to illustrate the complex relationships of these devices and figure out future directions for promoting clinical translations.

The ultimate goal of functional biomaterials is to stimulate tissue repair/regeneration following injury and damage [1]. When exposed to functional biomaterials, cells can sense signals from the material's features, such as its topography and chemical composition, and convert them into biological responses [7,8]. Zhang et al. (contribution 10) fabricated four different submicron-grooved polystyrene films using soft lithography. The results show that submicron groove structures can significantly transform the morphology and cytoskeleton of Schwann cells and upregulate the gene expression of Schwann cells responsible for axonal regeneration and myelination, which is promising for peripheral nerve repair. Xie and co-workers (contribution 11) fabricated a duplex film with an inner MnOOH layer and an outer FeOOH layer on PEO-treated AZ31 using a simple immersion processes. The oxyhydroxide film completely sealed the pores on the PEO surface, thus providing better corrosion resistance. The modified sample exhibited improved osteogenic activity in vitro and enhanced bone regeneration in vivo, indicating promising potential for orthopedic applications. Wang and co-workers (contribution 12) investigated the mechanism of magnesium in regulating osteogenic activity. The results reveal that the AMPK/mTOR signaling pathway is involved in the process of autophagy associated with magnesium-induced osteogenic differentiation of BMSCs. Their study advances our understanding of the link between autophagy induction and osteogenic differentiation. Recently, much evidence has suggested that tissue repair/regeneration is not simply regulated by local signals and does not occur independently of other organ functions; instead, it requires the precise coordination of different organ systems [9]. Zhao et al. (contribution 13) provide an overview of biomaterials-mediated immune responses that regulate bone regeneration, methods to assess the bone immunomodulatory properties of biomaterials, as well as the strategies that can be used for future bone tissue engineering applications.

The development of functional biomaterials is also highly reliant on the fine-tuning of material structures and properties, as well as on progress in fabrication techniques [10]. Surface modification has been well recognized as an economical and efficient approach to endow the surfaces of conventional biomaterials with bioactivity and multifunctionality for therapeutic purposes [11]. This can help retain the favorable bulk properties that determine whether a material can adapt to the human body. For example, plasma immersion ion implantation and deposition (PIII&D) can introduce metal or gas elements to objects with flexible shapes [12]. Typically, the depth of the modified layer is relatively shallow (<100 nm), and the amount of the doped element is relatively low. Thus, Ag-doped zirconia implant can exhibit good biocompatibility with human gingival fibroblasts (contribution 1). Among various surface modification techniques, plasma electrolyte oxidation (PEO) is extensively used to produce porous and thick metal oxide coatings on metals (e.g., titanium, magnesium, aluminum, and their alloys) in industrial and biomedical applications, especially in dental implants, bone fusion, bone fixation, etc. [13]. It is for this reason that this technique has been widely explored by researchers and used to develop a variety of functional biomaterials (contributions 11 and 12). As a high-throughput tool for surface patterning, soft lithography can create ordered topographical features (grooves, pillars, and pores) and benefits from ease of preparation, high efficiency, and reusability [14]. It plays an important role in exploring the relationship between surface topography and cellular behaviors (contribution 10). Distinct from surface modification, nanomaterials such as nanoparticles and MOFs are usually synthesized by one-pot, hydrothermal, ultrasonic-assisted, electrochemical, and mechanochemical methods [15] (contributions 7 and 8). The high cost of nanomaterials, as well as complications in their purification and stability, are the main factors influencing their future applications.

3. Conclusions

In summary, this Special Issue, entitled "State-of-the-Art Functional Biomaterials in China", presents a diverse collection of research papers and review articles on trailblazing antibacterial surfaces, advanced multifunctional biomaterials, and clinical studies on the applications of biomaterials. These collaborations of academic and clinical research will provide unique and essential insights into functional biomaterials and increase the efficiency of their clinical translation.

Author Contributions: Conceptualization, Y.Q. and H.C.; writing—original draft preparation, Y.Q. and H.C.; writing—review and editing, Y.Q. and H.C. All authors have read and agreed to the published version of the manuscript.

Funding: This work was jointly supported by grants from the National Natural Science Foundation of China (32271399 and 31870945), the Natural Science Foundation of Shanghai (21ZR1415700 and 20ZR1465100), and the Shanghai Committee of Science and Technology (23S31901700 and 20S31903300).

Data Availability Statement: Not applicable.

Conflicts of Interest: The authors declare no conflicts of interest.

List of Contributions:

1. Yang, Y.; Liu, M.Y.; Yang, Z.; Lin, W.S.; Chen, L.; Tan, J.G. Enhanced Antibacterial Effect on Zirconia Implant Abutment by Silver Linear-Beam Ion Implantation. *J. Funct. Biomater.* **2023**, *14*, 46.
2. Ye, K.; Huang, M.R.; He, X.J.; An, Z.Q.; Qin, H. Synergistic Antibacterial Effect of Zinc Oxide Nanoparticles and Polymorphonuclear Neutrophils. *J. Funct. Biomater.* **2022**, *13*, 25.
3. Li, W.L.; Hua, G.P.; Cai, J.F.; Zhou, Y.M.; Zhou, X.; Wang, M.; Wang, X.M.; Fu, B.Q.; Ren, L. Multi-Stimulus Responsive Multilayer Coating for Treatment of Device-Associated Infections. *J. Funct. Biomater.* **2022**, *13*, 24.
4. Chen, X.; Bi, Y.K.; Huang, M.R.; Cao, H.L.; Qin, H. Why Is Tantalum Less Susceptible to Bacterial Infection? *J. Funct. Biomater.* **2022**, *13*, 264.
5. Zhi, Q.; Zhang, Y.H.; Wei, J.X.; Lv, X.L.; Qiao, S.C.; Lai, H.C. Cell Responses to Calcium- and Protein-Conditioned Titanium: An In Vitro Study. *J. Funct. Biomater.* **2023**, *14*, 253.
6. Peng, T.; Shi, Q.; Chen, M.L.; Yu, W.Y.; Yang, T.T. Antibacterial-Based Hydrogel Coatings and Their Application in the Biomedical Field—A Review. *J. Funct. Biomater.* **2023**, *14*, 243.
7. Zhang, X.; Peng, F.; Wang, D. MOFs and MOF-Derived Materials for Antibacterial Application. *J. Funct. Biomater.* **2022**, *13*, 215.
8. Yang, Y.J.; Jiang, X.; Lai, H.C.; Zhang, X.M. Smart Bacteria-Responsive Drug Delivery Systems in Medical Implants. *J. Funct. Biomater.* **2022**, *13*, 173.
9. Cao, H.L.; Qiao, S.C.; Qin, H.; Jandt, K.D. Antibacterial Designs for Implantable Medical Devices: Evolutions and Challenges. *J. Funct. Biomater.* **2022**, *13*, 86.
10. Zhang, Z.; Lv, Y.L.; Harati, J.; Song, J.A.; Du, P.; Ou, P.Y.; Liang, J.Q.; Wang, H.Y.; Wang, P.Y. Submicron-Grooved Films Modulate the Directional Alignment and Biological Function of Schwann Cells. *J. Funct. Biomater.* **2023**, *14*, 238.
11. Xie, J.N.; Cheng, S.; Zhong, G.Q.; Zhou, R.X.; Zhang, C.; He, Y.; Peng, F.; Zhang, Y. Oxyhydroxide-Coated PEO-Treated Mg Alloy for Enhanced Corrosion Resistance and Bone Regeneration. *J. Funct. Biomater.* **2022**, *13*, 50.
12. Wang, G.F.; Luo, J.X.; Qiao, Y.Q.; Zhang, D.D.; Liu, Y.L.; Zhang, W.J.; Liu, X.Y.; Jiang, X.Q. AMPK/mTOR Pathway Is Involved in Autophagy Induced by Magnesium-Incorporated TiO_2 Surface to Promote BMSC Osteogenic Differentiation. *J. Funct. Biomater.* **2022**, *13*, 221.
13. Zhao, T.; Chu, Z.Z.; Ma, J.; Ouyang, L.P. Immunomodulation Effect of Biomaterials on Bone Formation. *J. Funct. Biomater.* **2022**, *13*, 103.

References

1. Gaharwar, A.K.; Singh, I.; Khademhosseini, A. Engineered biomaterials for in situ tissue regeneration. *Nat. Rev. Mater.* **2020**, *5*, 686–705. [CrossRef]
2. Kalelkar, P.P.; Riddick, M.; García, A.J. Biomaterial-based antimicrobial therapies for the treatment of bacterial infections. *Nat. Rev. Mater.* **2022**, *7*, 39–54. [CrossRef] [PubMed]
3. Darby, E.M.; Trampari, E.; Siasat, P.; Gaya, M.S.; Alav, I.; Webber, M.A.; Blair, J.M.A. Molecular mechanisms of antibiotic resistance revisited. *Nat. Rev. Microbiol.* **2023**, *21*, 280–295. [CrossRef] [PubMed]
4. Rosenberg, G.; Riquelme, S.; Prince, A.; Avraham, R. Immunometabolic crosstalk during bacterial infection. *Nat. Microbiol.* **2022**, *7*, 497–507. [CrossRef] [PubMed]
5. Cao, H.L.; Dauben, T.J.; Helbing, C.; Jia, Z.C.; Zhang, Y.C.; Huang, M.R.; Müller, L.; Gu, S.; Zhang, X.Y.; Qin, H.; et al. The antimicrobial effect of calcium-doped titanium is activated by fibrinogen adsorption. *Mater. Horiz.* **2022**, *9*, 1962–1968. [CrossRef] [PubMed]
6. Duan, S.; Wu, R.N.; Xiong, Y.H.; Ren, H.M.; Lei, C.Y.; Zhao, Y.Q.; Zhang, X.Y.; Xu, F.J. Multifunctional antimicrobial materials: From rational design to biomedical applications. *Prog. Mater. Sci.* **2022**, *125*, 100887. [CrossRef]

7. Sadtler, K.; Singh, A.; Wolf, M.T.; Wang, X.K.; Pardoll, D.M.; Elisseeff, J.H. Design, clinical translation and immunological response of biomaterials in regenerative medicine. *Nat. Rev. Mater.* **2016**, *1*, 16040. [CrossRef]
8. Rahmati, M.; Silva, E.A.; Reseland, J.E.; Heyward, C.A.; Haugen, H.J. Biological responses to physicochemical properties of biomaterial surface. *Chem. Soc. Rev.* **2020**, *49*, 5178–5224. [CrossRef]
9. Sun, F.; Poss, K.D. Inter-organ communication during tissue regeneration. *Development* **2023**, *150*, dev202166. [CrossRef] [PubMed]
10. Collins, M.N.; Ren, G.; Young, K.; Pina, S.; Reis, R.L.; Oliveira, J.M. Scaffold Fabrication Technologies and Structure/Function Properties in Bone Tissue Engineering. *Adv. Funct. Mater.* **2021**, *31*, 2010609. [CrossRef]
11. Hu, X.; Wang, T.; Li, F.Q.; Mao, X. Surface modifications of biomaterials in different applied fields. *RSC Adv.* **2023**, *13*, 20495–20511. [CrossRef] [PubMed]
12. Akdogan, E.; Sirin, H.T. Plasma surface modification strategies for the preparation of antibacterial biomaterials: A review of the recent literature. *Mater. Sci. Eng. C-Mater.* **2021**, *131*, 112474. [CrossRef] [PubMed]
13. Makurat-Kasprolewicz, B.; Ossowska, A. Recent advances in electrochemically surface treated titanium and its alloys for biomedical applications: A review of anodic and plasma electrolytic oxidation methods. *Mater. Today Commun.* **2023**, *34*, 105425. [CrossRef]
14. Tran, K.T.M.; Nguyen, T.D. Lithography-based methods to manufacture biomaterials at small scales. *J. Sci.-Adv. Mater. Dev.* **2017**, *2*, 1–14. [CrossRef]
15. Ge, X.Y.; Wong, R.; Anisa, A.; Ma, S.Q. Recent development of metal-organic framework nanocomposites for biomedical applications. *Biomaterials* **2022**, *281*, 121322. [CrossRef] [PubMed]

Disclaimer/Publisher's Note: The statements, opinions and data contained in all publications are solely those of the individual author(s) and contributor(s) and not of MDPI and/or the editor(s). MDPI and/or the editor(s) disclaim responsibility for any injury to people or property resulting from any ideas, methods, instructions or products referred to in the content.

Journal of
Functional Biomaterials

MDPI

Article

Multi-Stimulus Responsive Multilayer Coating for Treatment of Device-Associated Infections

Wenlong Li [1,†], Guanping Hua [1,†], Jingfeng Cai [1], Yaming Zhou [1], Xi Zhou [1], Miao Wang [1,*], Xiumin Wang [2], Baoqing Fu [3,*] and Lei Ren [1,4,*]

1 Higher Educational Key Laboratory for Biomedical Engineering of Fujian Province, Research Center of Biomedical Engineering of Xiamen, Department of Biomaterials, College of Materials, Xiamen University, 422 Siming Nan Road, Xiamen 361005, China; 20720190153834@stu.xmu.edu.cn (W.L.); hgpailber@163.com (G.H.); 20720201150033@stu.xmu.edu.cn (J.C.); blue_zzyymm@163.com (Y.Z.); xizhou@xmu.edu.cn (X.Z.)

2 School of Pharmaceutical Sciences, Xiamen University, Xiamen 361102, China; wangxm@xmu.edu.cn

3 Department of Laboratory Medicine, Xiang'an Hospital of Xiamen University, School of Medicine, Xiamen University, Xiamen 361102, China

4 State Key Laboratory of Physical Chemistry of Solid Surfaces, College of Chemistry and Chemical Engineering, Xiamen University, Xiamen 361005, China

* Correspondence: miaowang@xmu.edu.cn (M.W.); fbq@sina.com (B.F.); renlei@xmu.edu.cn (L.R.)

† These authors contributed equally to this work.

Abstract: Antibacterial coating with antibiotics is highly effective in avoiding device-associated infections (DAIs) which is an unsolved healthcare problem that causes significant morbidity and mortality rates. However, bacterial drug resistance caused by uncontrolled release of antibiotics seriously restricts clinical efficacy of antibacterial coating. Hence, a local and controlled-release system which can release antibiotics in response to bacterial infected signals is necessary in antibacterial coating. Herein, a multi-stimulus responsive multilayer antibacterial coating was prepared through layer-by-layer (LbL) self-assembly of montmorillonite (MMT), chlorhexidine acetate (CHA) and Poly(protocatechuic acid-polyethylene glycol 1000-bis(phenylboronic acid carbamoyl) cystamine) (PPPB). The coating can be covered on various substrates such as cellulose acetate membrane, polyacrylonitrile membrane, polyvinyl chloride membrane, and polyurethane membrane, proving it is a versatile coating. Under the stimulation of acids, glucose or dithiothreitol, this coating was able to achieve controlled release of CHA and kill more than 99% of Staphylococcus aureus and Escherichia coli (4×10^8 CFU/mL) within 4 h. In the mouse infection model, CHA releasing of the coating was triggered by infected microenvironment to completely kill bacteria, achieving wounds healing within 14 days.

Keywords: device-associated infections; multi-stimulus responsiveness; antibacterial coating; wound healing

Citation: Li, W.; Hua, G.; Cai, J.; Zhou, Y.; Zhou, X.; Wang, M.; Wang, X.; Fu, B.; Ren, L. Multi-Stimulus Responsive Multilayer Coating for Treatment of Device-Associated Infections. *J. Funct. Biomater.* **2022**, *13*, 24. https://doi.org/10.3390/jfb13010024

Academic Editors: James Kit-hon Tsoi, Huiliang Cao and Yuqin Qiao

Received: 30 January 2022
Accepted: 25 February 2022
Published: 28 February 2022

Publisher's Note: MDPI stays neutral with regard to jurisdictional claims in published maps and institutional affiliations.

Copyright: © 2022 by the authors. Licensee MDPI, Basel, Switzerland. This article is an open access article distributed under the terms and conditions of the Creative Commons Attribution (CC BY) license (https://creativecommons.org/licenses/by/4.0/).

1. Introduction

Nosocomial infection is acquired in healthcare facilities [1]. More than 50% nosocomial infections are related to medical devices [2]. Even with aseptic techniques, microbial communities from the patient's surface or external environment can be attached to medical devices leading to device-associated infections (DAIs) which is a major health care problem that has not been solved [3]. For instance, the substantial mortality caused by catheter-associated infection have reached 25% [4,5]. A study from the World Health Organization demonstrated that patients in the intensive care unit (ICU) are at a significantly higher risk of acquiring DAIs. About 30% of patients have experienced at least one DAIs with significant associated morbidity and mortality [6].

To prevent the occurrence of DAIs, a promising strategy is to modify the antibiotics on the devices surface [7]. Chlorhexidine acetate (CHA), a kind of common broad-spectrum antibiotic, carries a large number of positive groups that can bind to the anionic sites on the bacteria membrane [8]. This binding can disrupt bacteria membrane potential and cause varying degrees of membrane damage [9]. Therefore, it is difficult for bacteria to develop resistance to CHA [10]. Given the above advantages, CHA is often used as an antibacterial component of the antibacterial coating. For instance, Usha et al. modified CHA on the surface of fabrics and achieved the destruction of bacteria (almost 100%) within 24 h [11]. However, these kinds of antibacterial coatings have obvious defects. While killing bacteria, abundant surface positive charges will cause a large number of bacterial fragments to accumulate, and the antibacterial ability of the coatings will soon fail [12]. At the same time, excessive release of antibiotics also contributes to the development of bacterial resistance [13]. In order to solve the above problems, nano-lamellar MMT, as a material with large specific surface area, good absorbability, good biocompatibility, and drug carrying capability, was used to load CHA to complete slow drug release [14]. MMT can load positively charged substances such as CHA between the lamellar layers through cation exchange effect [15]. However, this drug delivery mode can only play a role of slow release, but cannot achieve the effect of on-demand release and sustained maintenance of antibacterial properties. Therefore, how to maintain the antibacterial ability of cationic antibacterial coating on the medical device surface and significantly reduce the toxicity and bacterial resistance is a major bottleneck in current research [16].

In recent years, stimulus-responsive antibacterial coatings have attracted great attention because of the reduction of the excessive release of antibiotics. This antibacterial coating can be used for topical drug delivery by responding to the bacterial infection microenvironment (e.g., weak acid environment (pH < 5), high peptidoglycan content, and high glutathione content) [17], which will inhibit the development of antibiotics resistance and achieve the long-term antibacterial effect. Most of the previous work on stimuli-responsive antibacterial coatings has focused on responding to a single stimulus (e.g., pH, electric field, light, temperature or bioactive molecules [18–22]). Although such single stimulus-responsive antibacterial coatings have been researched deeply in the treatment of DAIs, it is clear that antibacterial coatings with multi-stimulus responsive ability would be more advantageous and widely applied. Because multiple stimulus occurs simultaneously in a real biological and physical environment [7]. For example, Yang Zhou et al. [23] constructed a surface coating with a multi-stimulus response on a silicon substrate containing thermally responsive components poly(*N*-isopropylacrylamide) (PNIPAAm) and phenylboric acid. The coating can respond to changes of pH, sugar and temperature in the biological environment to release drugs and kill bacteria. Therefore, constructing multi-stimulus-responsive antibacterial coatings that are sensitive to the bacterial infection environment may serve as a good strategy for the treatment of DAIs.

However, most coatings are highly selective to the substrate and are limited to the variety of substrate materials [24]. Therefore, it is necessary to develop a kind of coating that can modify the surface of various substrate. In this process, we need to integrate antibacterial agents and responsive drug delivery systems into the coating material. However, it is not easy to load antibiotics into releasing systems because the weak binding between the matrix and small molecules [25], which would result in bacterial antibiotic resistance [26]. Layer-by-layer (LbL) technology provides a solution to this problem. LbL technology originally refers to a technique of forming self-assembled multilayers on charged substrates by alternating deposition of polyelectrolytes, which was first studied and proposed by R. K. Iler [27]. Since the 1990s, LbL technology has developed rapidly. LbL technology can be widely used in various forms (e.g., spraying or immersion) for surface modification of various materials (e.g., planar or particulate substrates) [28]. It has been proved that LbL assembly could provide a controlled carrier system in respect of releasing antibiotics [29]. LbL self-assembly technology has the ability to fully absorb and retain the biological activity of the drug, which is used to construct a good responsive drug releasing mechanism on the surface of the medical devices.

In this paper, a multi-stimulus responsive multilayer antibacterial coating $(MMT\text{-}PPPB\text{-}CHA)_n$ was prepared by the electrostatic adsorption between the inorganic nano-lamellar MMT, cationic antibacterial agent CHA and PPPB (Figure 1). We studied the drug loading capacity, drug release capacity, and antibacterial activity of $(MMT\text{-}PPPB\text{-}CHA)_n$ coating to prove that the coating can be a versatile material used to modify a variety of substrate, and the coating has good drug loading ability and responsive drug release ability. In addition, a mouse epidermal infection model was established to study the in vivo antibacterial activity of $(MMT\text{-}PPPB\text{-}CHA)_n$ coating. This coating can respond to the microenvironment of bacterial infection (such as weak acid environment (pH < 5), high peptidoglycan content, and high glutathione content) to release cationic CHA, providing a new strategy for solving DAIs.

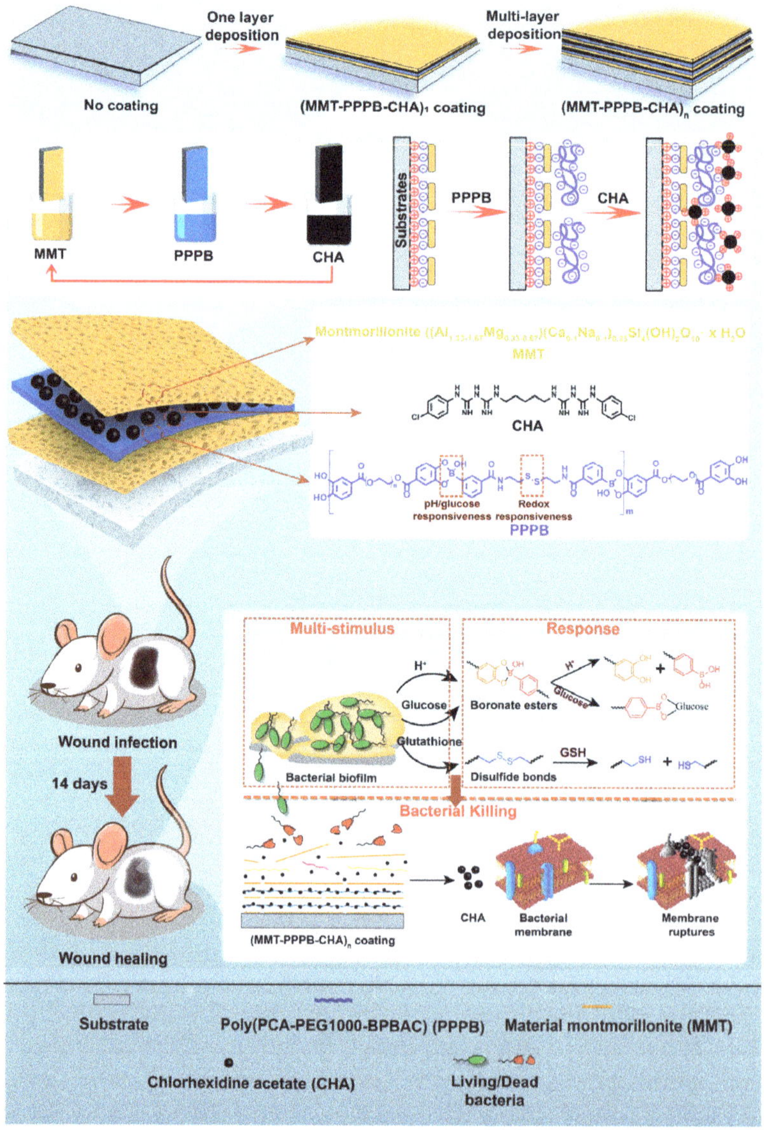

Figure 1. Schematic illustration of the multi-stimulus responsive multilayer antibacterial coating $(MMT\text{-}PPPB\text{-}CHA)_n$.

2. Materials and Methods

2.1. Materials and Instrumentation

Ethanol absolute (AR Grade, ≥99.7%) and chloroform (AR Grade, ≥99.0%) were purchased from Sinopharm Chemical Reagent Shanghai Co., Ltd. (Shanghai, China). Acetic acid (AR Grade, ≥99.5%), phosphate buffer saline (PBS buffer), and dimethyl sulfoxide (DMSO, molecular biology grade, ≥99.7%), 1-(3-Dimethylaminopropyl)-3-ethylcarbodiimide hydro (EDC, ≥95.0%, C600433-0025), glucose (≥99.8%, A100188-0005) were purchased from Sangon Biotechnology Co., Ltd. (Shanghai, China). *E. coli* (ATCC 43888) and *S. aureus* (ATCC BAA-1721) were purchased from China Center of Industrial Culture Collection, CICC. Chlorhexidine acetate (CHA, ≥99.0%, C107054), dithiothreitol (DTT, ≥99.0%, D104859), cysteamine dihydrochloride (CYS, ≥97.0%, C153647), protocatechuic acid (PCA, ≥97.0%, S30117), polyethylene glycol 1000 (PEG1000, AR Grade, 12803702) were purchased from Aladdin (Co. Ltd. Shanghai, China). MMT k-10 (M813515), 3-carboxyphenyl boric acid (CPBA, ≥99.0%, C804442), 4-Dimethylaminopyridine (DMAP, ≥97.0%, D807273), N-Hydroxysuccinimide (NHS, ≥97.0%, N811124) Ethylene imine polymer (PEI, ≥99.0%, E808878), Deuterium oxide (D_2O, ≥99.9%, D807644), Methanol-d4 (CD_3OD, ≥99.9%, M812876), DMSO-d6 (≥ 99.9%, D806935) were purchased from Macklin Biochemical Co., Ltd. (Shanghai, China).

2.2. Synthesis of PPPB

Firstly, 1.65 g CPBA (10.0 mmol), 1.53 g EDC (8.0 mmol), and 1.40 g NHS (12.2 mmol) were accurately weighed and dissolved in 100 mL PBS buffer (pH 6.86). After incubation for 2 h under full stirring at room temperature, 0.90 g CYS (4.0 mmol) was added, and then the reaction mixture was continuously stirred at room temperature for 12 h. The precipitate produced by the reaction was filtered, washed several times with PBS and ultrapure water, and then centrifuged twice at 6000 rpm. The product was vacuum dried at 60 °C, and finally recrystallized and purified with methanol to obtain white powder BPBAC.

Secondly, 1.54 g PCA and 2.50 g PEG1000 were accurately weighed and dissolved in 20.0 mL DMSO solution. 1.92 g EDC and 305 mg DMAP were added under stirring conditions, and the reaction was carried out at room temperature for 48 h. After the reaction, the precipitate was removed by filtration, and the filtrate was dialyzed by dialysis bag with molecular weight cut-off 1000. The dialysate was firstly dialyzed by DMSO for 12 h, and then by ultrapure water for 36 h. The dialysate was changed every 6 h during dialysis. After dialysis, the product was precipitated into dark yellow solid and dried in vacuum at 30 °C to obtain PCA-PEG1000.

Finally, bis(phenylboronic acid carbamoyl) cystamine (BPBAC) was dissolved in a small amount of methanol solution, then BPBAC methanol solution and PCA-PEG1000 were fully dissolved in 20.0 mL NaOH aqueous solution (pH 10.0), and stirred at 300 rpm for 12 h at room temperature. At the end of the reaction, dialysis bags with molecular weight cut-off 1000 were used in NaOH solution with pH 10.0 for 48 h, and dialysate was changed every 8 h. Finally, the solution was freeze-dried to obtain a dark yellow colloidal substance, namely PPPB.

2.3. Synthesis of (MMT-PPPB-CHA)$_n$ Coating

Nanosheet MMT was first prepared. 0.8 g MMT was dissolved in 20.0 mL ultrapure water and 20.0 mL methanol, respectively. After sealing and stirring at room temperature for 1 week, 40.0 mg/mL nanosheet MMT aqueous dispersion and 40.0 mg/mL nanosheet MMT methanol dispersion were prepared for standby.

The substrate was immersed in PEI solution (0.5 mg/mL) for 5 min to form a precursor layer. Then, the composite coating (MMT-PPPB-CHA)$_n$ was prepared according to the following steps: (1) The substrate was soaked in 1.0 mg/mL MMT dispersion for 30 s and then dried naturally. (2) Natural drying after immersion in 1.0 mg/mL PPPB solution for 30 s. (3) Natural drying after soaking in 2.0 mg/mL CHA solution for 15 s. In this preparation process, the substrate surface is covered with MMT layer, then PPPB layer, and finally CHA layer, forming a layer (MMT-PPPB-CHA) coating. Repeat the above steps "n" times, $(MMT\text{-}PPPB\text{-}CHA)_n$ coating was prepared on the surface of substrates. The substrate surface coatings only modified with MMT and MMT-CHA were prepared by the same method, denoted as $(MMT)_n$ and $(MMT\text{-}CHA)_n$, respectively, as the control groups. In order to ensure the accuracy and objectivity of performance characterization, the prepared coating was washed with pure water and PBS buffer (pH 7.4) for three times before the test, in order to remove the excess CHA, PPPB and MMT that did not form the coating during the preparation process.

2.4. Characterization of Polymers PPPB

The FT-IR spectra of BPBAC, PCA-PEG1000, and PPPB were recorded on a Nicolet 6700 FT-IR spectrometer (Thermo Scientific) in the 4000–600 cm^{-1} range. The ^1H nuclear magnetic resonance (^1H NMR) (AVANCE III 400 MHz, Bruker, Karlsruhe, Germany) of BPBAC, PCA-PEG1000, and PPPB was recorded at 25 °C using CD_3OD, DMSO-d6, and D_2O as solvent separately. The Ultraviolet−visible (UV−vis) absorption spectra of PPPB products with the three stimuli were determined by UV spectro-photometer (UV-2550, Shimadzu, Tokyo, Japan).

2.5. Characterization of $(MMT\text{-}PPPB\text{-}CHA)_n$ Coating

The morphology of $(MMT\text{-}PPPB\text{-}CHA)_n$ coating was characterized by a field emission scanning electron microscope (SEM, SU-70, Hitachi Nake high-tech enterprise, Tokyo, Japan) at an acceleration voltage of 5 kV. X-ray diffraction (XRD) was performed with the Axs D8-A25 advance (Bruker, Karlsruhe, Germany). The Fourier transform infrared (FTIR) spectra assay was performed using a Nicolet iS10 (Thermo Scientific, Waltham, MA, USA) with the KBr pellet technique. UV−vis absorption spectra were acquired with the UV spectrophotometer (UV-2550, Shimadzu, Karlsruhe, Germany, Japan). MMT, PPPB, and CHA were prepared into 0.5 mg/mL solution. $(MMT\text{-}CHA)_3$ and $(MMT\text{-}PPPB\text{-}CHA)_3$ coatings were scraped off the substrate and dispersed in water to form a suspension (0.5 mg/mL). The Zeta potential of samples was measured by Malvern Mastersizer 2000 particle size and potential analyzer (Mastersizer2000/MAL1012737, Spectris, Shanghai, China). First of all, the test mode was set as potential test in Zetasizer Software (test temperature: 25 °C, number of repeated tests: 3 times, and potential sample pool model: Dts1070). Then, the solution or suspension was fully shaken and injected at the calibration line of the sample pool. Then, the sample pool was put into the analyzer. Finally, the potential values measured in the three experiments were averaged as the final sample Zeta potential. The water contact angle of samples was measured by a dynamic water contact angle measuring instrument (OCA20, Dataphysics, Germany, Esslingen, Germany).

2.6. The Standard Curve of CHA

The concentration of CHA solution prepared was 0.1 mg/mL, 0.05 mg/mL, 0.025 mg/mL, 0.01 mg/mL, 0.005 mg/mL and 0.0025 mg/mL, respectively. The absorbance of CHA at 252 nm was measured by a UV-VIS spectrophotometer (UV-2550, Shimadzu, China) with three times to take the average value. Then the standard curve of CHA concentration and absorbance was drawn.

2.7. CHA Loading Capacity of $(MMT\text{-}PPPB\text{-}CHA)_n$ Coating

Five groups of $(MMT\text{-}PPPB\text{-}CHA)_n$ coatings (n = 2, 4, 8, 12, and 16) were constructed on the CA films with an area of 2 cm^2 according to the above steps. 5.0 mL CHA solution with the concentration of 5.0 mg/mL. To facilitate the calculation, 0.5 mL CHA (5.0 mg/mL) was used in preparation. After the preparation, we collected the remaining CHA solution and tested the absorbance of CHA at 252 nm. The relationship between drug loading ($\mu g/cm^2$) and layers number was further calculated. The calculation formula is:

$$\text{CHA content per unit area } (\mu g/cm^2) = \frac{\text{Total dose } (\mu g) - \text{Remaining dose } (\mu g)}{\text{Total area } (cm^2)}$$

2.8. Sustained Drug-Releasing Capacity of (MMT-PPPB-CHA)$_3$ Coating

The coating (1×2 cm^2) was put into 5.0 mL PBS buffer. 0.5 mL of the release medium was taken out successively at 1, 2, 4, 12, 24, 48, and 72 h, and the same amount of PBS buffer was immediately replaced after taking. The absorbance of CHA was measured at 252 nm by an UV-VIS spectrophotometer (UV-2550, Shimadzu, China), and the CHA release kinetics curves of the different coating were obtained within 72 h.

2.9. Multi-Stimulus Responsive Ability of (MMT-PPPB-CHA)$_3$ Coating

The (MMT-PPPB-CHA)$_3$ coating (1×2 cm^2) was placed in 5.0 mL release media with different pH values (5.5, 6.0, 7.0, and 8.0), different concentrations of glucose solutions (0, 0.5, and 1.0 mg/mL) and different concentrations of DTT solutions (0, 0.5, and 1.0 mg/mL), respectively. 0.5 mL release medium was taken out successively at 1, 2, 4, 12, 24, 48, and 72 h, and an equal amount of release medium with corresponding pH value was immediately replaced after taking. The absorbance of the CHA was measured at 252 nm by an UV-VIS spectrophotometer (UV-2550, Shimadzu, China), and the release kinetics curve of CHA was obtained according to the standard curve.

To further explore the synergistic drug release behavior of (MMT-PPPB-CHA)$_3$ coating, orthogonal test L_9 (3^4) (three factors and three levels) was used to study the synergistic drug release effect of (MMT-PPPB-CHA)$_3$ coating under the different conditions. The orthogonal test results were analyzed as shown in Table 1.

Table 1. The result analysis of $L_9(3^4)$ orthogonal test.

Groups	Factors			CHA Release Rate (%)
	pH	**Glucose (mg/mL)**	**DTT (mg/mL)**	
a	6.0	0	0	56.0 ± 3.2
b	6.0	0.5	0.5	60.1 ± 2.0
c	6.0	1.0	1.0	70.1 ± 1.3
d	7.0	0	0.5	56.7 ± 1.7
e	7.0	0.5	1.0	62.0 ± 2.3
f	7.0	1.0	0	53.9 ± 1.1
g	8.0	0	1.0	43.3 ± 1.6
h	8.0	0.5	0	34.3 ± 1.8
i	8.0	1.0	0.5	50.7 ± 2.8
K1	186.2	156.0	144.2	
K2	172.6	156.4	167.8	
K3	128.3	174.7	175.4	
k1	62.1	52.0	48.1	
k2	57.5	52.1	55.9	
k3	42.8	58.2	58.5	
R	19.3	6.2	10.4	

2.10. Preparation of Bacteria Suspensions

E. coli (ATCC 43888) and *S. aureus* (ATCC BAA-1721) were cultured in 10 mL LB broth at 37 °C for 6 h. After culturing, the bacterial sediments were resuspended and washed with normal saline (0.9%, pH 7.4). Finally, the bacteria suspension was diluted to an optical density of 0.1 at 600 nm (OD600 = 0.1)

2.11. In Vitro Antibacterial Activity of $(MMT\text{-}PPPB\text{-}CHA)_3$ Coating

The coated or uncoated substrate was co-cultured with bacteria suspension at different pH values (6.0, 7.0 and 8.0), different concentrations of DTT solutions (0, 0.5, and 1.0 mg/mL), and glucose solutions (0, 0.5, and1.0 mg/mL) for 2 h, respectively. Then, 100 µL of the bacteria suspension was taken out and inoculated on AGAR medium. After 16 h of static culture, the number of colonies was counted and digital photos of culture plates were taken.

Fluorescein isothiocyanate (FITC) was dissolved in DMSO to prepare the 1 mM FITC solution. Live bacteria were labeled with 100 µL FITC solution and incubated with the coated or uncoated substrate at different pH values (6.0, 7.0 and 8.0), different concentrations of DTT solutions (0, 0.5, and 1.0 mg/mL), and glucose solutions (0, 0.5, and 1.0 mg/mL) for 2 h, respectively. The bacteria suspension was dropped into the surface dish specially used for laser confocal microscopy, and the survival of the bacteria was further observed under a laser confocal microscopy (SP8-STED 3X, Leica, German).

The bacteria were fixed with 2% glutaraldehyde aqueous solution, and then the gradient dehydration was carried out with 10%, 20%, 30%, 40%, 50%, 70%, 90%, 100% ethanol aqueous solution successively. The growth and adhesion of bacteria on the surface of the coating material were observed under a SEM (SU-70, Hitachi Nake High-Tech Enterprise, Tokyo, Japan) after dehydration.

2.12. Long-Term Antibacterial Performance of $(MMT\text{-}PPPB\text{-}CHA)_3$ Coating

Sterilized CHA, $(MMT\text{-}CHA)_3$ and $(MMT\text{-}PPPB\text{-}CHA)_3$ coated and uncoated substrates were placed in 5.0 mL bacteria suspension, respectively. Take 100 µL of bacteria solution from each group every 4 h and inoculate them into the AGAR medium for colony counting. At the same time, the bacteria liquid of each group was discarded, cleaned with sterile normal saline, and 5.0 mL of bacteria suspension was added again to continue co-culture. Repeat the preceding steps six times.

2.13. In Vivo Antibacterial Activity and Full-Thickness Skin Defect Healing

Female BALB/c mice (8 weeks, 15–20 g) were obtained from Laboratory Animal Center of Xiamen University and divided into five groups with three mice in each group: (1) normal saline (control); (2) uncoated substrate; (3) CHA coated substrate; (4) $(MMT\text{-}CHA)_3$ coated substrate; (5) $(MMT\text{-}PPPB\text{-}CHA)_3$ coated substrate.

10 µL of *S. aureus* and *E. coli* solution with a concentration of 1.0×10^6 CFU/mL was subcutaneously injected into the back of mice. 48 h later, obvious suppurative infection symptoms could be observed at the bacteria inoculated site on the back of mice. The suppurative infection site was cut open with a sterilized scalpel to form an open resection wound with a diameter of about 6 mm. Corresponding materials were affixed to the wound of each mouse and fixed with medical tape. The materials were removed 24 h later.

All mice were weighed every 24 h, and the wound healing of the subcutaneous infected site was observed and photographed at certain intervals. ImageJ was used to measure the change of wound area at different periods to calculate the wound healing rate [30], which was calculated as follows:

$$\text{Wound healing rate } (\%) = \left(1 - \frac{\text{wound area on day n}}{\text{wound area on day 0}}\right) \times 100\%$$

Finally, the mice were killed, and the bacterial concentration of the infected tissue was observed on the AGAR plate. This in vivo experiment has been approved by the Institutional Animal Care and Use Committee of Xiamen University. Approval date: 26 February 2018; Approval code: XMULAC20180003.

2.14. Statistical Analysis

Data are expressed as mean ± S.D. At least three independent experiments were performed for in vitro experiments. Statistical analysis was carried out using SPSS version

18.0, Origin 2021 and GraphPad Prism 7.00. Data were analyzed by two-sided student t-tests for comparison of two groups and one-way ANOVA for multiple groups.

3. Results and Discussion

3.1. Synthesis of the Multi-Stimulus Responsive Polymer PPPB

We firstly synthesized a multi-stimulus responsive polymer PPPB based on phenylborate ester bond and disulfide bond (Figure 2a) to construct a multi-stimulus responsive antibacterial coating. PPPB was synthesized through dynamic covalent reactions between BPBAC (Figure S1) and PCA-PEG1000 (Figure S2) under alkaline conditions (pH 10.0). The structure of PPPB was identified based on IR and ^{1}H NMR spectrum. The position and peak area ratio (1:1) of the characteristic peaks (δ 2.9 ppm and δ 3.6 ppm) indicated that PPPB possessed the characteristic structure of both BPBAC and PCA-PEG1000 (Figure 2b). Moreover, IR spectrum of PPPB also showed the characteristic peaks of BPBAC and PCA-PEG1000 at 2919 cm^{-1} (C–H) and 1641 cm^{-1} (C=O) (Figure 2c). More importantly, the disappearance of characteristic peaks at 1347 cm^{-1} indicated the phenylboronic acid was transformed to boronate ester, further confirming the successful synthesis of PPPB.

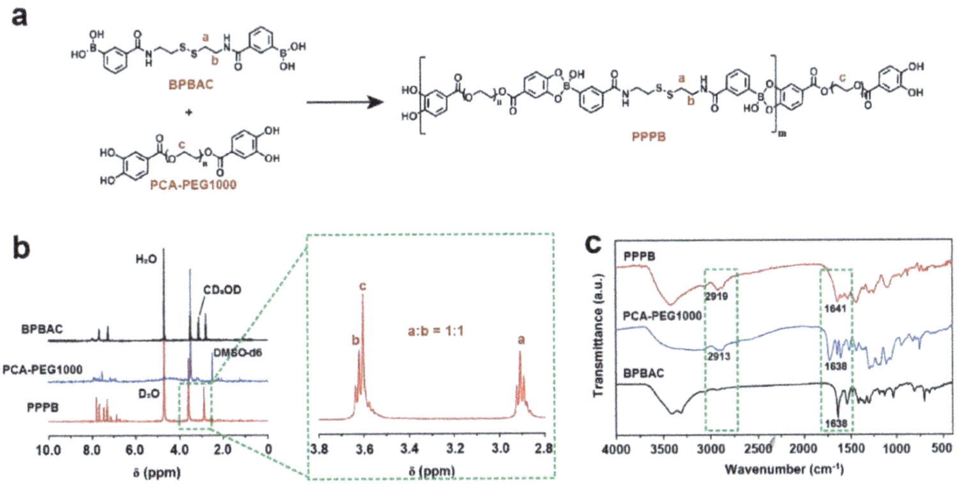

Figure 2. (**a**) Chemical structural formula of polymer PPPB; (**b**) ^{1}H-NMR spectrums of PPPB, PCA-PEG1000 and BPBAC; (**c**) FT-IR absorbance spectrums of PPPB, PCA-PEG1000, and BPBAC.

Due to the presence of boronate ester and disulfide bonds, polymer PPPB should exhibit pH-, glucose-, and redox-responsive features theoretically (Figure S3a). Therefore, the multi-stimulus responsive degradation behavior of PPPB was explained by UV-absorption, IR and ^{1}H NMR spectrum. As shown in Figure S3b, PPPB displayed a distinct corresponding UV peak with an absorption maximum at 300 nm. However, the intensity of the peak decreased with the addition of PBS buffer (pH 6.0), glucose aqueous solution (0.5 mg/mL) and DTT aqueous solution (0.5 mg/mL), indicating that the disulfide bond content in the product is reduced. In order to further analyze the stimulus responsiveness of PPPB, we purified the products by dialysis (molecular weight cut-off: 1000 Da), which means that the product obtained by dialysis should be the part containing PEG-1000 fragment. Disulfide bonds in polymer PPPB break when DTT stimulus responsiveness occurs. According to Figure 3, the intercepted part of the broken PPPB should contain part of the structure of BPBAC molecule, which means the existence of the characteristic peak of -S-CH_2- (δ 2.9 ppm) in the ^{1}H-NMR spectrum. On the contrary, the disappearance of the characteristic peak of-S-CH_2- (δ 2.9 ppm) and the existence of characteristic peak of -CH2-CH2 (δ 3.6 ppm) indicates the intercepted part of the broken PPPB is PCA-PEG1000, which means that pH/redox stimulus responsiveness have occurred.

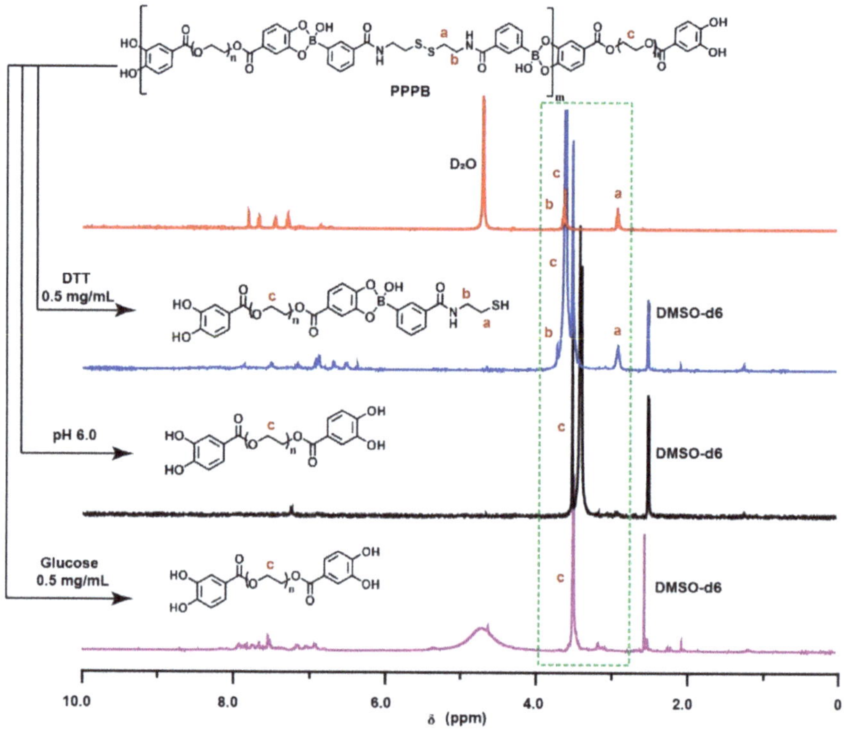

Figure 3. ^{1}H-NMR spectrums of PPPB in PBS buffer (pH 6.0), glucose aqueous solution (0.5 mg/mL) and DTT aqueous solution (0.5 mg/mL).

3.2. Construction of the of (MMT-PPPB-CHA)$_n$ Coating

Based on multi-stimulus responsive polymer PPPB, we designed a layer-by-layer assembly coating with the specificity of the response to the bacterial infection microenvironment (weak acid environment (pH < 5), high peptidoglycan content, and high glutathione content) (Figure 1). We chose cellulose acetate membrane (CA), polyacrylonitrile membrane (PAN), polyvinyl chloride membrane (PVC), and polyurethane membrane (PU) as substrates to prepare the coating (MMT-PPPB-CHA)$_3$ because they are common medical materials in the clinic [31]. According to the color change of the substrate surface, it is preliminarily concluded that (MMT-PPPPB-CHA)$_3$ coating could be prepared on various substrate materials. As shown in Figure S4, the surfaces of coated groups appeared even rougher than uncoated groups. Moreover, the change of water contact angle can prove that the coating can obviously improve the hydrophilicity of the substrate, which is helpful to improve the affinity of medical devices to human body (Figure S5). According to the cross-section of the coating (Figure 4a–c), (MMT-PPPB-CHA)$_3$ coating could present a thick and loose layer stacking state. The reasons could be explained by zeta potential (Figure 4d). After the coatings have been successfully prepared on the substrate, they are gently scraped off and configured to a 0.5 mg/mL suspension (the solvent is pure water). Zeta potential of the sample was measured by Malvern Mastersizer 2000. Both MMT and PPPB exhibit electronegativity, which was very favorable for loading the positively charged CHA with electrostatic interaction. In addition, (MMT-CHA)$_3$ exhibited a weak positive charge after LbL self-assembly of MMT and CHA, which also proved the possibility of PPPB participating in electrostatic interaction. Finally, (MMT-PPPB-CHA)$_3$ coatings were negatively charged, confirming the electrostatic interaction between MMT, PPPB, and CHA. This interaction could be beneficial for CHA loading and enlarge the layer spacing of the coating. It can be seen from Figure 3e that the (001) diffraction peak of $(MMT)_3$ and $(MMT\text{-}CHA)_3$ are very obvious, which are corresponded to an interlayer d spacing of 1.2357 nm and 1.3043 nm respectively, according to Bragg's equation. The (001) interlayer spacing of $(MMT)_n$ increased about 0.0686 nm after co-deposition with CHA, indicating that some of CHA was loaded between MMT layers by ion exchange [25,32]. However, the (001) diffraction peak of $(MMT\text{-}PPPB\text{-}CHA)_3$ can not be observed, indicating that the addition of PPPB completely destroyed the layered ordering structure of MMT. Finally, FT-IR absorbance spectrum demonstrated that (MMT-PPPB-CHA)3 coating was prepared successfully (Figure 4f).

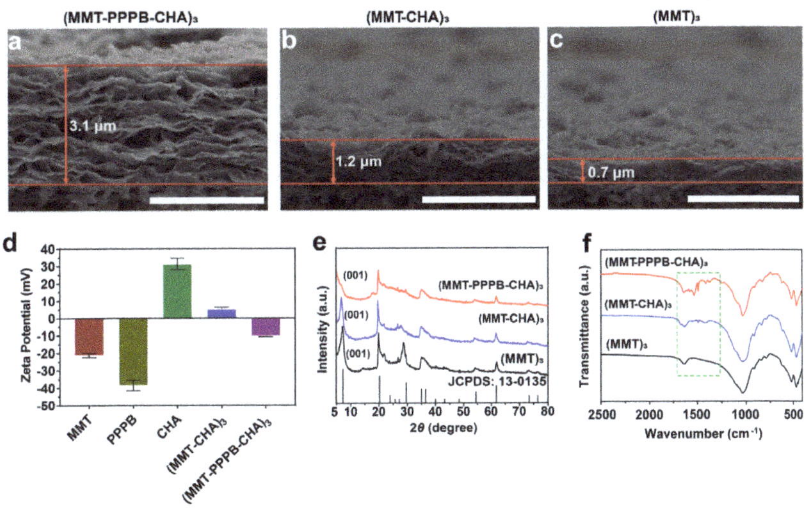

Figure 4. The SEM micrograph of cross-sectional of (**a**) MMT-PPPB-CHA$)_3$, (**b**) $(MMT\text{-}CHA)_3$, and (**c**) $(MMT)_3$ coating. Scale bar: 3 μm; (**d**) Zeta potential of MMT, PPPB, CHA, $(MMT\text{-}CHA)_3$ and $(MMT\text{-}PPPB\text{-}CHA)_3$; (**e**) XRD patterns and (**f**) FT-IR absorbance spectrums of $(MMT)_3$, $(MMT\text{-}CHA)_3$, $(MMT\text{-}PPPB\text{-}CHA)_3$.

3.3. CHA Loading Capacity and Releasing Kinetics of (MMT-PPPB-CHA)$_n$ Coating

Herein, CHA loading capacity and kinetics of CHA releasing in $(MMT\text{-}PPPB\text{-}CHA)_n$ coating were studied. As shown in Figure S6, the drug loading of $(MMT\text{-}PPPB\text{-}CHA)_n$ coating had a linear relationship with the number of layers (n). This result was in line with our expectation that number of layers could regulate the CHA loading dosages in $(MMT\text{-}PPPB\text{-}CHA)_n$ coating. According to Figure S7, the drug release rate of $(MMT\text{-}PPPB\text{-}CHA)_n$ coating was less than 30%, while the drug release rate of $(MMT\text{-}CHA)_n$ was more than 60% within 72 h. This indicated that $(MMT\text{-}PPPB\text{-}CHA)_n$ coating possessed good sustained release capacity.

3.4. Multistimulus-Responsive Drug Release Behavior of (MMT-PPPB-CHA)$_n$ Coating

The ability of sustained release alone does not play a role in inhibiting bacterial resistance. Prolonged exposure to drugs can encourage bacteria to develop drug resistance. Therefore, a controlled drug delivery system in response to bacterial infection microenvironment (weak acid environment (pH < 5), high peptidoglycan content, and high glutathione content) is very important. Considering the presence of multi-stimulus responsive PPPB, $(MMT\text{-}PPPB\text{-}CHA)_3$ coating should theoretically be capable of responsive drug release. Therefore, the multi-stimulus responsive drug release behavior of $(MMT\text{-}PPPB\text{-}CHA)_3$ coating was studied (Figure 5). As shown in Figure 5a, the release rate of CHA was only about

20% in the alkaline or neutral phase (pH 8.0–7.0). When pH reduced to 5.5, the CHA release rate increased to 40% within 72 h. The lower the pH value, the higher the CHA release rate. Moreover, according to Figure 5b,c, the CHA release rate increased with the increased glucose and DTT concentration. When glucose concentration was 1.0 mg/mL, the CHA release rate can reach about 35% and 65% within 72 h respectively. This responsive drug release behavior of the $(MMT\text{-}PPPB\text{-}CHA)_3$ coating was related to the dynamic boronate ester and disulfide bonds in the polymer PPPB. The dynamic covalent bonds in PPPB could be reversibly fractured in weak acid environment ($pH < 7.0$), high peptidoglycan solution (0.5 mg/mL), and high glutathione solution (0.5 mg/mL). When bond fracture occurs in PPPB, the layered structure of $(MMT\text{-}PPPB\text{-}CHA)_3$ coating gradually became loose and the gradual release of the CHA was started.

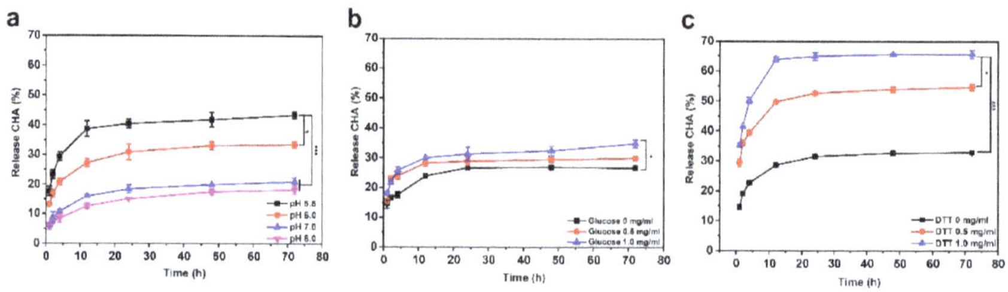

Figure 5. CHA release curves of $(MMT\text{-}PPPB\text{-}CHA)_3$ within 72 h in (**a**) different pH (5.5, 6.0, 7.0 and 8.0), (**b**) concentrations of glucose (0, 0.5, 1.0 mg/mL) and (**c**) DTT (0, 0.5, 1.0 mg/mL) solutions. * $p < 0.05$, *** $p < 0.001$.

The above results showed that $(MMT\text{-}PPPB\text{-}CHA)_3$ coating exhibited responsive drug release ability in different pH (5.5, 6.0, 7.0 and 8.0), concentrations of glucose (0, 0.5, 1.0 mg/mL) and DTT (0, 0.5, 1.0 mg/mL) solutions, respectively. However, these factors exist together in a real infectious environment. Therefore, we studied the influence of these three factors (pH, glucose, and DTT) on CHA release rate. Based on the investigation of single factors, the three factors, namely pH (factor A), glucose (factor B), and DTT (factor C) were selected. CHA release rate was regarded as investigating indicator for screening the optimal combination of three factors through the orthogonal test L_9 (3^4) (three factors and three levels) as shown in Table 1 [33]. The average drug release rate was the highest in group c and the lowest in group h, while the drug release levels of groups a, b, d, e, and f were similar. It indicated that the optimal combination of factors to promote CHA release was pH 6.0, glucose 1.0 mg/mL and DTT 1.0 mg/mL. However, the effects of the three factors on the drug release rate of the coating also have a certain order. To explore this order, we analyzed the extremum values (R) which indicates the importance degrees of three factors in Table [34]. By comparing the R-value, the significance order of the three factors on the average CHA release rate was pH > DTT > glucose. Therefore, it can be concluded that pH had the greatest influence on the drug release of the coating.

3.5. In Vitro Antibacterial Activity of $(MMT\text{-}PPPB\text{-}CHA)_n$ Coating

Given the results of orthogonal test, we further studied the antibacterial properties of $(MMT\text{-}PPPB\text{-}CHA)_3$ coating in vitro. As shown in Figure 6a,b, bacterial mortality rate of $(MMT\text{-}PPPB\text{-}CHA)_3$ group reached about 99% in 4 h at a normal environment against *S. aureus* and *E. coli*. When coating degradation triggered by weak acid environment (pH 6.0), glucose solution (1.0 mg/mL), and DTT solution (1.0 mg/mL), $(MMT\text{-}PPPB\text{-}CHA)_3$ coating could kill 99% of *S. aureus* and *E. coli* only in 2 h. We speculated that the reason for this might be the high CHA releasing rate of the $(MMT\text{-}PPPB\text{-}CHA)_3$ coating. According to Figures S6 and S7, the minimum release amount of CHA was 23 μg ($n = 3$). This dose (23 μg/mL) was still above the MIC of CHA (1.0 μg/mL) according to the literature we reviewed [35]. This caused the bacteria to be killed more quickly in higher concentrations of CHA.

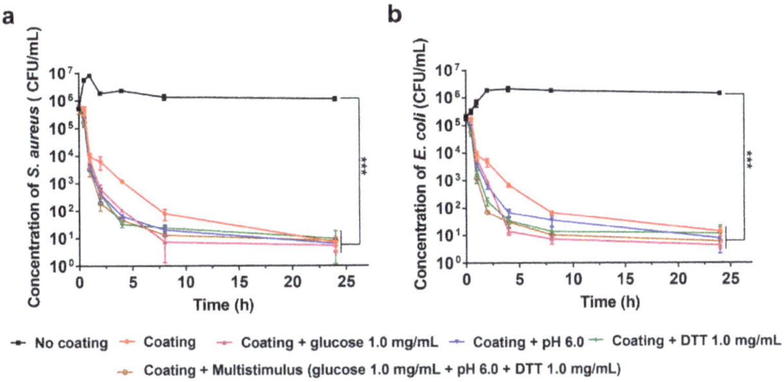

Figure 6. Stimulation-responsive killing effects of $(MMT\text{-}PPPB\text{-}CHA)_3$ on high concentrations of (**a**) *S. aureus* and (**b**) *E. coli* after incubating for 24 h. *** $p < 0.001$.

The efficient bactericidal ability can inhibit the development of bacterial resistance, but it does not mean that the medical device surface can be better protected. In particular, cationic antimicrobial modified surfaces are more likely to adsorb bacterial debris. If the dead bacteria can't be cleaned up quickly, they can provide an attachment site for other living bacteria. In this case, the antibacterial surface can quickly fail and the medical device becomes contaminated. Previous results have proved that the layered self-assembly $(MMT\text{-}PPPB\text{-}CHA)_3$ coating has strong electronegativity and hydrophilicity. According to the literature, these properties should endow the coating with good anti-adhesion capacity [18]. Herein, bacterial adhesion on $(MMT\text{-}PPPB\text{-}CHA)_3$ coating was observed in PBS buffer (pH 5.5, 6.0, 7.0 and 8.0), glucose aqueous solution (0.5 and 1.0 mg/mL) and DTT aqueous solution (0.5 and 1.0 mg/mL). FITC labelled *S. aureus* (2×10^6 CFU/mL) and *E. coli* (2×10^6 CFU/mL) were co-cultured with the $(MMT\text{-}PPPB\text{-}CHA)_3$ coating. As shown in Figure 7a, Figures S8a and S9a, $(MMT\text{-}PPPB\text{-}CHA)_3$ coating could effectively inhibit the growth of bacteria within 2 h in different solutions (PBS buffer (pH 5.5, 6.0, 7.0 and 8.0), glucose aqueous solution (0.5 and 1.0 mg/mL) and DTT aqueous solution (0.5 and 1.0 mg/mL)), respectively. The number of visible bacteria was significantly reduced (99% reduction) (Figures S10 and S11). Moreover, only a few bacteria fragments could be on the $(MMT\text{-}PPPB\text{-}CHA)_3$ coating (Figures 7b, S8b and S9b). By contrast, the bacteria survived well on the uncoated substrates (Figures 7c, S8c and S9c).

After proving the anti-adhesion property, we further explored the recyclability of the $(MMT\text{-}PPPB\text{-}CHA)_3$ coating. The substrate covered with different coatings was reused for six cycles (4 h as 1 cycle) (Figure 8). The concentration of the bacteria increased to 10^8–10^9 CFU/mL after 4 h in the uncoated group. For CHA coating (Figure 8b), the complete killing of *S. aureus* and *E. coli* was achieved in the first cycle and lost its antibacterial properties in the following cycle. As shown in Figure 8c, the $(MMT\text{-}CHA)_3$ coating achieved effect bacterial killing in two cycles but began to lose its antibacterial ability in the third cycle. According to Figure 8d, $(MMT\text{-}PPPB\text{-}CHA)_3$ coating completely killed *S. aureus* and *E. coli* in four cycles. In the fifth and sixth cycles, low concentrations (10^3 CFU/mL) of *S. aureus* and *E. coli* appeared on the surface of the coating, respectively. It could be thus concluded that $(MMT\text{-}PPPB\text{-}CHA)_3$ coating can maintain a long-term antibacterial effect.

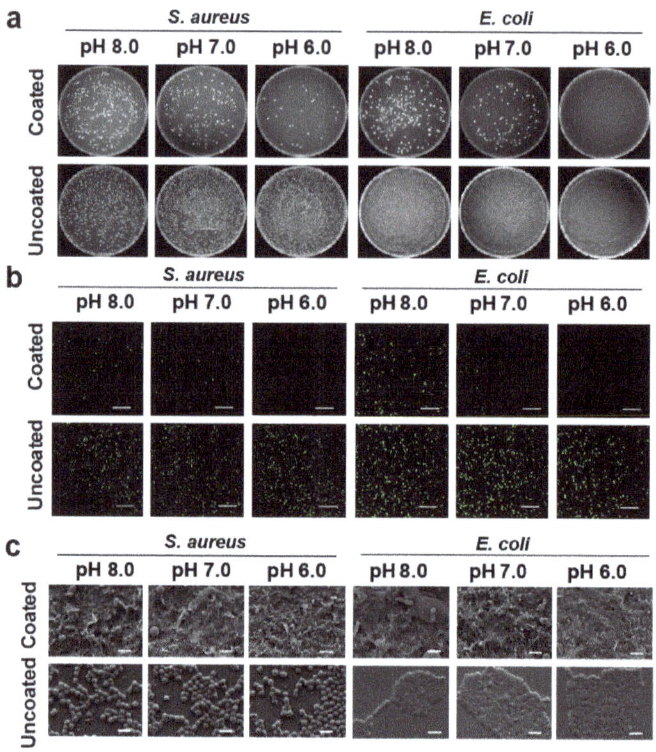

Figure 7. (**a**) Images of bacterial clones on the agar plates after treatment with $(MMT\text{-}PPPB\text{-}CHA)_3$ coating for 2 h under different pH values (6.0, 7.0, and 8.0); (**b**) CLSM diagrams and (**c**) SEM images of bacterial adhesion on the $(MMT\text{-}PPPB\text{-}CHA)_3$ coating at different pH values (6.0, 7.0 and 8.0) for 2 h. Scale bar: 20 μm and 1 μm.

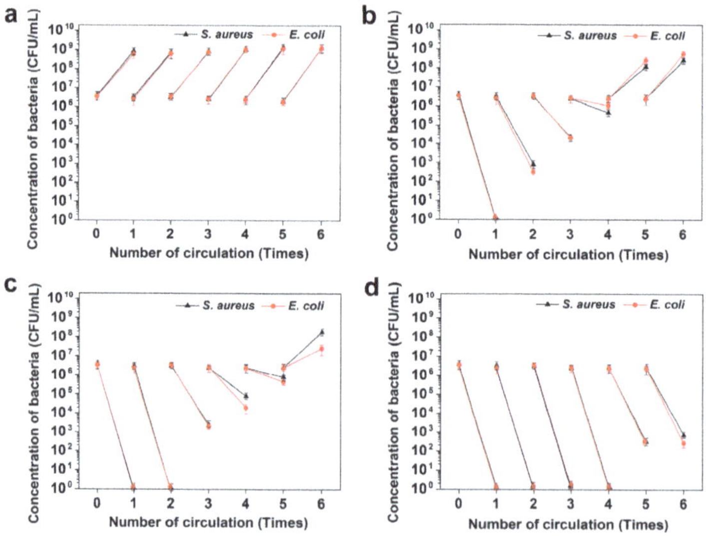

Figure 8. Bacteria cyclic killing effects of (**a**) Uncoated, (**b**) CHA coated, (**c**) $(MMT\text{-}CHA)_3$ coated and (**d**) $(MMT\text{-}PPPB\text{-}CHA)_3$ coated substrate against *S. aureus* and *E. coli*.

3.6. In Vivo Antibacterial Activity and Full-Thickness Skin Defect Healing of (MMT-PPPB CHA)$_n$ Coating

Given the highly effective antibacterial efficacy of (MMT-PPPB-CHA)$_3$ coating in vitro, we studied its in vivo antibacterial performance in an infected mouse model. The process of animal experimentation was depicted in Figure 9a. We inoculated the back of the mouse with bacteria, and cut it open with a scalpel to form an open resection of the infected wound where there were obvious signs of suppurative infection. Then, different coatings were used for treatment. The wound healing of the subcutaneous infected site in mice was observed and photographed on days 1, 5, 10, and 14, respectively, as shown in Figure 9b. The wounds treated with CHA, (MMT-CHA)$_3$ and (MMT-PPPB-CHA)$_3$ coatings showed no further deterioration and a good healing trend. As shown in Figure 9c,d, (MMT-PPPB-CHA)$_3$ coating treated wounds healed 91–97% after 14 days, respectively. However, the infected wounds in the untreated group and the blank substrate group were basically in a state of severe infection. Some symptoms of suppurative infection still existed after 14 days. The healing rates of *S. aureus* and *E. coli* infected wounds were 63% and 72%, respectively. After 14 days, the tissues of the infected site were fully soaked in sterile normal saline and inoculated on AGAR plates to observe the bacteria colony growth, as shown in Figure S11. It was observed that there were still many bacteria living in the infected sites of mice in the untreated group and blank group, indicating the infection did not been eliminated. By contrast, the wound tissue treated with CHA, (MMT-CHA)$_3$ and (MMT-PPPB-CHA)$_n$ coatings was free of any bacteria.

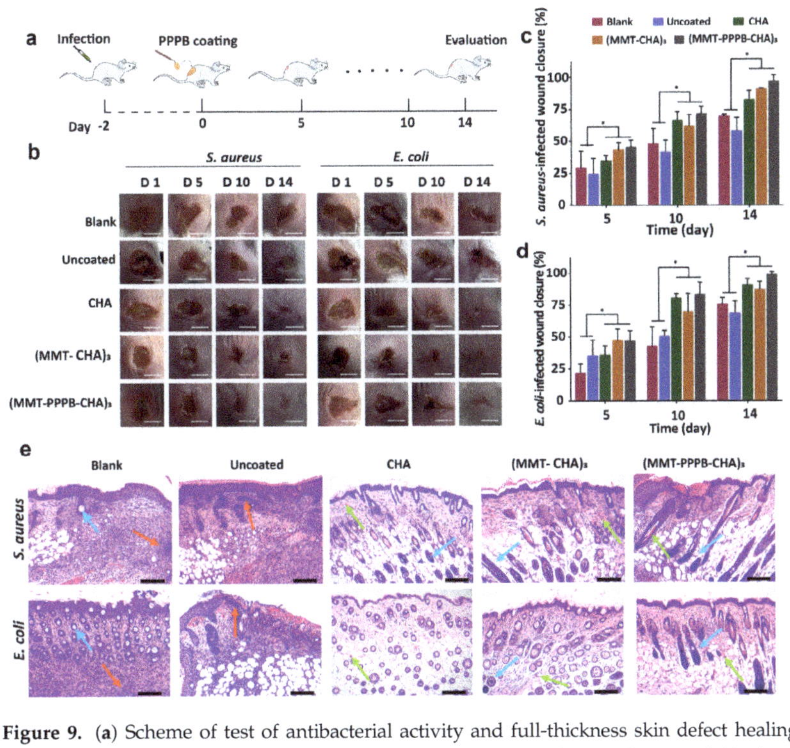

Figure 9. (**a**) Scheme of test of antibacterial activity and full-thickness skin defect healing in vivo; (**b**) Digital photographs of infected wound healing in vivo at days 1, 5, 10, and 14 after different treatments. Scale bar: 1 mm; (**c**) *S. aureus* and (**d**) *E. coli* infected wound closure rates during the 14 days in treatment; (**e**) H & E-staining results of healed skin tissues at determined times (blood vessels: green arrows, hair follicles: blue arrows, neutrophils: red arrows). Scale bar: 100 μm. * $p < 0.05$.

In clinical practice, medical instruments (e.g., peripherally inserted central catheter and urethral catheter) are exposed to wounds on the surface of the body for a long time [36]. During this time, bacteria have a chance to contaminate medical equipment and invade the body to cause infection [36]. Therefore, it is very important for the coating of medical devices to have high antibacterial properties and promote wound healing. Wound healing is a complex and orderly process consisting of four stages: hemostasis, inflammation, proliferation, and wound remodeling. Herein, H & E staining was applied to study the process of wound healing from a histological perspective (Figure 9e). The skin tissues in the untreated group and the blank substrate group exhibited acute inflammatory responses and displayed very little sprouting angiogenesis on the 14 days [37]. The results showed that the wound healing was in a transitional stage between inflammation and regeneration, indicating the infections in the wound were not resolved. Differently, the skin tissues treated $(MMT\text{-}PPPB\text{-}CHA)_3$ coatings had regenerated hair follicles and capillaries. It is worth noting that the tissue treated with $(MMT\text{-}PPPB\text{-}CHA)_3$ coating showed better hair follicle regeneration, while CHA and $(MMT\text{-}CHA)_n$ group showed more angiogenesis. Considering the wound can regenerate a large number of blood vessels during the regeneration stage, and complete the degeneration of some blood vessels and generate hair follicles during the remodeling stage. The presence of more hair follicles indicated the best wound recovery in $(MMT\text{-}PPPB\text{-}CHA)_3$ group. In conclusion, a comparison of wound healing of each group showed that $(MMT\text{-}PPPB\text{-}CHA)_3$ coating group healed best no matter *S. aureus* or *E. coli* infected wounds, indicating that $(MMT\text{-}PPPB\text{-}CHA)_n$ coating had good biocompatibility (Figure S12). Moreover, it had the best bactericidal effect on the subcutaneous infected site, thus accelerating the elimination of infection and promoting wound repair.

4. Conclusions

In this study, a multi-stimulus responsive multilayer coating with controlled release of CHA was constructed for the treatment of DAIs. The in vitro studies highlighted the pH-, glucose- and DTT-responsive abilities of $(MMT\text{-}PPPB\text{-}CHA)_n$ coating, and its antibacterial capacity. In medium with pH 6.0, glucose 1.0 mg/mL and DTT 1.0 mg/mL, the coating showed extremely high bactericidal ability (99% reduction). In real clinical scenario, the coating is intended to address bacterial infections that occur when medical devices (such as urinary catheters or peripheral venous catheters) are connected to open wounds in humans. The three components involved in the coating are biocompatible components. In vivo studies also show that the coating can kill bacteria quickly and promote wound healing within 14 days, which supports the coating as a non-toxic and highly bactericidal material. In addition, $(MMT\text{-}PPPB\text{-}CHA)_n$ coating could be reused 6 times with antibacterial ability, reducing the pain caused by frequent dressing changes for patients. However, it needs further research that whether the coating still has good antibacterial and biocompatibility in case applied for the stents and prosthetic devices in vivo.

Supplementary Materials: The following supporting information can be downloaded at: https://www.mdpi.com/article/10.3390/jfb13010024/s1. Chemical structural formula, ^{1}H-NMR spectrums, and FT-IR absorbance spectrums of BPBAC (Figure S1) and PCA-PEG1000 (Figure S2), the responsive mechanism of PPPB (Figure S3a), UV absorption spectrums of PPPB in different solutions (Figure S3b), SEM images (Figure S4) and water contact angles (Figure S5) of the surfaces of CA, PAN, PVC and PU substrates coated and uncoated with (MMT-PPPB-CHA)n,, the relationship between drug loading and the number the (MMT-PPPB-CHA)n bilayers (Figure S6), CHA release curves of (MMT-CHA)n and (MMT-PPPB-CHA)n coating (Figure S7), *S. aureus* and *E. coli* after treatment with (MMT-PPPB-CHA)n coating for 2 h (Figures S8 and S9), concentrations of *S. aureus* and *E. coli* after incubating with (MMT-PPPB-CHA)n coating (Figures S10 and S11), the growth of *S. aureus* and *E. coli* in infected wound tissues after 14 days of different treatments (Figure S12).

Author Contributions: Conceptualization, W.L., G.H. and L.R.; methodology, W.L., G.H. and J.C.; investigation, Y.Z., X.Z. and X.W.; resources, L.R.; data curation, W.L. and G.H.; writing—original draft preparation, W.L.; writing—review and editing, W.L., M.W., B.F. and L.R.; visualization, L.R.; supervision, L.R.; project administration, L.R.; funding acquisition, L.R. All authors have read and agreed to the published version of the manuscript.

Funding: This work was supported by the National Natural Science Foundation of China (Grant No. 31870994 and U1904206) and Fujian Provincial Science and Technology Cooperation Project (20210002).

Institutional Review Board Statement: The animal study protocol was approved by the Ethics Committee of the Institutional Animal Care and Use Committee of Xiamen University (protocol code: XMULAC20180003 and date of approval: 26 February 2018).

Informed Consent Statement: The study did not involve humans.

Conflicts of Interest: The authors declare no conflict of interest.

References

1. Brooks, B.; Olm, M.R.; Firek, B.A.; Baker, R.; Geller-McGrath, D.; Reimer, S.R.; Soenjoyo, K.R.; Yip, J.S.; Dahan, D.; Thomas, B.C.; et al. The developing premature infant gut microbiome is a major factor shaping the microbiome of neonatal intensive care unit rooms. *Microbiome* **2018**, *6*, 12. [CrossRef] [PubMed]
2. Burny, F.; Donkerwolcke, M.; Moulart, F.; Bourgois, R.; Puers, R.; Van Schuylenbergh, K.; Barbosa, M.; Paiva, O.; Rodes, F.; Begueret, J.B.; et al. Concept, design and fabrication of smart orthopedic implants. *Med. Eng. Phys.* **2000**, *22*, 469–479. [CrossRef]
3. Arciola, C.R.; Campoccia, D.; Montanaro, L. Implant infections: Adhesion, biofilm formation and immune evasion. *Nat. Rev. Microbiol.* **2018**, *16*, 397–409. [CrossRef] [PubMed]
4. Weiner-Lastinger, L.M.; Abner, S.; Edwards, J.R.; Kallen, A.J.; Karlsson, M.; Magill, S.S.; Pollock, D.; See, I.; Soe, M.M.; Walters, M.S.; et al. Antimicrobial-resistant pathogens associated with adult healthcare-associated infections: Summary of data reported to the National Healthcare Safety Network, 2015–2017. *Infect. Control. Hosp. Epidemiol.* **2020**, *41*, 1–18. [CrossRef]
5. Zhou, C.; Wu, Y.; Thappeta, K.R.V.; Subramanian, J.T.L.; Pranantyo, D.; Kang, E.T.; Duan, H.W.; Kline, K.; Chan-Park, M.B. In Vivo Anti-Biofilm and Anti-Bacterial Non-Leachable Coating Thermally Polymerized on Cylindrical Catheter. *ACS Appl. Mater. Interfaces* **2017**, *9*, 36269–36280. [CrossRef]
6. Wang, B.L.; Liu, H.H.; Wang, Z.F.; Shi, S.; Nan, K.H.; Xu, Q.W.; Ye, Z.; Chen, H. A self-defensive antibacterial coating acting through the bacteria-triggered release of a hydrophobic antibiotic from layer-by-layer films. *J. Mat. Chem. B* **2017**, *5*, 1498–1506. [CrossRef]
7. Wei, T.; Yu, Q.; Chen, H. Responsive and Synergistic Antibacterial Coatings: Fighting against Bacteria in a Smart and Effective Way. *Adv. Healthc. Mater.* **2019**, *8*, 1801381. [CrossRef]
8. Chen, R.; Shi, C.; Xi, Y.; Zhao, P.; He, H. Fabrication of cationic polymer surface through plasma polymerization and layer-by-layer assembly. *Mater. Manuf. Processes* **2020**, *35*, 221–229. [CrossRef]
9. Liu, P.; Fu, K.; Zeng, X.; Chen, N.; Wen, X. Fabrication and Characterization of Composite Meshes Loaded with Antimicrobial Peptides. *ACS Appl. Mater. Interfaces* **2019**, *11*, 24609–24617. [CrossRef]
10. Liu, Z.; Ma, S.; Duan, S.; Deng, X.; Sun, Y.; Zhang, X.; Xu, X.; Guan, B.; Wang, C.; Hu, M.; et al. Modification of Titanium Substrates with Chimeric Peptides Comprising Antimicrobial and Titanium-Binding Motifs Connected by Linkers To Inhibit Biofilm Formation. *ACS Appl. Mater. Interfaces* **2016**, *8*, 5124–5136. [CrossRef]
11. Bhaskara, U.R.; Warmoeskerken, M.; Nabers, M.G.D.; Nanomi, B.V. Antimicrobial Functionalization of beta-Cyclodextrin Treated Fabrics. *AATCC J. Res.* **2016**, *3*, 15–24. [CrossRef]
12. Zhou, C.; Wang, F.; Chen, H.; Li, M.; Qiao, F.; Liu, Z.; Hou, Y.; Wu, C.; Fan, Y.; Liu, L.; et al. Selective Antimicrobial Activities and Action Mechanism of Micelles Self-Assembled by Cationic Oligomeric Surfactants. *ACS Appl. Mater. Interfaces* **2016**, *8*, 4242–4249. [CrossRef]
13. Yao, Q.Q.; Ye, Z.; Sun, L.; Jin, Y.Y.; Xu, Q.W.; Yang, M.; Wang, Y.; Zhou, Y.L.; Ji, J.; Chen, H.; et al. Bacterial infection microenvironment-responsive enzymatically degradable multilayer films for multifunctional antibacterial properties. *J. Mat. Chem. B* **2017**, *5*, 8532–8541. [CrossRef]
14. Saha, K.; Butola, B.S.; Joshi, M. Synthesis and characterization of chlorhexidine acetate drug-montmorillonite intercalates for antibacterial applications. *Appl. Clay Sci.* **2014**, *101*, 477–483. [CrossRef]
15. Datta, S.M. Clay-polymer nanocomposites as a novel drug carrier: Synthesis, characterization and controlled release study of Propranolol Hydrochloride. *Appl. Clay Sci.* **2013**, *80–81*, 85–92. [CrossRef]
16. Fillion, M.; Goudreault, M.; Voyer, N.; Bechinger, B.; Auger, M. Amphiphilicity Is a Key Determinant in the Membrane Interactions of Synthetic 14-mer Cationic Peptide Analogues. *Biochemistry* **2016**, *55*, 6919–6930. [CrossRef]
17. Chen, H.; Jin, Y.Y.; Wang, J.J.; Wang, Y.Q.; Jiang, W.Y.; Dai, H.D.; Pang, S.Y.; Lei, L.; Ji, J.; Wang, B.L. Design of smart targeted and responsive drug delivery systems with enhanced antibacterial properties. *Nanoscale* **2018**, *10*, 20946–20962. [CrossRef]

18. Li, M.; Wang, H.Q.; Chen, X.M.; Jin, S.N.; Chen, W.; Meng, Y.C.; Liu, Y.; Guo, Y.S.; Jiang, W.Y.; Xu, X.; et al. Chemical grafting of antibiotics into multilayer films through Schiff base reaction for self-defensive response to bacterial infections. *Chem. Eng. J.* **2020**, *382*, 122973. [CrossRef]
19. Xu, Q.W.; Li, X.; Jin, Y.Y.; Sun, L.; Ding, X.X.; Liang, L.; Wang, L.; Nan, K.H.; Ji, J.; Chen, H.; et al. Bacterial self-defense antibiotics release from organic-inorganic hybrid multilayer films for long-term anti-adhesion and biofilm inhibition properties. *Nanoscale* **2017**, *9*, 19245–19254. [CrossRef]
20. Zhao, Y.Q.; Sun, Y.J.; Zhang, Y.D.; Ding, X.K.; Zhao, N.N.; Yu, B.R.; Zhao, H.; Duan, S.; Xu, F.J. Well-Defined Gold Nanorod/Polymer Hybrid Coating with Inherent Antifouling and Photothermal Bactericidal Properties for Treating an Infected Hernia. *ACS Nano* **2020**, *14*, 2265–2275. [CrossRef]
21. Wei, T.; Zhan, W.J.; Yu, Q.; Chen, H. Smart Biointerface with Photoswitched Functions between Bactericidal Activity and Bacteria-Releasing Ability. *ACS Appl. Mater. Interfaces* **2017**, *9*, 25767–25774. [CrossRef] [PubMed]
22. Nastyshyn, S.; Raczkowska, J.; Stetsyshyn, Y.; Orzechowska, B.; Bernasik, A.; Shymborska, Y.; Brzychczy-Wloch, M.; Gosiewski, T.; Lishchynskyi, O.; Ohar, H.; et al. Non-cytotoxic, temperature-responsive and antibacterial POEGMA based nanocomposite coatings with silver nanoparticles. *RSC Adv.* **2020**, *10*, 10155–10166. [CrossRef]
23. Zhou, Y.; Zheng, Y.J.; Wei, T.; Qu, Y.C.; Wang, Y.R.; Zhan, W.J.; Zhang, Y.X.; Pan, G.Q.; Li, D.; Yu, Q.; et al. Multistimulus Responsive Biointerfaces with Switchable Bioadhesion and Surface Functions. *ACS Appl. Mater. Interfaces* **2020**, *12*, 5447–5455. [CrossRef] [PubMed]
24. Chouirfa, H.; Bouloussa, H.; Migonney, V.; Falentin-Daudre, C. Review of titanium surface modification techniques and coatings for antibacterial applications. *Acta Biomater.* **2019**, *83*, 37–54. [CrossRef]
25. Wang, B.L.; Jin, T.W.; Xu, Q.W.; Liu, H.H.; Ye, Z.; Chen, H. Direct Loading and Tunable Release of Antibiotics from Polyelectrolyte Multilayers To Reduce Bacterial Adhesion and Biofilm Formation. *Bioconjugate Chem.* **2016**, *27*, 1305–1313. [CrossRef]
26. Manyi-Loh, C.; Mamphweli, S.; Meyer, E.; Okoh, A. Antibiotic Use in Agriculture and Its Consequential Resistance in Environmental Sources: Potential Public Health Implications. *Molecules* **2018**, *23*, 795. [CrossRef]
27. Iler, R.K. Multilayers of colloidal particles. *J. Colloid Interface Sci.* **1966**, *21*, 569–594. [CrossRef]
28. Richardson, J.J.; Cui, J.W.; Bjornmalm, M.; Braunger, J.A.; Ejima, H.; Caruso, F. Innovation in Layer-by-Layer Assembly. *Chem. Rev.* **2016**, *116*, 14828–14867. [CrossRef]
29. Zhang, X.H.; Ou-yang, S.; Wang, J.L.; Liao, L.; Wu, R.F.; Wei, J.C. Construction of Antibacterial Surface Via Layer-by-Layer Method. *Curr. Pharm. Des.* **2018**, *24*, 926–935. [CrossRef]
30. Singh, A.V.; Baylan, S.; Park, B.W.; Richter, G.; Sitti, M. Hydrophobic pinning with copper nanowhiskers leads to bactericidal properties. *PLoS ONE* **2017**, *12*, e0175428. [CrossRef]
31. Yu, H.-C.; Kim, Y.-K. Fractures of implant fixtures: A retroscccpective clinical study. *Maxillofac. Plast. Reconstr. Surg.* **2020**, *42*, 13. [CrossRef]
32. Lobato-Aguilar, H.; Uribe-Calderon, J.A.; Herrera-Kao, W.; Duarte-Aranda, S.; Baas-Lopez, J.M.; Escobar-Morales, B.; Cauich-Rodriguez, J.V.; Cervantes-Uc, J.M. Synthesis, characterization and chlorhexidine release from either montmorillonite or palygorskite modified organoclays for antibacterial applications. *J. Drug Deliv. Sci. Technol.* **2018**, *46*, 452–460. [CrossRef]
33. Song, X.P.; Tang, J. Extraction Optimization, Preliminary Characterization and Bioactivities in Vitro of Ligularia hodgsonii Polysaccharides. *Int. J. Mol. Sci.* **2016**, *17*, 788. [CrossRef]
34. Wang, W.T.; Wu, B.; Qin, H.; Liu, P.T.; Qin, Y.; Duan, G.W.; Hu, G.Q.; He, M.X. Genome shuffling enhances stress tolerance of Zymomonas mobilis to two inhibitors. *Biotechnol. Biofuels* **2019**, *12*, 288. [CrossRef]
35. Fan, W.; Sun, Q.; Li, Y.Y.; Tay, F.R.; Fan, B. Synergistic mechanism of Ag^{+}-Zn^{2+} in anti-bacterial activity against Enterococcus faecalis and its application against dentin infection. *J. Nanobiotechnol.* **2018**, *16*, 10. [CrossRef]
36. Li, W.T.; Zhang, H.X.; Li, X.L.; Yu, H.; Che, C.Y.; Luan, S.F.; Ren, Y.F.; Li, S.; Liu, P.; Yu, X.T.; et al. Multifunctional Antibacterial Materials Comprising Water Dispersible Random Copolymers Containing a Fluorinated Block and Their Application in Catheters. *ACS Appl. Mater. Interfaces* **2020**, *12*, 7617–7630. [CrossRef]
37. Joshi, K.M.; Shelar, A.; Kasabe, U.; Nikam, L.K.; Pawar, R.A.; Sangshetti, J.; Kale, B.B.; Singh, A.V.; Patil, R.; Chaskar, M.G. Biofilm inhibition in Candida albicans with biogenic hierarchical zinc-oxide nanoparticles. *Mater. Sci. Eng. C* **2021**, 112592. [CrossRef]

Journal of
Functional Biomaterials

Article

Synergistic Antibacterial Effect of Zinc Oxide Nanoparticles and Polymorphonuclear Neutrophils

Kai Ye [1,†], Moran Huang [1,2,†], Xiaojian He [3], Zhiquan An [1,*] and Hui Qin [1,*]

1 Department of Orthopedic Surgery, Shanghai Jiao Tong University Affiliated Sixth People's Hospital, NO.600 Yishan Road, Shanghai 200233, China; yekai706@163.com (K.Y.); huangmoran44@126.com (M.H.)
2 Department of Orthopaedics, Affiliated Anhui Provincial Hospital of Anhui Medical University, NO.17 Lujiang Road, Hefei 230001, China
3 Department of Orthopaedics, Qingpu Branch of Zhongshan Hospital, Fudan University, NO.1158 Gongyuandong Road, Shanghai 201700, China; hexiaojian1@126.com
* Correspondence: anzhiquan@126.com (Z.A.); qh197826@163.com (H.Q.)
† These authors contributed equally to this work.

Abstract: Zinc oxide nanoparticles (ZnONPs) are inorganic nano-biomaterials with excellent antimicrobial properties. However, their effects on the anti-infection ability of the innate immune system remains poorly understood. The aim of the present study was to explore the potential immunomodulatory effects of ZnONPs on the innate immune system, represented by polymorphonuclear leukocytes (PMNs), and determine whether they can act synergistically to resist pathogen infections. In vitro experiment showed that ZnONPs not only exhibit obvious antibacterial activity at biocompatible concentrations but also enhance the antibacterial property of PMNs. In vivo experiments demonstrated the antibacterial effect of ZnONPs, accompanied by more infiltration of subcutaneous immune cells. Further ex vivo and in vitro experiments revealed that ZnONPs enhanced the migration of PMNs, promoted their bacterial phagocytosis efficiency, proinflammatory cytokine (TNF-α, IL-1β, and IL-6) expression, and reactive oxygen species (ROS) production. In summary, this study revealed potential synergistic effects of ZnONPs on PMNs to resist pathogen infection and the underlying mechanisms. The findings suggest that attempts should be made to fabricate and apply biomaterials in order to maximize their synergy with the innate immune system, thus promoting the host's resistance to pathogen invasion.

Keywords: antimicrobial property; polymorphonuclear leukocyte; synergistic effect; zinc oxide nanoparticle

Citation: Ye, K.; Huang, M.; He, X.; An, Z.; Qin, H. Synergistic Antibacterial Effect of Zinc Oxide Nanoparticles and Polymorphonuclear Neutrophils. *J. Funct. Biomater.* **2022**, *13*, 35. https://doi.org/10.3390/jfb13020035

Academic Editors: Ik-Hwan Kim and John H.T. Luong

Received: 23 February 2022
Accepted: 16 March 2022
Published: 23 March 2022

Publisher's Note: MDPI stays neutral with regard to jurisdictional claims in published maps and institutional affiliations.

Copyright: © 2022 by the authors. Licensee MDPI, Basel, Switzerland. This article is an open access article distributed under the terms and conditions of the Creative Commons Attribution (CC BY) license (https://creativecommons.org/licenses/by/4.0/).

1. Introduction

Pathogenic bacterial infections in the orthopedic field, including implant and periprosthetic joint infections, constitute common clinical challenges that require repeated surgery and time-consuming therapy [1]. The incidence of orthopedic device infections varies from 2% to 5% clinically [2], and the dominating pathogens are *Staphylococcus* species, particularly *Staphylococcus aureus* (*S. aureus*) [3]. Devastating outcomes along with unacceptable economic burden are usually associated with implant and periprosthetic joint infections [4].

The immune system plays a crucial role in controlling pathogenic infections [5,6], as it is the body's main line of defense against pathogen invasions. Specifically, polymorphonuclear leukocytes (PMNs or neutrophils) are among the cell types that are mobilized as a part of the host's innate immune response in the initial stages, toward the site of infection [7]. PMNs can effectively neutralize pathogens through various mechanisms, including phagocytosis [8], chemotaxis [9], and superoxide production [10], thereby effectively preventing the development of infections. Hence, a fully functional PMN response represents a fundamental component of the host immune response to invading pathogens.

Recently, the emergence of nanoparticle-based antimicrobial agents has provided a new paradigm for antimicrobial treatments. These agents, including gold [11], silver [12], and copper [13] nanoparticles, possess good antibacterial properties in vitro. However, they are often criticized for their lack of corresponding effects in vivo [14]. Several studies have demonstrated limited in vivo antibacterial effects of implants coated with antimicrobial nanoparticles; thus, redundant efforts have been made for complicated modifications that are associated with high cost and erratic yield [15,16]. During bacterial infections, the phagocytic immune response can be suppressed by virulence factors liberated by bacteria [17,18]. Bacterial biofilms on implants can also impair microbial eradication by inhibiting phagocytosis of immune cells [19]. However, the present practice of nanoparticle-based antimicrobial treatment invariably neglects the restoration of immunosuppression, which may explain the inability of nanoparticles to maximize their antibacterial effects in vivo. Knowledge of interaction between nanoparticles and the immune system will help reduce in vivo applications of these agents. In other words, the development of agents that can mobilize or enhance the antibacterial ability of the body's immune system and simultaneously have superior antibacterial properties might be a breakthrough in the anti-infection research of nanomaterials [20–22].

Zinc oxide nanoparticles (ZnONPs) constitute a common type of nanoparticle with excellent antimicrobial properties [23,24]. They have the ability to mobilize immunoreaction. ZnONPs reportedly promote the proinflammatory response by activating monocytes [25] and releasing interleukin-8 by human bronchial epithelial cells [26]. There is also evidence that ZnONPs enhance the ability of PMNs to exert phagocytosis and degranulation [27–29]. The nanocomposites of ZnO/Ag nanoparticles armed in chitosan/polyethylene oxide nanofibers have been successfully applied to exert synergic antibacterial effect with fibroblast cells [30]. Nonetheless, the actual effect of the interaction between ZnONPs and the innate immune system in the context of protecting the host from pathogenic infection remains unclear [31–33].

Considering that the neutrophils are the first-line responders of the host's innate immune system [5–7], the exploration of the antibacterial effect under the interaction between ZnONPs and neutrophils may be a paradigm for the future application of nanomaterials that act synergistically with the host immune system to eliminate bacteria in early infection stages. In this study, in vitro nanoparticle–PMN–bacterial co-culture and in vivo *S. aureus*-infected air pouch models were utilized to explore the potential effects and underlying mechanisms of ZnONPs on neutrophils to resist pathogen infection.

2. Materials and Methods

2.1. Materials

2.1.1. Characterization of ZnONP Nanoparticles

ZnO nanoparticle dispersions with an average particle size of <20 nm were purchased from Maogon Nano (Shanghai, China). Particle morphology and size were characterized using TEM (JSM-IT200; JEOL, Tokyo, Japan). The chemical constituents and states of the elements were characterized by XPS (Escalab 250Xi; Thermo Fisher Scientific, Waltham, MA, USA).

2.1.2. Human PMN Isolation

This study was conducted in accordance with the principles of the Declaration of Helsinki, and ethical approval was granted by the review board of our institution. Heparinized venous blood was obtained from healthy volunteers after informed consent was obtained. PMNs were isolated by the neutrophil separating Kit (Solarbio, Beijing, China) using the gradient density centrifugation method [34]. Briefly, 4 mL of Solution A, 2 mL of Solution C, and 4 mL of heparinized human venous blood were successively added into a 15 mL centrifuge tube. Then, the tube was centrifuged at 610× *g* for 30 min with the lowest accelerated and retarded speeds. The neutrophils were carefully removed into another 15 mL centrifuge tube, resuspended with 5 mL Red Blood Cell Lysis Buffer, reacted for 5 min, and centrifuged at 250× *g* for 10 min. After discarding the supernatant, the neutrophils were then resuspended with 5 mL of PBS (Ca^{2+}/Mg^{2+} free) and centrifuged at 250× *g* for 10 min. The remaining neutrophils were finally resuspended with RPMI-1640 culture medium. Cell viability was determined by trypan blue exclusion method.

2.1.3. Bacterial Preparation

Freeze-dried gram-positive *S. aureus* (ATCC 43300) was acquired from the American Type Culture Collection (Manassas, VA, USA). Before each experiment, the bacteria were inoculated onto sheep blood agar plates and grown overnight at 37 °C. An individual colony was then selected, transferred into 4 mL of fresh trypticase soy broth (TSB) medium, and incubated on an orbital shaker overnight at 37 °C. The bacterial density was estimated by optical density (OD) at a wavelength of 600 nm and was finally maintained at 0.8–1.0. The bacteria were gradually diluted for subsequent experiments.

2.2. Biocompatibility Evaluation

Using a 96-well plate, PMN (1×10^5 per well) suspensions were incubated in RPMI-1640 complete culture medium (Hyclone Laboratories Inc., Logan, UT, USA) with ZnONPs included at various concentrations (0, 25, 50, 100, and 200 μg mL^{-1}) at 37 °C in a humidified atmosphere containing 5% CO_2. After co-culturing for 4 h, the medium was substituted with 0.1 mL RPMI-1640 culture medium. Next, 10 μL of CCK-8 solution (Dojindo, Kumamoto, Japan) was added to each well and incubated at 37 °C for 1 h. Finally, the plate was assessed with a microplate reader (Thermo Fisher Scientific, Waltham, USA) at a wavelength of 450 nm. An LDH leakage assay was also performed using an LDH assay kit (Beyotime, Shanghai, China), according to the manufacturer's protocol.

2.3. In Vitro Antibacterial Evaluation

To assess the antibacterial activity of ZnONPs, 1 mL of bacterial suspension (1×10^5 CFU mL^{-1}) was incubated with ZnONPs (0, 25, 50, and 100 μg mL^{-1}) in 24-well plates at 37 °C for 4 h. The SPM was then used to evaluate the bacterial count of the culture medium with different nanoparticle concentrations. To further assess the antibacterial activity of PMNs under the influence of ZnONPs, 1-mL PMN suspensions (1×105 cells mL^{-1}) were incubated with bacteria (1×10^5 CFU mL^{-1}) and ZnONPs (0, 25, 50, and 100 μg mL^{-1}) in 24-well plates at 37 °C for 4 h. SPM was also used to estimate the bacterial count, which was presented as $\log_{10}$ CFU.

2.4. In Vivo S. aureus-Infected Air Pouch Model

2.4.1. In Vivo *S. aureus* Infected-Air Pouch Model

The animal operations were approved by our Animal Care and Experiment Committee and performed in accordance with the NIH Guide for Care and Use of Laboratory Animals guidelines. Twenty-four 8-week-old female Sprague–Dawley (SD) rats were acclimated for 1 week under standard conditions. Before each operation, the rats were anesthetized intraperitoneally with 3% pentobarbital sodium. The operation area was located at the intra-scapular area, which was shaved and sterilized twice with 70% ethanol. Using a 20 mL syringe and a 0.2 μm sterile filter, 20 mL of sterile air was injected subcutaneously to generate an air pouch. Three days later, the rats were anesthetized again, and an additional 10 mL of sterile air was injected to maintain the air pouch. Six days after the first injection, the rats were randomly divided into four groups ($n = 6$): control group, *S. aureus*-infected group, 50 μg mL^{-1} ZnONPs + *S. aureus*-infected group, and 100 μg mL^{-1} ZnONPs + *S. aureus*-infected group. Next, 2 mL of PBS, *S. aureus* (1×10^7 CFU mL^{-1}) with ZnONPs (50 μg mL^{-1}), and *S. aureus* (1×10^7 CFU mL^{-1}) with ZnONPs (100 μg mL^{-1}) were injected into the air pouches of each group member.

2.4.2. In Vivo Microbiological Assessment

The exudate and the tissue of the air pouch was harvested 24 h after the final injection. Briefly, before six rats from each group were euthanized, 5 mL of sterile PBS was injected into the air pouch, and the contents were mixed by massaging them gently. A sagittal incision was made to collect the total volume of the exudate. The number of bacteria was calculated using the SPM.

2.4.3. Histological Study

Three rats from each group were euthanized, and partial full-thickness skin within the air pouch area was harvested and fixed with 10% paraformaldehyde solution at room temperature. After being washed with fresh PBS, the skin samples were dehydrated with an alcohol gradient and embedded in paraffin. Histological sections were then prepared using a sledge microtome (Leica, Hamburg, Germany). Finally, sample sections were deparaffinized with a xylene solution, subjected to H&E and Giemsa staining, and observed under an optical microscope (Nikon, Tokyo, Japan).

2.5. In Vitro Mechanism Study

2.5.1. Chemotactic Migration of PMNs

For the chemotactic migration assay, RPMI-1640 complete medium loaded with 1×10^5 PMNs and various concentrations of ZnONPs (0, 25, 50, and 100 μg mL^{-1}) in a volume of 0.2 mL were added into the upper chambers of 24-well transwell plates (pore size: 3 μm; Corning, NY, USA). The lower chamber was loaded with 0.5 mL of RPMI-1640 complete medium containing various concentrations of ZnONPs (0, 25, 50, and 100 μg mL^{-1}) and bacterial suspensions (1×10^6 CFU mL^{-1}). After being incubated for 2 h at 37 °C, the non-migrating PMNs were carefully removed using a swab. The migrated PMNs in the membrane were dipped in 4% paraformaldehyde for 10 min, transferred into 0.1% crystal violet for 15 min and gently washed thrice with PBS. The chambers were viewed under an optical microscope (Nikon, Tokyo, Japan).

2.5.2. Quantitative Real-Time PCR Detection of Proinflammatory Cytokines Released by PMNs In Vitro

To evaluate the expression of tumor necrosis factor (TNF)-α), interleukin (IL)-1β, and IL-6, 1 mL of 1×10^6 PMNs was incubated with various concentrations of ZnONPs (0, 25, 50, and 100 μg mL^{-1}) and a bacterial suspension (1×10^7 CFU mL^{-1}) in 24-well plates at 37 °C for 4 h. The culture medium was then centrifuged at $250\times$ *g* for 10 min and resuspended in TRIzol reagent (Invitrogen, Waltham, MA, USA). Total RNA from PMNs was extracted and quantified according to the manufacturer's protocol. Total RNA (1 μg) was reverse transcribed into cDNA using the PrimeScript RT reagent Kit (TaKaRa, Shiga, Japan). RT-qPCR was performed using SYBR Premix Ex Taq (Takara) with a LightCycler480 system (Roche, Mannheim, Germany).

2.5.3. Detection of Phagocytosis of PMNs Isolated from Human Whole Blood Using a Bacteria–Blood–ZnONPs Co-Culture Model

Whole blood (0.5 mL) was mixed with bacteria (1×10^7 CFU mL^{-1}) and ZnONPs (0, 25, 50, and 100 μg mL^{-1}), and subsequently incubated in a 24-well plate at 37 °C for 0, 5, 15, 30, 45, 60, and 90 min. At each time point, three random sample smears for each group were created for Wright–Giemsa staining and observed under an optical microscope (Nikon, Tokyo, Japan). The phagocytosis rate was calculated as follows. Six isolated microscopic fields were evaluated for each group at each time point. The number of phagocytosed, adherent, and dissociated bacteria within 10 μm of a PMN was counted. Phagocytosis rate of PMN (%) = (phagocytosed count/the sum of phagocytosed, bound, and dissociated bacteria count) × 100.

2.5.4. Cell Lysis of PMNs during the Antibacterial Process

A 10-μL suspension of bacteria (1×10^7 CFU mL^{-1}) was added to 0.1 mL of PMNs (1×10^5 cells mL^{-1}) and co-cultured with various concentrations of ZnONPs (0, 25, 50, and 100 μg mL^{-1}) in 96-well plates at 37 °C for 1.5 h. LDH release was detected by an LDH assay kit (Beyotime, Shanghai, China). The evaluation was based on the OD value at a wavelength of 490 nm. Cell lysis rate (%) = (experimental group–control group)/(cell disruption group–control group).

2.5.5. Reactive Oxygen Species (ROS) Production by PMNs In Vitro

ROS production was measured by a ROS assay kit (Beyotime, Shanghai, China). To evaluate the influence of ZnONPs on the capacity of PMNs to produce ROS, 0.1 mL of the latter (1×10^5 mL^{-1} cells mL^{-1}) was co-cultured with ZnONPs (0, 25, 50, and 100 μg mL^{-1}) in 96-well plates at 37 °C for 1.5 h. To evaluate the effect of ZnONPs on the capacity of PMNs to produce ROS during the antibacterial process, 10 μL of a bacterial suspension (1×10^7 CFU mL^{-1}) was added simultaneously. Next, 2′,7′-dichlorodihydrofluorescein diacetate (DCFH-DA) diluted with serum-free RPMI-1640 medium was added into each well at a final concentration of 10 μM. After an additional incubation time of 30 min in the dark, the cells were centrifuged and washed twice with PBS to remove the residual DCFH-DA. Fluorescence intensity was examined under an inverted fluorescence microscope, and quantitative measurements were conducted using a fluorescence microplate reader (Thermo Fisher Scientific, Waltham, MA, USA).

2.5.6. Superoxide Production by PMNs In Vitro

Following the manufacturer's protocol for the Superoxide Assay Kit (Beyotime, Shanghai, China), PMNs were collected, centrifuged, and resuspended in a 0.2 mL working solution at a final concentration of 1×10^5 cells mL^{-1}. The PMNs were then incubated with various concentrations of ZnONPs (0, 25, 50, and 100 μg mL^{-1}) in 96-well plates, with or without bacteria (1×10^6 CFU mL^{-1}), for 2 h. The OD value was measured at a wavelength of 450 nm.

2.5.7. Statistical Analysis

Experimental data were displayed as the mean ± standard deviation. Statistical analyses were conducted by one-way analysis of variance and two-tailed Student's t-test. SPSS software (version 25.0, IBM Corp, Armonk, NY, USA) was used to perform statistical analyses. Significance was set at a p value < 0.05.

3. Results

3.1. Characterization of ZnONPs

The transmission electron microscopy (TEM) results showed that the nanostructured ZnONPs had a particle size of approximately 20 nm (Figure 1A). The elemental composition of the ZnONPs was further determined by X-ray photoelectron spectroscopy (XPS; Figure 1B,C). Results showed that the binding energies of Zn2p were at 1044.6 eV for Zn2p1 and 1021.5 eV for Zn2p3, indicative of the characteristic Zn signals.

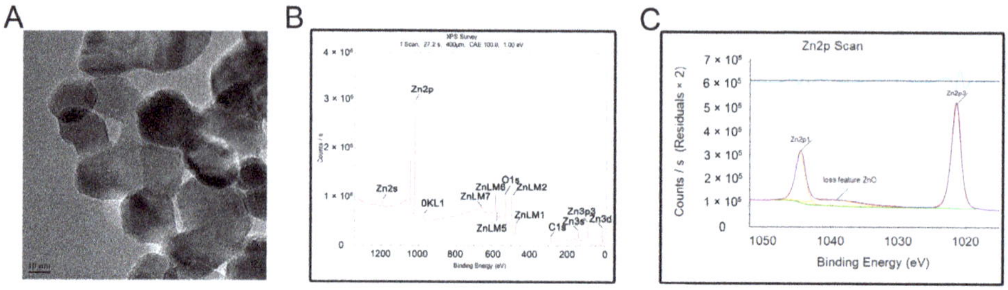

Figure 1. Material characterization of zinc oxide nanoparticles (ZnONPs). (**A**) Transmission electron microscopy of ZnONPs. (**B**,**C**) X-ray photoelectron spectroscopy showed the elementary composition of ZnONPs.

3.2. *In Vitro Biocompatibility Evaluation of ZnONPs*

PMN cell viability was detected following incubation with ZnONPs, using the CCK-8 assay (Figure 2A). After 4 h of incubation, ZnONPs at concentrations of 25, 50, and 100 μg mL^{-1} did not significantly affect the viability of PMNs compared with the control group, whereas 200 μg mL^{-1} resulted in significantly decreased viability. This was further verified by a lactate dehydrogenase (LDH) leakage assay (Figure 2B). LDH release was equivalent between the control group and the groups treated with ZnONPs at 25, 50, and 100 μg mL^{-1}; however, it was increased significantly in the group treated with 200 μg mL^{-1} ZnONP. Therefore, concentrations of 25, 50, and 100 μg mL^{-1} were used in subsequent experiments.

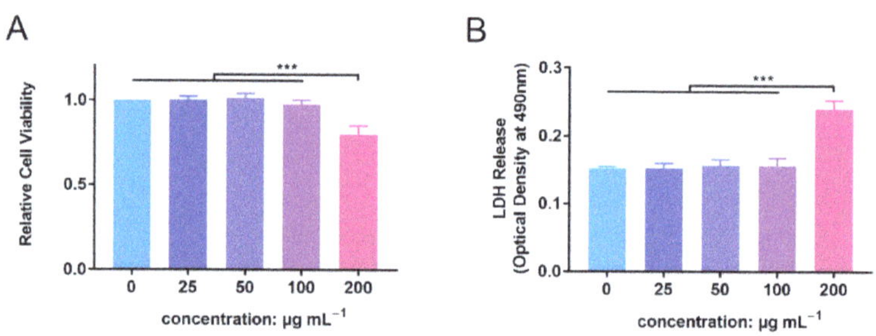

Figure 2. Cytotoxicity of zinc oxide nanoparticles (ZnONPs) on polymorphonuclear neutrophils (PMNs). (**A**) Relative cell viability of PMNs after co-culture with ZnONPs at various concentrations (0, 25, 50, 100, 200 μg mL^{-1}) for 4 h using CCK-8 assay. (**B**) Lactate dehydrogenase (LDH) release of PMNs treated with ZnONPs at various concentrations (0, 25, 50, 100, 200 μg mL^{-1}) for 4 h. *** $p < 0.001$.

3.3. *In Vitro Antimicrobial Evaluation of ZnONPs and PMNs*

Antibacterial activity was verified using the spread-plate method (SPM) and demonstrated using $\log_{10}$ CFU mL^{-1} values. After 4 h of incubation, the ZnONP-treated groups showed obvious antibacterial activity compared with the positive control group (Figure 3A,B). This effect was more obvious in the group treated with a ZnONP concentration of 100 μg mL^{-1} than in the other two groups (25 and 50 μg mL^{-1}). The bacterial count decreased in the PMN-treated group. Interestingly, the bacterial counts decreased further in the PMN-treated groups under the influence of ZnONPs at various concentrations and this decrease presented in a dose-dependent manner.

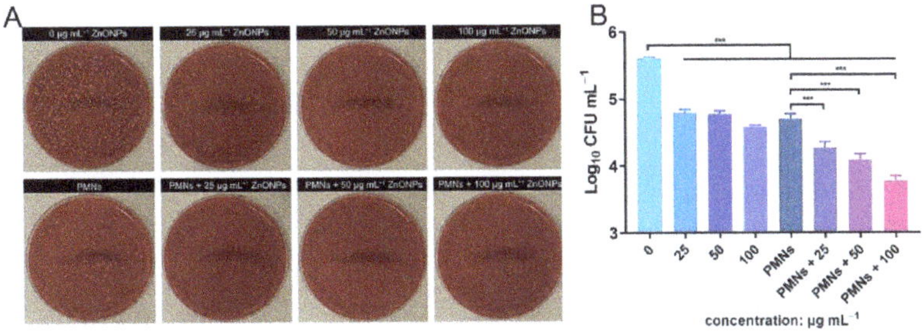

Figure 3. In vitro evaluation of the antibacterial activity elicited by zinc oxide nanoparticles (ZnONPs) alone and polymorphonuclear neutrophils (PMNs) treated with ZnONPs for 4 h. (**A**) Representative photos of the blood plate medium used for co-culturing of *Staphylococcus aureus* and ZnONPs (0, 25, 50, 100 μg mL^{-1}), with or without PMNs. (**B**) Quantitative evaluation using spread-plate method (SPM). *** $p < 0.001$.

3.4. In Vivo Antibacterial Evaluation of ZnONPs Using the Infected Air Pouch Model

The bacterial count of the lavage harvested from the air pouch was evaluated using SPM. The bacterial count decreased in the ZnONP groups in a concentration-dependent manner compared to the positive control group (Figure 4A). This was further verified by quantitative analysis using $\log_{10}$ CFU mL^{-1} values (Figure 4B). These results indicate that ZnONPs exhibit antibacterial activity in vivo.

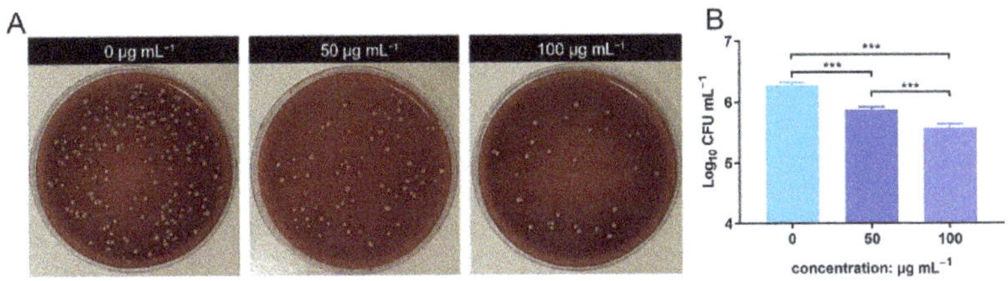

Figure 4. In vivo anti-bacteria activity of zinc oxide nanoparticles (ZnONPs) using the air pouch model infected with *Staphylococcus aureus* for 24 h. (**A**) Representative photos of the blood plate medium used for lavage harvest from the air pouch of groups treated with ZnONPs (0, 50, 100 μg mL^{-1}). (**B**) Quantitative evaluation of the bacteria counts. *** $p < 0.001$.

3.5. Histological Evaluation of the S. aureus-Infected Air Pouch Model

Hematoxylin and eosin (H&E) staining, and Giemsa staining, were performed for the histological examination of the harvested air pouch (Figure 5). In the H&E stained slices, immune cell infiltration was the lowest in the control group, while the an obvious immune cell infiltration was observed in the bacteria-infected group; this effect was promoted by ZnONPs treatment, in a concentration-dependent manner. Moreover, the Giemsa staining results also reflected the antibacterial function of ZnONPs, with more bacteria observed in the bacteria-infected group, fewer observed in the ZnONP-treated groups, and no bacteria detected in the control group.

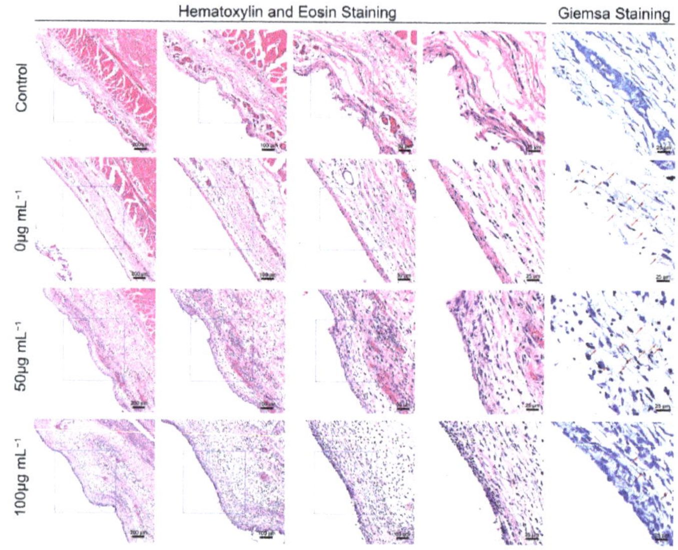

Figure 5. Histological evaluation of the full-thickness skin harvested from the air pouch after 24 h of the establishment of the model. Hematoxylin and eosin staining (50×, 100×, 200×, 400×) showed the neutrophil infiltration, and Giemsa staining (400×) mainly showed bacteria residual of the control group and infected groups treated with ZnONPs (0, 50, 100 μg mL^{-1}). Red arrows indicated the bacteria residual.

3.6. ZnONP Treatment Enhances PMN Migration and Cytokine Expression In Vitro

The transwell assay facilitated the detection of PMN migration activity under the influence of ZnONP nanoparticles. The number of PMNs that migrated through the membrane increased in following treatment with ZnONP, in a concentration-dependent manner (Figure 6A). The quantitative cell count in the transwell test also reflected this result (Figure 6B). Further, the ZnONP, PMN, and bacteria co-culture increased the mRNA expression of inflammation-related cytokines (TNF-α, IL-1β, and IL-6) in a dose-dependent manner (Figure 6C–E). These results suggest that ZnONPs promote PMN migration and proinflammatory cytokine expression.

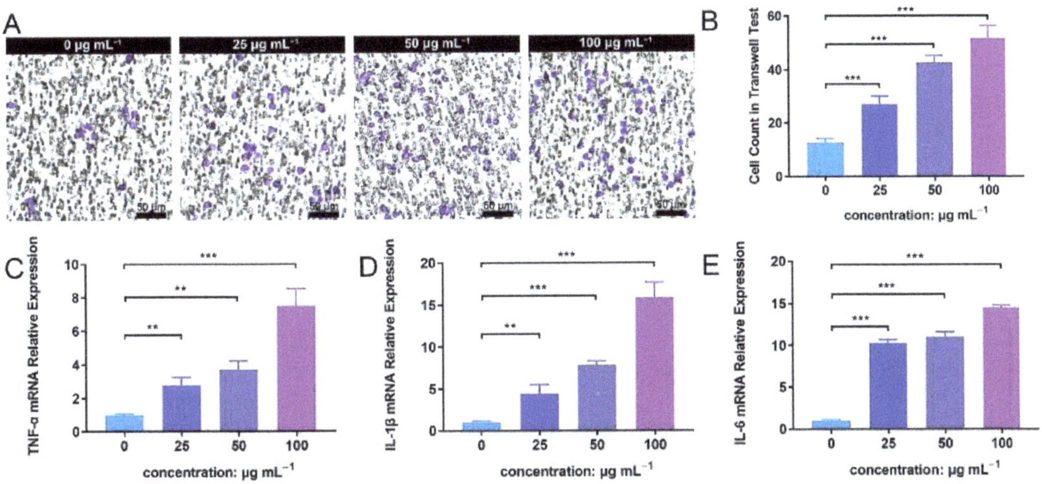

Figure 6. Cell migration and proinflammatory cytokine mRNA expression of polymorphonuclear neutrophils (PMNs) in the presence of bacteria treated with zinc oxide nanoparticles (ZnONPs). (**A**) Representative images of transwell tests demonstrating cell migration under the influence of *Staphylococcus aureus* and ZnONPs (0, 25, 50, 100 μg mL^{-1}) for 2 h. (**B**) Quantitative calculation of migrant PMNs. (**C–E**) Proinflammatory cytokine mRNA expression (TNF-α, IL-1β and IL-6) of PMNs in co-culture with *Staphylococcus aureus* and ZnONPs (0, 25, 50, 100 μg mL^{-1}) for 4 h. ** $p < 0.01$, *** $p < 0.001$.

3.7. ZnONPs Increase Phagocytosis of Bacteria by PMNs in Human Whole Blood

To further explore the effects of ZnONPs on PMNs, a phagocytosis assay was conducted with human whole blood (Figure 7A). In the established bacteria–blood–ZnONP co-culture model, the number of bacteria ingested by PMNs increased gradually with incubation time. Meanwhile, the number of bacteria phagocytosed by PMNs was significantly increased in co-cultures with ZnONPs.

The mean phagocytosis rates of the untreated group and ZnONPs treated groups (0, 25, 50, and 100 μg mL^{-1}) were 8%, 10%, 10%, and 11%, respectively, at 5 min; 25%, 26%, 30%, and 33%, respectively, at 15 min; 40%, 50%, 53%, and 62%, respectively, at 30 min; 47%, 61%, 72%, and 80%, respectively, at 45 min; 54%, 65%, 77%, and 88%, respectively, at 60 min; and 55%, 66%, 78%, and 89%, respectively, at 90 min. The overall trend of the phagocytosis rate was the lowest in the untreated group and higher in the ZnONP groups, in a concentration-dependent manner, with a more significant effect observed after 30 min of incubation (Figure 7B), indicating that ZnONPs demonstrated a synergistic immune regulatory function with PMNs by promoting phagocytosis.

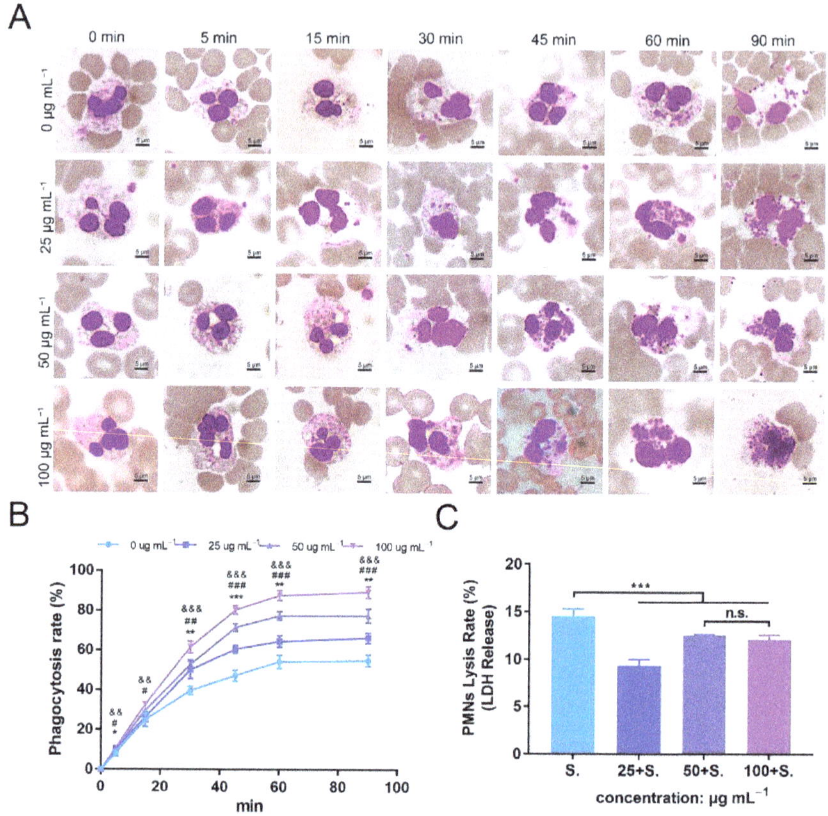

Figure 7. Phagocytosis of *Staphylococcus aureus* by polymorphonuclear neutrophils (PMNs) isolated from human whole blood, and PMN lysis rate, under co-culture with zinc oxide nanoparticles (ZnONPs). (**A**) Representative micrographs (1000×) of phagocytosis following Wright–Giemsa staining during continuous incubation (5, 15, 30, 45, 60, 90 min) with ZnONP (0, 25, 50, 100 μg mL^{-1}) treatment. (**B**) Phagocytosis rate of *Staphylococcus aureus* by PMNs. * 25 μg mL^{-1} group versus 0 μg mL^{-1} group, & 50 μg mL^{-1} group versus 0 μg mL^{-1} group, #100 μg mL^{-1} group versus 0 μg mL^{-1} group; *, &, # $p < 0.05$; **, ##, && $p < 0.01$; ***, &&&, ### $p < 0.001$. (**C**) Cell lysis rate of PMNs after co-culturing with *Staphylococcus aureus* and ZnONPs (0, 25, 50, 100 μg mL^{-1}). *** $p < 0.001$, n.s. represents no statistical significance.

3.8. ZnONPs Cause Slight Decrease in Lysis of PMNs Isolated from Human Whole Blood

The degree of PMN lysis was also determined with the average PMN lysis rate found to be 14.6%, 9.3%, 12.5%, and 12.1% in groups treated with ZnONP concentrations of 0, 25, 50, and 100 μg mL^{-1}, respectively (Figure 7C). Thus, ZnONPs slightly, yet significantly, attenuated the degree of PMN lysis.

3.9. ZnONPs Activate PMN ROS and Superoxide Production In Vitro

To further investigate the enhanced antibacterial effect of PMNs in the presence of ZnONPs, as well as the mechanism of their synergistic effects on the immune system, ROS and superoxide production were detected. Fluorescence levels were negligible in the untreated group, with a slight increase observed following incubation of PMNs with ZnONPs (Figure 8A). Co-culturing with bacteria also induced an increase in ROS production by PMNs; however, this effect was amplified in the presence of ZnONPs, in a concentration-dependent manner. These results were confirmed by quantitative measurement of ROS fluorescence (Figure 8B). Moreover, superoxide production exhibited a similar tendency to that observed for ROS production (Figure 8C). These results suggest that the enhanced antibacterial effect elicited by PMNs, following co-culture with ZnONPs, may be due to increased production of ROS and superoxide.

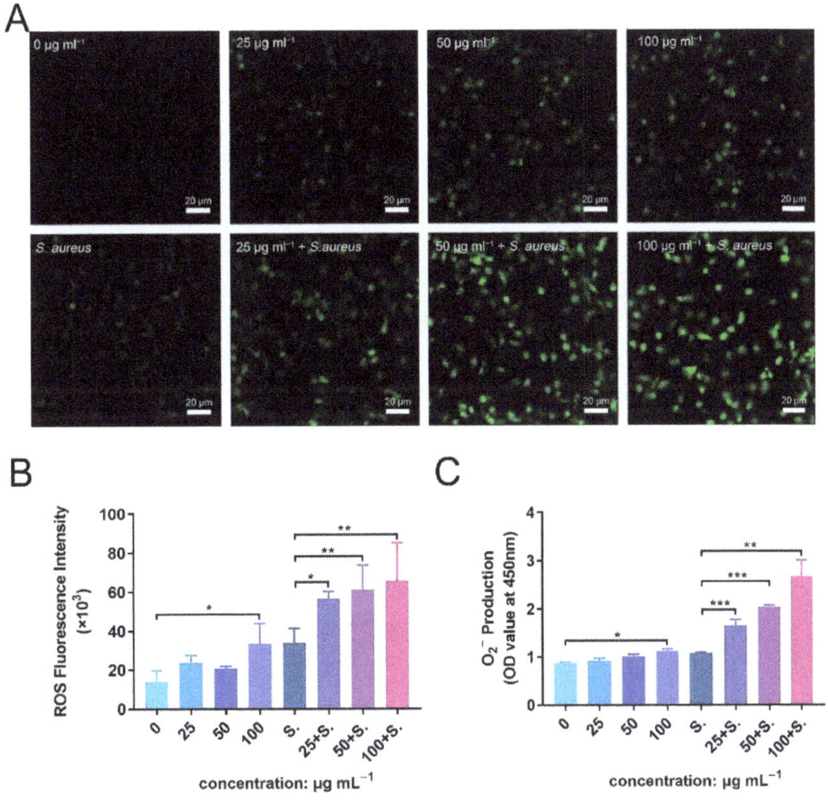

Figure 8. ROS and superoxide production by polymorphonuclear neutrophils (PMNs) after co-culturing with zinc oxide nanoparticles (ZnONPs), with or without bacteria. (**A**) Representative fluorescent images of ROS production in PMNs after 2 h of incubation with ZnONPs (0, 25, 50, 100 μg mL^{-1}), with or without *Staphylococcus aureus*. (**B**,**C**) Quantitative fluorescence measurement of ROS and superoxide (O^{2-}) production by PMNs after 2 h of incubation with ZnONPs (0, 25, 50, 100 μg mL^{-1}), with or without *Staphylococcus aureus*. * $p < 0.05$, ** $p < 0.01$, *** $p < 0.001$.

4. Discussion

In this study, we found that during bacterial infection, ZnONPs enhance the antibacterial property of PMNs, alleviate subcutaneous immune cell infiltration in vivo, enhance PMNs migration, improve bacterial phagocytosis efficiency of PMNs, promote PMNs to express the proinflammatory cytokines (TNF-α, IL-1β, and IL-6), and facilitate reactive oxygen species (ROS) production of PMNs. ZnONPs reportedly possess excellent antimicrobial properties, and they have been applied as antibacterial agents to prevent indwelling device infections [35,36]. Nanoparticles of various particle sizes, generally from 5 to 300 nm, have been reported to possess sufficient antimicrobial properties [37,38]. The antibacterial activity of ZnONPs is size dependent—ZnONPs of smaller particle sizes can easily penetrate bacterial membranes because of the large interfacial area [39]. However, decreased particle size inevitably affects biocompatibility [40]. The mechanism of cytotoxicity attributed to the dissolution of ZnONPs and the generation of ROS [20]. In this study, the particle size of ZnONPs was 20 nm, and these particles showed a good antibacterial efficacy in vitro. The chosen particle size is for the balance between antimicrobial properties and cell viability. Indeed, previous studies have confirmed that 20-nm ZnONPs at adequate concentrations possessed good biocompatibility [21]. The antimicrobial properties and biocompatibility are also influenced by nanoparticle concentration. Thus, we verified the cytocompatibility of ZnONPs and found no distinct cytotoxicity at a concentration gradient of below 100 μg mL^{-1}. This result is consistent with the findings of previous studies, which set 100 μg mL^{-1} as the default biosafety concentration for PMNs [27–29]. Thus, the antibacterial effect and biocompatibility determined in this study equilibrated at the chosen concentrations.

Neutrophils are among the cell types that are mobilized as a part of the host's innate immune response in the early stages; they act as first-line responders against pathogenic invasion [5–7]. Here, we found that the combination of ZnONPs and PMNs further reduced bacterial load in vitro. An in vitro study showed that ZnONPs enhance the phagocytosis of human neutrophils [27]. Another study reported that ZnONPs positively modulate the degranulation process in human neutrophils [28]. Several studies have also demonstrated the pro-inflammatory effect of ZnONPs on the innate immune system including macrophages [21,41]. Given the positive immunoregulation and pro-inflammatory effect of ZnONPs on neutrophils, they may exert a synergistic antibacterial effect, rather than exerting the effect independently. To further manifest the antibacterial effect of ZnONPs in vivo, we applied an *S. aureus*-infected air pouch model—a typical model to determine the antibacterial property of certain agents and reflect acute inflammation [42]. The results showed that ZnONPs significantly reduced bacterial count in the harvested exudate; Giemsa staining also showed a decrease in bacterial proportion in the subcutaneous tissue within the air pouch. These results were consistent with those of in vitro experiments. A previous study has also demonstrated that ZnONPs significantly attenuated the severity of infection in a localized soft tissue infection model owing to their antimicrobial property [43]. Interestingly, even though bacterial burden was relieved under the intervention of ZnONPs, the subcutaneous immune cell infiltration was increased following ZnONP intervention, which indicated a potential promotion effect of ZnONPs on the innate immune system. It is worth mentioning that although local bacterial burden was relieved by ZnONPs, substances released by multiple bacteria may still trigger immune cell migration. Nevertheless, the exact effect of ZnONPs on neutrophils during antibacterial process needs further verification.

When encountering local pathogen infection, the host exerts inflammatory responses by recruiting multiple immune cells from the blood, nearly 70% of which are neutrophils—the first leukocytes to be recruited to an infected site [7,44,45]. Excessive mobilization and recruitment of neutrophils are a prerequisite for local tissue cell infiltration [44,46]. The results of the Transwell assay in our study showed that ZnONPs significantly facilitated neutrophil recruitment. This effect was possibly mediated by the release of zinc ions that dissolved from ZnONPs in aqueous medium [47,48]. During the neutrophil recruitment cascade, neutrophils in the circulation are mobilized by the change in the surface of endothelium. This process results from the stimulation of inflammatory mediators, including cytokines that are released from tissue-resident neutrophils in contact with bacteria [7]. Using the ZnONP, PMN, and bacterial co-culture system, we proved that ZnONPs enhanced proinflammatory cytokine expression in PMNs at the mRNA level. Although a few studies have reported the proinflammatory effect of ZnONPs on PMNs during antibacterial process, several studies have demonstrated proinflammatory properties of ZnONPs on other types of immune cells. Poon et al. [25] proved that ZnONPs enhanced TNF-α and IL-1β expression in THP-1 (acute monocytic human leukemia) cells. ZnONPs were also found to stimulate macrophages and primary dendritic cells to release TNF-α and IL-6 [21]. Roy et al. [41] demonstrated that ZnONP-activated macrophages upregulated TNF-α, IL-1β, and IL-6 expression depending on MAPK signaling. Excessive production of proinflammatory cytokines could further trigger consecutive antibacterial actions of neutrophils including immune cell recruitment, phagocytosis, and ROS production [49].

After reaching the inflammatory site and experiencing multiple stimuli, neutrophils are more activated as phagocytic cells than circulating neutrophils [46]. Multitudinous of stimuli within the inflammatory site, such as lipopolysaccharide, chemokines, and cytokines, will facilitate neutrophils to mount an increased activation following successive stimuli [50]. However, in the orthopedic field, conditions become more complex. This is because in early-stage infections after orthopedic surgery, bleeding around indwelling devices caused by surgical operation invariably constitutes a unique microenvironment. In this circumstance, the blood becomes a medium where interaction occurs between implants, neutrophils, and pathogens. To further evaluate the effect of ZnONPs on phagocytic behavior of neutrophils, and to best simulate in vivo conditions, phagocytosis, one of the core functions of neutrophils in eliminating pathogens [50], was evaluated using the ex vivo bacteria–blood–ZnONP co-culture model. The results demonstrated that ZnONPs enhanced the efficiency of PMNs to phagocytose bacteria. This promoting effect is in accordance with a previous study finding that ZnONPs increase the ability of neutrophils to phagocytose sheep red blood cells and fluorescent latex beads [27]. Indeed, an increased phagocytic capacity of macrophages promoted by zinc ions is a common finding both in vitro and in vivo [41,51,52]. In this study, neutrophils gathered within the inflammatory site were activated and behaved as phagocytic cells. Zinc ions released from ZnONPs may concentrate within the phagolysosomes of activated neutrophils to form a high zinc concentration environment, thus exerting an enhanced bactericidal efficiency through specific mechanisms, for example, the destruction of the Fe-S cluster [53,54]. It is worth mentioning that the rapid phagosomal maturation and phagocytosis within neutrophils are not perfect processes; thus, oxidative products and cytolytic contents may be released from the cells, damaging the surrounding tissues. Unexpectedly, we confirmed the protective effect of ZnONPs on PMNs via a slight decrease in cell lysis rate during antibacterial process. This cytoprotective effect might be explained by the results of Girard et al. [29], who demonstrated a delayed neutrophil apoptotic effect under the influence of ZnONPs through a de novo protein synthesis-dependent mechanism.

After the encapsulation of bacteria in phagosomes, one of the essential mechanisms of immune cells is to kill them in a nicotinamide adenine dinucleotide phosphate (NADPH) oxygenase-dependent manner [46], also referred to as the respiratory burst. During this process, ROS are a vital weapon for neutrophils as they contribute to the destruction of bacteria [10], whereas superoxide anions serve as the initiating switch leading to the generation of ROS [49]. In this study, neither ZnONPs nor bacterial cells alone induced significantly enhanced ROS or superoxide production by PMNs. In this context, previous studies have proved that ZnONPs at bio-safe concentrations will not directly enhance ROS production in neutrophils [29]. We further observed that ROS and superoxide anions were mass-produced under the presence of both ZnONPs and bacteria. Under such circumstances, ZnONPs are more likely to function as priming agents, which might trigger certain pathways of PMNs, thereby increasing ROS production. It is worth noting that NOX2 functions as the catalytic unit of NADPH oxidase, which is an essential protein in ROS and superoxide production. [55] Hence, we speculated that ZnONPs might improve ROS production through NOX2-related mechanisms. In addition, NF-κB-related genes also play a role in stimulating ROS production, [56] and ZnONPs can reportedly activate NF-κB signaling in THP-1 cells [57]. However, further studies are required to clarify the precise mechanism underlying the promotion of ROS production.

This study has proved the potential synergistic effects of ZnONPs on PMNs to resist pathogen infection. Further research should be devoted to reducing the toxicity of ZnONPs through various methods, including forming complexes with substances such as chitosan [30], or embedding on implant surfaces [43]. These attempts may be expected to increase the concentration of ZnONPs within the biosafety range, thus achieving a better bactericidal effect. On the other hand, how to further increase immune cell infiltration in the infected site is also an urgent problem. ZnONPs were reportedly produce massive ROS under ultraviolet irradiation [24]. A possible paradigm is to use UV light irradiation at specific times to induce local neutrophil apoptosis, which is expected to further trigger immune cell recruitment [33]. In addition, ZnONPs applied on the animal model of subcutaneous soft tissue infection were proved to be effective in this study. We believe that these results will be equally valid in other models including humans; however, further research is needed. This study had certain limitations. First, neutrophils are not the only immune cells in the innate immune system. Indeed, many cells, including macrophages, play a crucial role in anti-infection immunity [41,52]. Hence, focusing exclusively on neutrophil function could have resulted in a lack of comprehensiveness. Second, the antibacterial effect of PMNs is not limited to phagocytosis and ROS production. Other mechanisms including neutrophil extracellular traps [8] and degranulation [28] are of importance, and they were not explored in this study.

5. Conclusions

ZnONPs have potential synergistic effects on neutrophils to resist pathogen infection along with promoting neutrophil migration, proinflammatory cytokine mRNA expression, phagocytosis, and ROS production. The findings help to fabricate and apply biomaterials in order to maximize their synergy with the innate immune system, thus promoting the host's resistance to pathogen invasion.

Author Contributions: K.Y. and M.H. contributed equally to this work. H.Q., Z.A. and X.H. conceived and designed this study. K.Y. and M.H. conducted the studies in antibacterial effects and mechanisms of ZnONPs and neutrophils. K.Y. and M.H. also performed the animal studies, collected data, performed data analysis, and drafted the manuscript. H.Q., Z.A. and X.H. critically reviewed the manuscript. All authors have read and agreed to the published version of the manuscript.

Funding: This research was funded by the National Natural Science Foundation of China, grant numbers 31771022.

Institutional Review Board Statement: The study was conducted in accordance with the Declaration of Helsinki, and approved by the Institutional Review Board of Shanghai Sixth People's Hospital (protocol code: 2021-KY-029(K) and date of approval: 12 May 2021). The animal study protocol was approved by the Institutional Review Board of Shanghai Sixth People's Hospital (protocol code: DWLL2022-0337 and date of approval: 3 March 2022).

Informed Consent Statement: Informed consent was obtained from all subjects involved in the study.

Data Availability Statement: Not applicable.

Conflicts of Interest: The authors declare no conflict of interest.

References

1. Zimmerli, W.; Trampuz, A.; Ochsner, P.E. Prosthetic-joint infections. *N. Engl. J. Med.* **2004**, *351*, 1645–1654. [CrossRef] [PubMed]
2. Darouiche, R.O. Treatment of infections associated with surgical implants. *N. Engl. J. Med.* **2004**, *350*, 1422–1429. [CrossRef] [PubMed]
3. Arciola, C.R.; An, Y.H.; Campoccia, D.; Donati, M.E.; Montanaro, L. Etiology of implant orthopedic infections: A survey on 1027 clinical isolates. *Int. J. Artif. Organs.* **2005**, *28*, 1091–1100. [CrossRef] [PubMed]
4. Mponponsuo, K.; Leal, J.; Puloski, S.; Chew, D.; Chavda, S.; Au, F.; Rennert-May, E. Economic burden of surgical management of surgical site infections following hip and knee replacements in Calgary, Alberta, Canada. *Infect. Control Hosp. Epidemiol.* **2021**, 1–8. [CrossRef] [PubMed]
5. Buvelot, H.; Posfay-Barbe, K.M.; Linder, P.; Schrenzel, J.; Krause, K.H. Staphylococcus aureus, phagocyte NADPH oxidase and chronic granulomatous disease. *FEMS Microbiol. Rev.* **2017**, *41*, 139–157. [CrossRef]
6. Alves-Filho, J.C.; Sonego, F.; Souto, F.O.; Freitas, A.; Verri, W.A., Jr.; Auxiliadora-Martins, M.; Basile-Filho, A.; McKenzie, A.N.; Xu, D.; Cunha, F.Q.; et al. Interleukin-33 attenuates sepsis by enhancing neutrophil influx to the site of infection. *Nat. Med.* **2010**, *16*, 708–712. [CrossRef]
7. Kolaczkowska, E.; Kubes, P. Neutrophil recruitment and function in health and inflammation. *Nat. Rev. Immunol.* **2013**, *13*, 159–175. [CrossRef]
8. Manfredi, A.A.; Ramirez, G.A.; Rovere-Querini, P.; Maugeri, N. The Neutrophil's Choice: Phagocytose vs Make Neutrophil Extracellular Traps. *Front. Immunol.* **2018**, *9*, 288. [CrossRef]
9. Petri, B.; Sanz, M.J. Neutrophil chemotaxis. *Cell Tissue Res.* **2018**, *371*, 425–436. [CrossRef]

10. El-Benna, J.; Hurtado-Nedelec, M.; Marzaioli, V.; Marie, J.C.; Gougerot-Pocidalo, M.A.; Dang, P.M. Priming of the neutrophil respiratory burst: Role in host defense and inflammation. *Immunol. Rev.* **2016**, *273*, 180–193. [CrossRef]
11. Wang, L.; Li, S.; Yin, J.; Yang, J.; Li, Q.; Zheng, W.; Liu, S.; Jiang, X. The Density of Surface Coating Can Contribute to Different Antibacterial Activities of Gold Nanoparticles. *Nano Lett.* **2020**, *20*, 5036–5042. [CrossRef] [PubMed]
12. Cao, C.; Ge, W.; Yin, J.; Yang, D.; Wang, W.; Song, X.; Hu, Y.; Yin, J.; Dong, X. Mesoporous Silica Supported Silver-Bismuth Nanoparticles as Photothermal Agents for Skin Infection Synergistic Antibacterial Therapy. *Small* **2020**, *16*, e2000436. [CrossRef] [PubMed]
13. Qiao, Y.; He, J.; Chen, W.; Yu, Y.; Li, W.; Du, Z.; Xie, T.; Ye, Y.; Hua, S.Y.; Zhong, D.; et al. Light-Activatable Synergistic Therapy of Drug-Resistant Bacteria-Infected Cutaneous Chronic Wounds and Nonhealing Keratitis by Cupriferous Hollow Nanoshells. *ACS Nano* **2020**, *14*, 3299–3315. [CrossRef] [PubMed]
14. Haase, H.; Fahmi, A.; Mahltig, B. Impact of silver nanoparticles and silver ions on innate immune cells. *J. Biomed. Nanotechnol.* **2014**, *10*, 1146–1156. [CrossRef]
15. Guan, M.; Chen, Y.; Wei, Y.; Song, H.; Gao, C.; Cheng, H.; Li, Y.; Huo, K.; Fu, J.; Xiong, W. Long-lasting bactericidal activity through selective physical puncture and controlled ions release of polydopamine and silver nanoparticles-loaded TiO2 nanorods in vitro and in vivo. *Int. J. Nanomed.* **2019**, *14*, 2903–2914. [CrossRef]
16. Brennan, S.A.; Ni Fhoghlu, C.; Devitt, B.M.; O'Mahony, F.J.; Brabazon, D.; Walsh, A. Silver nanoparticles and their orthopaedic applications. *Bone Jt. J.* **2015**, *97-B*, 582–589. [CrossRef]
17. Pinho-Ribeiro, F.A.; Baddal, B.; Haarsma, R.; O'Seaghdha, M.; Yang, N.J.; Blake, K.J.; Portley, M.; Verri, W.A.; Dale, J.B.; Wessels, M.R.; et al. Blocking Neuronal Signaling to Immune Cells Treats Streptococcal Invasive Infection. *Cell* **2018**, *173*, 1083–1097.e22. [CrossRef]
18. Zhao, Q.; Wang, J.; Yin, C.; Zhang, P.; Zhang, J.; Shi, M.; Shen, K.; Xiao, Y.; Zhao, Y.; Yang, X.; et al. Near-Infrared Light-Sensitive Nano Neuro-Immune Blocker Capsule Relieves Pain and Enhances the Innate Immune Response for Necrotizing Infection. *Nano Lett.* **2019**, *19*, 5904–5914. [CrossRef]
19. Heim, C.E.; Vidlak, D.; Scherr, T.D.; Kozel, J.A.; Holzapfel, M.; Muirhead, D.E.; Kielian, T. Myeloid-derived suppressor cells contribute to Staphylococcus aureus orthopedic biofilm infection. *J. Immunol.* **2014**, *192*, 3778–3792. [CrossRef]
20. Saptarshi, S.R.; Duschl, A.; Lopata, A.L. Biological reactivity of zinc oxide nanoparticles with mammalian test systems: An overview. *Nanomedicine* **2015**, *10*, 2075–2092. [CrossRef]
21. Heng, B.C.; Zhao, X.; Tan, E.C.; Khamis, N.; Assodani, A.; Xiong, S.; Ruedl, C.; Ng, K.W.; Loo, J.S. Evaluation of the cytotoxic and inflammatory potential of differentially shaped zinc oxide nanoparticles. *Arch. Toxicol.* **2011**, *85*, 1517–1528. [CrossRef] [PubMed]
22. Tsou, T.C.; Yeh, S.C.; Tsai, F.Y.; Lin, H.J.; Cheng, T.J.; Chao, H.R.; Tai, L.A. Zinc oxide particles induce inflammatory responses in vascular endothelial cells via NF-kappaB signaling. *J. Hazard. Mater.* **2010**, *183*, 182–188. [CrossRef] [PubMed]
23. Padmavathy, N.; Vijayaraghavan, R. Enhanced bioactivity of ZnO nanoparticles-an antimicrobial study. *Sci. Technol. Adv. Mater.* **2008**, *9*, 035004. [CrossRef] [PubMed]
24. Jin, S.E.; Jin, J.E.; Hwang, W.; Hong, S.W. Photocatalytic antibacterial application of zinc oxide nanoparticles and self-assembled networks under dual UV irradiation for enhanced disinfection. *Int. J. Nanomed.* **2019**, *14*, 1737–1751. [CrossRef]
25. Poon, W.L.; Alenius, H.; Ndika, J.; Fortino, V.; Kolhinen, V.; Mesceriakovas, A.; Wang, M.; Greco, D.; Lahde, A.; Jokiniemi, J.; et al. Nano-sized zinc oxide and silver, but not titanium dioxide, induce innate and adaptive immunity and antiviral response in differentiated THP-1 cells. *Nanotoxicology* **2017**, *11*, 936–951. [CrossRef]
26. Wu, W.; Samet, J.M.; Peden, D.B.; Bromberg, P.A. Phosphorylation of p65 is required for zinc oxide nanoparticle-induced interleukin 8 expression in human bronchial epithelial cells. *Environ. Health Perspect.* **2010**, *118*, 982–987. [CrossRef]
27. Babin, K.; Goncalves, D.M.; Girard, D. Nanoparticles enhance the ability of human neutrophils to exert phagocytosis by a Syk-dependent mechanism. *Biochim. Biophys. Acta* **2015**, *1850*, 2276–2282. [CrossRef]
28. Babin, K.; Antoine, F.; Goncalves, D.M.; Girard, D. TiO_2, CeO_2 and ZnO nanoparticles and modulation of the degranulation process in human neutrophils. *Toxicol. Lett.* **2013**, *221*, 57–63. [CrossRef]
29. Goncalves, D.M.; Girard, D. Zinc oxide nanoparticles delay human neutrophil apoptosis by a de novo protein synthesis-dependent and reactive oxygen species-independent mechanism. *Toxicol. Vitr.* **2014**, *28*, 926–931. [CrossRef]
30. Bagheri, M.; Validi, M.; Gholipour, A.; Makvandi, P.; Sharifi, E. Chitosan nanofiber biocomposites for potential wound healing applications: Antioxidant activity with synergic antibacterial effect. *Bioeng Transl. Med.* **2022**, *7*, e10254. [CrossRef]
31. Soares, T.; Ribeiro, D.; Proenca, C.; Chiste, R.C.; Fernandes, E.; Freitas, M. Size-dependent cytotoxicity of silver nanoparticles in human neutrophils assessed by multiple analytical approaches. *Life Sci.* **2016**, *145*, 247–254. [CrossRef] [PubMed]
32. Wang, C.; Liu, X.; Han, Z.; Zhang, X.; Wang, J.; Wang, K.; Yang, Z.; Wei, Z. Nanosilver induces the formation of neutrophil extracellular traps in mouse neutrophil granulocytes. *Ecotoxicol. Environ. Saf.* **2019**, *183*, 109508. [CrossRef] [PubMed]
33. Zhang, P.; Zhao, Q.; Shi, M.; Yin, C.; Zhao, Z.; Shen, K.; Qiu, Y.; Xiao, Y.; Zhao, Y.; Yang, X.; et al. Fe_3O_4@TiO_2-Laden Neutrophils Activate Innate Immunity via Photosensitive Reactive Oxygen Species Release. *Nano Lett.* **2020**, *20*, 261–271. [CrossRef] [PubMed]
34. Freitas, M.; Porto, G.; Lima, J.L.; Fernandes, E. Isolation and activation of human neutrophils in vitro. The importance of the anticoagulant used during blood collection. *Clin. Biochem.* **2008**, *41*, 570–575. [CrossRef]
35. Lin, M.H.; Wang, Y.H.; Kuo, C.H.; Ou, S.F.; Huang, P.Z.; Song, T.Y.; Chen, Y.C.; Chen, S.T.; Wu, C.H.; Hsueh, Y.H.; et al. Hybrid ZnO/chitosan antimicrobial coatings with enhanced mechanical and bioactive properties for titanium implants. *Carbohydr. Polym.* **2021**, *257*, 117639. [CrossRef]

36. McGuffie, M.J.; Hong, J.; Bahng, J.H.; Glynos, E.; Green, P.F.; Kotov, N.A.; Younger, J.G.; VanEpps, J.S. Zinc oxide nanoparticle suspensions and layer-by-layer coatings inhibit staphylococcal growth. *Nanomedicine* **2016**, *12*, 33–42. [CrossRef]
37. Lallo da Silva, B.; Caetano, B.L.; Chiari-Andreo, B.G.; Pietro, R.; Chiavacci, L.A. Increased antibacterial activity of ZnO nanoparticles: Influence of size and surface modification. *Colloids. Surf. B Biointerfaces* **2019**, *177*, 440–447. [CrossRef]
38. Dutta, R.K.; Sharma, P.K.; Bhargava, R.; Kumar, N.; Pandey, A.C. Differential susceptibility of Escherichia coli cells toward transition metal-doped and matrix-embedded ZnO nanoparticles. *J. Phys. Chem. B* **2010**, *114*, 5594–5599. [CrossRef]
39. Sirelkhatim, A.; Mahmud, S.; Seeni, A.; Kaus, N.H.M.; Ann, L.C.; Bakhori, S.K.M.; Hasan, H.; Mohamad, D. Review on Zinc Oxide Nanoparticles: Antibacterial Activity and Toxicity Mechanism. *Nanomicro. Lett.* **2015**, *7*, 219–242. [CrossRef]
40. Tang, F.; Li, L.; Chen, D. Mesoporous silica nanoparticles: Synthesis, biocompatibility and drug delivery. *Adv. Mater.* **2012**, *24*, 1504–1534. [CrossRef]
41. Roy, R.; Singh, S.K.; Das, M.; Tripathi, A.; Dwivedi, P.D. Toll-like receptor 6 mediated inflammatory and functional responses of zinc oxide nanoparticles primed macrophages. *Immunology* **2014**, *142*, 453–464. [CrossRef] [PubMed]
42. Duarte, D.B.; Vasko, M.R.; Fehrenbacher, J.C. Models of Inflammation: Carrageenan Air Pouch. *Curr. Protoc. Pharmacol.* **2016**, *72*, 561–569. [CrossRef] [PubMed]
43. Wang, J.; Zhou, H.; Guo, G.; Tan, J.; Wang, Q.; Tang, J.; Liu, W.; Shen, H.; Li, J.; Zhang, X. Enhanced Anti-Infective Efficacy of ZnO Nanoreservoirs through a Combination of Intrinsic Anti-Biofilm Activity and Reinforced Innate Defense. *ACS Appl. Mater. Interfaces* **2017**, *9*, 33609–33623. [CrossRef] [PubMed]
44. Nauseef, W.M.; Borregaard, N. Neutrophils at work. *Nat. Immunol.* **2014**, *15*, 602–611. [CrossRef]
45. Caielli, S.; Banchereau, J.; Pascual, V. Neutrophils come of age in chronic inflammation. *Curr. Opin. Immunol.* **2012**, *24*, 671–677. [CrossRef]
46. Borregaard, N. Neutrophils, from marrow to microbes. *Immunity* **2010**, *33*, 657–670. [CrossRef]
47. Hujanen, E.S.; Seppa, S.T.; Virtanen, K. Polymorphonuclear leukocyte chemotaxis induced by zinc, copper and nickel in vitro. *Biochim. Biophys. Acta* **1995**, *1245*, 145–152. [CrossRef]
48. Ibs, K.H.; Rink, L. Zinc-altered immune function. *J. Nutr.* **2003**, *133*, 1452S–1456S. [CrossRef]
49. Nauseef, W.M. How human neutrophils kill and degrade microbes: An integrated view. *Immunol. Rev.* **2007**, *219*, 88–102. [CrossRef]
50. Mayadas, T.N.; Cullere, X.; Lowell, C.A. The multifaceted functions of neutrophils. *Annu. Rev. Pathol.* **2014**, *9*, 181–218. [CrossRef]
51. Nowak, J.E.; Harmon, K.; Caldwell, C.C.; Wong, H.R. Prophylactic zinc supplementation reduces bacterial load and improves survival in a murine model of sepsis. *Pediatr. Crit. Care Med.* **2012**, *13*, e323–e329. [CrossRef] [PubMed]
52. Lastra, M.D.; Pastelin, R.; Camacho, A.; Monroy, B.; Aguilar, A.E. Zinc intervention on macrophages and lymphocytes response. *J. Trace Elem. Med. Biol.* **2001**, *15*, 5–10. [CrossRef]
53. Botella, H.; Peyron, P.; Levillain, F.; Poincloux, R.; Poquet, Y.; Brandli, I.; Wang, C.; Tailleux, L.; Tilleul, S.; Charriere, G.M.; et al. Mycobacterial p(1)-type ATPases mediate resistance to zinc poisoning in human macrophages. *Cell Host Microbe* **2011**, *10*, 248–259. [CrossRef] [PubMed]
54. Lemire, J.A.; Harrison, J.J.; Turner, R.J. Antimicrobial activity of metals: Mechanisms, molecular targets and applications. *Nat. Rev. Microbiol.* **2013**, *11*, 371–384. [CrossRef] [PubMed]
55. Belambri, S.A.; Rolas, L.; Raad, H.; Hurtado-Nedelec, M.; Dang, P.M.; El-Benna, J. NADPH oxidase activation in neutrophils: Role of the phosphorylation of its subunits. *Eur. J. Clin. Investig.* **2018**, *48* (Suppl. S2), e12951. [CrossRef] [PubMed]
56. Morgan, M.J.; Liu, Z.G. Crosstalk of reactive oxygen species and NF-kappaB signaling. *Cell Res.* **2011**, *21*, 103–115. [CrossRef]
57. Senapati, V.A.; Kumar, A.; Gupta, G.S.; Pandey, A.K.; Dhawan, A. ZnO nanoparticles induced inflammatory response and genotoxicity in human blood cells: A mechanistic approach. *Food Chem. Toxicol.* **2015**, *85*, 61–70. [CrossRef]

Journal of
Functional Biomaterials

MDPI

Article

Oxyhydroxide-Coated PEO–Treated Mg Alloy for Enhanced Corrosion Resistance and Bone Regeneration

Juning Xie [1,2,†], Shi Cheng [2,†], Guoqing Zhong [2,3], Ruixiang Zhou [2,4], Chi Zhang [2], Yue He [1,2], Feng Peng [2,*] and Yu Zhang [1,2,*]

1 School of Medicine, South China University of Technology, Guangzhou 510006, China; juningxie@163.com (J.X.); raul86@126.com (Y.H.)
2 Medical Research Center, Department of Orthopedics, Guangdong Provincial People's Hospital, Guangdong Academy of Medical Sciences, Guangzhou 510080, China; chengshi@gdph.org.cn (S.C.); gqzhong@foxmail.com (G.Z.); 15616228413@163.com (R.Z.); zhangchi@gdph.org.cn (C.Z.)
3 Medical College, Shantou University, Shantou 515041, China
4 Medical Research Center, Jinzhou Medical University, Jinzhou 121001, China
* Correspondence: pengfeng@gdph.org.cn (F.P.); zhangyu@gdph.org.cn (Y.Z.)
† These authors contributed equally to this work.

Abstract: Plasma electrolytic oxidation (PEO) is widely used as a surface modification method to enhance the corrosion resistance of Mg alloy, the most likely applied biodegradable material used in orthopedic implants. However, the pores and cracks easily formed on the PEO surface are unfavorable for long-term corrosion resistance. In this study, to solve this problem, we used simple immersion processes to construct Mn and Fe oxyhydroxide duplex layers on the PEO-treated AZ31 (PEO–Mn/Fe). As control groups, single Mn and Fe oxyhydroxide layers were also fabricated on PEO (denoted as PEO–Mn and PEO–Fe, respectively). PEO–Mn showed a similar porous morphology to the PEO sample. However, the PEO–Fe and PEO–Mn/Fe films completely sealed the pores on the PEO surfaces, and no cracks were observed even after the samples were immersed in water for 7 days. Compared with PEO, PEO–Mn, and PEO–Fe, PEO–Mn/Fe exhibited a significantly lower self-corrosion current, suggesting better corrosion resistance. In vitro C3H10T1/2 cell culture showed that PEO–Fe/Mn promoted the best cell growth, alkaline phosphatase activity, and bone-related gene expression. Furthermore, the rat femur implantation experiment showed that PEO–Fe/Mn–coated Mg showed the best bone regeneration and osteointegration abilities. Owing to enhanced corrosion resistance and osteogenesis, the PEO–Fe/Mn film on Mg alloy is promising for orthopedic applications.

Keywords: biomedical magnesium alloy; corrosion resistance; bone repair; surface modification

Citation: Xie, J.; Cheng, S.; Zhong, G.; Zhou, R.; Zhang, C.; He, Y.; Peng, F.; Zhang, Y. Oxyhydroxide-Coated PEO–Treated Mg Alloy for Enhanced Corrosion Resistance and Bone Regeneration. *J. Funct. Biomater.* **2022**, *13*, 50. https://doi.org/10.3390/jfb13020050

Academic Editors: Anderson de Oliveira Lobo and Serena Danti

Received: 19 March 2022
Accepted: 22 April 2022
Published: 1 May 2022

Publisher's Note: MDPI stays neutral with regard to jurisdictional claims in published maps and institutional affiliations.

Copyright: © 2022 by the authors. Licensee MDPI, Basel, Switzerland. This article is an open access article distributed under the terms and conditions of the Creative Commons Attribution (CC BY) license (https://creativecommons.org/licenses/by/4.0/).

1. Introduction

To date, the most widely and successfully used clinical orthopedic biomaterials are metals such as Ti, Ti alloys, Co–Cr alloy, and 316 stainless steel, owing to their good mechanical properties and biocompatibility [1–3]. However, long-term clinical studies have revealed several drawbacks to these non-degradable metals, including stress-shielding effects, host response, and inflammation caused by elastic modulus mismatch, permanent implantation, and wear particles, respectively [4–6]. To avoid these drawbacks, researchers have focused on biodegradable Mg, which is considered a next-generation biomedical metal; its elastic modulus is close to that of natural bone (41 vs. 7 to 30 GPa) [7], its low standard electrode potential (−2.37 V vs. normal hydrogen electrode) leads to complete degradation in fluids [8], and its mechanical strength is suitable for orthopedic implants [9,10].

However, Mg-based orthopedic implants are not applied on a large scale, mainly due to their rapid degradation [11–13], which not only results in the reduction of mechanical strength but also causes the accumulation of excessive OH^- and H_2. The former might lead to implantation failure, and the latter inhibits bone regeneration. Therefore, enhancing the corrosion resistance of Mg is of great importance for its orthopedic applications.

Surface modification is effective in improving the corrosion resistance of Mg. Commonly used technologies for surface modification of Mg include plasma electrolytic oxidation (PEO), hydrothermal treatment, spray coating, fluoride treatment, etc. [14–16]. Among these technologies, PEO coating, performed under high voltage, is widely accepted in the industry. The main component of PEO coating is metal oxide; therefore, it can provide favorable corrosion protection for the substrate. Moreover, high temperatures (over 2000 °C) on the substrate surface can melt the coating and result in strong binding forces between the coating and substrate [17,18]. Numerous studies have treated Mg with PEO and investigated its orthopedic applications [19,20]. Rendenbach et al. modified WE43 Mg alloy with PEO treatment and found that the PEO-treated Mg alloy showed enhanced corrosion resistance and osteointegration upon implantation in Gottingen miniature pigs [21]. However, the dielectric breakdown effect caused pore formation and cracks easily formed and spread over the PEO coating, which is unfavorable for long-term corrosion resistance. A study by Fischerauer et al. revealed that PEO-coated ZX50 Mg alloy vanished completely after 12–16 weeks of implantation in rat femurs [22]. Therefore, a manner by which pores and cracks on the PEO coating can be avoided remains a challenge for PEO-treated Mg applications in orthopedic implants.

Mn and Fe are trace elements that participate in numerous physiological reactions. In addition, both the ions can be adsorbed on alkaline surfaces to form hydroxide. Therefore, in this study, we first used a simple immersing method to fabricate MnOOH and FeOOH coatings on Mg alloy. Interestingly, we found that MnOOH did not change the surface morphology of the PEO coating. However, for the FeOOH film, a layer of nano-sheet-like structures formed on the PEO coating and totally sealed the pores. Hence, to obtain a more protective oxyhydroxide coating on the Mg alloy, we designed and fabricated a duplex Mn/Fe oxyhydroxide (with an inner MnOOH layer and an outer FeOOH layer) on the top of the PEO coating. The corrosion resistance of the newly designed film was investigated. Moreover, the in vitro and in vivo osteogenesis performances of the coated Mg alloy were studied via rat bone marrow stem cell (rBMSCs) cultivation and bone implantation experiments, respectively.

2. Materials and Methods

2.1. Sample Preparation and Characterization

AZ31 magnesium alloy sheets (with 3% Al, 0.8% Zn, 0.4% Mn, and the balance Mg) were purchased from Suzhou plain metal materials Co., Ltd. (Suzhou, China) and cut into pieces of 10 mm in diameter and 2 mm in length for in vitro tests and pieces of 2 mm in diameter and 8 mm in length for in vivo tests. The AZ31 specimens were ground with 800# silicon carbide abrasive paper and then ultrasonically cleaned in ethanol. The PEO process was conducted in the electrolyte containing 10 g/L $C_3H_7Na_2O_6P$ and 12.5 g/L KOH. The constant current, frequency, duty cycle, and stop voltage were 0.8 A, 1000 Hz, 10%, and 340 V, respectively. The PEO-treated specimens were then immersed in 12 g/L of $MnCl_2 \cdot 4H_2O$ for 9 h, and the obtained samples were labelled as PEO–Mn. Similarly, after immersion in 2 g/L of $FeCl_2 \cdot 4H_2O$ for 4 h, the samples were denoted as PEO–Fe. The specimens first immersed in 12 g/L of $MnCl_2 \cdot 4H_2O$ for 9 h and then in 2 g/L of $FeCl_2 \cdot 4H_2O$ for 4 h were denoted as PEO–Mn/Fe. The surface views were determined using scanning electron microscopy (SEM; S-3400N, HITACHI, Tokyo, Japan) with a working voltage of 15 kV and an emission current of 0.16 mA. The phase compositions were determined using X-ray diffraction (XRD; D2PHASE, Bruker, Billerica, MA, USA) with a scanning rate of 5°/min and a step size of 0.02°. Element compositions were determined using energy dispersive spectrometry (EDS; IXRF-550i, IXRF SYSTEMS, Austin, TX, USA) with a working voltage of 15 kV and a detection time of 15 min. X-ray photoelectron spectroscopy (XPS; RBD upgraded PHI-5000C ESCA system, Perkin Elmer, Waltham, MA, USA) was conducted with an energy step size of 1 eV, a working voltage of 12 kV, and a filament current of 6 mA.

2.2. Corrosion Evaluation

A potentiodynamic polarization test was conducted in phosphate buffered saline (PBS) using an electrochemical analyzer (CHI760C, Shanghai, China). The samples were kept in PBS to obtain a stable open circuit potential and the potentiodynamic polarization test was performed from −2 to 0 V at a scan rate of 10 mV/s.

The samples were placed in a 24-well plate and then 1 mL of α minimum essential medium culture medium (α-MEM) was added to each well. After incubation at 37 °C for 7 days, the samples were collected and rinsed with ultrapure water. The corrosion morphology of the samples was observed using SEM. Moreover, all the culture medium changed from reddish to dark red, indicating that the pH value of culture medium increased.

2.3. Live/Dead Staining

C3H10T1/2 cells were purchased from the Type Culture Collection of the Chinese Academy of Sciences and cultured in Modified Eagle's medium (MEM; Gibco, Waltham, MA, USA) containing 10% FBS, 2 mM L-glutamine (Sigma Aldrich, Missouri, MO, USA) and 1% sodium pyruvate (Leagene Biotechnology, Beijing, China). The conditions of the cell incubator were 37 °C, 5% CO_2, and 95% humidity [23]. All the in vitro cell experiments were conducted using C3H10T1/2.

Before all experiments, the materials were sterilized by ethylene oxide gas and maintained for 7 days at room temperature to clear remnants. C3H10T1/2 cells were used to evaluate in vitro biocompatibility of the samples. The cells were seeded on the surface of each sample at a density of 5×10^4 cells/mL and cultured for 3 days. Thereafter, the samples were rinsed with PBS and 500 μL of fetal bovine serum–free medium containing calcein-AM (2 μM) and propidium iodide (5 μM) was added to each well, followed by incubation in the dark for 15 min. Ultimately, the samples were rinsed with PBS and the live or dead cells were observed using a fluorescence microscope (Olympus IX 71, Olympus, Tokyo, Japan) with an excitation wavelength of 490 nm for living cells and an excitation wavelength of 545 nm for dead cells.

2.4. Cell Proliferation

The cells (5×10^4 cells/mL) were seeded on the sample surfaces and cultured for 1, 3, and 5 days. At each time point, the samples were rinsed with PBS and moved to a new 24-well plate. Immediately, 0.5 mL of AlamarBlue assay (AbD Serotec Ltd., Kidlington, UK; diluted 10-fold with culture medium was added to each well and cultured for another 2 h. Soon after that, 100 μL culture medium from each well was transferred to a black 96-well plate and measured using an enzyme-labeling instrument (BIO-TEK, ELX 800) with an absorption wavelength of 560 nm and a scattering wavelength of 590 nm).

2.5. Alkaline Phosphatase (ALP) Activity Assay

The samples were immersed in culture medium (1.25 cm^2/mL) for 24 h and the extracts were collected for osteogenic differentiation evaluation assays. The cells (5×10^4 cells/mL) were seeded in a 24-well plate and cultured with extracts from different samples supplied with 10 mM β-glycerophosphate, 100 nM dexamethasone, and 50 mM ascorbate and glutamine for 3 and 7 days. At the predetermined time, the BCIP/NBT ALP Color Development Kit (Beyotime, Shanghai, China) was used to stain ALP in the cells in line with the instructions of manufacturer. For quantitative detection, the intracellular ALP activity was quantitated using Alkaline Phosphatase Kit (Beyotime, Shanghai, China) and the total protein was measured using BCA protein quantitation kit (Themo, Waltham, MA, USA). The ALP activity was normalized with total protein content.

2.6. Quantitative Real Time Polymerase Chain Reaction (qRT-PCR) Assay

The cells were cultured as described in Section 2.5. Total cellular RNA in each well was collected using Total RNA Kit I (Omega R6834-01, Omega, Guangzhou, China) following the manufacturer's instructions and the concentration of acquired RNA was measured using a NanoDrop^{TM} 2000 spectrophotometer (ThermoFisher, Waltham, MA, USA). The RNA from each group was then reverse transcribed into complementary DNA (cDNA) using TransScript II All-in-one Fist-Strand cDNA Synthesis SuperMix. The cDNA was then amplified and analyzed by qRT-PCR (TransGen Biotech, Beijing, China) with TransStart Green qPCR SuperMix (TransGen Bioteh, Beijing, China) and primers. The relative expressions of osteoblastic differentiation-related genes, including ALP, osteopontin (OPN), RUNX family transcription factor 2 (RUNX2), collagen-I (COL-I), and osteocalcin (OCN) were quantified using the cycle threshold value and $2^{-\Delta\Delta}$CT method. The expression of housekeeping gene glyceraldehyde-3-phosphate dehydrogenase (GAPDH) was used as an endogenous control for normalization. PCR primers sequences are provided in the Table S1.

2.7. Bone Implantation Evaluation

All the procedures of animal experiments were performed in accordance with the Guidelines for Care and Use of Laboratory Animals of South China University of Technology and approved by the Animal Ethics Committee of Guangdong Provincial People's Hospital (KY2020-018-01-01).

Twelve Sprague–Dawley rats (male, 250–300 g) were purchased from Hunan SJA Laboratory Animal Co., Ltd. (Hunan, China) and randomly divided into four groups. After the skin on the bilateral legs was disinfected and lanced, a 2 mm hand-operated drill was used to create a cylindrical hole in the trochlear groove of the femur to reach the marrow cavity, and the materials (2 mm in diameter and 8 mm in length) were implanted in a direction parallel to the longitudinal axis of the femur. The muscles and skin were carefully sutured. All rats were intraperitoneally injected with penicillin for 3 days for protection from postoperative infection. Thereafter, all rats were euthanatized with an overdose injection of pentobarbital sodium at 8 weeks post-surgery, and the bilateral femurs were collected for micro-computed tomography (micro-CT) scanning. The femurs were embedded and cut into sections. The sections were ground, polished, and stained with van Gieson's (VG) solution. Two femurs for each group were used for VG staining and two random regions for each stained section were pictured. In total, four random regions for each group were pictured to evaluate the area of newly formed bone (the red-stained area) and the distance between the newly formed bone layer and the implant.

2.8. Statistical Analysis

Data are presented as the mean ± standard deviation. Differences among groups were analyzed by two-way analysis of variance followed by Tukey's post hoc test using the GraphPad Prism 8.3.0 (GraphPad Software; San Diego, CA, USA). Statistical significance was set at $p < 0.05$.

3. Results and Discussion

The surface views and colors of the various samples are shown in Figure 1a and Figure S1, respectively. PEO samples exhibited a porous structure, the typical morphology of the PEO-treated surface [24,25]. After immersion in $MnCl_2$ solution, the surface view of the PEO-Mn sample was similar to that of the PEO sample; however, the grey color changed to black. Upon immersing the PEO-treated sample in $FeCl_2$ solution, a layer of nanoflower-like structures which completely sealed the pores was formed. The color of the PEO-Fe sample was dark brown. The surface view of the PEO–Mn/Fe sample showed nanoflower-like structures and was black in color. Figure 1b shows corresponding element distributions of the coated samples. Mn and Fe were uniformly distributed on the PEO–Mn and PEO–Fe samples, respectively. Both Mn and Fe were uniformly distributed on the surface of PEO-Mn/Fe sample. Although no difference was observed on the SEM images of the PEO and PEO–Mn samples, element composition and surface color analysis confirmed that a Mn layer was formed on top of the PEO layer for the PEO–Mn sample. Moreover, the cross-sectional views of the PEO–Fe and PEO–Mn/Fe samples showed a significant layer formed on top of the PEO layer, which is consistent with the surface morphology results in Figure 1a.

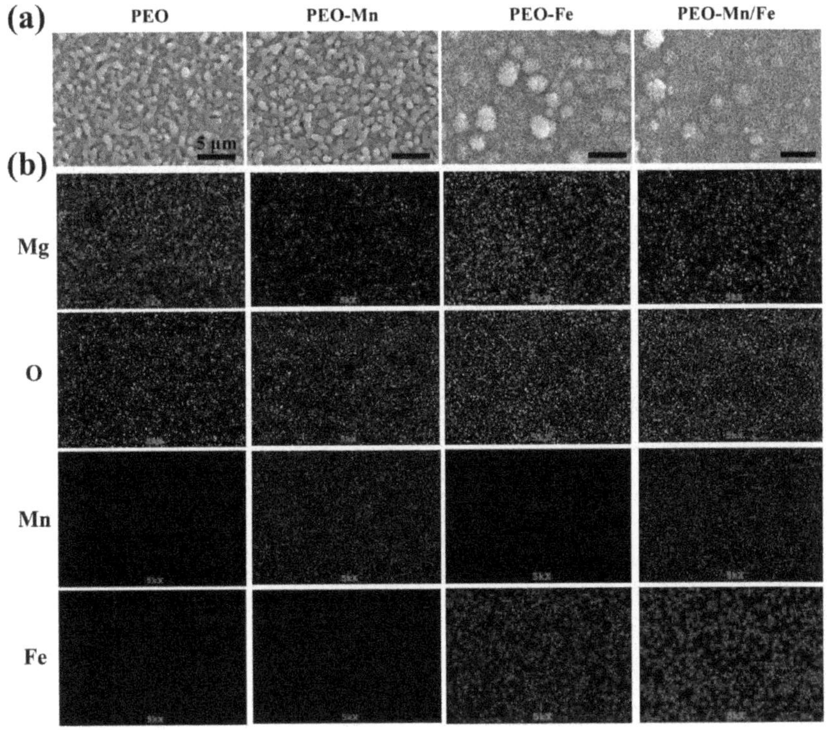

Figure 1. Surface views (**a**) and element distribution (**b**) of PEO, PEO–Mn, PEO–Fe, and PEO–Mn/Fe samples.

To analyze the phase compositions of the films, XPS and XRD were performed. As shown in Figure 2a, C, O, and Mn were detected in the full XPS spectrum of the PEO–Mn sample, whereas C, O, and Fe were detected in the PEO–Fe and PEO–Mn/Fe samples, further confirming the element compositions of the films. The Mn oxidation state was calculated from the following formula: $\triangle E = 7.88–0.85n$ ($2 \leq n \leq 4$), where $\triangle E$ is the separation energy of Mn 3s, and n represents the valence state of Mn [26–28]. As shown in Figure 2b, the $\triangle E$ value is approximately 0.53 and hence the calculated n value is approximately 3, which corresponds with the valence state of Mn in MnOOH. High-resolution spectra of O 1s are shown in Figure 2c. O 1s of the PEO–Mn spectrum can be divided into two peaks centered at 529.3 and 530.7 eV, which correspond with the structures of Mn–O–Mn and Mn–OH, respectively. This further confirmed that the outer layer of PEO–Mn sample was MnOOH. Moreover, the PEO–Fe and PEO–Mn/Fe O1s spectra can be fitted into two peaks centered at 529.2 and 530.5 eV, which attributed to Fe–O–Fe and Fe–OH structures. This confirmed that the outer layer of both samples was FeOOH [29–31]. Figure 3 shows the XRD patterns of various samples. Obvious peaks at 44° representing the MgO phase were detected for all samples. Minute peaks were observed at 27° for the PEO–Fe and PEO–Mn/Fe samples, which indicated the formation of FeOOH. Barring Mg and MgO peaks, no other characteristic peaks were observed in the PEO–Mn group, possibly because the MnOOH film was amorphous or too thin for detection, as shown in Figure S2, no obvious MnOOH layer was observed on the cross section of PEO–Mn sample.

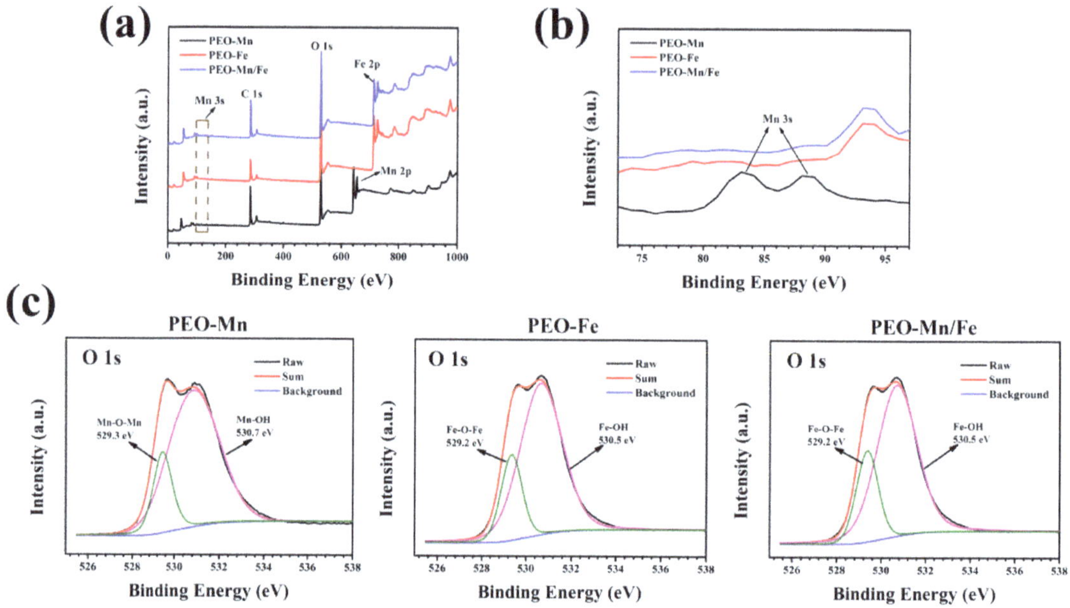

Figure 2. Full XPS spectra (**a**) and enlarged spectra (**b**) of PEO–Mn, PEO–Fe, and PEO–Mn/Fe samples. High-resolution XPS spectra of O 1 s of PEO–Mn, PEO–Fe, and PEO–Mn/Fe samples (**c**).

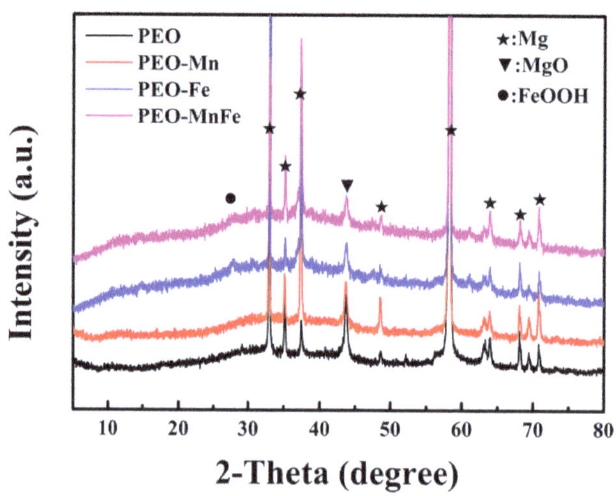

Figure 3. XRD patterns of PEO, PEO–Mn, PEO–Fe, and PEO–Mn/Fe samples.

Both the formation of MnOOH and FeOOH can be ascribed to an alkaline-microenvironment-induced self-assembly process, as described in our previous studies [28,31]. Briefly, MgO, the main component of PEO film, is an alkaline substance that produces OH^- near the coating. Generated OH^- ions react with M^{2+} ions (M represents Fe and Mn) to form unstable $M(OH)_2$. Finally, the $M(OH)_2$ layers are oxidized by oxygen in the solution and MOOH is generated.

The corrosion resistance of the films was evaluated using electrochemical and immersion tests. As shown in Figure 4a, although the PEO–Mn sample exhibited a higher self-corrosion potential than did the PEO sample, their self-corrosion currents were at a similar level. However, the self-corrosion currents of PEO–Fe and PEO–Mn/Fe were significantly lower than those of the PEO and PEO–Mn samples, suggesting better corrosion resistance. Notably, the PEO–Mn/Fe sample exhibited the highest self-corrosion potential, revealing that it was most difficult to corrode. Figure 4b shows the corrosion morphology of various samples after immersion in α-MEM for 7 days. Numerous cracks were observed on the surface of the PEO sample. After MnOOH coating, the cracks were still present on the PEO–Mn surface; however, they were narrower than those on the PEO surface. This might be because the MnOOH layer on the PEO coating is too thin to prevent the corrosive fluid from permeating into the PEO layer (Figures 1a and S2). Interestingly, no cracks were observed on the PEO–Fe and PEO–Mn/Fe samples, indicating the favorable corrosion resistance of the films. As shown in Figure 1a, a layer of nanosheet-like structures completely covered the PEO layer in both the groups, which could be sufficient to prevent the contact between the liquid and the PEO layer, and thus greatly improved the corrosion resistance of the samples. On the other hand, oxyhydroxide is the precursor of layered double hydroxide (LDHs) and would gradually transfer to LDHs in fluid [28]. LDHs are widely considered to be biodegradable materials for biomedical applications [32]. Therefore, the as-prepared films on the Mg alloy would not inhibit the advantage of the biodegradable ability of the magnesium alloy implants.

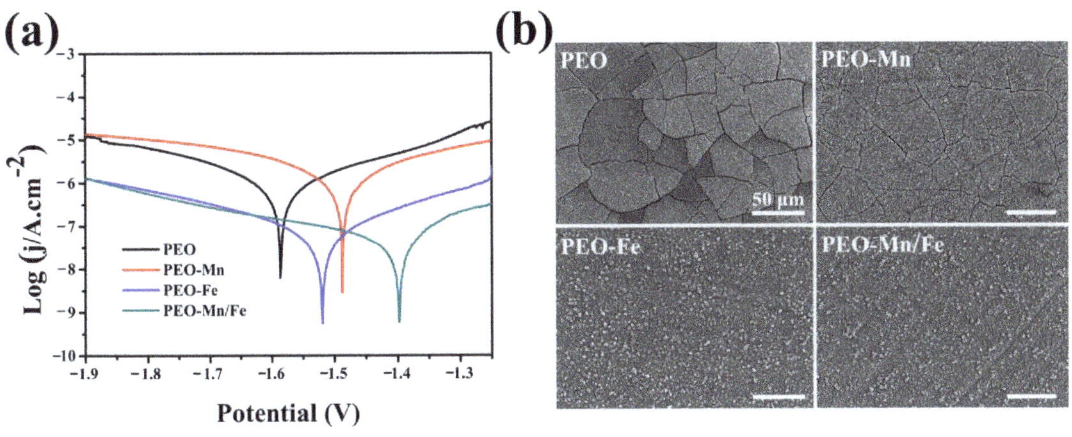

Figure 4. Potentiodynamic polarization curves (**a**) and corrosion morphology (**b**) of PEO, PEO–Mn, PEO–Fe, and PEO–Mn/Fe samples.

Cytocompatibility of the samples was evaluated by culturing C3H10T1/2 cells directly on sample surfaces. Figure 5a shows the live/dead staining results. Only a few living cells were observed on the PEO sample surface. After modifying the PEO with a MnOOH layer, more living cells were detected. Significantly larger numbers of living cells and more connected pseudopods were observed on the PEO–Fe and PEO–Mn/Fe samples than on the PEO–Mn sample. The Alamar Blue results are presented in Figure 5b. The PEO–Fe and PEO–Mn/Fe samples showed the highest proliferation rate among the four groups, consistent with the live/dead staining results. It should be noted that there were no significant differences for the live/dead staining and cell proliferation results between the PEO–Fe and PEO–Mn/Fe samples. These data suggest that the FeOOH film greatly improved the cytocompatibility of the substrate.

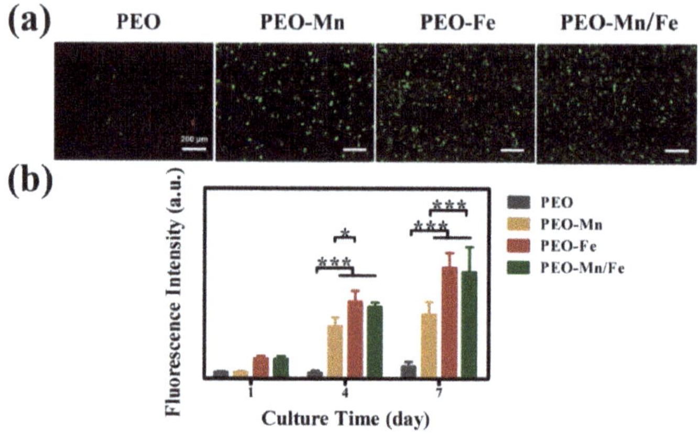

Figure 5. Live/dead staining (**a**) and cell proliferation (**b**) of C3H10T1/2 cells cultured on PEO, PEO–Mn, PEO–Fe, and PEO–Mn/Fe sample surfaces. (*: $p < 0.05$; ***: $p < 0.001$, and n = 4).

Because the cells cultured on sample surfaces showed poor viability, we used extracts to evaluate the osteogenesis induction ability of various groups, according to the ISO 10993-5 standard [33]. The ALP staining and corresponding quantitative results are shown in Figure 6a and b, respectively. The PEO–Fe and PEO–Mn/Fe groups exhibited higher ALP activity than did the other two groups at both time points. In particular, on day 7, the PEO–Mn/Fe sample exhibited the highest ALP activity. At the molecular level, the PEO–Mn/Fe group showed the highest OPN gene expression when cultured for 3 days (Figure 7). Intriguingly, after extending the culture time to 7 days, ALP, OPN, RUNX2, and COL-I expression was highest in cells cultured in the PEO–Mn/Fe extract. On day 7, OCN expression was similar for the PEO–Fe and PEO–Mn/Fe groups, but was still significantly higher than that in the other two groups. Combining the results of ALP activity and the gene expression analyses, it can be concluded that the PEO–Mn/Fe sample is most favorable for the osteogenic differentiation of bone stem cells.

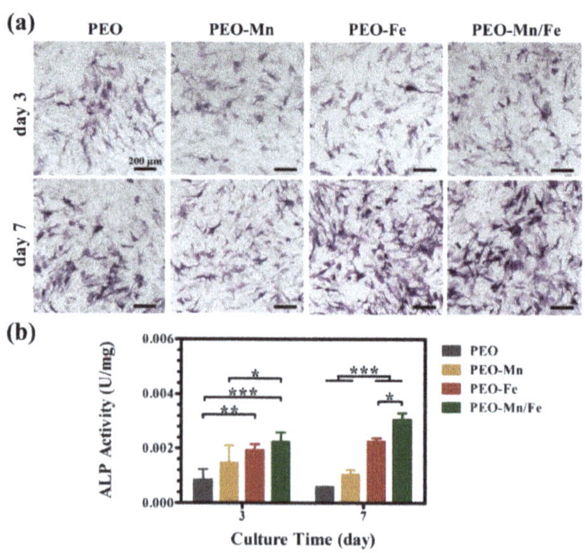

Figure 6. ALP activity in C3H10T1/2 cells cultured in PEO, PEO–Mn, PEO–Fe, and PEO–Mn/Fe extracts (**a**) and corresponding quantitative results (**b**). (*: $p < 0.05$; **: $p < 0.01$; ***: $p < 0.001$, and n = 4).

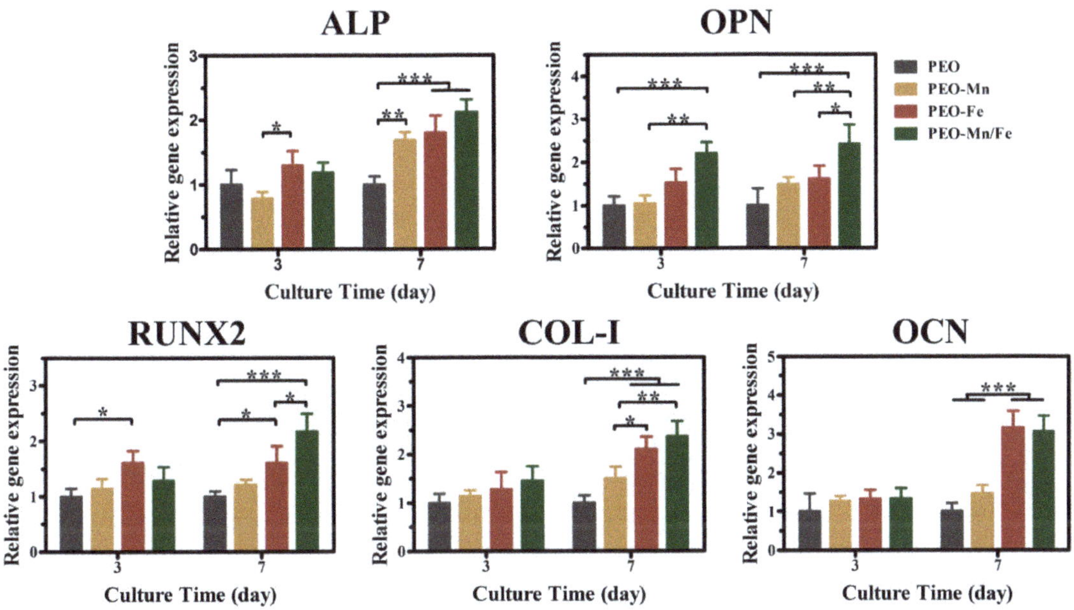

Figure 7. The expression of bone-related genes in C3H10T1/2 cells cultured in PEO, PEO–Mn, and PEO–Fe extracts. (*: $p < 0.05$; **: $p < 0.01$; ***: $p < 0.001$, and n = 4).

To further investigate in vivo osteogenesis ability, all the samples were implanted in rat femurs for 8 weeks. The Micro-CT results shown in Figure 8a suggest that the structures of all femurs were normal and no bone resorption or osteonecrosis was present. The collected femurs were stained with VG solution and the results are shown in Figure 8b and the corresponding quantitative analysis is shown in Figure 8c. Large gaps between the newly formed bone and the implants were observed for the PEO and PEO–Mn groups. Notably, the narrowest gap was observed for the PEO–Mn/Fe group; additionally some newly formed bone closely adhered to the implant surface (indicated by the yellow arrow). Moreover, the largest amount of newly formed bone was observed in the PEO–Mn/Fe group. This suggested that the PEO–Mn/Fe implant was the most favorable for bone regeneration and osteointegration.

The PEO–Mn/Fe sample showed the best osteogenesis performance owing to its corrosion resistance being the best and its sustained release of Mg, Fe, and Mn ions. On the one hand, the corrosion products H_2 and OH^- hugely decreased with improved corrosion resistance, thus reducing the damage to the bone remodeling process. On the other hand, all metal ions mentioned above are bioactive and essential for new bone formation. The Mg ions released from Mg implants upregulate calcitonin gene–related peptide (CGRP) and then promote osteogenic differentiation of bone stem cells [34]. Fe is involved in vitamin D metabolism and collagen synthesis, thus influencing the bone formation process [35,36]. Mn can bind to integrin and trigger integrin-mediated signaling cascades to enhance osteogenesis process [37,38]; additionally, it participates in the synthesis of chondroitin sulfate and glycosyltransferases, which play critical roles in the formation of skeletal and cartilage matrices [39]. Furthermore, there is evidence that these bioactive ions have synergistic effects on the improvement of bone regeneration [40]. Therefore, the PEO–Mn/Fe implant possessed the best bone regeneration capability was expected.

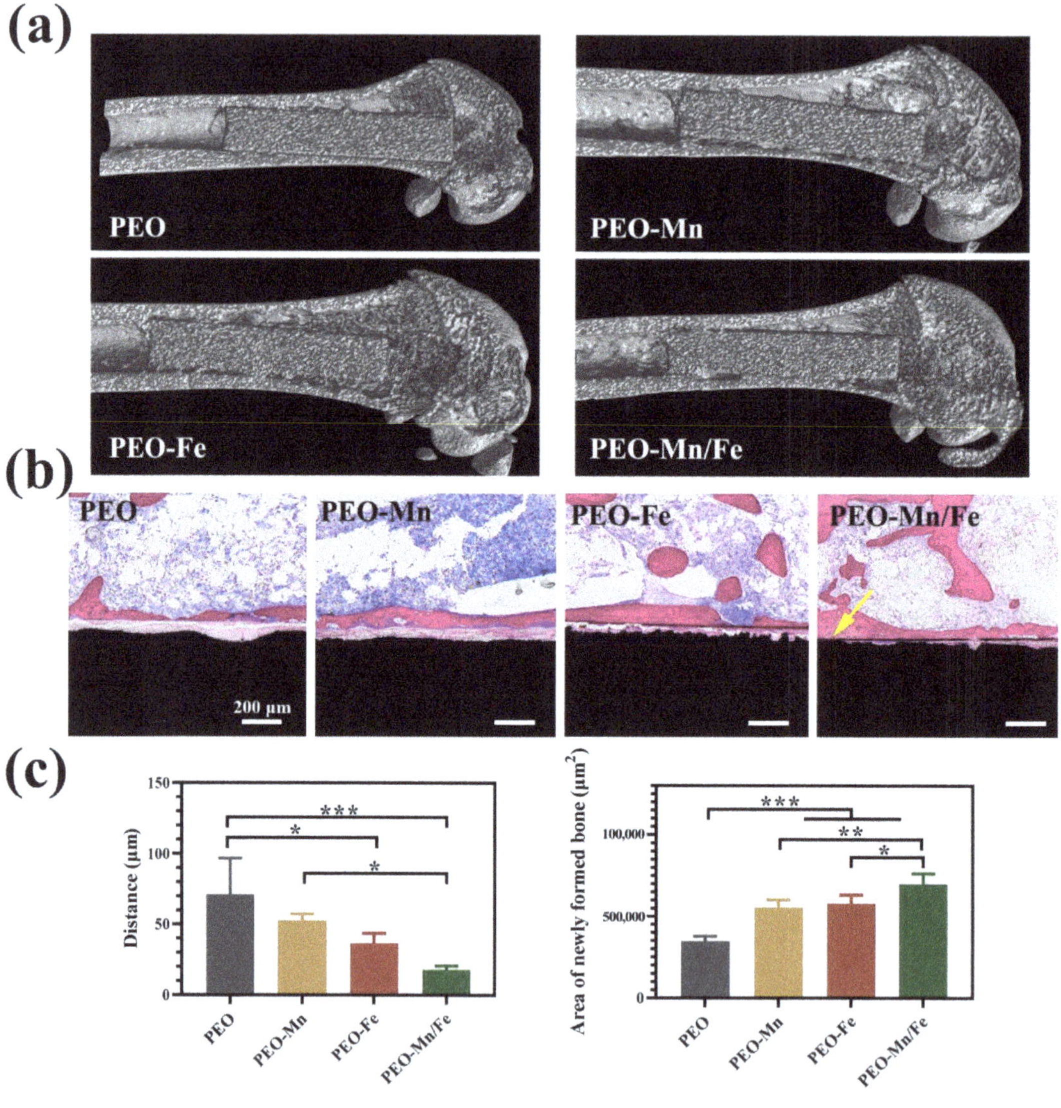

Figure 8. Representative Micro-CT images (**a**) and VG staining (**b**) of PEO, PEO–Mn, and PEO–Fe implants after 8 weeks of implantation. The quantitative analysis of gap distance between newly formed bone and the implants, and the area of newly formed surrounding the implants (**c**). (*: $p < 0.05$; **: $p < 0.01$; ***: $p < 0.001$). The yellow arrow in b indicates the closely adhesion between the newly formed bone and the implant.

4. Conclusions

In conclusion, the current study successfully fabricated a duplex film with an inner MnOOH layer and an outer FeOOH layer on PEO-coated Mg alloy. The oxyhydroxide film completely sealed the pores on the PEO surface. Therefore, it inhibited the occurrence and development of cracks on the PEO layer. The modified sample exhibited improved osteogenesis induction ability in vitro and enhanced bone regeneration in vivo, owing to its better corrosion resistance and sustained release of bioactive ions, including Mg, Fe, and

Mn. The novel design and fabricated oxyhydroxide-modified PEO film formed on Mg alloy shows promising potential for orthopedic applications.

Supplementary Materials: The following supporting information can be downloaded at: https://www.mdpi.com/article/10.3390/jfb13020050/s1, Figure S1: Representative optical images of PEO, PEO–Mn, PEO–Fe, and PEO–Mn/Fe samples; Figure S2: Cross-sectional images of PEO, PEO–Mn, PEO–Fe, and PEO–Mn/Fe samples; Table S1: Primer sequences of the osteogenesis-related genes used in this study.

Author Contributions: Conceptualization and funding acquisition: F.P. and S.C.; investigation: J.X., G.Z. and R.Z.; methodology: J.X., Y.H. and C.Z.; methodology project administration and supervision: Y.Z.; writing—review and editing: F.P. and J.X. All authors have read and agreed to the published version of the manuscript.

Funding: The authors acknowledge the High-level (Full-time) Talents Project of Guangdong Provincial People's Hospital (KY012021462), National Natural Science Foundation of China (52001076), NSFC Incubation Project of Guangdong Provincial People–s Hospital (No. KY0120220044), and Guangdong Basic and Applied Basic Research Foundation (No. 2021A1515110135).

Institutional Review Board Statement: The study was conducted according to the guidelines of the South China University of Technology and approved by Animal Ethics Committee of Guangdong Provincial People's Hospital (KY2020-018-01-01).

Informed Consent Statement: Not applicable.

Data Availability Statement: All raw data from the characterizations are available from the corresponding author upon request.

Conflicts of Interest: The authors declare no conflict of interest.

References

1. Geetha, M.; Singh, A.K.; Asokamani, R.; Gogia, A.K. Ti based biomaterials, the ultimate choice for orthopaedic implants—A review. *Prog. Mater. Sci.* **2009**, *54*, 397–425. [CrossRef]
2. Okazaki, Y.; Gotoh, E.; Nishimori, M.; Katsuda, S.; Manabe, T.; Kobayashi, K. Osteocompatibility of stainless steel, Co-Cr-Mo, Ti-6Al-4V and Ti-15Zr-4Nb-4Ta alloy implants in rat bone tissue. *Mater. Trans.* **2005**, *46*, 1610–1617. [CrossRef]
3. Niinomi, M.; Nakai, M.; Hieda, J. Development of new metallic alloys for biomedical applications. *Acta Biomater.* **2012**, *8*, 3888–3903. [CrossRef] [PubMed]
4. Lah, N.A.C.; Hussin, M.H. Titanium and Titanium Based Alloys as Metallic Biomaterials in Medical Applications—Spine Implant Case Study. *Pertanika J. Sci. Technol.* **2019**, *27*, 459–472.
5. Zhou, Z.; Shi, Q.; Wang, J.; Chen, X.; Hao, Y.; Zhang, Y.; Wang, X. The unfavorable role of titanium particles released from dental implants. *Nanotheranostics* **2021**, *5*, 321–332. [CrossRef]
6. Sumner, D.R. Long-term implant fixation and stress-shielding in total hip replacement. *J. Biomech.* **2015**, *48*, 797–800. [CrossRef]
7. Zhao, D.W.; Witte, F.; Lu, F.Q.; Wang, J.L.; Li, J.L.; Qin, L. Current status on clinical applications of magnesium-based orthopaedic implants: A review from clinical translational perspective. *Biomaterials* **2017**, *112*, 287–302. [CrossRef]
8. Zheng, Y.F.; Gu, X.N.; Witte, F. Biodegradable metals. *Mat. Sci. Eng. R Rep.* **2014**, *77*, 1–34. [CrossRef]
9. Dargusch, M.S.; Balasubramani, N.; Yang, N.; Johnston, S.; Ali, Y.; Wang, G.; Venezuela, J.; Carluccio, J.; Lau, C.; Allavena, R.; et al. In vivo performance of a rare earth free Mg-Zn-Ca alloy manufactured using twin roll casting for potential applications in the cranial and maxillofacial fixation devices. *Bioact. Mater.* **2022**, *12*, 85–96. [CrossRef]
10. Li, C.; Deng, B.; Dong, L.; Liu, X.; Du, K.; Shi, B.; Dong, Y.; Peng, F.; Zhang, Z. Effect of Zn addition on the microstructure and mechanical properties of as-cast BCC Mg-11Li based alloys. *J. Alloys Compd.* **2022**, *895*, 162718. [CrossRef]
11. Zhang, D.D.; Peng, F.; Liu, X.Y. Protection of magnesium alloys: From physical barrier coating to smart self-healing coating. *J. Alloys Compd.* **2021**, *853*, 157010. [CrossRef]
12. Yu, Y.; Lu, H.; Sun, J. Long-term in vivo evolution of high-purity Mg screw degradation—Local and systemic effects of Mg degradation products. *Acta Biomater.* **2018**, *71*, 215–224. [CrossRef] [PubMed]
13. Lee, J.W.; Han, H.S.; Han, K.J.; Park, J.; Jeon, H.; Ok, M.R.; Seok, H.K.; Ahn, J.P.; Lee, K.E.; Lee, D.H.; et al. Long-term clinical study and multiscale analysis of in vivo biodegradation mechanism of Mg alloy. *Proc. Natl. Acad. Sci. USA* **2016**, *113*, 716–721. [CrossRef] [PubMed]
14. Quan, P.H.; Antoniac, I.; Miculescu, F.; Antoniac, A.; Paltanea, V.M.; Robu, A.; Bița, A.-I.; Miculescu, M.; Saceleanu, A.; Bodog, A.D.; et al. Fluoride Treatment and In Vitro Corrosion Behavior of Mg-Nd-Y-Zn-Zr Alloys Type. *Materials* **2022**, *15*, 566. [CrossRef]
15. Peng, F.; Li, H.; Wang, D.H.; Tian, P.; Tian, Y.X.; Yuan, G.Y.; Xu, D.M.; Liu, X.Y. Enhanced Corrosion Resistance and Biocompatibility of Magnesium Alloy by Mg-Al-Layered Double Hydroxide. *ACS Appl. Mater. Interfaces* **2016**, *8*, 35033–35044. [CrossRef]

16. Liu, Y.; Zhang, Y.; Wang, Y.-L.; Tian, Y.-Q.; Chen, L.-S. Research progress on surface protective coatings of biomedical degradable magnesium alloys. *J. Alloys Compd.* **2021**, *885*, 161001. [CrossRef]
17. Khaselev, O.; Weiss, D.; Yahalom, J. Structure and composition of anodic films formed on binary Mg-Al alloys in KOH-aluminate solutions under continuous sparking. *Corros. Sci.* **2001**, *43*, 1295–1307. [CrossRef]
18. Narayanan, T.S.N.S.; Park, I.S.; Lee, M.H. Strategies to improve the corrosion resistance of microarc oxidation (MAO) coated magnesium alloys for degradable implants: Prospects and challenges. *Prog. Mater. Sci.* **2014**, *60*, 1–71. [CrossRef]
19. Lin, Z.S.; Wang, T.L.; Yu, X.M.; Sun, X.T.; Yang, H.Z. Functionalization treatment of micro-arc oxidation coatings on magnesium alloys: A review. *J. Alloys Compd.* **2021**, *879*, 160453. [CrossRef]
20. Esmaeili, M.; Tadayonsaidi, M.; Ghorbanian, B. The effect of PEO parameters on the properties of biodegradable Mg alloys: A review. *Surf. Innov.* **2021**, *9*, 184–198. [CrossRef]
21. Rendenbach, C.; Fischer, H.; Kopp, A.; Schmidt-Bleek, K.; Kreiker, H.; Stumpp, S.; Thiele, M.; Duda, G.; Hanken, H.; Beck-Broichsitter, B.; et al. Improved in vivo osseointegration and degradation behavior of PEO surface-modified WE43 magnesium plates and screws after 6 and 12 months. *Mat. Sci. Eng. C Mater.* **2021**, *129*, 112380. [CrossRef] [PubMed]
22. Fischerauer, S.F.; Kraus, T.; Wu, X.; Tangl, S.; Sorantin, E.; Hanzi, A.C.; Loffler, J.F.; Uggowitzer, P.J.; Weinberg, A.M. In vivo degradation performance of micro-arc-oxidized magnesium implants: A micro-CT study in rats. *Acta Biomater.* **2013**, *9*, 5411–5420. [CrossRef] [PubMed]
23. Li, M.; Chu, X.; Wang, D.; Jian, L.; Liu, L.; Yao, M.; Zhang, D.; Zheng, Y.; Liu, X.; Zhang, Y.; et al. Tuning the surface potential to reprogram immune microenvironment for bone regeneration. *Biomaterials* **2022**, *282*, 121408. [CrossRef]
24. Pavarini, M.; Moscatelli, M.; Candiani, G.; Tarsini, P.; Cochis, A.; Rimondini, L.; Najmi, Z.; Rocchetti, V.; De Giglio, E.; Cometa, S.; et al. Influence of frequency and duty cycle on the properties of antibacterial borate-based PEO coatings on titanium for bone-contact applications. *Appl. Surf. Sci.* **2021**, *567*, 150811. [CrossRef]
25. Rahmati, M.; Raeissi, K.; Toroghinejad, M.R.; Hakimizad, A.; Santamaria, M. Effect of Pulse Current Mode on Microstructure, Composition and Corrosion Performance of the Coatings Produced by Plasma Electrolytic Oxidation on AZ31 Mg Alloy. *Coatings* **2019**, *9*, 688. [CrossRef]
26. Zhang, Y.; Hu, Z.; An, Y.; Guo, B.; An, N.; Liang, Y.; Wu, H. High-performance symmetric supercapacitor based on manganese oxyhydroxide nanosheets on carbon cloth as binder-free electrodes. *J. Power Sources* **2016**, *311*, 121–129. [CrossRef]
27. Fang, H.; Zhang, S.; Wu, X.; Liu, W.; Wen, B.; Du, Z.; Jiang, T. Facile fabrication of multiwalled carbon nanotube/alpha-MnOOH coaxial nanocable films by electrophoretic deposition for supercapacitors. *J. Power Sources* **2013**, *235*, 95–104. [CrossRef]
28. Zhang, D.; Peng, F.; Qiu, J.; Tan, J.; Zhang, X.; Chen, S.; Qian, S.; Liu, X. Regulating corrosion reactions to enhance the anti-corrosion and self-healing abilities of PEO coating on magnesium. *Corros. Sci.* **2021**, *192*, 109840. [CrossRef]
29. Qian, X.; Wu, Y.; Kan, M.; Fang, M.; Yue, D.; Zeng, J.; Zhao, Y. FeOOH quantum dots coupled g-C3N4 for visible light driving photo- Fenton degradation of organic pollutants. *Appl. Catal. B Environ.* **2018**, *237*, 513–520. [CrossRef]
30. Zhang, D.D.; Peng, F.; Tan, J.; Liu, X.Y. In-situ growth of layered double hydroxide films on biomedical magnesium alloy by transforming metal oxyhydroxide. *Appl. Surf. Sci.* **2019**, *496*, 143690. [CrossRef]
31. Zhang, D.; Peng, F.; Tan, J.; Zhang, Y.; Wang, F.; Xie, J.; Xu, R.; Du, H.; Qian, S.; Qiao, Y.; et al. Self-assembled ferric oxyhydroxide nanosheet on PEO-coated magnesium alloy with photocatalytic/photothermal antibacterial and enhanced osteogenesis activities. *Chem. Eng. J.* **2022**, *437*, 135257. [CrossRef]
32. Cao, Z.B.; Zhang, L.; Liang, K.; Cheong, S.S.; Boyer, C.; Gooding, J.J.; Chen, Y.; Gu, Z. Biodegradable 2D Fe-Al Hydroxide for Nanocatalytic Tumor-Dynamic Therapy with Tumor Specificity. *Adv. Sci.* **2018**, *5*, 1801155. [CrossRef] [PubMed]
33. Lei, Z.; Jia, P.; Wang, H.; Shi, Y.; Niu, J.; Feng, Y.; Hua, H.; Hua, Z.; Yuan, G. Facile Preparation of Poly(lactic acid)/Brushite Bilayer Coating on Biodegradable Magnesium Alloys with Multiple Functionalities for Orthopedic Application. *ACS Appl. Mater. Interfaces* **2017**, *9*, 9437–9448.
34. Zhang, Y.; Xu, J.; Ruan, Y.C.; Yu, M.K.; O'Laughlin, M.; Wise, H.; Chen, D.; Tian, L.; Shi, D.; Wang, J.; et al. Implant-derived magnesium induces local neuronal production of CGRP to improve bone-fracture healing in rats. *Nat. Med.* **2016**, *22*, 1160–1169. [CrossRef] [PubMed]
35. Balogh, E.; Paragh, G.; Jeney, V. Influence of Iron on Bone Homeostasis. *Pharmaceuticals* **2018**, *11*, 107. [CrossRef] [PubMed]
36. Laura, T.; Vaquero, M. Chronic Iron Deficiency as an Emerging Risk Factor for Osteoporosis: A Hypothesis. *Nutrients* **2015**, *7*, 2324–2344.
37. Mould, A.P.; Akiyama, S.K.; Humphries, M.J. Regulation of integrin $\alpha5\beta1$-fibronectin interactions by divalent cations: Evidence for distinct classes of bind sites for Mn^{2+}, Mg^{2+}, and Ca^{2+}. *J. Biol. Chem.* **1995**, *270*, 26270–26277. [CrossRef]
38. Siebers, M.C.; Walboomers, X.F.; van den Dolder, J.; Leeuwenburgh, S.C.G.; Wolke, J.G.C.; Jansen, J.A. The behavior of osteoblast-like cells on various substrates with functional blocking of integrin-beta 1 and integrin-beta 3. *J. Mater. Sci. Mater. Med.* **2008**, *19*, 861–868. [CrossRef]
39. Wu, T.; Shi, H.; Liang, Y.; Lu, T.; Lin, Z.; Ye, J. Improving osteogenesis of calcium phosphate bone cement by incorporating with manganese doped beta-tricalcium phosphate. *Mat. Sci. Eng. C Mater.* **2020**, *109*, 110481. [CrossRef]
40. Du, Z.; Leng, H.; Guo, L.; Huang, Y.; Zheng, T.; Zhao, Z.; Liu, X.; Zhang, X.; Cai, Q.; Yang, X. Calcium silicate scaffolds promoting bone regeneration via the doping of Mg^{2+} or Mn^{2+} ion. *Compos. Part B Eng.* **2020**, *190*, 107937. [CrossRef]

Journal of
Functional Biomaterials

MDPI

Review

Antibacterial Designs for Implantable Medical Devices: Evolutions and Challenges

Huiliang Cao [1,2,3,*], Shichong Qiao [4,5,6,*], Hui Qin [7,*] and Klaus D. Jandt [3,8,9,*]

1 Interfacial Electrochemistry and Biomaterials, School of Materials Science and Engineering, East China University of Science and Technology, Shanghai 200237, China
2 Lab of Low-Dimensional Materials Chemistry, Key Laboratory for Ultrafine Materials of Ministry of Education, East China University of Science & Technology, Shanghai 200237, China
3 Chair of Materials Science, Otto Schott Institute of Materials Research (OSIM), Friedrich Schiller University Jena, 07743 Jena, Germany
4 Department of Implant Dentistry, Shanghai Ninth People's Hospital, College of Stomatology, Shanghai Jiao Tong University School of Medicine, Shanghai 200011, China
5 National Clinical Research Center for Oral Diseases, Shanghai 200011, China
6 Shanghai Key Laboratory of Stomatology & Shanghai Research Institute of Stomatology, Shanghai 200011, China
7 Department of Orthopaedics, Shanghai Jiaotong University Affiliated Sixth People's Hospital, Shanghai 200233, China
8 Jena Center for Soft Matter (JCSM), Friedrich Schiller University Jena, 07743 Jena, Germany
9 Jena School for Microbial Communication (JSMC), Neugasse 23, 07743 Jena, Germany
* Correspondence: huiliang.cao@ecust.edu.cn (H.C.); shichong_qiao@hotmail.com (S.Q.); qh197826@163.com (H.Q.); K.Jandt@uni-jena.de (K.D.J.)

Abstract: The uses of implantable medical devices are safer and more common since sterilization methods and techniques were established a century ago; however, device-associated infections (DAIs) are still frequent and becoming a leading complication as the number of medical device implantations keeps increasing. This urges the world to develop instructive prevention and treatment strategies for DAIs, boosting the studies on the design of antibacterial surfaces. Every year, studies associated with DAIs yield thousands of publications, which here are categorized into four groups, i.e., antibacterial surfaces with long-term efficacy, cell-selective capability, tailored responsiveness, and immune-instructive actions. These innovations are promising in advancing the solution to DAIs; whereas most of these are normally quite preliminary *"proof of concept"* studies lacking exact clinical scopes. To help identify the flaws of our current antibacterial designs, clinical features of DAIs are highlighted. These include unpredictable onset, site-specific incidence, and possibly involving multiple and resistant pathogenic strains. The key point we delivered is antibacterial designs should meet the specific requirements of the primary functions defined by the *"intended use"* of an implantable medical device. This review intends to help comprehend the complex relationship between the device, pathogens, and the host, and figure out future directions for improving the quality of antibacterial designs and promoting clinical translations.

Keywords: implantable antibacterial surfaces; polymicrobial infections; surface modification; biocompatibility; tissue integration; bacterial charging; cell-selective surfaces; antibiotic resistance; antimicrobials; protein adsorption

Citation: Cao, H.; Qiao, S.; Qin, H.; Jandt, K.D. Antibacterial Designs for Implantable Medical Devices: Evolutions and Challenges. *J. Funct. Biomater.* **2022**, *13*, 86. https://doi.org/10.3390/jfb13030086

Academic Editor: James Kit-hon Tsoi

Received: 22 May 2022
Accepted: 17 June 2022
Published: 21 June 2022

Publisher's Note: MDPI stays neutral with regard to jurisdictional claims in published maps and institutional affiliations.

Copyright: © 2022 by the authors. Licensee MDPI, Basel, Switzerland. This article is an open access article distributed under the terms and conditions of the Creative Commons Attribution (CC BY) license (https://creativecommons.org/licenses/by/4.0/).

1. Introduction

It was estimated that over 500,000 types of medical devices, such as dental implants, vascular graft/endograft, orthopedic prosthetics, catheters, etc., are currently marketing globally for medical applications [1]. Every year, there are about 10,000,000 dental implants and more than 1,000,000 cardiovascular electronic devices inserted around the world [2,3]. It has been estimated that 100 million urinary catheters are sold worldwide each year [4]. As the population of the aged increases, procedures for implantable medical devices are expected to increase rapidly in the coming years. In the United States of America (USA), the primary total knee arthroplasty (TKA) is going to grow by 85%, to 1.26 million procedures by 2030 [5]. In Germany, by 2040, the total number of TKA is expected to increase by 45% to over 244,000 procedures; and the incidence rate of total hip arthroplasty (THA) is projected to increase to 437 per 100,000 inhabitants [6]. In the United Kingdom, the volume of hip and knee joint replacement is expected to increase by almost 40% by 2060 [7]. Bacterial infections are one of the most frequent and severe complications associated with the clinical application of implantable medical devices [1]. It was reported that device-associated infections (DAIs), including ventilator-associated pneumonia, catheter-associated urinary tract infection, and central-catheter-associated bloodstream infection), accounted for approximately 26% of all healthcare-associated infections (HAIs) in the USA [8]. The annual number of HAIs in European Union countries is about 3.2 million, including 37,000 registered mortalities [9]. The financial burden for the treatment of a DAI is also extraordinarily high. For instance, the average revision costs in the USA for infected hip and knee arthroplasty were approximately USD 80 and 60 thousand, respectively [10]. Additionally, by 2030, the estimated combined annual hospital costs related to arthroplasty infection will rise to USD 1.85 billion in the USA alone [11]. This urges the world to develop instructive prevention and treatment strategies for DAIs.

Accordingly, fundamental research on the development of various antibacterial surfaces has dramatically increased in recent years. Screening for "*antibacterial surface*" or "*antibacterial coating*" in the topic of the articles included in the Web of Science (*www.webofscience.com*; accessed on 14 February 2022) can hit more than 50,000 records between the years 1996 and 2021. Around 80% of these records were published during the last decade (between 2012 and 2021), and over 67% of them were published during the last five years (between 2017 and 2021), identifying a boom in developing antibacterial surfaces or coatings. Developing antibacterial surfaces for implantable medical devices also is currently a hot direction among the Chinese communities focusing on biomaterials science and engineering. Typical designs published in the first half of 2022 include copper-bearing titanium [12], surface charge and wettability control in lysozyme [13], light-activatable carbon monoxide gas generation by triiron dodecacarbonyl loaded polydopamine [14], clickable peptide engineered surface [15], calcium-doped titanium targeting blood protein adsorption [16], puncture and ROS (reactive oxygen species) release by nanorod zinc oxide patterns [17], light-stimulated ROS generation by rare-earth elements-doped titanium dioxide coating [18], on-demand antibiotics release by responsive polymers [19,20], and bacteriophage-modified alginate hydrogels [21]. This trend demonstrates that the academic community has already realized the urgency of solving the DAI problem, whereas only a limited number of these innovations have entered clinical applications or clinical studies around the world. A very small number of registered records (we found merely eight) concerning antibacterial surfaces were found in *ClinicalTrials.gov* (accessed on 22 May 2022) by searching for "*device infection*" in the "*Condition or disease*" field. As shown in Table 1, silver in metallic or ionic forms is the most popular active ingredient in developing antibacterial medical devices. Currently, a handful of antibacterial surfaces have been branded for clinical uses, which are commonly silver-based and normally custom-made (available on request). These include *Acticoat* using magnetron sputtering synthesized nanosilver coatings for wound care [22], *MUTARS* prosthesis reducing infections by electroplating a metallic-silver surface, METS prosthesis acting against pathogenic bacteria by absorption of ionic silver to anodized titanium implants [23], *PorAg* prosthesis taking advantage of a controlled electrochemical reaction (do not directly release silver ions) in a titanium-silver alloy for disinfection [23], and *PROtect* nails administrating gentamicin for prevention of infections in complex open fractures [24]. These commercial promotions have set examples for the development of antibacterial surfaces for implantable medical devices (here we coin them "*implantable antibacterial surfaces*"); however, it is still a challenge to improve the quality and efficiency of translational research over those "*antibacterial surface*" or "*antibacterial coating*" reports.

Table 1. Antibacterial surface registered for clinical studies *.

Active Ingredients	Devices	Phase	Locations	First Posted
Silver coating	Intravenous catheters	Not applicable	United States	25 August 2009
Antibiotics (minocycline and rifampin)	Antibacterial envelope for a cardiac implantable electronic device	Not applicable	United States	7 January 2010
Silver-based coating	Urinary catheter	Not applicable	United States	10 September 2012
Ionic silver	Wound dressings for a cardiac implantable electronic device	Phase 4	United States	24 May 2016
Silver-doped hydroxyapatite coating	Orthopedic implants (hip joint prostheses, intramedullary nails, and external fixator implants)	Not applicable	Turkey	17 November 2017
Gold-silver-palladium coating	Invasive devices (endotracheal tube, central venous catheter, and urinary catheter)	Phase 1, 2	Brazil	11 March 2019
Iodine	Barrier dressing for a cardiac implantable electronic device	Not applicable	Canada	19 October 2020
Antibiotic (gentamycin)	Platform wound device	Phase 4	United States	15 February 2021

* Data were obtained by searching for *"device infection"* in the *"Condition or disease"* field of the registered clinical studies conducted around the world on *ClinicalTrials.gov* (plus manual exclusion, as of 31 March 2022).

Herein we firstly analyze the cases associated with device-associated infections (DAIs) by highlighting the clinical features and challenges in DAIs prevention and treatment, then present the state-of-art research by identifying the evolutions in developing antibacterial surfaces for implantable medical devices, i.e., implantable antibacterial surfaces and, finally, illuminate the flaws in reporting of the findings in fundamental researchers to advance the development and translation of innovative designs against bacterial infections and promote the success of implantable medical devices.

2. Clinical Features of Device-Associated Infections

2.1. Site-Specific Incidence

Infection is a common and frequent complication associated with all types of biomedical materials, despite the infection rate varying greatly among different intended uses of various implantable devices (Table 2) [25–61]. Orthopedic implants, such as the ankle, hip, knee, elbow, shoulder, and finger joint prosthetics, are made of metals (titanium alloys, stainless steel, cobalt-chromium alloy, etc.) and are expected to serve long periods (>10 years) in patients' bodies. Infections of these devices are extremely troublesome [1]. Ankle arthroplasty has higher infection rates (2.4–8.9%) than hip (0.4–2.4%) and knee (1–2%) arthroplasty, although they are normally made of the same materials (Table 2). This is remarkably related to wound dehiscence (or prolonged drainage) developed due to the frail soft tissue surrounding ankles and increased chance of delayed wound healing following ankle arthroplasty [26,62]. The infection situation will be even more serious in revision cases. For example, the incidence of infection for primary hip and knee arthroplasty is around 2% (Table 2), yet this will be possibly as high as 12% and 22% for revision hip and knee arthroplasty, respectively [63]. Moreover, the number of infection cases is expected to increase progressively because the number of arthroplasty surgeries is going to grow in the coming years. In Taiwan, China, for instance, a total of 728 hip and knee infection cases were recorded in 2013 and this number was expected to increase markedly to over 3500 by 2035 [10]. Not only these metallic implants are connected to bacterial infection, but also polymer devices are susceptible to this complication (Table 2). Examples include breast implants, vascular graft/endograft, cardiovascular electronic devices, and cochlear implants, which are made of silicone, polytetrafluoroethylene, plastics, or Teflon, and have infection incidence high up to 10.2% [32], 6% [33], 7% [37], and 8% [40], respectively. Additionally, the DAIs may occur due to the device design. As in brain stimulation implants, the battery of the pulse generator should be replaced typically every 2 years, and such multiple replacements increase the risk of DAIs [46]. Furthermore, the incidence of infection is highly determined by the site a device is placed in. As shown in Table 2, the infection rates in urinary catheters (up to 13.7 cases per 1000 catheter-days), cerebrospinal fluid shunts (27%), internal fixation devices (32%), and dental implants (47%) are high. This is because these devices are highly challenged by bacterial adhesion and biofilm formation during their insertion and the subsequent service period. For example, urinary catheters provide routes for the entry of pathogenic bacteria, increasing the risk of acquiring infections [51]. Investigations of the bacterial sources in infected shunts also demonstrate that a majority of harmful microbes gained entry from the skin of the patients themselves [64]. The risk of complications in fixation of fractures is highly in connection to the low blood supply and elder people are susceptible to infection [59]. Additionally, there are more than 500 bacterial species associated with commensals or pathogens within the oral cavity [65]. This situation makes the prevention of infections in dental implants extremely complicated. The reported incidence rates for dental implants serving of over 3 and 5 years are 9.25% and 9.6%, respectively, and this rate for implants with service periods of over 10 years is up to 26% [61]. More importantly, the prevalence of the pathogenic strains is also associated with specific anatomical locations. Although *Staphylococcus* spp. is the most prevalent microbe associated with all types of bacterial infections, other pathogens can be involved in specific sites. Gram-negative microbes are involved in 10–40%, 20%, and 35–55% of vertebral, trauma/fracture, and foot/ankle-related infections [66]. Additionally, 15–30%, 20–30%, and 30–80% of polymicrobial infections occur in vertebral, trauma/fracture, and foot/ankle, respectively [66]. Different bacterial strains may have different metabolisms and pathogenic mechanisms that require specifically tailored treatments. This is especially critical to cure infections involving multiple pathogenic strains; as a result, developing an all-around antibacterial solution for all medical devices is hardly possible.

Table 2. Incidence of typical device-associated infections.

Device	Materials	Incidence	Reference
Ankle arthroplasty	Metals (titanium alloys), Ceramic, Polyethylene	2.4–8.9%	[25,26]
Hip arthroplasty	Metals (titanium alloys, stainless steel), Ceramics (alumina, zirconia), Polymers (polyethylene, polyetheretherketone), Composites	0.4–2.4%	[10,27,28]
Knee arthroplasty	Metals (titanium alloys, cobalt-chromium alloy), Ceramics (zirconia, titanium nitride), Polymers (polyethylene,)	1–2%	[10,29]
Breast implants	Silicone	1–10.2%	[30–32]
Vascular graft/endograft	Polytetrafluoroethylene, Polyethylene Terephthalate, Nitinol	0.16–6%	[33]
Cardiovascular electronic devices	Plastic polymers, Titanium, Teflon, Gold, Copper	0.9–7%	[34–38]
Cochlear implant	Teflon, Platinum-iridium alloy, Silicone, Titanium, Ceramics	1–8%	[39–43]
Brain stimulation implant	Stainless steel, Platinum, Titanium oxide, Iridium oxide	2–10%	[44–46]
Urinary catheters *	Natural rubber, Polyisoprene, Polymer ethylene vinyl acetate, Polytetrafluoroethylene, Hydrogel	0.1–13.7 cases per 1000 catheter-days	[47–52]
Cerebrospinal fluid shunts	Silicone rubber	1.9–27%	[53–57]
Internal fixation devices	Stainless steel, Cobalt-chromium alloys, Titanium alloys	7–32%	[58,59]
Dental implants	Titanium, Ceramics (zirconia, alumina)	6–47%	[60,61]

* The incidence of catheter-associated urinary tract infection is typically expressed as the number of infections per 1000 urinary catheter-days [52].

2.2. The Unpredictable Onset

Device-associated infections become even stickier because of those host-specific, transient, or resident factors (Table 3) [67–79]. The onset of DAI is not predictable, it can onset years after implantation (Cases 1 through 6 in Table 3). The soft tissue envelope around an implant likely degenerates with aging and can be disrupted by an occasional scratch, which may have promoted the infection of an alloplastic chin implant 45 years after placement [67]. Breast implants significantly risk bacterial contamination from hematogenous spread of distant antecedent infections. It was reported that the *Achromobacter xylosoxidans* (lives in wet soil) from a chronic footsore and *streptococcus viridans* (lives in the oral cavity) from recurrent periodontitis can cause infection of breast implants even 7 and 25 years after the implantation [68]. *Staphylococcus epidermidis* (*S. epidermidis*) can colonize various biomedical implants and escape from the immune clearance and antibiotic treatments, hence possibly causing symptom-free (such as pain, redness, or fever) chronic infection lasting even for 30 years before being identified by clinical approaches [69]. *Cutibacterium acnes* (previously known as *propionibacterium acnes*), a common conjunctival inhabitant, are slow-growing, anaerobic Gram-positive rods, and can manifest several years or even decades before leading to late infections in orbital implants made of silicone or tantalum [70,71]. The sources of the pathogens of the DAIs can be host-specific (Cases 7 through 9 in Table 3). DAIs can be initiated by acute illness (e.g., diarrhea developed during a holiday journey [31]), penetration of contaminated water during participating in outdoor activities [45], or even when the patients play with their pets (bacterial contamination from zoonotic sources) [72]. Moreover, the occurrence of DAIs is commonly associated with a compromised immune system in the hosts (Cases 10 and 11 in Table 3). Methotrexate, a folate antagonist, can affect neutrophil chemotaxis and induce apoptosis of T cells and reactivation of opportunistic pathogens; hence chronic treatment of rheumatoid arthritis with this kind of drug significantly increases the risk of infections around the battery for brain stimulation [73]. *Nocardia nova* is a common environmental pathogen and rarely affects immunocompetent hosts; however, this species successfully colonized a tibia implant placed in an immunocompetent patient [74]. *Listeria monocytogenes*, a common organism associated with unpasteurized dairy products (e.g., deli meats and unpasteurized cheeses), can induce a periprosthetic joint infection in a patient with a history of diabetes mellitus, asthma, and psoriatic arthritis [75]. *Anaerobiospirillum succiniciproducens*, a common settler in the gastrointestinal tract of cats and dogs, can induce a prosthetic hip joint infection in an immunocompromised patient [76]. DAIs are normally initiated by bacterial seeding and as a result tissue integration will be impaired quickly; however, some cases failed to identify any organism by cultures [77,78] and tissue integration was intact after being infected [79]. These situations add difficulties to the prevention, diagnosis, and treatment of DAIs.

Table 3. Representative cases showing the latent period of DAIs.

Case	Devices	Latent Period (Post Insertion)	Pathogens	Causes	Reference
1	Alloplastic chin implant	45 years	/	After scratching herself (soft tissue degeneration due to aging)	[67]
2	Breast implant	Seven years	*Achromobacter xylosoxidans* (a pathogen that lives in wet soil)	Development of a chronic footsore (hematogenous spread from distant bacterial infection sites)	[68]
3	Breast implant	25 years	*Streptococcus viridans* (a pathogen that lives in the oral cavity)	After extensive dental treatment (hematogenous spread from distant bacterial infection sites)	[68]
4	Alloplastic implant	30 years	*Staphylococcus epidermidis*	Bacterial contamination years before identifying the infection (a symptom-free chronic infection; the pathogen escaped immune clearance and antibiotic treatments)	[69]
5	Orbital implant	30 years	*Cutibacterium acnes* (previously known as *Propionibacterium acnes*)	Bacterial contamination during the primary implantation (the pathogen can manifest for several decades)	[70]
6	Orbital implant	26 years (implant exposure 10 years before the presentation was documented)	*Propionibacterium acnes (renamed Cutibacterium acnes)*	Bacterial contamination during the primary implantation or implant exposure during scleral patch graft repair	[71]
7	Breast Implant	Five months	*Salmonella serogroup C*	Hematogenous seeding due to developing of diarrhea during a holiday travel	[31]
8	Generator for brain stimulation	Four months	Multispecies including the rare *Cupriavidus pauculus species* (an environmental Pathogen in "*water*")	Penetration of contaminated water during participating in outdoor activities	[45]
9	Breast implant	Seven months	*Pasteurella canis* (a pathogen normally lives in the oropharyngeal commensal flora of cats and dogs)	Bacterial contamination from a patient-owned cat	[72]
10	Battery for brain stimulation	Two cases (Two years or 10 years)	*Staphylococcus aureus*	Chronic treatment of rheumatoid arthritis with methotrexate	[73]

Table 3. *Cont.*

Case	Devices	Latent Period (Post Insertion)	Pathogens	Causes	Reference
11	Tibia Tenodesis Implant	Four and half months	*Nocardia nova* (a common environmental pathogen, rarely affects immunocompetent hosts)	Contamination of his tibial wound by the outside facility	[74]
12	Knee arthroplasty	4 months	*Listeria monocytogenes* (a facultative intracellular organism; commonly associated with deli meats and unpasteurized cheeses)	Consuming unpasteurized dairy products (an immunocompromised patient)	[75]
13	Hip arthroplasty	10 years	*Anaerobiospirillum succiniciproducens* (lives in the gastrointestinal tract of cats and dogs)	Breeding a dog (an immunocompromised patient)	[76]
14	Knee arthroplasty	Eight years	*Bartonella henselae* (a pathogen that induces acute infections but is hard to be diagnosed by culture)	A cat scratch	[77]
15	Cranioplasty implant	Two years and three months	No bacteria were cultured, but the infection was clinically evident	/	[78]
16	Shoulder prosthesis	Three years	*Staphylococcus* spp.	/	[79]

2.3. Diversity of Relevant Pathogens

Infections associated with medical devices with the same intended use (the same device category) but placed in different individuals are possibly connected with different bacterial strains. As shown in Table 3, the infection of breast implants can result from *achromobacter xylosoxidans* (Gram-negative rod) [68], *streptococcus viridans* [68], and *salmonella serogroup C* [31], or *Pasteurella canis* [72]. Polymicrobial infections become more prevalent in DAIs [66,80]. Even a single infection in a specific individual often has a polymicrobial composition [81]. Multispecies including the rare *Cupriavidus pauculus* species were isolated in an infection associated with the generator for brain stimulation [45]. Since the bacteria associated with an infection of a medical device may have diverse morphologies and arrangements, an effective antibacterial strategy must be capable of eliminating multiple pathogenic species. *Cocci* cells (spherical bacteria) range from 0.5 to 2.0 μm in diameter, rods are approximate 0.5–1.0 μm in width and 1–10 μm in length, and spiral bacteria are up to 20 μm in length and 0.1–1 μm in diameter [82]. Moreover, bacterial morphology varies with the growth environments (medium, surfaces, etc.) and growth phase (normally smallest in the logarithmic phase) [83,84]. These facts add additional difficulties to developing a competent antibacterial surface for implantable devices. On account of these features of DAIs, antibacterial surfaces only have a pore-size-based cell selectivity [85], or those peptide-loaded surfaces merely have specific actions to Gram-positive or Gram-negative strains [86] and are not likely adequate to prevent infection of implantable medical devices.

2.4. Prevalence of Antibiotic Resistance

The uses of internal implants in humans are safer and more common since sterilization methods and techniques were established at the end of the 19th century [87], and the commercialization of antibiotics especially penicillin in the first half of the 20th century [88]. Antibiotics have become an integral component of contemporary biomedical practice, producing a serious follow-up threat: antibiotic resistance in bacteria [89,90]. Clinical cases in orthopedic practice have shown that infections of antibiotic-resistant bacteria, such as methicillin-resistant *Staphylococcus aureus* (MRSA), are closely related to high morbidity and mortality [91]. Antibiotic resistance in bacteria even multidrug-resistant (MDR) bacteria is now a worldwide challenge [91]. Antibiotic-resistant infections were frequently reported all over the world, including in both developing and developed countries (Table 4) [92–110]. During an infection, *Staphylococcus aureus* (*S. aureus*) often forms biofilms on implantable devices, which dramatically increases the ability of the species to acquire resistance via horizontal plasmid transfer [111]. This is why *S. aureus* has high rates of resistance. As shown by the typical cases reported in recent years (Table 4), MRSA has become the most common strain causing infections of various implantable medical devices, including cardiac devices [93,95,99,103,106], orthopedic prosthetics [96,97], cochlear implants [98], breast implants [100], laryngeal implants [101], and stent grafts [109]. In addition, there is an alarming increase in antibiotic resistance in other strains, such as *Acinetobacter baumannii* [92], *Mycobacterium chelonae* [94], *Enterobacter cloacae complex* [102], *S. epidermidis* [104,110], *Klebsiella pneumoniae* [105], *Staphylococcus haemolyticus* [107], and *Staphylococcal endophthalmitis* [108], are also involved in various resistant DAIs. Those resistant DAIs impacted patients have to experience prolonged hospital stays, bear high medical costs, and risk increased mortality (references in Table 4). Antibiotic recalcitrance is a worldwide threat that likely causes substantial global economic costs ranging from USD 21,832 per individual case to over USD 3 trillion in gross domestic product (GDP) loss by 2050 [112]. In the USA alone, at least 2 million infections and 23,000 deaths per year were caused by antibiotic-resistant bacteria, costing USD 55–70 billion [90]. Currently, antibiotic-loaded materials are important complements to modular medical practices for the prevention of recurrent infections in various medical devices, such as wound dressings, bone cement, bone plates, nails, or prostheses [24,113,114]. However, applications of these surfaces in *"uninfected tissues"* to prevent DAIs should be careful and in strict guidance, because the prolonged release of prophylactic antibiotics possibly contributes to arising resistant mutants [115]. Silver-based surfaces also have attractive efficacy in the prevention of DAIs [116], improper use of this material may also pose bacterial-resistant problems [117,118]. In addition, pathogenic bacteria have many defensive actions resistant to antimicrobial challenges [91,119,120]: (a) express polymer biofilms to protect themselves from antibiotic attacks; (b) remodel their outer surface to reduce antibiotic uptake; (c) synthesize precursors to modify the target of antimicrobials; (d) produce enzymes to detoxify dangerous drugs. Therefore, antibacterial surfaces, especially those release-killing ones, should be designed to bypass these actions of bacterial cells.

Table 4. Epidemiology of antibiotic-resistant DAIs.

Case	Resistant Pathogens	Implant	Latent Period	Reference
1	*Multidrug-resistant Acinetobacter baumannii*	Hip arthroplasty	12–25 days	[92]
2	*Methicillin-resistant Staphylococcus aureus (MRSA)*	Cardiac pacemaker	Nine years	[93]
3	*Clarithromycin-resistant Mycobacterium chelonae*	Breast implant	Four days	[94]
4	*MRSA*	Transvenous lead	Four years	[95]
5	*MRSA*	Ankle fracture fixation	Eight weeks	[96]
6	*MRSA*	Cranial implant	Three months	[97]
7	*MRSA*	Cochlear implant	Five months	[98]
8	*MRSA*	Pacemaker	Two months	[99]
9	*MRSA*	Breast Implant	Two days	[100]
10	*MRSA*	Laryngeal implant	More than one year	[101]
11	*Carbapenem-resistant Acinetobacter baumannii; Fluoroquinolone-resistant Enterobacter cloacae complex (AmpC overexpression)*	Internal fixation for an open proximal tibial fracture	Two months	[102]
12	*MRSA*	Pacemaker	Two years	[103]
13	*Multidrug-resistant Staphylococcus epidermidis*	Plates and wire cerclages for periprosthetic fractures	Three months	[104]
14	*Carbapenem-resistant Klebsiella pneumoniae*	Lumbar instruments,	Seven days	[105]
15	*MRSA*	The ventricular lead of an implanted defibrillator	Eight weeks	[106]
16	*Methicillin-resistant Staphylococcus haemolyticus*	Hip joint	Two years	[107]
17	*Ofloxacin-resistant staphylococcal endophthalmitis*	Intravitreal ozurdex implant	Three days	[108]
18	*MRSA*	Stent graft	Three days	[109]
19	*Methicillin-resistant Staphylococcus epidermidis*	Spinal instrumentation	7–88 days	[110]

3. Innovative Designs to Mitigate Device-Associated Infections

Based on the reports we screened, the innovations of implantable antibacterial surfaces can be categorized into four groups, namely surfaces with prolonged or cell-selective bactericidal efficacy and responsive or immune-instructive surfaces. Prolonged antibacterial efficacy (or long-term antibacterial activity) can be realized by taking advantage of the degradation or surface structures of the supporting materials for orchestrated release or immobilizing the antimicrobials yielding contact-killing surfaces. Cell selectivity of antibacterial surfaces can be obtained by doping of multifunctional metals, the combinational release of ingredients that are respectively good for antibacterial and tissue integration promotion or applying the electrochemical reactions evoked by the host's physiological fluids to recognize bacteria from mammalian cells. Responsive antibacterial surfaces can deliver services over the stimulation of external light irradiation, or internally by the bacterial charge or bacterial infection-associated pH shifts. Since the key players (neutrophils and macrophages) in the immune system can be regulated by proper surface chemistry, topography, wettability, or stiffness, immune-instructive antibacterial surfaces are expected to be produced by control of these parameters. These evolutions in the development of implantable antibacterial surfaces help us rethink those complex interactions among device surface, host, and the pathogen (Figure 1), advancing the solution of DAIs.

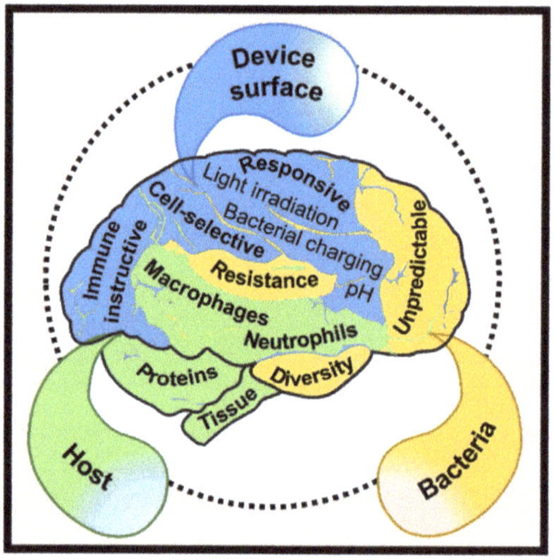

Figure 1. Rethinking the interplay among device surface, host, and pathogen.

3.1. Prolonged Antibacterial Efficacy

As shown by Tables 3 and 4, the latent period of a DAI can be days after implant placement [92,94,105,108–110], years after the surgery [78,79,93,95], or even decades later [67,68]. This feature of DAIs lays the basis for the development of antibacterial surfaces with long active durations. As shown by the representative reports on the development of *"long-term"* antibacterial surfaces (Table 5) [121–139], various ingredients such as commercial antibiotics (tigecycline, vancomycin, amoxicillin, etc.) [121,132], metals or metal ions (silver, copper, or zinc) [124,125,129], and other drugs [127,128] were taken to equip implantable biomaterials (titanium, silicone, ceramics, etc.) with prolonged antibacterial efficacy, ranging from days [127,131,133] to months [122,134]. Extending the release period of the antimicrobials is currently a major pathway leading to *"long-term"* antibacterial surfaces. Calcium phosphate cement (CPC) has proved an effective carrier to retain vancomycin (effective for the treatment of MRSA) to local sites [140,141], ensuring the antibiotic has a 24-week release profile in vivo [122]. Proper antibiotic concentration is a key factor that determines the mechanical strength of vancomycin-impregnated CPC and influences the effective antibacterial period of the composite [140]. Electrochemical oxidation, namely micro-arc oxidation (also known as plasma electrolytic oxidation) and anodic oxidation, is a well-known class of approaches that can produce porous surface layers on implant materials and, in the meantime, load antibacterial agents on the material's surface. Shivaram et al. demonstrated that the silver loaded in an anodized titanium substrate had a release period of a minimum of 6 months [134]. The titanium substrates were fabricated with 25 vol% porosity by using a powder-based additive manufacturing technique [142]. Then electrochemical anodization was applied to the porous titanium in a hydrofluoric acid electrolyte to produce a surface layer of nanotube arrays with a thickness of 375 $\pm$ 35 nm and diameter of 105 nm $\pm$ 30 nm, which facilitated the loading of silver from a 0.1 M silver nitrate ($AgNO_3$) solution via direct current deposition [134]. After heating at 500 °C, tightly adhered silver particles with a coverage of 13.5% were detected on the nanotube-structured surface. The 27-week cumulative release profiles demonstrated that silver release was within 10 ppm (mg/mL), which ensured good early-stage osseointegration of the porous titanium implants, along with good antibacterial activities [134]. Micro-arc oxidation is another technique that can produce a porous titanium surface which may facilitate the control of antimicrobial release. Very recently, Tsutsumi et al. reported that silver and zinc load micro-arc oxidation layer on titanium exhibited excellent activity against *Escherichia coli* (*E. coli*) after a six-month immersion in physiological saline [124]. Another way to prolong the effective period of antimicrobials is to immobilize (or embed) them in the non-degradable implant surfaces and prevent release. Cao et al. developed a silver plasma immersion ion implantation and deposition (Ag PIII&D) procedure to in situ synthesis and immobilize silver nanoparticles (Ag NPs) on titanium surface [143]. The process is generally carried out in a vacuum chamber of about 2.5 $\times$ 10^{-3} Pa and takes a pure silver rod (10 mm in diameter) as a cathode to produce cathodic arcs, which serve as sources of positively charged silver ions (Ag^{n+}) (Figure 2a). The silver arcs are filtered by a curved magnetic duct to remove the macro-particles produced. The energetic silver ions in a plasma form are accelerated and injected in a non-line-of-sight manner onto the titanium surfaces, which are negatively biased by a pulsed high voltage synchronizing with the cathodic arc current under a certain frequency. The positively charged silver ions become neutral atoms when they reach the titanium surfaces. As the process continues, the neutral atoms are further condensed and nanoparticles precipitate. By using this process, well-distributed Ag NPs were synthesized and immobilized on titanium. Figure 2b shows a group of Ag NPs average of 5 nm in diameter was produced by Ag PIII&D under a 30 kV bias for 30 min. Cross-sectional TEM images also confirmed that those Ag PIII&D produced nanoparticles were metallic silver (Figure 2c) [144]. The antibacterial efficacy of these nanoparticles was found independent of silver release [143]. As shown in Figure 2d, the grain boundaries of the titanium substrate were exposed after the material (Ag PIII&D treated for 30 min under a 30 kV bias) have defeated the colonization of S. aureus (cultured for 24 h at 37 °C with a bacteria concentration of 10^8 CFU/mL), indicating the antibacterial efficacy of those immobilized Ag NPs is associated with the corrosion of the titanium substrate. Since the standard electrode potential of titanium, -1.63 V, is more negative than that of silver at 0.80 V, the Ag PIII&D treated titanium surface embedded with a proper number of well-distributed Ag NPs likely evoked micro-galvanic corrosion, in which the cathodic reactions on Ag NPs may establish proton depleted regions on the titanium surface that possibly disrupt the proton electrochemical gradient (i.e., proton-motive force) in the intermembrane space of the microbes and get them killed (Figure 2e) [143]. Followed-up studies demonstrated that Ag PIII&D treated titanium implants have a long activity duration (60 days, the longest time point studied) against bacterial colonization [138]. The effectiveness of this micro-galvanic-associated antibacterial mechanism in copper-bearing stainless steels was also evidenced recently [145]. It was found that the contact killing of copper-bearing stainless steel was manipulated by the potential difference of the microdomains (the copper-rich phase and the matrix) in the material, which also produced proton depletion in bacteria and as a result cell death [145].

Table 5. Representative reports on long-term antibacterial surfaces.

Active Ingredients	Intended Use (Substrates)	Effective Period	Reference
Tigecycline, Copper ions	Treatment for osteomyelitis (Alginate aerogel)	18 days	[121]
Vancomycin	Cement (Calcium phosphate)	168 days	[122]
(Z-)-4-bromo-5-(bromomethylene)-2(5H)-furanone	Dental implants (Titanium)	60 days	[123]
Silver/Zinc ions	An orthopedic and dental implant (Titanium)	180 days	[124]
Nanosilver	Bone implant (Polylactic acid fiber)	11 days	[125]
Honokiol	Remineralization of demineralized enamel (Poly(amido amine) (PAMAM) (Dendrimer)	24 days	[126]
Patchouli Essential Oil	Wound Dressing (Polyvinyl alcohol and chitosan)	2 days	[127]
Cetylpyridinium chloride	Endodontic sealers (Polyhydroxyethyl methacrylate trimethylolpropanetrimethacrylate)	48 days	[128]
Metallic silver	Hard tissue replacements (Titanium)	84 days	[129]
Copper	Orthopedics (Titanium)	14 days	[130]
Zinc/Copper	Cement (dicalcium silicate)	3 days	[131]
Amoxicillin	Wound dressing (Poly (e-caprolactone))	7 days	[132]
Chlorhexidine	Medical devices (not clear, 316L)	3 days	[133]
Silver ions	Orthopedic implants (Titanium)	189 days (silver release)	[134]
Nanosilver	Biomedicine (not clear)	7 days	[135]
Nanogold/Titania	Orthopedic implants (Titanium)	6 days	[136]
Nanosilver	Orthopedic implants (Titanium)	60 days	[137]
Silver nanoparticles	Orthopedic implants (Titanium)	60 days	[138]
Poly (poly (ethylene glycol) dimethacrylate)	Peritoneal dialysis catheters (Silicone)	30 days	[139]

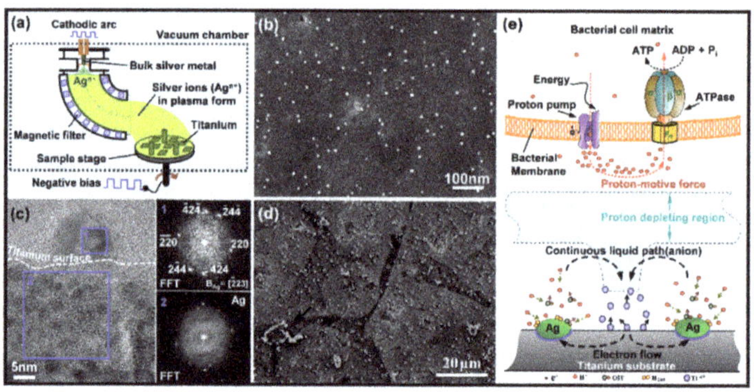

Figure 2. Contact killing of silver nanoparticles synthesized and immobilized on titanium by ion implantation: (**a**) schematic representation of the silver plasma immersion ion implantation and deposition (Ag PIII&D) process; (**b**) SEM image of the silver nanoparticles synthesized and immobilized on titanium by Ag PIII&D under a 30 kV bias for 30 min; (**c**) cross-sectional TEM of the silver nanoparticles synthesized and immobilized on titanium by Ag PIII&D, with corresponding fast Fourier transform patterns (FFT, 1 and 2) inserted; (**d**) SEM image of the Staphylococcus aureus cells cultured on an Ag PIII&D treated (treated for 30 min under a 30 kV bias) titanium for 24 h at 37 °C with a bacteria concentration of 10^8 CFU/mL; (e) possible antibacterial mechanism of the Ag PIII&D treated titanium; (**b**, **d**, and **e**) reused with permission from Elsevier [143]; (**c**) reused with permission from American Chemical Society [144].

These studies provide insights into solving the problem of DAIs; however, they are still far from meeting all the clinical requirements. This applies, especially for those late hematogenous cases, which may suddenly come years or decades after surgeries [68]. At present, there is no clear definition of the time of effectiveness of "*long-term*" antibacterial surfaces. Our opinion on determining "*long-term*" is to clarify the "*intended use*" of the surface first and take into account the time for a specific tissue repair process. The skin healing process consists of ordered stages, which are inflammation (15 min to 6 days), proliferation (2–3 days to 2–3 weeks), and maturation (3 weeks to 2 years) [132]. In this respect, we believe that the effective period for a "*long-term*" antibacterial wound dressing should exceed 3 weeks. Generally, the healing of a bone fracture involves three distinct but overlapping stages, i.e., the early inflammatory period (a few hours to days), the repairing period (weeks), and the late remodeling period (months to years) [146,147]. As a result, the typical time for a new bone to achieve adequate strength is 3 to 6 months [146]. Therefore, we believe the effective period for "*long-term*" antibacterial bone implants should exceed 3 months, and this time required for elder patients should be longer because their bone healing process is relatively slow. Although studies on "*long-term*" antibacterial surfaces are increasing these years, most of the reports did not consider the "*matching*" problem between antibacterial duration and tissue integration, yielding barriers to translational research. Since the incidence of DAI is site-specific, this aspect should be considered in future studies.

3.2. Response to pH Shifts

It is known that the pH shift is a common phenomenon of bacterial infections [148–150], laying the basis for the development of pH-responsive antibacterial surfaces. The antibacterial activity of pH-responsive films or coatings can be triggered by the protonation or deprotonation of their ionic groups. The thiazole and triazole groups, for example, in polymer PS54-b-PTTBM23 (on porous polystyrene surfaces) can be protonated under acidic pH levels, which increased the positive charge density on the materials surface to act against bacterial adhesion (Figure 3a) [151]. In addition, the killed bacteria can be further removed by increasing the pH levels (pH 7.4, for instance), which induced deprotonation of the thiazole and triazole groups in the materials [151]. Normally, pH-responsive surfaces are designed for the controllable release of antibacterial agents by manipulating the materials' pH-associated swelling or shrinking processes. By shifting the environmental pH, the protons of the carboxyl repeat units in poly(methacrylic acid) can be removed to make the material swell, which can control the release of antibacterial drugs [152]. In this manner, Wei et al. developed a pH-responsive surface capable of loading bacteriolytic lysozyme at acidic pH levels and releasing it under neutral or basic pH [152]. A pH-sensitive fibrous membrane was also developed to control the release of antibacterial gatifloxacin hydrochloride and silver nanoparticles [153]. The backbone (hydrophobic)-attached amino groups (weak basic moieties) of chitosan adapt to a deprotonated state above pH 6 while becoming protonated and positively charged at low pH, demonstrating a pH-dependent extension of the colloid chains and consequently swelling of the material [154]. Accordingly, chitosan was crosslinked with hydroxypropyl methylcellulose and 2-hydroxypropyl-β-cyclodextrin to produce a superabsorbent hydrogel for controllable delivery of antibacterial 3,4-dihydroxy cinnamic acid in response to pH changes [154]. Similarly, the structure of keratin hydrogel was reorganized by manipulating the protonation and deprotonation process of carboxyl groups in the material, yielding pH-dependent shrinking and swelling at low and high pH levels, respectively [155]. This behavior of the keratin hydrogel was taken to control the release of biocidal zinc oxide nanoplates in a pH-dependent manner, which can be a potential therapy response to a bacteria-contaminated media with biased pH and treatment of chronic wounds [156].

In addition, acid-labile bonds can also be used to program the release of antibacterial agents. Antibacterial gentamicin was conjugated with an alginate dialdehyde Schiff base reaction between the aldehyde groups (-CHO) and amino groups (-NH2) from the polymer, and was released due to the acidic environment triggered the disintegration of the Schiff base bonds (Figure 3b) [157]. Similarly, antimicrobial 6-Chloropurine was conjugated to 4-(vinyloxy) butyl methacrylate (VBMA) to produce 4-(1-(6-chloro-7H-purin-7-yl) ethoxy) butyl methacrylate (CPBMA), which contained a hemiaminal ether linkage can be hydrolyzed in mildly acidic conditions and allowed the controllable release of the antibacterial agent (Figure 3c) [158]. Moreover, pH-induced material structural evolutions, such as degradation, disintegration, and conformational changes, are also applied for the controllable delivery of biocides. Polyacetal-based polymers are degradable under acidic pH levels and possess a relatively low critical solution temperature (LCST) which allows wettability control by shifting the temperature (between LCST and room temperature) [159]. On account of this, a film-forming triple polymer-gel matrix containing polyacetal-based polymer was prepared by De Silva et al. to control the topical release of silver sulfadiazine, which was highly efficient against wound pathogens, such as *S. aureus*, *Pseudomonas aeruginosa* (*P. aeruginosa*), and *Candida albicans* (*C. Albicans*) [159]. The Schiff base structure between the amino groups (-NH_2) in dopamine and the aldehyde groups (-CHO) in oxidized dextran can be formed at pH 7.0 under the protection of nitrogen (N_2) [160]. The Schiff base bonds were disintegrated due to exposure to acidic bacteria-infected diabetic wounds, which was the mechanism used by Hu et al. to control the release of antibacterial silver nanoparticles by dopamine-conjugated oxidized dextran polymers (Figure 3d) [160]. The pH-induced conformational changes in silk fibroin (in a nanocapsule structure) were also applied to control the delivery of eugenol, which exhibited strong efficacy against both Gram-positive *S. aureus* and Gram-negative *E. coli* [161].

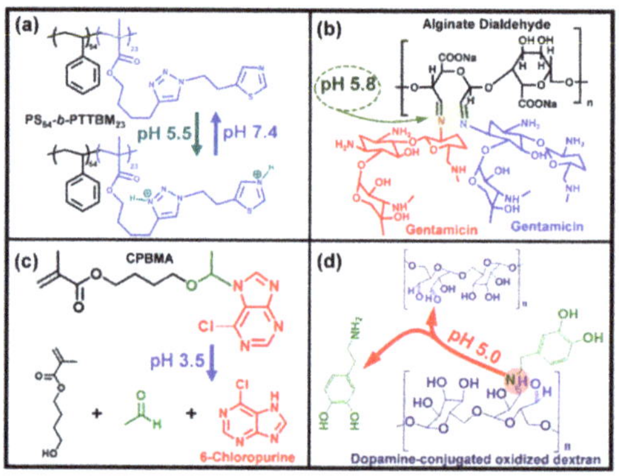

Figure 3. Typical methods toward pH-responsive surfaces: (**a**) protonation of polystyrene-*b*-poly(4-(1-(2-(4-methylthiazol-5-yl)ethyl)-1*H*-1,2,3-triazol-4-yl)butyl methacrylate) (PS_{54}-*b*-$PTTBM_{23}$) at acidic pH levels and increase of the positive charge density on the surfaces [151]; (**b**) breaking the Schiff base bonds between antibacterial gentamicin and alginate dialdehyde by acidic environments [157]; (**c**) hydrolyzation of the hemiaminal ether linkage of antimicrobial 6-Chloropurine in 4-(1-(6-chloro-7H-purin-7-yl) ethoxy) butyl methacrylate (CPBMA) by mild acidic conditions [158]; (**d**) destruction of dopamine-conjugated oxidized dextran polymer to release the contained silver nanoparticles by disintegration the Schiff base structures in the polymer [160]. (**a**,**c**) reused with permission from John Wiley and Sons and American Chemical Society, respectively; (**b**,**d**) reused with permission from Elsevier.

3.3. Response to Bacterial Charging

Membrane-bound respiratory electron transfer in bacteria plays a critical role in the synthesis of adenosine triphosphate and bacterial maintenance [162]; therefore, it can be a potential target for antibacterial surfaces. Extracellular electron transfer is a general mechanism required for bacterial growth [163–166]. The microbial cell envelope is not electrically conductive; hence bacteria have evolved strategies to exchange electrons with extracellular substances [167], including direct electron transfer via physical contacts (through the bacterial envelope or pili) between a microbe and a material surface, and redox-active compounds mediating electron shuttle between bacteria and the material's surface serve as electron acceptors [168].

Accordingly, Cao et al. proposed to construct antibacterial coatings targeting the extracellular electron transfer process in pathogenic bacteria (Figure 4) [169,170]. Ag NPs in various sizes (4–25 nm) were in situ synthesized and immobilized onto plasma-spraying-prepared titanium oxide coatings by manipulating the atomic-scale heating effect in silver plasma immersion ion implantation. The antibacterial efficacy of the resulting composite coatings was dependent on the particle size and interparticle space of the immobilized silver, i.e., large particles (5–25 nm) induced fatal cytosolic content leakage and lysis of both Gram-negative and Gram-positive bacteria while small ones (~4 nm) did not [169]; and a relatively large interparticle space was superior to a small interparticle space is disrupting the integrity of the adherent bacterial cells [170]. Similar results were also reported in follow-up studies by using silver nanoparticles decorated with tantalum oxide coatings [171,172]. By using plasma spraying, graphene nanomaterials decorated with titania coatings were prepared for antibacterial applications [173]. The coatings can collect the electrons extruded by adherent bacterial cells due to the rectification of the Schottky-like graphene-titania boundaries [173]. In vitro evidence showed that cobalt-titanium dioxide and cobalt oxide (CoO or Co_3O_4)-titanium dioxide nanoscale heterojunctions can downregulate the expression of respiratory gene levels in bacteria and cause oxidative damage to bacterial surfaces [174]. In another study, Wang et al. also found that the antibacterial efficacy of tungsten-incorporated titanium dioxide coatings (prepared by micro-arc oxidization) was related to their strong capability in the storage of bacteria-extruded electrons and accumulation of sufficient valence-band holes inducing oxidative damages to the microbes [175]. These findings have opened up new avenues for taking advantage of the intrinsic feature of biological systems to design and control the antibacterial actions of biomaterials.

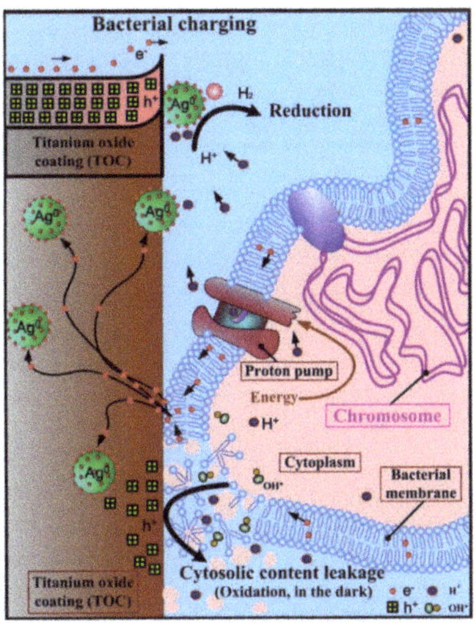

Figure 4. Silver nanoparticle decorated titanium oxide coating acting against bacterial colonization by taking advantage of extracellular electron transfer in bacteria: collection and storage of bacteria-extruded electrons on the immobilized silver nanoparticles ("bacterial charging"), accumulation of valence-band hole (h^+) at the titanium oxide side of the silver–titanium oxide boundaries, and disruption of bacterial cell walls (cytosolic content leakage) by those accumulated valence-band holes (oxidation) [169]. Reused with permission from Elsevier.

3.4. Response to Light Irradiation

Sterilizing materials' surfaces with ultraviolet (UV) light is a well-established standard method that has been around for decades. Materials converting light energy to heat for local disruption of bacterial colonization, i.e., photothermal therapy, are promising alternatives to antibiotics that possibly circumvent the problem of resistance [176]. Typical reports in this direction are listed in Table 6. Gold nanoparticles have been studied widely because of their high efficiency of photothermal conversion via surface plasmon resonance in the near-infrared (NIR) region (in the range of 700–1100 nm) [177]. It was reported that a gold nanoparticle and phase-transitioned lysozyme hybrid film was able to kill 99% of adherent bacteria within 5 min under the illumination of a NIR laser [177]. Composite thin films were produced by coordinative assembly of tannic acid (TA) and iron ions (Fe^{3+}) and yielded Au-TA/Fe (Figure 5a; the support can be other materials, rather than gold) [178]. These films exhibited high absorption and efficient light-to-heat convention under NIR irradiation (Figure 5b), as a result, they had efficient photothermal killing effects that disrupted 99% of adherent microbes, including both Gram-negative *E. coli* and Gram-positive MRSA strains (Figure 5c).

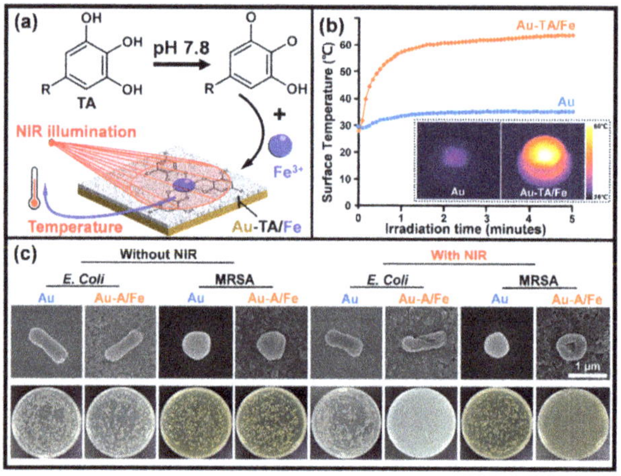

Figure 5. A photothermal antibacterial surface: (**a**) schematic illustration of the coordinated assembly of tannic acid (TA) and Fe^{3+} ions (iron chloride hexahydrate) on gold (can be other materials), yielding the Au-TA/Fe; (**b**) near-infrared (NIR) irradiation (808 nm, 2.2 $W \cdot cm^{-2}$) induced temperature changes on the material surface immersed in phosphate-buffered saline (PBS), with corresponding thermal images inserted; (**c**) SEM images of adherent bacteria (*E. coli* or MRSA) on materials surface with/without NIR irradiation (5 min), together with the typical photographs of bacterial colonies re-cultured from materials surface of different processing histories. Adapted from reference [178] with permission from the American Chemical Society.

Materials absorbing light energy to produce reactive oxygen species (ROS) for oxidization of bacterial cell walls, yielding photodynamic therapies [179], are also developed for potential biomedical applications. Quantum dots are sensitive to blue light illumination and capable of producing singlet oxygen, which is a powerful ROS able to disintegrate bacterial cell walls [180]. It was reported that hydrophobic carbon quantum dots incorporated into various polymer matrices in form of thin films had significant activity against *S. aureus* under blue light irradiation [180]. An antibacterial coating composed of black phosphorus nanosheets (BPS) and poly (4-pyridonemethylstyrene) (PPMS) was coated on a titanium surface (PPMS/BPS) (Figure 6a) [181]. Under the stimulation of visible light (660 nm, 0.5 $W \cdot cm^{-2}$), the contained photosensitizer BPS released ROS (singlet oxygen), which was evidenced by the gradual decrease of UV absorption at 415 nm as the irradiation duration was prolonged to 50 min (Figure 6b). This was monitored by using 1, 3-diphenylisobenzofuran, which reacts with singlet oxygen to decrease the absorption around 415 nm [181]. In addition, the coating was good at the storage of the illumination-generated ROS via transforming PPMS to poly (4-pyridonemethylstyrene) endoperoxide (PPMS-EPO) (Figure 6a). The stored ROS can be released in the dark, mediating the *"killing without light"* capability of the coating. By illuminating in presence of oxygen gas (O_2) for 40 min, the PPMS/BPS group was transferred into PMMS-EPO/BPS, which was able to release ROS even after being contained in the dark at 37 °C for 24 h (the insert in Figure 6b). A reverse transformation between pyridone and endoperoxide was evidenced by the arising 1H nuclear magnetic resonance (1H NMR) peaks corresponding to the endoperoxide ring and the proton of endoperoxide in PMMS-EPO (Figure 6c). Tan et al. demonstrated that the PPMS-EPO/BPS coating had an antibacterial rate (against *E. coli* and *S. aureus*) of over 99.0% under light illumination and around 70.0% without light irradiation [181]. Such designs can compensate those photo-based therapies for applications in implantable medical devices that lack light reach.

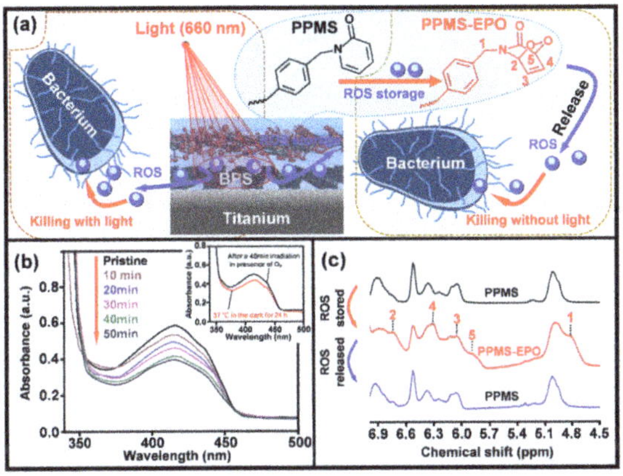

Figure 6. A photodynamic antibacterial material surface: (**a**) schematic illustration of the killing actions of the composite coating composed of black phosphorus nanosheets (BPS) and poly (4-pyridonemethylstyrene) (PPMS). Under light irradiation (660 nm, 0.5 W·cm^{-2}), BPSs generate reactive oxygen species (ROS), which can directly act on bacterial cells or are stored by the coating itself through the transfer of PPMS into poly (4-pyridonemethylstyrene) endoperoxide (PPMS-EPO), yielding antibacterial activity in the dark (killing without light). (**b**) UV-vis spectra show the capability of ROS production in PPMS/BPS with the increasing irradiation duration in the air (20 °C, 660 nm, 0.5 W·cm^{-2}). The insert shows the capability of ROS production by a PMMS-EPO/BPS (fabricated by illuminating the PPMS/BPS group for 40 min in presence of oxygen gas (O_2) after being contained in the dark at 37 °C for 24 h. (**c**) ^{1}H NMR spectra show the reversible structure change of PPMS and PPMS-EPO. Peaks corresponding to the endoperoxide ring and proton of endoperoxide were detected. Adapted from reference [181] with permission from John Wiley and Sons.

Materials having both photothermal and photodynamic activities are also developed for the disinfection of biomedical materials. Recently, a hydrothermally prepared nanorod array of titanium dioxide was demonstrated as an efficient photosensitizer for antibacterial applications [179]. In response to a NIR light (808 nm), the nanorods had both efficient light-to-heat conversion and ROS production properties, showing their excellent in vitro and in vivo antibacterial efficacy [179]. To endow titanium surface with both photothermal and photodynamic activities, the near-infrared fluorescent dye IR780 modified red phosphorus (has high photothermal conversion efficiency) films were developed [182]. In addition to the excellent compatibility with mammalian cells and normal tissue, the composite coatings demonstrated the synergistic effect of thermal and singlet oxygen in the eradication of *S. aureus* biofilms in vitro and in vivo [182]. Moreover, the thermal conversion activity of photosensitizers also can be applied to control the release of antibacterial agents for disinfection. Nano-structural molybdenum sulfide coating alone has high photothermal conversion efficacy that may induce hyperthermia capable of disintegrating bacterial envelopes [183]. This property of molybdenum sulfide can be used to control the release of antibacterial nitric oxide [184,185]. Typically, nano molybdenum sulfide assembled with heat-sensitive N,N′-di-sec-butyl-N,N′-dinitroso-1,4-phenylenediamine (as a nitric oxide donor) demonstrated a rapid antibacterial activity depending on nitric oxide release, yielding promising treatments for bacterial infections, even heat-resistant strain associated [184].

Table 6. Representative reports on photo-responsive antibacterial surfaces.

Action: Active Ingredient	Light Parameter	Pathogens Tested	Intended Use	Reference
Heat: gold	NIR light	*E. coli, MRSA*	In vitro (not specific)	[177]
Heat: tannic acid and iron	NIR light	*E. coli, MRSA*	Not specific	[178]
Heat: titanium dioxide	NIR light	*E. coli, S. aureus*	Orthopedic/dental implants	[179]
Heat: carbon dots	Blue light	*S. aureus*	Not specific	[180]
ROS: black phosphorus	Visible light	*E. coli, S. aureus*	Implantable materials/device (not specific)	[181]
Heat and ROS: fluorescent modified red phosphorus	NIR light	*S. aureus*	Treatment for joint implants	[182]
Heat and Nitric oxide: molybdenum sulfide assembled with a nitric oxide donor	NIR light	*Ampicillin-resistant E. coli, heat-resistant E. faecalis,* and *S. aureus*	Wound repair (not specific)	[184]

Moreover, antibacterial coatings responsive to other stimuli, such as temperature, electricity, and oxidative species, are also developed. Environmental temperature-responsive films based on poly(N-isopropylacrylamide) were developed to control the release of Ag NPs (at 37 °C) for antibacterial applications [186]. Triggered by an external electric field, a polypyrrole-doped polydimethylsiloxane coating was capable to release an antimicrobial drug loaded (crystal violet), making it a promising candidate for responsive antibacterial surfaces [187]. A branched poly(ethylene glycol)-poly(propylene sulfide) (PEG-PPS) polymer coating was found capable of actively releasing antibiotics (tigecycline or vancomycin) in response to an oxidative environment (ROS), which would occur adjacent to the infection site of a periprosthetic joint [188]. However, current reports on responsive antibacterial coatings normally are quite preliminary *"proof of concept"* studies; hence a lot of further efforts are needed to confirm their specific clinical scopes.

3.5. Cell-Selective Materials Surfaces

As aforementioned, an implant surface can be contaminated by pathogenic bacteria during surgery or during serving in the host. This requires the device to be highly selective over bacterial and mammalian cells, i.e., toxic to the adhesion of bacteria while compatible with the host cells. Here *"compatible"* includes two aspects, to be inert for a temporary device (for example a titanium bone plate) that does not stimulate rejections and to be bioactive for a permanent device (for example the implant-bone boundary for an artificial joint) that actively orchestrate tissue repair and integration in the host. We already know many surface features, such as surface composition (ion release), nanostructures, and wettability, that can determine the adhesion of cells to implantable medical device surface [189]. However, it is still hard to engineer a cell-selective surface on implantable medical devices because bacterial and mammalian cells share many mechanisms in adhesion, and the defense systems in the host are normally perturbed by placing the device. In the 1980s, Gristina first proposed the *"race for the surface"* concept [190], which suggested that the fate of an implantable device is a contest between bacterial adhesion and tissue integration to the device's surface. If the race is won by tissue cells, then the device surface is normally in host protection from bacterial colonization and infections. Since then, various in vitro and in vivo methods were developed to simultaneously study biofilm formation and tissue integration on the same surface. Subbiahdoss et al. developed an in vitro method that allows the growth of both S. epidermidis (ATCC 35983) and osteosarcoma cells (U2OS) in a parallel plate flow chamber [191]. So the *"race for the surface"* can be evaluated by determining the number of adhering cells and the area per spread cell. By using this protocol, the race between *S. epidermidis* and U2OS cells on various polymers with different wettability was studied [192]. The results demonstrated that the interactions of U2OS cells with biomaterials were hampered by biofilm formation on the materials, and neither hydrophobic nor hydrophilic surfaces have the potential to help U2OS win the race. In contrast, the presence of integrin-active arginine-glycine-aspartic acid (RGD) peptide on biomaterials significantly compromised the negative effects of biofilm presence by increasing surface coverage of U2OS but detaching bacterial biofilms at elevated flow shear (5.6 s-1, phosphate-buffered saline) [193]. The competition for a poly(methylmethacrylate) surface between U2OS cells and highly virulent *S. aureus* or *P. aeruginosa* in presence of murine macrophages was also described by the research group [194]. The presence of *S. aureus* decreased the adherence of human osteogenic sarcoma (SaOS-2) or primary osteoblast (hOB) cells to the surfaces of titanium, polydimethylsiloxane, and polystyrene; on the other hand, the presence of either type of these human cells was also associated with a reduction of bacterial colonization to the material's surface [195]. Martinez-Perez et al. described an in vitro approach for the study of the adherence of *S. aureus* and *S. epidermidis* in the presence of pre-osteoblastic cells (MC3T3-E1), which can be used for assessing the effects of surface coatings with antibacterial potentials [196]. By using a bilateral intramedullary rat model and injecting bacteria (*S. aureus*) into the tail vein after implant placement, the temporal interplay between host-cell adhesion and bacterial colonization was examined [197]. To determine the effects of hematogenous spreading of bacteria on infection subcutaneous implants (after healing of the implantation wound), rats were intravenously injected either with *S. aureus*, *S. epidermidis*, or *P. aeruginosa* 4 weeks after subcutaneous implantation of various biomaterials, including silicone rubber, polyethylene, polypropylene, poly(tetrafluoroethylene), poly(ethylene terephthalate), poly(methyl methacrylate), polyurethane, or glass [198]. The results demonstrated that late hematogenous infection of subcutaneous biomaterials does not occur in the rat, hence those reported late infections in humans were likely caused by perioperatively introduced pathogens [198], which is worthy of further studies. Reports covering biofilm formation on biomaterials surfaces lack often the differentiation between biofilm reducing and biofilm inhibitory effects [199]. Hence, the current biofilm methodologies used for judging the antibacterial effects of implant surfaces need to be critically revisited and if necessary revised and standardized. These reports show the significance of constructing cell-selective surfaces for implantable medical devices, as well as methods to evaluate the property.

Metallic ingredients are performing multiple functions in humans that can be building blocks leading to single-element-release mediated cell-selective surfaces. Zinc is known as a stimulus to the osteogenic function of bone cells and also an inhibitor of bacterial growth. Accordingly, zinc was loaded onto various titanium surfaces by using micro-arc oxidation [200], hydrothermal treatment [201], and ion implantation [202], demonstrating excellent antimicrobial and osteogenic properties. It is known that antimicrobial cobalt ions can induce hypoxia-like conditions [203]. By applying this feature of cobalt ions, antibacterial wound dressings with excellent capability for the promotion of angiogenesis and epithelialization were fabricated by Shi et al. [203]. Copper, in addition to its broad-spectrum bactericidal activity, was found to be capable of promoting osteoblast proliferation and bone formation [204]. It was reported that proper control of the content and release of copper in Ti-Cu alloys are capable of balancing the antibacterial and osteogenic properties of the metal implants [204].

The combinational release of antimicrobials and tissue-integration promoters is another pathway extensively studied to develop cell-selective implants. Low-molecular-weight polyethyleneimine is a cationic antimicrobial agent, and alendronate is a stimulus for new bone formation and osteointegration improvement [205]. They can be covalently conjugated onto ethanediamine-functionalized poly (glycidyl methacrylate) to construct titanium implants of both antibacterial and osteogenic properties [205]. Silver and hydroxyapatite are two typical materials with antibacterial and osteogenic properties, respectively. Recently, Fazel et al. developed a duplex process that sequentially employed micro-arc oxidation and hydrothermal treatment to decorate Ag NPs and hydroxyapatite nanocrystals on the surface of a porous Ti6Al4V substrate (fabricated by using selective laser melting), creating surfaces of both osteogenic and antibacterial properties [206]. Bacterial infection in burn wounds is common and fatal [207]. Many works have been done to develop wound dressings preventing bacterial infection and promoting wound healing. Porphyrin photosensitizer sinoporphyrin sodium has two macrocycles that show high efficiency against pathogenic bacteria via the production of ROS [208]. This dimeric photosensitizer was chosen to work with fibroblast growth factor in a carboxymethyl chitosan-sodium alginate matrix that has successfully suppressed the growth of bacteria and simultaneously accelerated the healing of bacteria-contaminated burn wounds in mice under mild photoirradiation (30 J$\cdot$cm^{-2}, 5 min) [208]. Due to inferior vascularization, poor re-epithelialization, and increased infection risk, treatment of diabetic wounds is considerably challenging and becomes a focus of cell-selective surfaces. The (11-mercaptoundecyl)-N,N,N-trimethylammonium (MTA) contains a quaternary ammonium cation that interacts strongly with the negatively charged cell membrane of microbes [209]. Together with the vascular endothelial growth factor, MTA can be conjugated into gold nanoparticles to produce dual-functional (antimicrobial and proangiogenic) dressings for treatments of diabetic wounds [209]. Ag NPs and pro-angiogenic deferoxamine were encapsulated in a pH-responsive hydrogel developed via double-crosslinking between chitosan quaternary ammonium salt and oxidized dextran-dopamine, achieving pH-sensitive feature in drug release and accelerated healing of infected diabetic wounds [160]. The antibacterial efficacy of hydroxypropyltrimethyl ammonium chloride chitosan is related to the substitution degree of its quaternary ammonium group [210]. This chitosan derivative cooperates with magnesium ions (magnesium chloride) in calcium alginate, yielding an antibacterial and angiogenic dressing for the treatment of infected diabetic wounds [210].

Normally, the surface of a medical device will be in contact with body fluids, which are electrolytes that facilitate electrochemical reactions on the implant surfaces. These reactions can intervene in the adjacent microscale biological environments and subsequently determine the fate of the implants, giving alternative pathways to construct cell-selective surfaces. To be specific, the dissimilar phases in a metal likely have different electrode potentials, as a result, electrochemical corrosion (also known as internal galvanic corrosion) will occur when the material comes into contact with an electrolyte [211,212], just as the case that a metallic implant is in contact with the physiological fluid. Based on this mechanism, Cao et al. firstly proposed to control the antibacterial activity and improve the biocompatibility of Ag NPs by taking advantage of the chemical reactions stemming between nanosilver precipitates and the titanium matrix (Figure 2e) [143]. The antimicrobial activity of these immobilized (or embedded) Ag NPs was well retained in addition to their excellent compatibility with the functions of bone cells and bone formation [138,213,214]. The biological basis for such a cell-selective property is likely the difference in size and structure between the prokaryotic (bacteria) and eukaryotic cells (bone cells), which makes bacterial and mammalian cells respond differently to the proton-depleting reactions over Ag NPs decorated materials surface [143], i.e., proton depletion mediated by immobilized Ag NPs likely disrupts the transmembrane proton electrochemical gradient and inactivate the adenosine triphosphate synthesis, ion transport, and metabolite sequestration, ultimately lead the bacterial death (Figure 2e), while it catalyzes the activation of an integrin-mediated cascade of osteoblast differentiation in rat bone marrow stem cells and improving osteointegration of metal implants [144]. Such cell-selective effects can be further boosted via the co-doping of silver and calcium into titanium surfaces [215]. Silver and calcium were in situ introduced into the titanium surface by two synchronously operating cathodic arcs as aforementioned in Figure 2a. The dopped silver was condensed into nanoparticles (Ag NPs) on the substrate surface, while the calcium was intermixed with the titanium matrix, yielding a modified surface layer of approximately 30 nm in thickness (designated as Ti-Ag/Ca) [216]. *E. coli* (ATCC 25922) at a concentration of 10^6 cfu$\cdot$ml^{-1} and rat bone marrow stem cells (BMSCs) at a density of 10^4 cells per ml were seeded onto the material surface and incubated at 37 °C for 24 h and 1 h, respectively. Serious bacterial cell disruption and distortion (Figure 7a) while accelerated spreading and coverage of BMSCs were detected (Figure 7b), demonstrating that Ti-Ag/Ca favored the functions of BMSCs and simultaneously acted against pathogenic bacteria, i.e., cell selectivity [215]. The cathodic and anodic reactions on Ti-Ag/Ca possibly disrupted the transmembrane proton-motive force (PMF) and the administration of calcium further stressed or even disordered the metabolic processes critical to bacterial maintenance (Figure 7c), showing the antibacterial activity as shown by Figure 7a. Additionally, the electrochemical reactions on Ti-Ag/Ca also possibly accelerated the proton extrusion of sodium-proton exchanger 1 (NHE1) and the calcium influx of sodium-calcium exchanger 1 (NCX1) in BMSCs (Figure 7d), showing enhanced membrane bleb nucleation, growth, and retraction as shown in Figure 7b. Moreover, the growth and retraction of membrane blebs can modulate cellular mechanics and promote the osteogenic differentiation of BMSCs, which is good to improve the osseointegration of titanium [215]. Such antibacterial surfaces are promising for orthopedic devices and dental implants, which are intended to be integrated into bone tissues.

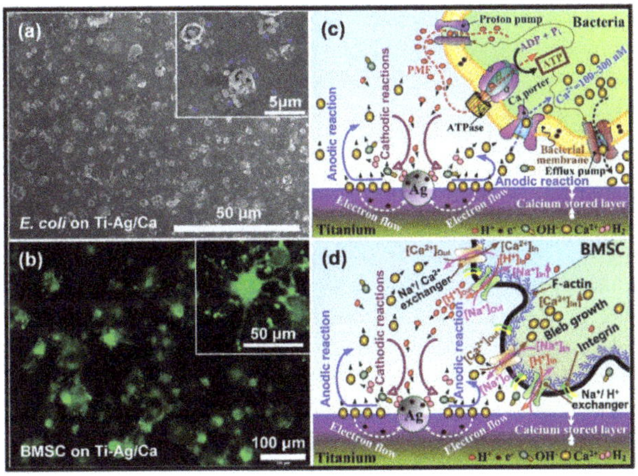

Figure 7. A cell-selective titanium surface: (**a**) SEM surface morphology of the microbes (*E. coli*) cultured for 24 h on titanium doped with both calcium and silver (Ti-Ag/Ca), with a high magnification image, inserted; (**b**) typical morphology of rat bone marrow stem cells (BMSCs) cultured for 1 h on Ti-Ag/Ca, with a high magnification image inserted; (**c**,**d**) potential mechanism underlying the actions of Ti-Ag/Ca on microbes and mammalian cells, respectively [215]. Reused with permission from the Royal Society of Chemistry.

In addition, topographical nanostructures, such as nanopillars [216–218], nanosheets [219], nanorod [220,221], and nano-roughness [222,223], with the efficacy of physical sterilization represent an innovative pathway toward antibacterial surfaces. Among these designs, nanopillars received the most attention because their puncture-based biocidal actions are material composition and bacterial species independent. The bactericidal mechanisms of such structures were already highlighted in a very recent review [224]. Additionally, there are several studies that demonstrated that titanium nanopillars with proper diameter, spacing, and height yield cell-selective surfaces. Hasan et al. fabricated nanopillars on a commercially pure titanium substrate (the surface appears black) using a reactive ion etching process [225]. They found that titanium nanopillars of about 1 μm in height have maximal bactericidal efficiency without compromising attachment and proliferation of human mesenchymal stem cells. Similarly, Ganjian et al. reported that titanium nanopillars with diameters of about 26 nm and lengths of about 1.1 μm have biocidal activity against Staphylococcus aureus and Escherichia coli, but murine preosteoblasts (*MC3T3-E1*) can attach and spread well [226]. Additionally, Modaresifar et al. recently concluded that the height and spatial organization are key factors contributing to the cell selectivity of the titanium nanopillars between *MC3T3-E1* and staphylococcus aureus [227]. These studies show that black titanium with nanopillars is a very promising material surface for orthopedic implants. However, further efforts are looking forward to confirming the cell selectivity of such a design in vivo.

3.6. Immune-Instructive Materials Surfaces

It is acknowledged that the high susceptibility of biomaterials to infections is not only related to bacterial contamination but also owed to the undesirable host responses which compromise the intrinsic immune capability in bacterial clearance [228,229]. The research of Zimmerli et al. revealed that 100 colony-forming units (CFU) of *S. aureus* were sufficient to cause infection to 95% of subcutaneous implants in pigs, whereas 10^5-fold higher CFU did not produce any infection in the same subcutaneous model without alien implants [230]. Southwood et al. found that the surgical site of an arthroplasty implant became infected by contamination with 50 CFU of *S. aureus*. This concentration was 200 times lower than that causing infection in a surgical site without any foreign device [231]. Given this, the inherent immunomodulatory effects of implantable devices and their interactions with the host's immunity system should be highlighted in developing advanced anti-infective biomaterials [232]. The immune system is generally divided into innate and adaptive arms, which cooperatively protect the host from bacterial infections [233]. The nonspecific nature of the former indicates its modulation potential yielding a broad spectrum against pathogenic bacteria [234]. The neutrophils and macrophages are key players mediating the innate immune response of the host at the cellular level. They have multiple actions against pathogenic bacteria, such as the production of reactive oxygen species (ROS) and reactive nitrogen species (nitric oxide), the release of granule proteins and neutrophil extracellular traps (NETs), cytokine expression of interleukins, and phagocytosis [235–238]; therefore, they are especially concerning to the biomaterial community.

The behaviors of neutrophil adhesion to biomaterials are determined by surface chemistry [239], topography [240], wettability [241], and stiffness [242]. Polystyrene and woven Dacron or Silastic induced neutrophil release of granule antibacterial products (are cationic peptides known as defensins) to create an environment hostile to phagocytic killing by neutrophils [243]. Polytetrafluoroethylene and Dacron promoted the production of ROS and mediated premature neutrophil death while polystyrene did not [239]. Furthermore, after being manually scratched with forceps, the polystyrene also induced the production of more reactive oxygen intermediates and rapid neutrophil death [240]. A rapid decrease in expression of L-selectin was detected within 16 min of neutrophil adhesion to titanium, and Fc gamma III receptor (CD16) expression dominated the initial adhesion (within 30 min) [244]. Human neutrophils rapidly adhered to sandblasted large-grit acid-etched titanium surfaces and released NETs [245], which are efficient actions of neutrophils limiting pathogenic spreading in a microbe-size-dependent manner [246]. Rough-hydrophilic titanium surfaces (produced by sandblasting and acid etching, and stored in a nitrogen environment) decreased the production of pro-inflammatory cytokines and enzymes as well as the formation of extracellular traps in adherent neutrophils [241]. In addition to the release of reactive oxygen radicals, neutrophils possibly enhance nitric oxide production in response to acute and chronic inflammation. The reaction of nitric oxide and reactive oxygen species readily generates peroxynitrite, which is a potent cytotoxic mediator [247]. It was demonstrated adherent neutrophils on polyethylene oxide-modified polyurethane produced lower amounts of nitric oxide; however, peroxynitrite formation did occur upon bacterial stimulation (*S. epidermidis*). This indicated that biomaterials can compromise neutrophil generation of nitric oxide, possibly diminishing the bacterial clearance capacity of the immune system and increasing the risk of DAIs [248]. Surface structures and stiffness have modulatory effects on macrophages. Surface-topography-induced changes in macrophages were examined in terms of polymer parallel gratings [248]. Such gratings, particularly of larger size, affected the adhesion, morphology, and cytokine secretion of macrophages. Adherent macrophages on both micro- and nanostructured silicon dioxide films did not increase the production of interleukin IL-6 or alter membrane mobility but had significantly greater phagocytic capacity than those on flat surfaces [249]. Macrophages are master regulators orchestrating host immune responses to biomaterials. Micropatterned surfaces (microgrooves/ridges and micropillars) did induce distinct gene expression profiles in human macrophages [250]. To be specific, micropillars (5–10 μm in diameter) were dominant in driving macrophage attachment to a polystyrene chip, and pillar size and spacing were critical in priming anti-inflammatory phenotype [251]. Hotchkiss et al. studied the surface roughness and wettability effects of titanium on macrophage activation and cytokine production. They found that smooth titanium induced inflammatory macrophage (M1) activation to express high levels of interleukins IL-1β, IL-6, and tumor necrosis factor alpha (TNF-α), while hydrophilic and rough titanium induced anti-inflammatory macrophage (M2) activation to release high levels of interleukins IL-4 and IL-10 [252]. Poly(ethylene glycol) based hydrogels with lower stiffness induced mild macrophage activation and a more typical mild foreign body reaction [253]. Similar results were found on polyacrylamide gels, i.e., stiff gels (323 kPa) prime pro-inflammatory macrophages with impaired phagocytosis while soft (11 kPa) and medium-stiff (88 kPa) gels prime anti-inflammatory and highly phagocytic phenotype [254]. These results were consistent with that reported by Previtera et al., who demonstrated that the production of pro-inflammatory mediators by macrophages was mechanically regulatable, namely stiff substrates enhanced proinflammation [255].

The immunomodulatory effects of antimicrobials attract extensive attention in addition to their biocidal activities. Silver-based coatings are commonly proposed for antibacterial applications; however, Croes et al. demonstrated that the electroplated silver on porous titanium was cytotoxic to neutrophils via releasing an excessive amount of silver ions and diminishing neutrophil phagocytic activity [256]. Diffusive Ag NPs were able to rapidly penetrate inside neutrophils and induce atypical cell death [257], which possibly inhibited neutrophil reactive oxygen production and subsequently impair the antibacterial efficacy of the innate immune system [258]. Other studies concluded that the immunomodulatory efficacy of diffusive Ag NPs was correlated with their ability to release silver ions [259]. Moreover, the immunomodulatory effects of engineered nanomaterials are chemistry-dependent. It was demonstrated that macrophages responded to diffusive nanoparticles by significantly increasing the generation of IL-6, nuclear translocation of nuclear factor-kappa B, induction of cyclooxygenase-2, and expression of TNF-α, with maximum prominent such pro-inflammatory responses detected in cells treated by diffusive Ag NPs, followed by aluminum, carbon black, and carbon-coated Ag NPs [260]. Moreover, zinc and copper are branded immunomodulatory and antimicrobial activities. Zinc plays multiple roles in the innate immune system, and zinc deficiency normally reduces the chemotaxis of neutrophils and the phagocytosis of macrophages [261]. Zinc oxide nanoparticles were coated on titanium by magnetron sputtering. This surface improved the antibacterial efficacy of macrophages and neutrophils in terms of phagocytosis and inflammatory cytokine secretion [262]. Copper-doped titanium oxide coatings (fabricated by micro-arc oxidation technique) induced high levels of inducible nitric oxide synthase activity and IL-6 release but low levels of IL-4 and IL-10 in macrophages to prime M1 phenotype, which exhibited enhanced phagocytosis and antibacterial efficacy [263]. Copper nanoparticles coated with polyetheretherketone (with a porous microstructure produced by sulfonation) by magnetron sputtering technique also were capable of polarizing macrophages to a pro-inflammatory phenotype with improved phagocytic ability toward MRSA [264].

This evidence gave us a comprehensive understanding of the biological actions of various silver-, zinc-, and copper-based materials, which are extensively concerned with developing antibacterial surfaces for medical devices [116,265]. However, they are normally passive studies that just demonstrate the immune-interfering actions of synthetic antibacterial materials, rather than active studies directly taking advantage of immunomodulatory biomaterials to construct antibacterial activity. In this respect, the immunomodulatory effects of essential metals, such as magnesium and calcium, should be appreciated. It was reported that high extracellular magnesium concentration can attenuate neutrophil activation by inhibiting the generation of superoxide radicals [266]. Calcium also plays a critical role in the regulation of pro-inflammatory functions of neutrophils, such as the release of superoxide anions, secretion of cytokine, formation of NETs, and phagocytosis [267]. Previously, we fabricated calcium-doped titanium (designated as Ti-Ca) by using a calcium plasma immersion ion implantation technique [268], which is similar to that shown in Figure 2a (just replace the silver cathode with pure calcium, and treat for 90 min with a 30 kV bias). Although in vitro tests demonstrated Ti-Ca was poor against bacterial colonization, the Ti-Ca implants survived the challenge of MRSA (ATCC 43300) in the tibia of rabbits and promoted osseointegration of titanium, while the pure titanium control (Ti) failed [256]. Very recently, we found that the locally delivered calcium by titanium can react with carboxy-terminal regions of the Aα chains and influence their interaction with the N-termini of Bβ chains in fibrinogen (a blood protein), which facilitates the exposure of the protein's antimicrobial motifs, showing the surprising antimicrobial efficacy of calcium-doped titanium (Figure 8) [16]. This finding validates that the antibacterial surfaces can address the functions of the host rather than targeting directly the pathogenic bacteria, breaking the existing paradigm on minimizing DAIs. Fibrinogen adsorption is an essential process in the intrinsic immune responses of the human body to implantation operations, the aforementioned effect of calcium-doped titanium on fibrinogen adsorption brings further insight into the design of immunomodulatory biomaterials. In addition, the increased release ratio of magnesium/calcium from a magnesium alloy was found to prime the M2 phenotype of macrophages [269]. Since magnesium serves as a natural calcium antagonist [270], such synergistic effects of magnesium and calcium indicate a fruitful direction for the development of immunomodulatory antibacterial surfaces.

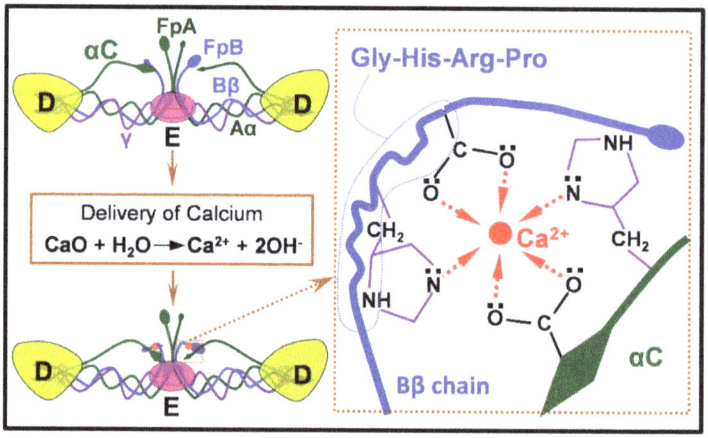

Figure 8. An antibacterial surface targeting the adsorption of fibrinogen: the calcium released by titanium turns the intramolecular interactions between αC regions and the amino-terminal of Bβ chains, and subsequently contributes to the exposure of the antibacterial peptide in fibrinogen. The Gly-His-Arg-Pro (Gly: glycine; His: histidine; Pro: proline; Arg: arginine) are the start sequences of the antibacterial peptide Bβ15–42 which locates at the N-terminal end of the β chain [16]. Reused with permission from the Royal Society of Chemistry.

4. Directions to Improve the Quality of Antibacterial Reports

As demonstrated in Section 3, tremendous antibacterial designs have been proposed to treat DAIs. Despite a large number of studies carried out systematical in vitro and in vivo tests, clinical translation of these designs is limited. All the devices have their own "*intended use*", which defines a primary function of an implantable medical device. For example, the primary functions of a wound dressing and dental implant are to help wound healing and promote osseointegration, respectively. Since not all the implantable medical devices are bound to a bacterial infection, antibacterial function, in our opinion, shall serve as a property secondary to those primary ones. This requires that the design, synthesis, and evaluation of implantable antibacterial surfaces should fit the needs of a specific "*intended use*", which will help to clarify the exact biological environment that the material is intended to integrate [116,271]. This is crucial to choose proper routes for material synthesis, applicable parameters for material characterizations, and the right strategies for biological evaluations, enabling reproducible, comparable, and reusable results that will be consistent in clinical translation [271]. Unfortunately, many previous reports were at the stage of *"proof of concept"*; they did not conform strictly to a specific application. According to the papers we have screened, major flaws in our current reporting on developing antibacterial surfaces for implantable medical devices are (Table 7 includes some typical examples [18,272–278]) (1) many reports are lack of comprehensive understanding of the requirements of a specific application or even have no clear indication of use (Cases 2, 3, 7, and 8 in Table 7); as a result, it is hard to ensure the effectiveness and safety of such designs because the incidence and pathogens of DAIs are site-specific [66] and the biocompatibility of biomaterials is referred to specific applications [279]. (2) Some reports choose testing assays that do not closely relate to the intended use. For example, in Case 4, the authors reported a light-responsive material for antibacterial wound dressings; however, the effects of light illumination on mammalian cells are not considered either in vitro or in vivo. In Case 6, the intended application for the study is *"dental implant"*, which requires good osseointegration; nevertheless, only human gingival fibroblasts have been tested in vitro. (3) Some of the reports have flaws that possibly undermine the confidence in clinical applications. Typical features of these flaws are using unidentified bacterial sources (Cases 2 and 4 in Table 7), and changing experimental conditions during the study (for example, in Cases 5 and 6, the light irradiation parameters, power, duration, and onset changed between different in vitro and in vivo tests, which may mislead the follow-up studies). For the sake of safety, in addition to antibacterial tests, the tissue integration or compatibility of the designs shall be tested; however, this aspect was not considered in many current reports (in Case 8, for example, a light-responsive surface was proposed for disinfection; however, the effects of light illumination on tissue integration were not evaluated). Many studies are *"proof of concept"* reports (like Case 1), and some of them seem inconsistent according to their results. For example, the authors in Case 3 tried to report an antimicrobial surface with long-term efficacy, but the antibacterial effect was considered 5 days post-operation using a subcutaneous implant model in mice. Moreover, although a bacterial infection is possibly associated with multiple species (Case 8, Table 3), normally one bacterial strain is involved in tests; co-culture of mammalian cells with bacteria or co-culture of different bacterial strains are rare in current studies.

Table 7. Typical flaws in our reports on antibacterial surfaces.

Case	Antibacterial Designs	Bacterial Strain (In Vitro)	Mammalian Cells Line (In Vitro)	In Vivo Tests	Intended Use	Reference
1	**Cell-selective:** Coating titanium nanowires with poly (ethyl acrylate) to organize fibronectin and deliver BMP-2	*P. aeruginosa* (ATCC 27853); cultured for 24 h	Primary human mesenchymal stem cell (MSCs); co-culture with bacteria	*None*	Orthopedic implant	[272]
2	**Cell-selective:** Ion release by Magnesium hydroxide	*S. aureus (unidentified source)*; E. coli *(unidentified source)*	Mouse *MC3T3-E1* pre-osteoblasts	Rat femoral condyle defect model; *Placed in for 7 days to examine the dis-infective effects. Placed in for 4 weeks to evaluate the osteogenic property*	*Not specific*	[273]
3	**Long-term efficacy:** salt-responsive polyzwitterionic brushes on a nanopatterned surface	*P. aeruginosa* (BNCC 337005); *Escherichia coli* (ATCC 25922)	Rabbit red blood cells (2 h- incubation); *L929* fibroblasts (cultured for 24 h)	Subcutaneous implant model in mice; *Placed in for 5 days*	*Not specific*	[274]
4	**Light-responsive** (808 nm laser irradiation, 1 W/cm^2, 5 min): Photosensitive gelatin methacryloyl incorporated with 4-octyl itaconate bearing black phosphorus	*S. aureus (unidentified source)*; E. coli *(unidentified source); The onset of light irradiation is not clear*	Human umbilical vein endothelial cells; *The effect of light illumination on the cell function was not clear (No data presented)*	Rat type I diabetes model (14 days); *The onset for light irradiation is not clear*	Wound dressing	[275]

Table 7. *Cont.*

Case	Antibacterial Designs	Bacterial Strain (In Vitro)	Mammalian Cells Line (In Vitro)	In Vivo Tests	Intended Use	Reference
5	**Light-responsive** (1060 nm laser, 0.3 W/cm^2, 0.6 W/cm^2, and 0.9 W/cm^2): Yb and Er-doped titanium dioxide nano-shovel/quercetin/L-arginine coatings	*S. aureus* (ATCC 29213);Light illumination (*0.6 W/cm^2*,15 min)	Osteosarcoma cells (Saos-2, *light irradiation at 0.9 W/cm^2 for 10 min*); Human umbilical vein endothelium cells (*light irradiation at 0.6 W/cm^2 for 10 min*); Bone marrow mesenchymal stem cells (*light irradiation at 0.6 W/cm^2 for 10 min*)	Tumor-bearing mouse model (*light irradiation at 0.9 W/cm^2 for 10 min and performed every other day*); Mice tibia infection model (*light irradiation at 0.6 W/cm^2 for 15 min and performed one day after surgery*); Mice tibia osteogenic model (*light irradiation at 0.6 W/cm^2 for 15 min and performed one day after surgery; samples collected 4 weeks after surgery*)	Bone implants	[18]
6	**Light-responsive** (808 nm laser irradiation): TiO_2/TiO_{2-x} metasurface	*E. coli* (ATCC 25922); *S. aureus* (ATCC 43300); *illuminated at 0.5 W/cm^2 for 10 min*	Human gingival fibroblasts; *light illumination at 0.5 W/cm^2 for 10 min*	Subcutaneous model in rats; *light illumination at 1.4 W/cm^2 for 10 min after surgery*	Dental implant	[276]
7	**Immune-instructive:** polydopamine functioned and antimicrobial peptide plasmid (LL37 plasmid) loaded porous zeolitic imidazolate framework-8 (ZIF8) in 3D-Printed Scaffolds	MRSA (ATCC 43300); *E. coli* (ATCC 25922)	MC3T3 cell; *The material effects on immune systems are not considered*	Murine quadriceps muscle infection model (MRSA injected after scaffold placement)	*Not specific*	[277]
8	**Light-responsive** (808 nm laser irradiation, 2 W/cm^2, 10 min): self-assembly of copper sulfide nanoparticle and reduced graphene oxide on anodized titanium	*S. aureus* (ATCC 29213); *E. coli* (ATCC 25922); *Light irradiation after inoculation*	Mice bone marrow stromal cells; *The effect of light illumination on the cell function is not clear (No data presented)*	Disinfection in rats (7 days); Osteogenic property in rats (8 weeks); *The effect of light illumination on osteogenesis was not clear (No data presented)*	*Not specific*	[278]

5. Summary and Outlook

Every DAI involves three participants, i.e., a device surface, pathogenic bacteria, and the host, which interact and interplay with each other and transform time dependently. That is why the onset of DAIs is uncertain and more DAIs are resistant to antibiotic treatments. Based on our growing knowledge of DAIs, the design paradigm toward implantable antibacterial surfaces is shifting to pursuing prolonged efficacy and being actively responsive, cell-selective, and immune instructive, experiencing a boom in advancing antibacterial surfaces (coatings) rapidly for various medical devices; nevertheless, clinical translation of these techniques is still rare. Since the incidence and associated bacterial strains of DAIs are site-specific, antibacterial designs for implantable medical devices shall conform with a specific intended use, which possibly promotes the clinical application of these technologies. Furthermore, the following aspects, in our opinion, are important to improve the quality of our reports in fundamental research. (1) Deepen interdisciplinary collaboration. The development of biomedical materials requires close collaboration in multiple disciplines, including materials sciences and engineering, biological sciences, medical sciences, etc. Researchers with a materials science and engineering background normally do not exactly know the experimental and reporting standards in biological and medical sciences; a closer collaboration will help to choose proper experimental assays and reduce the flaws in our publications, ensuring the impacts of our findings. (2) Publish in journals focusing on biomaterials science and engineering. Many academic journals are publishing biomaterials-associated studies; however, only those concentrated in biomaterials science and engineering are well equipped with experienced referees that can identify the flaws in the manuscript during the peer review process.

In addition, material designs targeting the adsorption processes of host proteins are a fruitful direction in developing implantable antibacterial surfaces. Spontaneous adsorption of proteins onto a biomedical device occurs seconds after its contact with body fluids, such as blood plasma, extracellular fluid, tears, saliva, and urine, depending on the specific intended use. Current efforts normally started both antibacterial tests and compatibility examinations at the cellular level regardless of the critical conditioning role of protein adsorption on the subsequent cell function and tissue integration. Blood plasma contains thousands of proteins that play key roles in diverse life activities, including signaling, transport, development, restoration, and disinfection [280]; however, plasma protein (fibrinogen) adsorbed to biomaterial surfaces is likely denatured into a pro-inflammatory state [281], which mediates foreign body reactions that contribute to DAIs [282]. Accordingly, control of protein adsorption in the host via material designs is possible to produce innovative antibacterial designs (reference No. 16 is one example) with a bright prospective for clinical applications, which is worthy of further efforts in the future.

Author Contributions: Conceptualization, writing—original draft preparation, visualization, H.C.; writing—review and editing, S.Q., H.Q., and K.D.J. All authors have read and agreed to the published version of the manuscript.

Funding: This work was funded by the National Natural Science Foundation of China (31870945 and 31771022) and the Natural Science Foundation of Shanghai (21ZR1415700). H.C. would like to thank the Alexander von Humboldt Foundation for awarding a Humboldt Research Fellowship.

Institutional Review Board Statement: Not applicable.

Informed Consent Statement: Not applicable.

Conflicts of Interest: The authors declare no conflict of interest.

References

1. Arciola, C.R.; Campoccia, D.; Montanaro, L. Implant infections: Adhesion, biofilm formation and immune evasion. *Nat. Rev. Microbiol.* **2018**, *16*, 397–409. [CrossRef] [PubMed]
2. Andersen, O.Z.; Offermanns, V.; Sillassen, M.; Almtoft, K.P.; Andersen, I.H.; Sørensen, S.; Jeppesen, C.S.; Kraft, D.C.E.; Bøttiger, J.; Rasse, M.; et al. Accelerated bone ingrowth by local delivery of strontium from surface functionalized titanium implants. *Biomaterials* **2013**, *34*, 5883–5890. [CrossRef]
3. Mond, H.G.; Proclemer, A. The 11th world survey of cardiac pacing and implantable cardioverter-defibrillators: Calendar year 2009-a World Society of Arrhythmia's project. *Pacing Clin. Electrophysiol.* **2011**, *34*, 1013–1027. [CrossRef] [PubMed]
4. Saint, S.; Wiese, J.; Amory, J.K.; Bernstein, M.L.; Patel, U.D.; Zemencuk, J.K.; Bernstein, S.J.; Lipsky, B.A.; Hofer, T.P. Are physicians aware of which of their patients have indwelling urinary catheters. *Am. J. Med.* **2000**, *109*, 476–480. [CrossRef]
5. Sloan, M.; Premkumar, A.; Sheth, N.P. Projected volume of primary total joint arthroplasty in the U.S., 2014 to 2030. *J. Bone Jt. Surg. Am.* **2018**, *100*, 1455–1460. [CrossRef] [PubMed]
6. Rupp, M.; Lau, E.; Kurtz, S.M.; Alt, V. Projections of primary TKA and THA in Germany from 2016 through 2040. *Clin. Orthop. Relat. Res.* **2020**, *478*, 1622–1633. [CrossRef] [PubMed]
7. Matharu, G.S.; Culliford, D.J.; Blom, A.W.; Judge, A. Projections for primary hip and knee replacement surgery up to the year 2060: An analysis based on data from the national joint registry for England, Wales, Northern Ireland and the Isle of Man. *Ann. R. Coll. Surg. Engl.* **2021**, *104*, 443–448. [CrossRef] [PubMed]
8. Magill, S.S.; Edwards, J.R.; Bamberg, W.; Beldavs, Z.G.; Dumyati, G.; Kainer, M.A.; Lynfield, R.; Maloney, M.; McAllister-Hollod, L.; Nadle, J.; et al. Emerging infections program healthcare-associated infections and antimicrobial use prevalence survey team. multistate point-prevalence survey of health care-associated infections. *N. Engl. J. Med.* **2014**, *370*, 1198–1208. [CrossRef]
9. Dadi, N.C.T.; Radochová, B.; Vargová, J.; Bujdáková, H. Impact of healthcare-associated infections connected to medical devices-an update. *Microorganisms* **2021**, *9*, 2332. [CrossRef]
10. Chang, C.H.; Lee, S.H.; Lin, Y.C.; Wang, Y.C.; Chang, C.J.; Hsieh, P.H. Increased periprosthetic hip and knee infection projected from 2014 to 2035 in Taiwan. *J. Infect. Public Health* **2020**, *13*, 1768–1773. [CrossRef]
11. Premkumar, A.; Kolin, D.A.; Farley, K.X.; Wilson, J.M.; McLawhorn, A.S.; Cross, M.B.; Sculco, P.K. Projected economic burden of periprosthetic joint infection of the hip and knee in the United States. *J. Arthroplast.* **2021**, *36*, 1484–1489. [CrossRef]
12. Zhang, S.; Yu, Y.; Wang, H.; Ren, L.; Yang, K. Study on mechanical behavior of Cu-bearing antibacterial titanium alloy implant. *J. Mech. Behav. Biomed. Mater.* **2022**, *125*, 104926. [CrossRef]
13. Chen, T.; Wang, Y.; Xie, J.; Qu, X.; Liu, C. Lysozyme amyloid fibril-integrated PEG injectable hydrogel adhesive with improved antiswelling and antibacterial capabilities. *Biomacromolecules* **2022**, *23*, 1376–1391. [CrossRef]
14. Yuan, Z.; Wu, J.; Fu, Z.; Meng, S.; Dai, L.; Cai, K. Polydopamine-Mediated Interfacial Functionalization of Implants for Accelerating Infected Bone Repair through Light-Activatable Antibiosis and Carbon Monoxide Gas Regulated Macrophage Polarization. *Adv. Funct. Mater.* **2022**, 2200374. [CrossRef]
15. Mou, X.; Zhang, H.; Qiu, H.; Zhang, W.; Wang, Y.; Xiong, K.; Huang, N.; Santos, H.A.; Yang, Z. Mussel-inspired and bioclickable peptide engineered surface to combat thrombosis and infection. *Research* **2022**, *2022*, 9780879. [CrossRef]
16. Cao, H.; Dauben, T.J.; Helbing, C.; Jia, Z.; Zhang, Y.; Huang, M.; Müller, L.; Gu, S.; Zhang, X.; Qin, H.; et al. The antimicrobial effect of calcium-doped titanium is activated by fibrinogen adsorption. *Mater. Horiz.* **2022**. [CrossRef]

17. Ye, J.; Li, B.; Li, M.; Zheng, Y.; Wu, S.; Han, Y. Formation of a ZnO nanorods-patterned coating with strong bactericidal capability and quantitative evaluation of the contribution of nanorods-derived puncture and ROS-derived killing. *Bioact Mater* **2022**, *11*, 181–191. [CrossRef]
18. Zhang, G.; Wu, Z.; Yang, Y.; Shi, J.; Lv, J.; Fang, Y.; Shen, Z.; Lv, Z.; Li, P.; Yao, X.; et al. A multifunctional antibacterial coating on bone implants for osteosarcoma therapy and enhanced osteointegration. A multifunctional antibacterial coating on bone implants for osteosarcoma therapy and enhanced osteointegration. *Chem. Eng. J.* **2022**, *428*, 131155. [CrossRef]
19. Yang, L.; Wang, C.; Li, L.; Zhu, F.; Ren, X.; Huang, Q.; Cheng, Y.; Li, Y. Bioinspired integration of naturally occurring molecules towards universal and smart antibacterial coatings. *Adv. Funct. Mater.* **2022**, *32*, 2108749. [CrossRef]
20. Li, W.; Hua, G.; Cai, J.; Zhou, Y.; Zhou, X.; Wang, M.; Wang, X.; Fu, B.; Ren, L. Multi-stimulus responsive multilayer coating for treatment of device-associated infections. *J. Funct. Biomater.* **2022**, *13*, 24. [CrossRef]
21. Shiue, S.; Syu, F.; Lin, H. Two types of bacteriophage-modified alginate hydrogels as antibacterial coatings for implants. Two types of bacteriophage-modified alginate hydrogels as antibacterial coatings for implants. *J. Taiwan Inst. Chem. Eng.* **2022**, *134*, 104353. [CrossRef]
22. Tredget, E.E.; Shankowsky, H.A.; Groenveld, A.; Burrell, R. A matched-pair, randomized study evaluating the efficacy and safety of Acticoat silver-coated dressing for the treatment of burn wounds. *J. Burn Care Rehabil.* **1998**, *19*, 531–537. [CrossRef] [PubMed]
23. Fiore, M.; Sambri, A.; Zucchini, R.; Giannini, C.; Donati, D.M.; De Paolis, M. Silver-coated megaprosthesis in prevention and treatment of peri-prosthetic infections: A systematic review and meta-analysis about efficacy and toxicity in primary and revision surgery. *Eur. J. Orthop. Surg. Traumatol.* **2021**, *31*, 201–220. [CrossRef] [PubMed]
24. Metsemakers, W.J.; Reul, M.; Nijs, S. The use of gentamicin-coated nails in complex open tibia fracture and revision cases: A retrospective analysis of a single centre case series and review of the literature. *Injury* **2015**, *46*, 2433–2437. [CrossRef]
25. Takakura, Y.; Tanaka, Y.; Kumai, T.; Sugimoto, K.; Ohgushi, H. Ankle arthroplasty using three generations of metal and ceramic prostheses. *Clin. Orthop. Relat. Res.* **2004**, *424*, 130–136. [CrossRef]
26. El-Sayed, D.; Nouvong, A. Infection protocols for implants. *Clin. Podiat.r Med. Surg.* **2019**, *36*, 627–649. [CrossRef]
27. Merola, M.; Affatato, S. Materials for hip prostheses: A review of wear and loading considerations. *Materials* **2019**, *12*, 495. [CrossRef]
28. Henderson, R.A.; Austin, M.S. Management of periprosthetic joint infection: The more we learn, the less we know. *J. Arthroplast.* **2017**, *32*, 2056–2059. [CrossRef]
29. Mihalko, W.M.; Haider, H.; Kurtz, S.; Marcolongo, M.; Urish, K. New materials for hip and knee joint replacement: What's hip and what's in kneed? *J. Orthop. Res.* **2020**, *38*, 1436–1444. [CrossRef]
30. Perry, D.; Frame, J.D. The history and development of breast implants. *Ann. R. Coll. Surg. Engl.* **2020**, *10*, 478–482. [CrossRef]
31. Hall, B.R.; Billue, K.L.; Sanders, S.E.; Meyer, B.R.; Johnson, P.J. Salmonella infection of breast implant associated with traveler's diarrhea: A case report. *JPRAS Open* **2018**, *18*, 59–64. [CrossRef]
32. Franchelli, S.; Pesce, M.; Savaia, S.; Marchese, A.; Barbieri, R.; Baldelli, I.; De Maria, A. Clinical and microbiological characterization of late breast implant infections after reconstructive breast cancer surgery. *Surg. Infect.* **2015**, *16*, 636–644. [CrossRef]
33. Chakfé, N.; Diener, H.; Lejay, A.; Assadian, O.; Berard, X.; Caillon, J.; Fourneau, I.; Glaudemans, A.W.J.M.; Koncar, I.; Lindholt, J.; et al. Editor's Choice—European Society for Vascular Surgery (ESVS) 2020 Clinical practice guidelines on the management of vascular graft and endograft infections. *Eur. J. Vasc. Endovasc. Surg.* **2020**, *59*, 339–384. [CrossRef]
34. Viola, G.M.; Darouiche, R.O. Cardiovascular implantable device infections. *Curr. Infect. Dis. Rep.* **2011**, *13*, 333–342. [CrossRef]
35. Zheng, Q.; Tang, Q.; Wang, Z.L.; Li, Z. Self-powered cardiovascular electronic devices and systems. *Nat. Rev. Cardiol.* **2021**, *18*, 7–21. [CrossRef]
36. Zerbo, S.; Perrone, G.; Bilotta, C.; Adelfio, V.; Malta, G.; Di Pasquale, P.; Maresi, E.; Argo, A. Cardiovascular implantable electronic device infection and new insights about correlation between pro-inflammatory markers and heart failure: A systematic literature review and meta-analysis. *Front. Cardiovasc. Med.* **2021**, *8*, 602275. [CrossRef]
37. Tarakji, K.G.; Chan, E.J.; Cantillon, D.J.; Doonan, A.L.; Hu, T.; Schmitt, S.; Fraser, T.G.; Kim, A.; Gordon, S.M.; Wilkoff, B.L. Cardiac implantable electronic device infections: Presentation, management, and patient outcomes. *Heart Rhythm.* **2010**, *7*, 1043–1047. [CrossRef]
38. Korkerdsup, T.; Ngarmukos, T.; Sungkanuparph, S.; Phuphuakrat, A. Cardiac implantable electronic device infection in the cardiac referral center in Thailand: Incidence, microbiology, risk factors, and outcomes. *J. Arrhythm.* **2018**, *34*, 632–639. [CrossRef] [PubMed]
39. Stöver, T.; Lenarz, T. Biomaterials in cochlear implants. *GMS Curr. Top. Otorhinolaryngol. Head Neck Surg.* **2009**, *8*, Doc10.
40. Lodhi, F.; Coelho, D.H. Non-tuberculous mycobacterial cochlear implant infection: An emerging pathogen. *Cochlear Implant. Int.* **2015**, *16*, 237–240. [CrossRef]
41. Sharma, S.; Gupta, A.; Bhatia, K.; Lahiri, A.K.; Singh, S. Salvaging cochlear implant after wound infection: Well worth a try. *Cochlear Implant. Int.* **2017**, *18*, 230–234. [CrossRef] [PubMed]
42. Tawfik, K.O.; Golub, J.S.; Roland, J.T.; Samy, R.N. Recurrent cochlear implant infection treated with exteriorization and partial mastoid obliteration. *Cochlear Implant. Int.* **2016**, *17*, 58–61. [CrossRef]
43. Vaid, N.; Vaid, S.; Manikoth, M. Case report-Biofilm infection of a cochlear implant. *Cochlear Implant. Int.* **2013**, *14*, 117–120. [CrossRef] [PubMed]
44. Zarrintaj, P.; Saeb, M.R.; Ramakrishna, S.; Mozafari, M. Biomaterials selection for neuroprosthetics. *Curr. Opin. Biomed. Eng.* **2018**, *6*, 99–109. [CrossRef]

45. Shenai, M.B.; Falconer, R.; Rogers, S. A cupriavidus pauculus infection in a patient with a deep brain stimulation implant. *Cureus* **2019**, *11*, e6104. [CrossRef]
46. Wei, Z.; Gordon, C.R.; Bergey, G.K.; Sacks, J.M.; Anderson, W.S. Implant site infection and bone flap osteomyelitis associated with the neuropace responsive neurostimulation system. *World Neurosurg.* **2016**, *88*, 687.e1–687.e6. [CrossRef]
47. Lawrence, E.L.; Turner, I.G. Materials for urinary catheters: A review of their history and development in the UK. *Med. Eng. Phys.* **2005**, *27*, 443–453. [CrossRef] [PubMed]
48. Huang, W.C.; Wann, S.R.; Lin, S.L.; Kunin, C.M.; Kung, M.H.; Lin, C.H.; Hsu, C.W.; Liu, C.P.; Lee, S.S.; Liu, Y.C.; et al. Catheter-associated urinary tract infections in intensive care units can be reduced by prompting physicians to remove unnecessary catheters. *Infect. Control Hosp. Epidemiol.* **2004**, *25*, 974–978. [CrossRef]
49. Lo, E.; Nicolle, L.E.; Coffin, S.E.; Gould, C.; Maragakis, L.L.; Meddings, J.; Pegues, D.A.; Pettis, A.M.; Saint, S.; Yokoe, D.S. Strategies to prevent catheter-associated urinary tract infections in acute care hospitals: 2014 update. *Infect. Control Hosp. Epidemiol.* **2014**, *35*, 464–479. [CrossRef]
50. Luzum, M.; Sebolt, J.; Chopra, V. Catheter-associated urinary tract infection, clostridioides difficile colitis, central line-associated bloodstream infection, and methicillin-resistant staphylococcus aureus. *Med. Clin. North Am.* **2020**, *104*, 663–679. [CrossRef]
51. Li, F.; Song, M.; Xu, L.; Deng, B.; Zhu, S.; Li, X. Risk factors for catheter-associated urinary tract infection among hospitalized patients: A systematic review and meta-analysis of observational studies. *J. Adv. Nurs.* **2019**, *75*, 517–527. [CrossRef]
52. Shuman, E.K.; Chenoweth, C.E. Urinary catheter-associated infections. *Infect. Dis. Clin. North Am.* **2018**, *32*, 885–897. [CrossRef]
53. Del Bigio, M.R. Biological reactions to cerebrospinal fluid shunt devices: A review of the cellular pathology. *Neurosurgery* **1998**, *42*, 319–326. [CrossRef]
54. Canadian Nosocomial Infection Surveillance Program. Device-associated infections in Canadian acute-care hospitals from 2009 to 2018. *Can. Commun. Dis. Rep.* **2020**, *46*, 387–397. [CrossRef]
55. Shibamura-Fujiogi, M.; Ormsby, J.; Breibart, M.; Warf, B.; Priebe, G.P.; Soriano, S.G.; Sandora, T.J.; Yuki, K. Risk factors for pediatric surgical site infection following neurosurgical procedures for hydrocephalus: A retrospective single-center cohort study. *BMC Anesthesiol.* **2021**, *21*, 124. [CrossRef]
56. Benachinmardi, K.K.; Ravikumar, R.; Indiradevi, B. Role of biofilm in cerebrospinal fluid shunt infections: A study at tertiary neurocare center from South India. *J. Neurosci. Rural. Pract.* **2017**, *8*, 335–341. [CrossRef]
57. Fernández-Méndez, R.; Richards, H.K.; Seeley, H.M.; Pickard, J.D.; Joannides, A.J. UKSR collaborators, Current epidemiology of cerebrospinal fluid shunt surgery in the UK and Ireland (2004–2013). *J. Neurol. Neurosurg. Psychiatry* **2019**, *90*, 747–754. [CrossRef]
58. Deshmukh, R.M.; Kulkarni, S.S. A review on biomaterials in orthopedic bone plate application. *International J. Curr. Eng. Technol.* **2015**, *5*, 2587–2591.
59. Toro-Aguilera, Á.; Zuriarrain, S.W.; Masdeu, M.G.; Sayol, R.R.; Billi, A.M.; Carrera, I.; de Caso, J. Risk factors for infection in fixation of distal tibia fractures. *Injury* **2021**, *52* (Suppl. 4), S104–S108. [CrossRef]
60. Guillaume, B. Dental implants: A review. *Morphologie* **2016**, *100*, 189–198. [CrossRef]
61. Neely, A.L.; Maalhagh-Fard, A. Successful management of early peri-implant infection and bone loss using a multidisciplinary treatment approach. *Clin. Adv. Periodontics* **2018**, *8*, 5–10. [CrossRef]
62. Patton, D.; Kiewiet, N.; Brage, M. Infected total ankle arthroplasty: Risk factors and treatment options. *Foot Ankle Int.* **2015**, *36*, 626–634. [CrossRef] [PubMed]
63. Gbejuade, H.O.; Lovering, A.M.; Webb, J.C. The role of microbial biofilms in prosthetic joint infections. *Acta Orthop.* **2015**, *86*, 147–158. [CrossRef] [PubMed]
64. Bayston, R.; Lari, J. A study of the sources of infection in colonised shunts. *Dev. Med. Child Neurol.* **1974**, *16*, 16–22. [CrossRef] [PubMed]
65. Reynolds-Campbell, G.; Nicholson, A.; Thoms-Rodriguez, C.A. Oral bacterial infections: Diagnosis and management. *Dent. Clin. N. Am.* **2017**, *61*, 305–318. [CrossRef] [PubMed]
66. Masters, E.A.; Ricciardi, B.F.; de Mesy Bentley, K.L.; Moriarty, T.F.; Schwarz, E.M.; Muthukrishnan, G. Skeletal infections: Microbial pathogenesis, immunity and clinical management. *Nat. Rev. Microbiol.* **2022**, *20*, 385–400. [CrossRef]
67. Bain, C.J.; Odili, J. Late infection of an alloplastic chin implant masquerading as squamous cell carcinoma. *J. Plast. Reconstr. Aesthet. Surg.* **2012**, *65*, e151–e152. [CrossRef]
68. Chang, J.; Lee, G.W. Late hematogenous bacterial infections of breast implants: Two case reports of unique bacterial infections. *Ann. Plast. Surg.* **2011**, *67*, 14–16. [CrossRef]
69. Beidas, O.E.; Rabb, C.H.; Sawan, K.T.; Tan, B.K. The pseudomeningocoele that wasn't: Case report of an adult who presented with a late infection of an implant. *J. Plast. Reconstr. Aesthet. Surg.* **2011**, *64*, 1228–1231. [CrossRef]
70. Vichitvejpaisal, P.; Dalvin, L.A.; Lally, S.E.; Shields, C.L. Delayed implant infection with Cutibacterium acnes (Propionibacterium acnes) 30 years after silicone sheet orbital floor implant. *Orbit* **2020**, *39*, 139–142. [CrossRef]
71. Coden, D.J.; Hornblass, A. Propionibacterium acnes orbital abscess. *Arch. Ophthalmol.* **1990**, *108*, 481. [CrossRef]
72. Hannouille, J.; Belgrado, J.P.; Vankerchove, S.; Vandermeeren, L. Breast implant infection with pasteurella canis: First case-report. *JPRAS Open* **2019**, *21*, 86–88. [CrossRef]
73. Oses, M.; Ordás, C.M.; Feliz, C.; Del Val, J.; Ayerbe, J.; García-Ruiz, P.J. Disease-modifying anti-rheumatic drugs as a risk factor for delayed DBS implant infection. *Parkinsonism Relat. Disord.* **2018**, *55*, 143–144. [CrossRef]

74. Young, P.; Riga, A.; Brunelli, J. Nocardia nova infection of tibia tenodesis implant after anterior cruciate ligament reconstruction in an immunocompetent patient. *J. Am. Acad. Orthop. Surg. Glob. Res. Rev.* **2020**, *4*, e19.00167. [CrossRef]
75. Paziuk, T.; Levicoff, E.; Tan, T.; Good, R. Periprosthetic joint infection with listeria monocytogenes: A case report. *JBJS Case Connect* **2020**, *10*, e1900489. [CrossRef]
76. Madden, G.R.; Poulter, M.D.; Crawford, M.P.; Wilson, D.S.; Donowitz, G.R. Case report: Anaerobiospirillum prosthetic joint infection in a heart transplant recipient. *BMC Musculoskelet. Disord.* **2019**, *20*, 301. [CrossRef]
77. Haimes, M.A.; Nelms, N.J. Total knee bartonella henselae infection: An unusual manifestation of cat scratch disease: A case report. *JBJS Case Connect* **2019**, *9*, e0081. [CrossRef]
78. Posti, J.P.; Piitulainen, J.M.; Hupa, L.; Fagerlund, S.; Frantzén, J.; Aitasalo, K.M.J.; Vuorinen, V.; Serlo, W.; Syrjänen, S.; Vallittu, P.K. A glass fiber-reinforced composite—bioactive glass cranioplasty implant: A case study of an early development stage implant removed due to a late infection. *J. Mech. Behav. Biomed. Mater.* **2016**, *55*, 191–200. [CrossRef]
79. Wahl, P.; Sprecher, C.M.; Brüning, C.; Meier, C.; Milz, S.; Gautier, E.; Moriarty, T.F. Successful bony integration of a porous tantalum implant despite longlasting and ongoing infection: Histologic workup of an explanted shoulder prosthesis. *J. Biomed. Mater. Res. B Appl. Biomater.* **2018**, *106*, 2924–2931. [CrossRef]
80. Hurdle, J.G.; O'Neill, A.J.; Chopra, I.; Lee, R.E. Targeting bacterial membrane function: An underexploited mechanism for treating persistent infections. *Nat. Rev. Microbiol.* **2011**, *9*, 62–75. [CrossRef]
81. Dowd, S.E.; Sun, Y.; Secor, P.R.; Rhoads, D.D.; Wolcott, B.M.; James, G.A.; Wolcott, R.D. Survey of bacterial diversity in chronic wounds using pyrosequencing, DGGE, and full ribosome shotgun sequencing. *BMC Microbiol.* **2008**, *8*, 43. [CrossRef] [PubMed]
82. Ryan, K.J.; Ahmad, N.; Alspaugh, J.A.; Drew, W.L. *Sherris Medical Microbiology*, 7th ed.; McGraw-Hill Education: New York, NY, USA, 2018; pp. 381–735.
83. Männik, J.; Driessen, R.; Galajda, P.; Keymer, J.E.; Dekker, C. Bacterial growth and motility in sub-micron constrictions. *Proc. Natl. Acad. Sci. USA* **2009**, *106*, 14861–14866. [CrossRef] [PubMed]
84. Pianetti, A.; Battistelli, M.; Citterio, B.; Parlani, C.; Falcieri, E.; Bruscolini, F. Morphological changes of Aeromonas hydrophila in response to osmotic stress. *Micron* **2009**, *40*, 426–433. [CrossRef] [PubMed]
85. Vargas-Alfredo, N.; Santos-Coquillat, A.; Martínez-Campos, E.; Dorronsoro, A.; Cortajarena, A.L.; Del Campo, A.; Rodríguez-Hernández, J. Highly efficient antibacterial surfaces based on bacterial/cell size selective microporous supports. *ACS Appl. Mater. Interfaces* **2017**, *9*, 44270–44280. [CrossRef]
86. Costa, F.; Carvalho, I.F.; Montelaro, R.C.; Gomes, P.; Cristina, M.; Martins, L. Covalent immobilization of antimicrobial peptides (AMPs) onto biomaterial surfaces. *Acta Biomater.* **2011**, *7*, 1431–1440. [CrossRef]
87. Lister, J. On a new method of treating compound fracture, abscess, etc.: With observations on the conditions of suppuration. *Lancet* **1867**, *89*, 326–329. [CrossRef]
88. Durand, G.A.; Raoult, D.; Dubourg, G. Antibiotic discovery: History, methods and perspectives. *Int. J. Antimicrob. Agents* **2019**, *53*, 371–382. [CrossRef]
89. Bryson, D.J.; Morris, D.L.J.; Shivji, F.S.; Rollins, K.R.; Snape, S.; Ollivere, B.J. Antibiotic prophylaxis in orthopaedic surgery: Difficult decisions in an era of evolving antibiotic resistance. *Bone Joint J.* **2016**, *98*, 1014–1019. [CrossRef]
90. Li, B.; Webster, T.J. Bacteria antibiotic resistance: New challenges and opportunities for implant-associated orthopedic infections. *J. Orthop. Res.* **2018**, *36*, 22–32. [CrossRef]
91. Makabenta, J.M.V.; Nabawy, A.; Li, C.H.; Schmidt-Malan, S.; Patel, R.; Rotello, V.M. Nanomaterial-based therapeutics for antibiotic-resistant bacterial infections. *Nat. Rev. Microbiol.* **2021**, *19*, 23–36. [CrossRef]
92. Vasiliadis, A.V.; Poutoglidou, F.; Chatziravdeli, V.; Metaxiotis, D.; Beletsiotis, A. Acute periprosthetic hip joint infection caused by multidrug-resistant acinetobacter baumannii: Is debridement, antibiotics, irrigation, and implant retention a viable treatment option? *Cureus* **2021**, *13*, e13090. [CrossRef]
93. Okada, A.; Shoda, M.; Tabata, H.; Kobayashi, H.; Shoin, W.; Okano, T.; Yoshie, K.; Kato, K.; Motoki, H.; Kuwahara, K. Simultaneous infection of abandoned leads and newly implanted leadless cardiac pacemaker: Why did this occur? *J. Cardiol. Cases* **2020**, *23*, 35–37. [CrossRef]
94. Jhaveri, V.V.; Singhal, D.; Riedel, S.; Rowley, C.F.; Nathavitharana, R.R. Surgical cure of clarithromycin resistant Mycobacterium chelonae breast implant infection: A case report and review of the literature. *J. Clin. Tuberc. Other Mycobact. Dis.* **2020**, *21*, 100183. [CrossRef]
95. El-Zein, R.S.; Stelzer, M.; Hatanelas, J.; Goodlive, T.W.; Amin, A.K. A ghost left behind after transvenous lead extraction: A finding to be feared. *Am. J. Case Rep.* **2020**, *21*, e924243. [CrossRef]
96. Palacios, L.; de Nova, A.A.; Pardo, M.G. Conservative multimodal management of osteosynthesis material in surgical wounds with polymicrobial superinfection, including methicillin-resistant Staphylococcus aureus, Clinical case. *Rev. Española Cirugía Ortopédica Traumatol. (Engl. Ed.)* **2020**, *64*, 125–129. [CrossRef]
97. Hwang, S.O.; Chang, L.S. Salvage of an exposed cranial prosthetic implant using a transposition flap with an indwelling antibiotic irrigation system. *Arch. Craniofac. Surg.* **2020**, *21*, 73–76. [CrossRef]
98. Fukushima, S.; Komune, N.; Kamizono, K.; Matsumoto, N.; Takaiwa, K.; Nakagawa, T.; Kadota, H. Use of negative pressure wound therapy to treat a cochlear implant infection around the auricle: A case report. *J. Wound Care* **2020**, *29*, 568–571. [CrossRef]
99. Bajaj, T.; Karapetians, A.; Karapetians, N.; Duong, H.; Heidari, A. Methicillin resistant Staphylococcus aureus infective endocarditis presenting as neutrophilic meningoencephalitis. *AME Case Rep.* **2020**, *4*, 4. [CrossRef]

100. Hisanaga, K.; Kadota, H.; Fukushima, S.; Inatomi, Y.; Shimamoto, R.; Kamizono, K.; Hanada, M.; Yoshida, S. Toxic shock syndrome caused by staphylococcal infection after breast implant surgery: A case report and literature review. *Ann. Plast. Surg.* **2019**, *83*, 359–362. [CrossRef]
101. Meleca, J.B.; Bryson, P.C. Delayed laryngeal implant infection and laryngocutaneous fistula after medialization laryngoplasty. *Am. J. Otolaryngol.* **2019**, *40*, 462–464. [CrossRef]
102. Siebenbürger, G.; Grabein, B.; Schenck, T.; Kammerlander, C.; Böcker, W.; Zeckey, C. Eradication of acinetobacter baumannii/enterobacter cloacae complex in an open proximal tibial fracture and closed drop foot correction with a multidisciplinary approach using the taylor spatial frame®: A case report. *Eur. J. Med. Res.* **2019**, *24*, 2. [CrossRef] [PubMed]
103. Nozoe, M.; Yoshida, D.; Nagatomo, D.; Suematsu, N.; Kubota, T.; Okabe, M.; Yamamoto, Y. Successful percutaneous retrieval of a micra transcatheter pacing system at 8 weeks after implantation. *J. Arrhythm.* **2018**, *34*, 653–655. [CrossRef] [PubMed]
104. Bonacker, J.; Darowski, M.; Haar, P.; Westphal, T.; Bergschmidt, P. Periprosthetic tibial fracture with nonunion and ascending prosthetic joint infection: A case report of an individual treatment strategy. *J. Orthop. Case Rep.* **2018**, *8*, 3–8. [PubMed]
105. Rico-Nieto, A.; Moreno-Ramos, F.; Fernández-Baillo, N. Lumbar arthrodesis infection by multi-resistant Klebsiella pneumoniae, successfully treated with implant retention and ceftazidime/avibactam. *Rev. Española Cirugía Ortopédica Traumatol. (Engl. Ed.)* **2018**, *62*, 471–473. [CrossRef] [PubMed]
106. Vaidya, G.N.; Deam, A.G. Simultaneous suction debulking of lead vegetation prior to percutaneous lead extraction. *J. Cardiol. Cases.* **2018**, *18*, 17–19. [CrossRef] [PubMed]
107. Sebastian, S.; Malhotra, R.; Pande, A.; Gautam, D.; Xess, I.; Dhawan, B. Staged reimplantation of a total hip prosthesis after co-infection with candida tropicalis and staphylococcus haemolyticus: A case report. *Mycopathologia* **2018**, *183*, 579–584. [CrossRef] [PubMed]
108. Mahalingam, P.; Topiwalla, T.T.; Ganesan, G. Drug-resistant coagulase-negative staphylococcal endophthalmitis following dexamethasone intravitreal implant. *Indian J. Ophthalmol.* **2017**, *65*, 634–636. [CrossRef]
109. Gharacholou, S.M.; Dworak, M.; Dababneh, A.S.; Palraj, R.V.; Roskos, M.C.; Chapman, S.C. Acute infection of viabahn stent graft in the popliteal artery. *J. Vasc. Surg. Cases Innov. Tech.* **2017**, *3*, 69–73. [CrossRef]
110. Takizawa, T.; Tsutsumimoto, T.; Yui, M.; Misawa, H. Surgical site infections caused by methicillin-resistant staphylococcus epidermidis after spinal instrumentation surgery. *Spine* **2017**, *42*, 525–530. [CrossRef]
111. Savage, V.J.; Chopra, I.; O'Neill, A.J. Staphylococcus aureus biofilms promote horizontal transfer of antibiotic resistance. *Antimicrob. Agents Chemother.* **2013**, *57*, 1968–1970. [CrossRef]
112. Naylor, N.R.; Atun, R.; Zhu, N.; Kulasabanathan, K.; Silva, S.; Chatterjee, A.; Knight, G.M.; Robotham, J.V. Estimating the burden of antimicrobial resistance: A systematic literature review. *Antimicrob. Resist. Infect. Control* **2018**, *7*, 58. [CrossRef]
113. Zilberman, M.; Elsner, J.J. Antibiotic-eluting medical devices for various applications. *J. Control Release* **2008**, *130*, 202–215. [CrossRef]
114. Freischmidt, H.; Armbruster, J.; Reiter, G.; Grützner, P.A.; Helbig, L.; Guehring, T. Individualized techniques of implant coating with an antibiotic-loaded, hydroxyapatite/calcium sulphate bone graft substitute. *Ther. Clin. Risk Manag.* **2020**, *16*, 689–694. [CrossRef] [PubMed]
115. Campoccia, D.; Montanaro, L.; Speziale, P.; Arciola, C.R. Antibiotic-loaded biomaterials and the risks for the spread of antibiotic resistance following their prophylactic and therapeutic clinical use. *Biomaterials* **2010**, *31*, 6363–6377. [CrossRef]
116. Cao, H.; Qin, H.; Li, Y.; Jandt, K.D. The action-networks of nanosilver: Bridging the gap between material and biology. *Adv. Healthc. Mater.* **2021**, *26*, e2100619. [CrossRef]
117. Percival, S.L.; Bowler, P.G.; Russell, D. Bacterial resistance to silver in wound care. *J. Hosp. Infect.* **2005**, *60*, 1–7. [CrossRef]
118. Panáček, A.; Kvítek, L.; Smékalová, M.; Večeřová, R.; Kolář, M.; Röderová, M.; Dyčka, F.; Šebela, M.; Prucek, R.; Tomanec, O.; et al. Bacterial resistance to silver nanoparticles and how to overcome it. *Nat. Nanotechnol.* **2018**, *13*, 65–71. [CrossRef]
119. Nadeem, S.F.; Gohar, U.F.; Tahir, S.F.; Mukhtar, H.; Pornpukdeewattana, S.; Nukthamna, P.; Moula Ali, A.M.; Bavisetty, S.C.B.; Massa, S. Antimicrobial resistance: More than 70 years of war between humans and bacteria. *Crit. Rev. Microbiol.* **2020**, *46*, 578–599. [CrossRef]
120. Ciofu, O.; Moser, C.; Jensen, P.Ø.; Høiby, N. Tolerance and resistance of microbial biofilms. *Nat. Rev. Microbiol.* **2022**. [CrossRef]
121. Wu, X.X.; Zhang, Y.; Hu, T.; Li, W.X.; Li, Z.L.; Hu, H.J.; Zhu, S.R.; Chen, W.Z.; Zhou, C.S.; Jiang, G.B. Long-term antibacterial composite via alginate aerogel sustained release of antibiotics and Cu used for bone tissue bacteria infection. *Int. J. Biol. Macromol.* **2021**, *167*, 1211–1220. [CrossRef]
122. Mukai, M.; Uchida, K.; Sugo, K.; Nakasu, M.; Nakajima, T.; Takata, K.; Takaso, M.; Urabe, K. Long-term antibacterial activity of vancomycin from calcium phosphate cement in vivo. *Biomed. Mater. Eng.* **2022**, *33*, 41–50. [CrossRef] [PubMed]
123. Cheng, Y.; Mei, S.; Kong, X.; Liu, X.; Gao, B.; Chen, B.; Wu, J. Long-term antibacterial activity of a composite coating on titanium for dental implant application. *J. Biomater. Appl.* **2021**, *35*, 643–654. [CrossRef] [PubMed]
124. Tsutsumi, H.; Tsutsumi, Y.; Shimabukuro, M.; Manaka, T.; Chen, P.; Ashida, M.; Ishikawa, K.; Katayama, H.; Hanaw, T. Investigation of the long-term antibacterial properties of titanium by two-step micro-arc oxidation treatment. *Coatings* **2021**, *11*, 798. [CrossRef]
125. Liu, F.; Cheng, X.; Xiao, L.; Wang, Q.; Yan, K.; Su, Z.; Wang, L.; Ma, C.; Wang, Y. Inside-outside Ag nanoparticles-loaded polylactic acid electrospun fiber for long-term antibacterial and bone regeneration. *Int. J. Biol. Macromol.* **2021**, *167*, 1338–1348. [CrossRef]

126. Tao, S.; Yang, X.; Liao, L.; Yang, J.; Liang, K.; Zeng, S.; Zhou, J.; Zhang, M.; Li, J. A novel anticaries agent, honokiol-loaded poly(amido amine) dendrimer, for simultaneous long-term antibacterial treatment and remineralization of demineralized enamel. *Dent. Mater.* **2021**, *37*, 1337–1349. [CrossRef]
127. Jia, J.; Duan, S.; Zhou, X.; Sun, L.; Qin, C.; Li, M.; Ge, F. Long-term antibacterial film nanocomposite incorporated with patchouli essential oil prepared by supercritical co 2 cyclic impregnation for wound dressing. *Molecules* **2021**, *26*, 5005. [CrossRef]
128. Kitagawa, H.; Kitagawa, R.; Tsuboi, R.; Hirose, N.; Thongthai, P.; Sakai, H.; Ueda, M.; Ono, S.; Sasaki, J.; Ooya, T.; et al. Development of endodontic sealers containing antimicrobial-loaded polymer particles with long-term antibacterial effects. *Dent. Mater.* **2021**, *37*, 1248–1259. [CrossRef]
129. Wan, R.; Chu, S.; Wang, X.; Lei, L.; Tang, H.; Hu, G.; Dong, L.; Li, D.; Gu, H. Study on the osteogenesis of rat mesenchymal stem cells and the long-term antibacterial activity of Staphylococcus epidermidis on the surface of silver-rich TiN/Ag modified titanium alloy. *J. Biomed. Mater. Res. B Appl. Biomater.* **2020**, *108*, 3008–3021. [CrossRef]
130. Liang, T.; Wang, Y.; Zeng, L.; Liu, Y.; Qiao, L.; Zhang, S.; Zhao, R.; Li, G.; Zhang, R.; Xiang, J.; et al. Copper-doped 3D porous coating developed on Ti-6Al-4V alloys and its *in vitro* long-term antibacterial ability. *Appl. Surf. Sci.* **2020**, *509*, 144717. [CrossRef]
131. Zhang, F.; Zhou, M.; Gu, W.; Shen, Z.; Ma, X.; Lu, F.; Yang, X.; Zheng, Y.; Gou, Z. Zinc-/copper-substituted dicalcium silicate cement: Advanced biomaterials with enhanced osteogenesis and long-term antibacterial properties. *J. Mater. Chem. B.* **2020**, *8*, 1060–1070. [CrossRef]
132. Yu, H.; Chen, X.; Cai, J.; Ye, D.; Wu, Y.; Liu, P. Dual controlled release nanomicelle-in-nanofiber system for long-term antibacterial medical dressings. *J. Biomater. Sci. Polym. Ed.* **2019**, *30*, 64–76. [CrossRef] [PubMed]
133. Daud, N.M.; Masri, N.A.; Malek, N.A.N.N.; Razak, S.I.A.; Saidin, S. Long-term antibacterial and stable chlorhexidine-polydopamine coating on stainless steel 316L. *Prog. Org. Coat.* **2018**, *122*, 147–153. [CrossRef]
134. Shivaram, A.; Bose, S.; Bandyopadhyay, A. Understanding long-term silver release from surface modified porous titanium implants. *Acta Biomater.* **2017**, *58*, 550–560. [CrossRef]
135. Zhao, R.; Lv, M.; Li, Y.; Sun, M.; Kong, W.; Wang, L.; Song, S.; Fan, C.; Jia, L.; Qiu, S.; et al. Stable nanocomposite based on pegylated and silver nanoparticles loaded graphene oxide for long-term antibacterial activity. *ACS Appl. Mater. Interfaces* **2017**, *9*, 15328–15341. [CrossRef]
136. Wang, G.; Feng, H.; Jin, W.; Gao, A.; Peng, X.; Li, W.; Wu, H.; Li, Z.; Chu, P.K. Long-term antibacterial characteristics and cytocompatibility of titania nanotubes loaded with Au nanoparticles without photocatalytic effects. *Appl. Surf. Sci.* **2017**, *414*, 230–237. [CrossRef]
137. Uhm, S.; Kwon, J.; Song, D.; Lee, E.; Jeong, W.; Oh, S.; Kim, K.; Choi, E.H.; Kim, K. Long-Term antibacterial performance and bioactivity of plasma-engineered Ag-NPs/TiO_2. *J. Biomed. Nanotechnol.* **2016**, *12*, 1890–1906. [CrossRef]
138. Qin, H.; Cao, H.; Zhao, Y.; Zhu, C.; Cheng, T.; Wang, Q.; Peng, X.; Cheng, M.; Wang, J.; Jin, G.; et al. In vitro and in vivo anti-biofilm effects of silver nanoparticles immobilized on titanium. *Biomaterials* **2014**, *35*, 9114–9125. [CrossRef]
139. Li, M.; Neoh, K.G.; Xu, L.Q.; Wang, R.; Kang, E.; Lau, T.; Olszyna, D.P.; Chiong, E. Surface modification of silicone for biomedical applications requiring long-term antibacterial, antifouling, and hemocompatible properties. *Langmuir* **2012**, *28*, 16408–16422. [CrossRef]
140. Urabe, K.; Naruse, K.; Hattori, H.; Hirano, M.; Uchida, K.; Onuma, K.; Park, H.J.; Itoman, M. In vitro comparison of elution characteristics of vancomycin from calcium phosphate cement and polymethylmethacrylate. *J. Orthop. Sci.* **2009**, *14*, 784–793. [CrossRef]
141. Uchida, K.; Sugo, K.; Nakajima, T.; Nakawaki, M.; Takano, S.; Nagura, N.; Takaso, M.; Urabe, K. *In vivo* release of vancomycin from calcium phosphate cement. *Biomed. Res. Int.* **2018**, *2018*, 4560647. [CrossRef]
142. Bandyopadhyay, A.; Shivaram, A.; Tarafder, S.; Sahasrabudhe, H.; Banerjee, D.; Bose, S. *In vivo* response of laser processed porous titanium implants for load-bearing implants. *Ann. Biomed. Eng.* **2017**, *45*, 249–260. [CrossRef] [PubMed]
143. Cao, H.; Liu, X.; Meng, F.; Chu, P.K. Biological actions of silver nanoparticles embedded in titanium controlled by micro-galvanic effects. *Biomaterials* **2011**, *32*, 693–705. [CrossRef] [PubMed]
144. Cao, H.; Zhang, W.; Meng, F.; Guo, J.; Wang, D.; Qian, S.; Jiang, X.; Liu, X.; Chu, P.K. Osteogenesis catalyzed by titanium-supported silver nanoparticles. *ACS Appl. Mater. Interfaces* **2017**, *9*, 5149–5157. [CrossRef] [PubMed]
145. Zhang, X.; Yang, C.; Yang, K. Contact killing of Cu-bearing stainless steel based on charge transfer caused by the microdomain potential difference. *ACS Appl. Mater. Interfaces* **2020**, *12*, 361–372. [CrossRef]
146. Kalfas, I.H. Principles of bone healing. *Neurosurg. Focus* **2001**, *10*, E1. [CrossRef]
147. Marsell, R.; Einhorn, T.A. The biology of fracture healing. *Injury* **2011**, *42*, 551–555. [CrossRef]
148. Ono, S.; Imai, R.; Ida, Y.; Shibata, D.; Komiya, T.; Matsumura, H. Increased wound pH as an indicator of local wound infection in second degree burns. *Burns* **2015**, *41*, 820–824. [CrossRef]
149. Sawyer, R.G.; Spengler, M.D.; Adams, R.B.; Pruett, T.L. The peritoneal environment during infection. The effect of monomicrobial and polymicrobial bacteria on pO2 and pH. *Ann. Surg.* **1991**, *213*, 253–260. [CrossRef]
150. Stassen, W.N.; McCullough, A.J.; Bacon, B.R.; Gutnik, S.H.; Wadiwala, I.M.; McLaren, C.; Kalhan, S.C.; Tavill, A.S. Immediate diagnostic criteria for bacterial infection of ascitic fluid. Evaluation of ascitic fluid polymorphonuclear leukocyte count, pH, and lactate concentration, alone and in combination. *Gastroenterology* **1986**, *90*, 1247–1254. [CrossRef]
151. Del Campo, A.; Echeverría, C.; Martín, M.S.; Cuervo-Rodríguez, R.; Fernández-García, M.; Muñoz-Bonilla, A. Porous microstructured surfaces with ph-triggered antibacterial properties. *Macromol. Biosci.* **2019**, *19*, 1900127. [CrossRef]

152. Wei, T.; Yu, Q.; Zhan, W.; Chen, H. A smart antibacterial surface for the on-demand killing and releasing of bacteria. *Adv. Healthc. Mater.* **2016**, *5*, 449–456. [CrossRef]
153. Wei, Z.; Yang, J.; Long, S.; Zhang, G.; Wang, X. Smart and in-situ formation electrospun fibrous membrane for the control of antimicrobial efficacy. *Smart Mater. Med.* **2021**, *2*, 87–95. [CrossRef]
154. Pinho, E.; Machado, S.; Soares, G. Smart hydrogel for the ph-selective drug delivery of antimicrobial compounds. *Macromol. Symp.* **2019**, *385*, 1800182. [CrossRef]
155. Ramos, M.L.P.; González, J.A.; Fabian, L.; Pérez, C.J.; Villanueva, M.E.; Copello, G.J. Sustainable and smart keratin hydrogel with pH-sensitive swelling and enhanced mechanical properties. *Mater. Sci. Eng. C Mater. Biol. Appl.* **2017**, *78*, 619–626. [CrossRef]
156. Villanueva, M.E.; Cuestas, M.L.; Pérez, C.J.; Dall Orto, V.C.; Copello, G.J. Smart release of antimicrobial ZnO nanoplates from a pH-responsive keratin hydrogel. *J. Colloid. Interface Sci.* **2019**, *536*, 372–380. [CrossRef]
157. Tao, B.; Deng, Y.; Song, L.; Ma, W.; Qian, Y.; Lin, C.; Yuan, Z.; Lu, L.; Chen, M.; Yang, X.; et al. BMP2-loaded titania nanotubes coating with pH-responsive multilayers for bacterial infections inhibition and osteogenic activity improvement. *Colloids Surf. B Biointerfaces* **2019**, *177*, 242–252. [CrossRef]
158. Phoungtawee, P.; Seidi, F.; Treetong, A.; Warin, C.; Klamchuen, A.; Crespy, D. Polymers with hemiaminal ether linkages for ph-responsive antibacterial materials. *ACS Macro. Lett.* **2021**, *10*, 365–369. [CrossRef]
159. De Silva, C.C.; Israni, N.; Zanwar, A.; Jagtap, A.; Leophairatana, P.; Koberstein, J.T.; Modak, S.M. "*Smart*" polymer enhances the efficacy of topical antimicrobial agents. *Burns* **2019**, *45*, 1418–1429. [CrossRef]
160. Hu, C.; Long, L.; Cao, J.; Zhang, S.; Wang, Y. Dual-crosslinked mussel-inspired smart hydrogels with enhanced antibacterial and angiogenic properties for chronic infected diabetic wound treatment via pH-responsive quick cargo release. *Chem. Eng.* **2021**, *411*, 128564. [CrossRef]
161. Quartinello, F.; Tallian, C.; Auer, J.; Schön, H.; Vielnascher, R.; Weinberger, S.; Wieland, K.; Weihs, A.M.; Herrero-Rollett, A.; Lendl, B.; et al. Smart textiles in wound care: Functionalization of cotton/PET blends with antimicrobial nanocapsules. *J. Mater. Chem. B.* **2019**, *7*, 6592–6603. [CrossRef]
162. Kaila, V.R.I.; Wikström, M. Architecture of bacterial respiratory chains. *Nat. Rev. Microbiol.* **2021**, *19*, 319–330. [CrossRef]
163. Reguera, G.; McCarthy, K.D.; Mehta, T.; Nicoll, J.S.; Tuominen, M.T.; Lovley, D.R. Extracellular electron transfer via microbial nanowires. *Nature* **2005**, *435*, 1098–1101. [CrossRef]
164. Harris, H.W.; El-Naggar, M.Y.; Bretschger, O.; Ward, M.J.; Romine, M.F.; Obraztsova, A.Y.; Nealson, K.H. Electrokinesis is a microbial behavior that requires extracellular electron transport. *Proc. Natl. Acad. Sci. USA* **2010**, *107*, 326–331. [CrossRef] [PubMed]
165. Kato, S.; Hashimoto, K.; Watanabe, K. Microbial interspecies electron transfer via electric currents through conductive minerals. *Proc. Natl. Acad. Sci. USA* **2012**, *109*, 10042–10046. [CrossRef] [PubMed]
166. Light, S.H.; Méheust, R.; Ferrell, J.L.; Cho, J.; Deng, D.; Agostoni, M.; Iavarone, A.T.; Banfield, J.F.; D'Orazio, S.E.F.; Portnoy, D.A. Extracellular electron transfer powers flavinylated extracellular reductases in Gram-positive bacteria. *Proc. Natl. Acad. Sci. USA* **2019**, *116*, 26892–26899. [CrossRef] [PubMed]
167. Shi, L.; Dong, H.; Reguera, G.; Beyenal, H.; Lu, A.; Liu, J.; Yu, H.; Fredrickson, J.K. Extracellular electron transfer mechanisms between microorganisms and minerals. *Nat. Rev. Microbiol.* **2016**, *14*, 651–662. [CrossRef] [PubMed]
168. Pankratova, G.; Hederstedt, L.; Gorton, L. Extracellular electron transfer features of Gram-positive bacteria. *Anal. Chim. Acta* **2019**, *1076*, 32–47. [CrossRef] [PubMed]
169. Cao, H.; Qiao, Y.; Liu, X.; Lu, T.; Cui, T.; Meng, F.; Chu, P.K. Electron storage mediated dark antibacterial action of bound silver nanoparticles: Smaller is not always better. *Acta. Biomater.* **2013**, *9*, 5100–5110. [CrossRef]
170. Cao, H.; Qiao, Y.; Meng, F.; Liu, X. Spacing-dependent antimicrobial efficacy of immobilized silver nanoparticles. *J. Phys. Chem. Lett.* **2014**, *5*, 743–748. [CrossRef]
171. Wang, M.; Cao, H.; Meng, F.; Zhao, X.; Ping, Y.; Lü, X.; Liu, X. Schottky barrier dependent antimicrobial efficacy of silver nanoparticles. *Mater. Lett.* **2016**, *179*, 1–4. [CrossRef]
172. Cao, H.; Meng, F.; Liu, X. Antimicrobial activity of tantalum oxide coatings decorated with Ag nanoparticles. *J. Vac. Sci. Technol. A* **2016**, *34*, 04C102. [CrossRef]
173. Yang, M.; Liu, H.; Qiu, C.; Iatsunskyi, I.; Coy, E.; Moya, S.; Wang, Z.; Wu, W.; Zhao, X.; Wang, G. Electron transfer correlated antibacterial activity of biocompatible graphene Nanosheets-TiO_2 coatings. *Carbon* **2020**, *166*, 350–360. [CrossRef]
174. Li, J.; Wang, J.; Wang, D.; Guo, G.; Yeung, K.W.K.; Zhang, X.; Liu, X. Band gap engineering of titania film through cobalt regulation for oxidative damage of bacterial respiration and viability. *ACS Appl. Mater. Interfaces* **2017**, *9*, 27475–27490. [CrossRef] [PubMed]
175. Wang, R.; Zhou, T.; Liu, J.; Zhang, X.; Long, F.; Liu, L. Bilayer microstructure of antibacterial TiO_2 coating on Ti6Al4V fabricated via micro-arc oxidation in W-containing electrolytes. *Surf. Coat. Technol.* **2021**, *413*, 127094. [CrossRef]
176. Ray, P.C.; Khan, S.A.; Singh, A.K.; Senapati, D.; Fan, Z. Nanomaterials for targeted detection and photothermal killing of bacteria. *Chem. Soc. Rev.* **2012**, *41*, 3193–3209. [CrossRef] [PubMed]
177. Qu, Y.; Wei, T.; Zhao, J.; Jiang, S.; Yang, P.; Yu, Q.; Chen, H. Regenerable smart antibacterial surfaces: Full removal of killed bacteria via a sequential degradable layer. *J. Mater. Chem. B.* **2018**, *6*, 3946–3955. [CrossRef]
178. Wang, Y.; Wei, T.; Qu, Y.; Zhou, Y.; Zheng, Y.; Huang, C.; Zhang, Y.; Yu, Q.; Chen, H. Smart, photothermally activated, antibacterial surfaces with thermally triggered bacteria-releasing properties. *ACS Appl. Mater. Interfaces* **2020**, *12*, 21283–21291. [CrossRef]

179. Zhang, X.; Zhang, G.; Chai, M.; Yao, X.; Chen, W.; Chu, P.K. Synergistic antibacterial activity of physical-chemical multi-mechanism by TiO_2 nanorod arrays for safe biofilm eradication on implant. *Bioact. Mater.* **2020**, *6*, 12–25. [CrossRef]
180. Kováčová, M.; Kleinová, A.; Vajd'ák, J.; Humpolíček, P.; Kubát, P.; Bodík, M.; Marković, Z.; Špitálský, Z. Photodynamic-active smart biocompatible material for an antibacterial surface coating. *J. Photochem. Photobiol. B.* **2020**, *211*, 112012. [CrossRef]
181. Tan, L.; Li, J.; Liu, X.; Cui, Z.; Yang, X.; Yeung, K.W.K.; Pan, H.; Zheng, Y.; Wang, X.; Wu, S. In situ disinfection through photo inspired radical oxygen species storage and thermal-triggered release from black phosphorous with strengthened chemical stability. *Small* **2018**, *14*, 1703197. [CrossRef]
182. Tan, L.; Li, J.; Liu, X.; Cui, Z.; Yang, X.; Zhu, S.; Li, Z.; Yuan, X.; Zheng, Y.; Yeung, K.W.K.; et al. Rapid biofilm eradication on bone implants using red phosphorus and near-infrared light. *Adv. Mater.* **2018**, *30*, e1801808. [CrossRef]
183. Yuan, Z.; Tao, B.; He, Y.; Liu, J.; Lin, C.; Shen, X.; Ding, Y.; Yu, Y.; Mu, C.; Liu, P.; et al. Biocompatible MoS_2/PDA-RGD coating on titanium implant with antibacterial property via intrinsic ROS-independent oxidative stress and NIR irradiation. *Biomaterials* **2019**, *217*, 119290. [CrossRef]
184. Gao, Q.; Zhang, X.; Yin, W.; Ma, D.; Xie, C.; Zheng, L.; Dong, X.; Mei, L.; Yu, J.; Wang, C.; et al. Functionalized MoS_2 nanovehicle with near-infrared laser-mediated nitric oxide release and photothermal activities for advanced bacteria-infected wound therapy. *Small* **2018**, *14*, e1802290. [CrossRef]
185. Liu, L.; Pan, X.; Liu, S.; Hu, Y.; Ma, D. Near-infrared light-triggered nitric oxide release combined with low-temperature photothermal therapy for synergetic antibacterial and antifungal. *Smart Mater. Med.* **2021**, *2*, 302–313. [CrossRef]
186. Yang, H.; Li, G.; Stansbury, J.W.; Zhu, X.; Wang, X.; Nie, J. Smart antibacterial surface made by photopolymerization. *ACS Appl. Mater. Interfaces* **2016**, *8*, 28047–28054. [CrossRef]
187. Děkanovský, L.; Elashnikov, R.; Kubiková, M.; Vokatá, B.; Švorčík, V.; Lyutakov, O. Dual-action flexible antimicrobial material: Switchable self-cleaning, antifouling, and smart drug release. *Adv. Funct. Mater.* **2019**, *29*, 1901880. [CrossRef]
188. Stavrakis, A.I.; Zhu, S.; Hegde, V.; Loftin, A.H.; Ashbaugh, A.G.; Niska, J.A.; Miller, L.S.; Segura, T.; Bernthal, N.M. In vivo efficacy of a "*smart*" antimicrobial implant coating. *J. Bone Joint Surg. Am.* **2016**, *98*, 1183–1189. [CrossRef]
189. Cao, H.; Liu, X. Plasma sprayed ceramic coatings for osseointegration. *Int. J. Appl. Ceram. Technol.* **2013**, *10*, 1–10. [CrossRef]
190. Gristina, A.G. Biomaterial-centered infection: Microbial adhesion versus tissue integration. *Science* **1987**, *237*, 1588–1595. [CrossRef]
191. Subbiahdoss, G.; Kuijer, R.; Grijpma, D.W.; van der Mei, H.C.; Busscher, H.J. Microbial biofilm growth vs. tissue integration: "*the race for the surface*" experimentally studied. *Acta Biomater.* **2009**, *5*, 1399–1404. [CrossRef]
192. Subbiahdoss, G.; Grijpma, D.W.; van der Mei, H.C.; Busscher, H.J.; Kuijer, R. Microbial biofilm growth vs. tissue integration on biomaterials with different wettabilities and a polymer-brush coating. *J. Biomed. Mater. Res. A* **2010**, *94*, 533–538. [PubMed]
193. Subbiahdoss, G.; Pidhatika, B.; Coullerez, G.; Charnley, M.; Kuijer, R.; van der Mei, H.C.; Textor, M.; Busscher, H.J. Bacterial biofilm formation versus mammalian cell growth on titanium-based mono- and bi-functional coating. *Eur. Cells Mater.* **2010**, *19*, 205–213. [CrossRef] [PubMed]
194. Subbiahdoss, G.; Fernandez, I.C.; Domingues, J.F.; Kuijer, R.; van der Mei, H.C.; Busscher, H.J. In vitro interactions between bacteria, osteoblast-like cells and macrophages in the pathogenesis of biomaterial-associated infections. *PLoS ONE* **2011**, *6*, e24827. [CrossRef] [PubMed]
195. Perez-Tanoira, R.; Han, X.; Soininen, A.; Aarnisalo, A.A.; Tiainen, V.M.; Eklund, K.K.; Esteban, J.; Kinnari, T.J. Competitive colonization of prosthetic surfaces by staphylococcus aureus and human cells. *J. Biomed. Mater. Res. A* **2017**, *105*, 62–72. [CrossRef] [PubMed]
196. Martinez-Perez, M.; Perez-Jorge, C.; Lozano, D.; Portal-Nuñez, S.; Perez-Tanoira, R.; Conde, A.; Arenas, M.A.; Hernandez-Lopez, J.M.; de Damborenea, J.J.; Gomez-Barrena, E.; et al. Evaluation of bacterial adherence of clinical isolates of Staphylococcus sp. using a competitive model: An in vitro approach to the "*race for the surface*" theory. *Bone Joint Res.* **2017**, *6*, 315–322. [CrossRef] [PubMed]
197. Shiels, S.M.; Mangum, L.H.; Wenke, J.C. Revisiting the "*race for the surface*" in a pre-clinical model of implant infection. *Eur. Cells Mater.* **2020**, *39*, 77–95. [CrossRef] [PubMed]
198. Gottenbos, B.; Klatter, F.; Van Der Mei, H.C.; Busscher, H.J.; Nieuwenhuis, P. Late hematogenous infection of subcutaneous implants in rats. *Clin. Diagn. Lab. Immunol.* **2001**, *8*, 980–983. [CrossRef] [PubMed]
199. Thieme, L.; Hartung, A.; Tramm, K.; Klinger-Strobel, M.; Jandt, K.D.; Makarewicz, O.; Pletz, M.W. MBEC Versus MBIC: The lack of differentiation between biofilm reducing and inhibitory effects as a current problem in biofilm methodology. *Biol. Proced. Online* **2019**, *21*, 18. [CrossRef]
200. Hu, H.; Zhang, W.; Qiao, Y.; Jiang, X.; Liu, X.; Ding., C. Antibacterial activity and increased bone marrow stem cell functions of Zn-incorporated TiO_2 coatings on titanium. *Acta Biomater.* **2012**, *8*, 904–915. [CrossRef]
201. Huo, K.; Zhang, X.; Wang, H.; Zhao, L.; Liu, X.; Chu, P.K. Osteogenic activity and antibacterial effects on titanium surfaces modified with Zn-incorporated nanotube arrays. *Biomaterials* **2013**, *34*, 3467–3478. [CrossRef]
202. Jin, G.; Cao, H.; Qiao, Y.; Meng, F.; Zhu, H.; Liu, X. Osteogenic activity and antibacterial effect of zinc ion implanted titanium. *Colloids Surf. B Biointerfaces* **2014**, *117*, 158–165. [CrossRef]
203. Shi, Q.; Luo, X.; Huang, Z.; Midgley, A.C.; Wang, B.; Liu, R.; Zhi, D.; Wei, T.; Zhou, X.; Qiao, M.; et al. Cobalt-mediated multi-functional dressings promote bacteria-infected wound healing. *Acta Biomater.* **2019**, *86*, 465–479. [CrossRef]

204. Zhang, W.; Zhang, S.; Liu, H.; Ren, L.; Wang, Q.; Zhang, Y. Effects of surface roughening on antibacterial and osteogenic properties of Ti-Cu alloys with different Cu contents. *J. Mater. Sci. Technol.* **2021**, *88*, 158–167. [CrossRef]
205. Sun, Y.; Zhao, Y.; Zeng, Q.; Wu, Y.; Hu, Y.; Duan, S.; Tang, Z.; Xu, F. Dual-functional implants with antibacterial and osteointegration-promoting performances. *ACS Appl. Mater. Interfaces* **2019**, *11*, 36449–36457. [CrossRef]
206. Fazel, M.; Salimijazi, H.R.; Shamanian, M.; Minneboo, M.; Modaresifar, K.; van Hengel, I.A.J.; Fratila-Apachitei, L.E.; Apachitei, I.; Zadpoor, A.A. Osteogenic and antibacterial surfaces on additively manufactured porous Ti-6Al-4V implants: Combining silver nanoparticles with hydrothermally synthesized HA nanocrystals. *Mater. Sci. Eng. C* **2021**, *120*, 111745. [CrossRef]
207. Wang, Y.; Beekman, J.; Hew, J.; Jackson, S.; Issler-Fisher, A.C.; Parungao, R.; Lajevardi, S.S.; Li, Z.; Maitz, P.K.M. Burn injury: Challenges and advances in burn wound healing, infection, pain and scarring. *Adv. Drug Deliv. Rev.* **2018**, *123*, 3–17. [CrossRef]
208. Mai, B.; Jia, M.; Liu, S.; Sheng, Z.; Li, M.; Gao, Y.; Wang, X.; Liu, Q.; Wang, P. Smart hydrogel-based DVDMS/bFGF nanohybrids for antibacterial phototherapy with multiple damaging sites and accelerated wound healing. *ACS Appl. Mater. Interfaces* **2020**, *12*, 10156–10169. [CrossRef]
209. Wei, S.; Chang, L.; Huang, C.; Chang, H. Dual-functional gold nanoparticles with antimicrobial and proangiogenic activities improve the healing of multidrug-resistant bacteria-infected wounds in diabetic mice. *Biomater. Sci.* **2019**, *7*, 4482–4490. [CrossRef]
210. Wang, M.; Yang, Y.; Yuan, K.; Yang, S.; Tang, T. Dual-functional hybrid quaternized chitosan/Mg/alginate dressing with antibacterial and angiogenic potential for diabetic wound healing. *J. Orthop. Translat.* **2021**, *30*, 6–15. [CrossRef]
211. Song, G.; Atrens, A. Corrosion mechanisms of magnesium alloys. *Adv. Eng. Mater.* **1999**, *1*, 11–33. [CrossRef]
212. Deshpande, K.B. Numerical modeling of micro-galvanic corrosion. *Electrochim. Acta* **2011**, *56*, 1737–1745. [CrossRef]
213. Qin, H.; Cao, H.; Zhao, Y.; Jin, G.; Cheng, M.; Wang, J.; Jiang, Y.; An, Z.; Zhang, X.; Liu, X. Antimicrobial and osteogenic properties of silver-ion-implanted stainless steel. *ACS Appl. Mater. Interfaces* **2015**, *7*, 10785–10794. [CrossRef] [PubMed]
214. Qiao, S.; Cao, H.; Zhao, X.; Lo, H.; Zhuang, L.; Gu, Y.; Shi, J.; Liu, X.; Lai, H. Ag-plasma modification enhances bone apposition around titanium dental implants: An animal study in Labrador dogs. *Int. J. Nanomed.* **2015**, *10*, 653–664.
215. Cao, H.; Tang, K.; Liu, X. Bifunctional galvanics mediated selective toxicity on titanium. *Mater. Horiz.* **2018**, *5*, 264–267. [CrossRef]
216. Hazell, G.; May, P.W.; Taylor, P.; Nobbs, A.H.; Welch, C.C.; Su, B. Studies of black silicon and black diamond as materials for antibacterial surfaces. *Biomater. Sci.* **2018**, *6*, 1424–1432. [CrossRef] [PubMed]
217. Ivanova, E.P.; Hasan, J.; Webb, H.K.; Gervinskas, G.; Juodkazis, S.; Truong, V.K.; Wu, A.H.F.; Lamb, R.N.; Baulin, V.A.; Watson, G.S.; et al. Bactericidal activity of black silicon. *Nat. Commun.* **2013**, *4*, 2838. [CrossRef] [PubMed]
218. Roy, A.; Chatterjee, K. Bactericidal anisotropic nanostructures on titanium fabricated by maskless dry etching. *ACS Appl. Nano Mater.* **2022**, *5*, 4447–4461. [CrossRef]
219. Tang, K.; Wang, L.; Geng, H.; Qiu, J.; Cao, H.; Liu, X. Molybdenum disulfide (MoS_2) nanosheets vertically coated on titanium for disinfection in the dark. *Arab. J. Chem.* **2020**, *13*, 1612–1623. [CrossRef]
220. Luo, Q.; Cao, H.; Wang, L.; Ma, X.; Liu, X. ZnO@ZnS nanorod-array coated titanium: Good to fibroblasts but bad to bacteria. *J. Colloids Interface Sci.* **2020**, *579*, 50–60. [CrossRef]
221. Li, J.; Tan, L.; Liu, X.; Cui, Z.; Yang, X.; Yeung, K.W.K.; Chu, P.K.; Wu, S. Balancing bacteria-osteoblast competition through selective physical puncture and biofunctionalization of ZnO/Polydopamine/Arginine-Glycine-Aspartic Acid-Cysteine nanorods. *ACS Nano* **2017**, *11*, 11250–11263. [CrossRef]
222. Lüdecke, C.; Roth, M.; Yu, W.; Horn, U.; Bossert, J.; Jandt, K.D. Nanorough titanium surfaces reduce adhesion of Escherichia coli and Staphylococcus aureus via nano adhesion points. *Colloids Surf. B Biointerfaces* **2016**, *145*, 617–625. [CrossRef]
223. Dauben, T.J.; Dewald, C.; Firkowska-Boden, I.; Helbing, C.; Peisker, H.; Roth, M.; Bossert, J.; Jandt, K.D. Quantifying the relationship between surfaces' nano-contact point density and adhesion force of Candida albicans. *Colloids Surf. B Biointerfaces.* **2020**, *194*, 111177. [CrossRef]
224. Hawi, S.; Goel, S.; Kumar, V.; Pearce, O.; Ayre, W.N.; Ivanova, E.P. Critical review of nanopillar-based mechanobactericidal systems. *ACS Appl. Nano Mater.* **2022**, *5*, 1–17. [CrossRef]
225. Hasan, J.; Jain, S.; Chatterjee, K. Nanoscale topography on black titanium imparts multi-biofunctional properties for orthopedic applications. *Sci. Rep.* **2017**, *7*, 41118. [CrossRef]
226. Ganjian, M.; Modaresifar, K.; Zhang, H.; Hagedoorn, P.; Fratila-Apachitei, L.E.; Zadpoor, A.A. Reactive ion etching for fabrication of biofunctional titanium nanostructures. *Sci. Rep.* **2019**, *9*, 18815. [CrossRef]
227. Modaresifar, K.; Ganjian, M.; Angeloni, L.; Minneboo, M.; Ghatkesar, M.K.; Hagedoorn, P.; Fratila-Apachitei, L.E.; Zadpoor, A.A. On the use of black Ti as a bone substituting biomaterial: Behind the scenes of dual-functionality. *Small* **2021**, *17*, e2100706. [CrossRef]
228. Zimmerli, W.; Lew, P.D.; Waldvogel, F.A. Pathogenesis of foreign body infection. Evidence for a local granulocyte defect. *J. Clin. Invest.* **1984**, *73*, 1191–1200. [CrossRef]
229. Zimmerli, W.; Sendi, P. Pathogenesis of implant-associated infection: The role of the host. *Semin. Immunopathol.* **2011**, *33*, 295–306. [CrossRef]
230. Zimmerli, W.; Waldvogel, F.A.; Vaudaux, P.; Nydegger, U.E. Pathogenesis of foreign body infection: Description and characteristics of an animal model. *J. Infect. Dis.* **1982**, *146*, 487–497. [CrossRef]
231. Southwood, R.T.; Rice, J.L.; McDonald, P.J.; Hakendorf, P.H.; Rozenbilds, M.A. Infection in experimental hip arthroplasties. *J. Bone Joint Surg. Br.* **1985**, *67*, 229–231. [CrossRef]

232. Yavari, S.A.; Castenmiller, S.M.; van Strijp, J.A.G.; Croes, M. Combating implant infections: Shifting focus from bacteria to host. *Adv. Mater.* **2020**, *32*, e2002962. [CrossRef] [PubMed]
233. Medzhitov, R. Recognition of microorganisms and activation of the immune response. *Nature* **2007**, *449*, 819–826. [CrossRef] [PubMed]
234. Hancock, R.; Nijnik, A.; Philpott, D. Modulating immunity as a therapy for bacterial infections. *Nat. Rev. Microbiol.* **2012**, *10*, 243–254. [CrossRef] [PubMed]
235. Fang, F. Antimicrobial reactive oxygen and nitrogen species: Concepts and controversies. *Nat. Rev. Microbiol.* **2004**, *2*, 820–832. [CrossRef] [PubMed]
236. Papayannopoulos, V. Neutrophil extracellular traps in immunity and disease. *Nat. Rev. Immunol.* **2018**, *18*, 134–147. [CrossRef] [PubMed]
237. Russell, D.G.; Huang, L.; VanderVen, B.C. Immunometabolism at the interface between macrophages and pathogens. *Nat. Rev. Immunol.* **2019**, *19*, 291–304. [CrossRef]
238. André, A.C.; Laborde, M.; Marteyn, B.S. The battle for oxygen during bacterial and fungal infections. *Trends. Microbiol.* **2022**, *30*, 643–653. [CrossRef]
239. Nadzam, G.S.; De La Cruz, C.; Greco, R.S.; Haimovich, B. Neutrophil adhesion to vascular prosthetic surfaces triggers nonapoptotic cell death. *Ann. Surg.* **2000**, *231*, 587–599. [CrossRef]
240. Chang, S.; Popowich, Y.; Greco, R.S.; Haimovich, B. Neutrophil survival on biomaterials is determined by surface topography. *J. Vasc. Surg.* **2003**, *37*, 1082–1090. [CrossRef]
241. Abaricia, J.O.; Shah, A.H.; Musselman, R.M.; Olivares-Navarrete, R. Hydrophilic titanium surfaces reduce neutrophil inflammatory response and NETosis. *Biomater. Sci.* **2020**, *8*, 2289–2299. [CrossRef]
242. Erpenbeck, L.; Gruhn, A.L.; Kudryasheva, G.; Günay, G.; Meyer, D.; Busse, J.; Neubert, E.; Schön, M.P.; Rehfeldt, F.; Kruss, S. Effect of adhesion and substrate elasticity on neutrophil extracellular trap formation. *Front. Immunol.* **2019**, *10*, 2320. [CrossRef]
243. Kaplan, S.S.; Heine, R.P.; Simmons, R.L. Defensins impair phagocytic killing by neutrophils in biomaterial-related infection. *Infect. Immun.* **1999**, *67*, 1640–1645. [CrossRef]
244. Eriksson, C.; Nygren, H. Adhesion receptors of polymorphonuclear granulocytes on titanium in contact with whole blood. *J. Lab. Clin. Med.* **2001**, *137*, 56–63. [CrossRef]
245. Vitkov, L.; Krautgartner, W.; Obermayer, A.; Stoiber, W.; Hannig, M.; Klappacher, M.; Hartl, D. The initial inflammatory response to bioactive implants is characterized by NETosis. *PLoS ONE* **2015**, *10*, e0121359. [CrossRef]
246. Branzk, N.; Lubojemska, A.; Hardison, S.E.; Wang, Q.; Gutierrez, M.G.; Brown, G.D.; Papayannopoulos, V. Neutrophils sense microbe size and selectively release neutrophil extracellular traps in response to large pathogens. *Nat. Immunol.* **2014**, *15*, 1017–1025. [CrossRef] [PubMed]
247. Patel, J.D.; Krupka, T.; Anderson, J.M. iNOS-mediated generation of reactive oxygen and nitrogen species by biomaterial-adherent neutrophils. *J. Biomed. Mater. Res A* **2007**, *80*, 381–390. [CrossRef]
248. Chen, S.; Jones, J.A.; Xu, Y.; Low, H.; Anderson, J.M.; Leong, K.W. Characterization of topographical effects on macrophage behavior in a foreign body response model. *Biomaterials* **2010**, *31*, 3479–3491. [CrossRef]
249. Makaremi, S.; Luu, H.; Boyle, J.P.; Zhu, Y.; Cerson, C.; Bowdish, D.M.E.; Moran-Mirabal, J.M. The topography of silica films modulates primary macrophage morphology and function. *Adv. Mater. Interfaces* **2019**, *6*, 1900677. [CrossRef]
250. Singh, S.; Awuah, D.; Rostam, H.M.; Emes, R.D.; Kandola, N.K.; Onion, D.; Htwe, S.S.; Rajchagool, B.; Cha, B.; Kim, D.; et al. Unbiased analysis of the impact of micropatterned biomaterials on macrophage behavior provides insights beyond predefined polarization states. *ACS Biomater. Sci. Eng.* **2017**, *3*, 969–978. [CrossRef]
251. Vassey, M.J.; Figueredo, G.P.; Scurr, D.J.; Vasilevich, A.S.; Vermeulen, S.; Carlier, A.; Luckett, J.; Beijer, N.R.M.; Williams, P.; Winkler, D.A.; et al. Immune modulation by design: Using topography to control human monocyte attachment and macrophage differentiation. *Adv Sci* **2020**, *7*, 1903392. [CrossRef]
252. Hotchkiss, K.M.; Reddy, G.B.; Hyzy, S.L.; Schwartz, Z.; Boyan, B.D.; Olivares-Navarrete, R. Titanium surface characteristics, including topography and wettability, alter macrophage activation. *Acta Biomater.* **2016**, *31*, 425–434. [CrossRef] [PubMed]
253. Blakney, A.K.; Swartzlander, M.D.; Bryant, S.J. The effects of substrate stiffness on the in vitro activation of macrophages and in vivo host response to poly(ethylene glycol)-based hydrogels. *J. Biomed. Mater. Res. A* **2012**, *100*, 1375–1386. [CrossRef] [PubMed]
254. Sridharan, R.; Cavanagh, B.; Cameron, A.R.; Kelly, D.J.; O'Brien, F.J. Material stiffness influences the polarization state, function and migration mode of macrophages. *Acta Biomater.* **2019**, *89*, 47–59. [CrossRef] [PubMed]
255. Previtera, M.L.; Sengupta, A. Substrate stiffness regulates proinflammatory mediator production through tlr4 activity in macrophages. *PLoS ONE* **2015**, *10*, e0145813. [CrossRef]
256. Croes, M.; Bakhshandeh, S.; van Hengel, I.A.J.; Lietaert, K.; van Kessel, K.P.M.; Pouran, B.; van der Wal, B.C.H.; Vogely, H.C.; Van Hecke, W.; Fluit, A.C.; et al. Antibacterial and immunogenic behavior of silver coatings on additively manufactured porous titanium. *Acta Biomater.* **2018**, *81*, 315–327. [CrossRef]
257. Liz, R.; Simard, J.; Leonardi, L.B.A.; Girard, D. Silver nanoparticles rapidly induce atypical human neutrophil cell death by a process involving inflammatory caspases and reactive oxygen species and induce neutrophil extracellular traps release upon cell adhesion. *Int. Immunopharmacol.* **2015**, *28*, 616–625. [CrossRef]

258. Huang, M.; Ye, K.; Hu, T.; Liu, K.; You, M.; Wang, L.; Qin, H. Silver nanoparticles attenuate the antimicrobial activity of the innate immune system by inhibiting neutrophil-mediated phagocytosis and reactive oxygen species production. *Int. J. Nanomed.* **2021**, *16*, 1345–1360. [CrossRef]
259. Haase, H.; Fahmi, A.; Mahltig, B. Impact of silver nanoparticles and silver ions on innate immune cells. *J. Biomed. Nanotechnol.* **2014**, *10*, 1146–1156. [CrossRef]
260. Nishanth, R.P.; Jyotsna, R.G.; Schlager, J.J.; Hussain, S.M.; Reddanna, P. Inflammatory responses of RAW 264.7 macrophages upon exposure to nanoparticles: Role of ROS-NFκB signaling pathway. *Nanotoxicology* **2011**, *5*, 502–516. [CrossRef]
261. Kim, B.; Lee, W. Regulatory role of zinc in immune cell signaling. *Mol. Cells* **2021**, *44*, 335–341. [CrossRef]
262. Wang, J.; Zhou, H.; Guo, G.; Tan, J.; Wang, Q.; Tang, J.; Liu, W.; Shen, H.; Li, J.; Zhang, X. Enhanced anti-infective efficacy of zno nanoreservoirs through a combination of intrinsic anti-biofilm activity and reinforced innate defense. *ACS Appl. Mater. Interfaces* **2017**, *9*, 33609–33623. [CrossRef]
263. Huang, Q.; Li, X.; Elkhooly, T.A.; Liu, X.; Zhang, R.; Wu, H.; Feng, Q.; Li, Y. The Cu-containing TiO_2 coatings with modulatory effects on macrophage polarization and bactericidal capacity prepared by micro-arc oxidation on titanium substrates. *Colloids Surf. B Biointerfaces* **2018**, *170*, 242–250. [CrossRef]
264. Liu, W.; Li, J.; Cheng, M.; Wang, Q.; Qian, Y.; Yeung, K.W.K.; Chu, P.K.; Zhang, X. A surface-engineered polyetheretherketone biomaterial implant with direct and immunoregulatory antibacterial activity against methicillin-resistant Staphylococcus aureus. *Biomaterials* **2019**, *208*, 8–20. [CrossRef]
265. Lemire, J.; Harrison, J.; Turner, R. Antimicrobial activity of metals: Mechanisms, molecular targets and applications. *Nat. Rev. Microbiol.* **2013**, *11*, 371–384. [CrossRef]
266. Bussière, F.I.; Gueux, E.; Rock, E.; Girardeau, J.; Tridon, A.; Mazur, A.; Rayssiguier, Y. Increased phagocytosis and production of reactive oxygen species by neutrophils during magnesium deficiency in rats and inhibition by high magnesium concentration. *Br. J. Nutr.* **2002**, *87*, 107–113. [CrossRef]
267. Hann, J.; Bueb, J.-L.; Tolle, F.; Bréchard, S. Calcium signaling and regulation of neutrophil functions: Still a long way to go. *J. Leukoc. Biol.* **2020**, *107*, 285–297. [CrossRef]
268. Cao, H.; Qin, H.; Zhao, Y.; Jin, G.; Lu, T.; Meng, F.; Zhang, X.; Liu, X. Nano-thick calcium oxide armed titanium: Boosts bone cells against methicillin-resistant Staphylococcus aureus. *Sci. Rep.* **2016**, *6*, 21761. [CrossRef]
269. Hou, Y.; Witte, F.; Li, J.; Guan, S. The increased ratio of Mg^{2+}/Ca^{2+} from degrading magnesium alloys directs macrophage fate for functionalized growth of endothelial cells. *Smart Mater. Med.* **2022**, 3, 188–198. [CrossRef]
270. Iseri, L.T.; French, J.H. Magnesium: Nature's physiologic calcium blocker. *Am. Heart J.* **1984**, *108*, 188–193. [CrossRef]
271. Cao, H.; Liu, X.; Jandt, K.D. The action-network of nanomaterials *in* On the issue of transparency and reproducibility in nanomedicine. *Nat. Nanotechnol.* **2019**, *14*, 629–635.
272. Damiati, L.A.; Tsimbouri, M.P.; Hernandez, V.; Jayawarna, V.; Ginty, M.; Childs, P.; Xiao, Y.; Burgess, K.; Wells, J.; Sprott, M.R.; et al. Materials-driven fibronectin assembly on nanoscale topography enhances mesenchymal stem cell adhesion, protecting cells from bacterial virulence factors and preventing biofilm formation. *Biomaterials* **2022**, *280*, 121263. [CrossRef] [PubMed]
273. Yuan, B.; Chen, H.; Zhao, R.; Deng, X.; Chen, G.; Yang, X.; Xiao, Z.; Aurora, A.; Iulia, B.A.; Zhang, K.; et al. Construction of a magnesium hydroxide/graphene oxide/hydroxyapatite composite coating on Mg–Ca–Zn–Ag alloy to inhibit bacterial infection and promote bone regeneration. *Bioact. Mater.* **2022**, *18*, 354–367. [CrossRef] [PubMed]
274. Liu, Z.; Yi, Y.; Song, L.; Chen, Y.; Tian, L.; Zhao, J.; Ren, L. Biocompatible mechano-bactericidal nanopatterned surfaces with salt-responsive bacterial release. *Acta Biomater.* **2022**, *141*, 198–208. [CrossRef] [PubMed]
275. Ding, Q.; Sun, T.; Su, W.; Jing, X.; Ye, B.; Su, Y.; Zeng, L.; Qu, Y.; Yang, X.; Wu, Y.; et al. Bioinspired multifunctional black phosphorus hydrogel with antibacterial and antioxidant properties: A stepwise countermeasure for diabetic skin wound healing. *Adv. Healthc. Mater.* **2022**, *11*, 2102791. [CrossRef]
276. Yang, M.; Qiu, S.; Coy, E.; Li, S.; Załęski, K.; Zhang, Y.; Pan, H.; Wang, G. NIR-Responsive TiO_2 Biometasurfaces: Toward in situ photodynamic antibacterial therapy for biomedical implants. *Adv. Mater.* **2022**, *34*, 2106314. [CrossRef]
277. Qu, X.; Wang, M.; Wang, M.; Tang, H.; Zhang, S.; Yang, H.; Yuan, W.; Wang, Y.; Yang, J.; Yue, B. Multi-mode antibacterial strategies enabled by gene-transfection and immunomodulatory nanoparticles in 3D-printed scaffolds for synergistic exogenous and endogenous treatment of infections. *Adv. Mater.* **2022**, *34*, 2200096. [CrossRef]
278. Zhang, Z.; Wang, Y.; Teng, W.; Zhou, X.; Ye, Y.; Zhou, H.; Sun, H.; Wang, F.; Liu, A.; Lin, P.; et al. An orthobiologics-free strategy for synergistic photocatalytic antibacterial and osseointegration. *Biomaterials* **2021**, *274*, 120853. [CrossRef]
279. Williams, D.F. There is no such thing as a biocompatible material. *Biomaterials* **2014**, *35*, 10009–10014. [CrossRef]
280. Sun, B.B.; Maranville, J.C.; Peters, J.E.; Stacey, D.; Staley, J.R.; Blackshaw, J.; Burgess, S.; Jiang, T.; Paige, E.; Surendran, P.; et al. Genomic atlas of the human plasma proteome. *Nature* **2018**, *558*, 73–79. [CrossRef]
281. Hu, W.J.; Eaton, J.W.; Ugarova, T.P.; Tang, L. Molecular basis of biomaterial-mediated foreign body reactions. *Blood* **2001**, *98*, 1231–1238. [CrossRef]
282. Takahashi, J.; Ebara, S.; Kamimura, M.; Kinoshita, T.; Misawa, H.; Shimogata, M.; Tozuka, M.; Takaoka, K. Pro-inflammatory and anti-inflammatory cytokine increases after spinal instrumentation surgery. *J. Spinal Disord. Tech.* **2002**, *15*, 294–300. [CrossRef] [PubMed]

Journal of *Functional Biomaterials*

MDPI

Review

Immunomodulation Effect of Biomaterials on Bone Formation

Tong Zhao [1,2,†], Zhuangzhuang Chu [1,2,†], Jun Ma [3,*] and Liping Ouyang [1,*]

1 Hongqiao International Institute of Medicine, Tongren Hospital, Shanghai Jiao Tong University School of Medicine, Shanghai 200336, China; tongzhao1218@163.com (T.Z.); czzlylotus@163.com (Z.C.)
2 Jiangsu Key Laboratory of Oral Diseases, Nanjing Medical University, Nanjing 210029, China
3 Department of General Practitioners, Tongren Hospital, Shanghai Jiao Tong University School of Medicine, Shanghai 200336, China
* Correspondence: lpouyang@shsmu.edu.cn (L.O.); jma@shsmu.edu.cn (J.M.); Tel.: +86-21-52039999 (L.O.); +86-21-52039999 (J.M.)
† The authors contributed equally to this work.

Abstract: Traditional bone replacement materials have been developed with the goal of directing the osteogenesis of osteoblastic cell lines toward differentiation and therefore achieving biomaterial-mediated osteogenesis, but the osteogenic effect has been disappointing. With advances in bone biology, it has been revealed that the local immune microenvironment has an important role in regulating the bone formation process. According to the bone immunology hypothesis, the immune system and the skeletal system are inextricably linked, with many cytokines and regulatory factors in common, and immune cells play an essential role in bone-related physiopathological processes. This review combines advances in bone immunology with biomaterial immunomodulatory properties to provide an overview of biomaterials-mediated immune responses to regulate bone regeneration, as well as methods to assess the bone immunomodulatory properties of bone biomaterials and how these strategies can be used for future bone tissue engineering applications.

Keywords: osteoimmunomodulation; bone regeneration; macrophages; biomaterials

Citation: Zhao, T.; Chu, Z.; Ma, J.; Ouyang, L. Immunomodulation Effect of Biomaterials on Bone Formation. *J. Funct. Biomater.* **2022**, *13*, 103. https://doi.org/10.3390/jfb13030103

Academic Editor: Masakazu Kawashita

Received: 26 June 2022
Accepted: 22 July 2022
Published: 25 July 2022

Publisher's Note: MDPI stays neutral with regard to jurisdictional claims in published maps and institutional affiliations.

Copyright: © 2022 by the authors. Licensee MDPI, Basel, Switzerland. This article is an open access article distributed under the terms and conditions of the Creative Commons Attribution (CC BY) license (https://creativecommons.org/licenses/by/4.0/).

1. Introduction

Bone loss in many situations, including aging, pathological fracture, periodontitis, and osteomyelitis, can lead to poor physical conditions, most of which require bone replenishment and timely surgical repair using implantation materials [1–3]. Nearly all dominant bone biomaterials have been developed and produced following the principle of great physicochemical properties and biocompatibility. Using this principle, candidate materials are usually subjected to the in vitro simulation of osteogenesis, which is intrinsically driven by osteoblastic activities in vivo [4,5]. In addition, owing to the advancement in material science, the fabrication of implants can meet the practical demands of patients.

Nevertheless, owing to the inconsistent outcomes of candidate materials in vitro and in vivo, it is not easy for these existing biomaterials to transform into clinically applicable implant materials in the human body. The reasons are that on the one hand, the physicochemical, biological, and mechanical properties of the candidate materials must certainly be optimized to meet specific demands in our complexed organic environment. On the other hand, as we obtain an increasingly profound understanding of bone biology and the underlying mechanisms of osteogenesis, we understand that the musculoskeletal system is not the sole contributor to this process. Osteogenesis is a composite bioprocess that collaborates with diverse molecular events. These concerns cause limitations in the dependence and efficiency of traditional bone biomaterials for practical purposes [6,7], and novel insights into the fabrication principle and operating mechanisms of bone biomaterials are urgently needed.

Osteoimmunology, an emerging theory derived from the latest findings in bone biology, hypothesizes that immune responses play essential roles in bone formation and homeostasis. Many functional proteins, signaling molecules, and cytokines have been confirmed to participate synchronously in reactive immune events and osteogenesis. Thus, immune signaling pathways are closely related to bone formation [8,9]. An immune response is defined as the body's defense against foreign matter or mutated autologous components. Triggered by the recognition of antigens, the entire process can generate an immune response to an antigen, including immune induction by antigens to the body, interactions between immune cells, and multiple effects mediated by immune effectors (e.g., sensitized lymphocytes, antibodies) [10]. Naturally recognized as a foreign body in this environment, the implant could provoke a series of adverse and even fatal events derived from the immune system, which can determine the destiny of bone biomaterials. In this type of situation, the irrevocable damage caused by various immune effector cells significantly changes the physicochemical and biological properties that are meant to help with bone formation in vivo, and the unexpected disturbance makes it difficult for biomaterials to work successfully in vitro. For example, the inappropriate immune responses to foreign implants which elicit excessive inflammation could lead to the formation of a fibrous capsule. The immune-related alterations significantly weaken the osteogenesis capability of biomaterials by preventing them from contacting and coordinating with bone cells. To reverse the impact, Chen and colleagues attached an SZS coating to Ti–6Al–4V and endowed the implant with immunomodulatory properties, considerably alleviating incorrect immune responses and favoring osteogenesis [11]. The immune system plays a central role in coordinating the repair and regeneration of damaged tissue after infection or injury. In the inflammatory phase, the regulatory function of the immune response in the bone healing, repair, and regeneration induced by biomaterials has been demonstrated [12]. Biomaterial-induced immunomodulation can provide space for osteoblast growth and maturation during the repair and remodeling phase. Thus, the theory of osteoimmunology has eventually emerged.

The pivotal advantage of osteoimmunology is that we can grant bone material-specific biological properties to artificially modulate the local immune environment so that it inversely favors the process of osteointegration of the implant and osteogenesis [13]. To fulfill this key goal, there has been an impressive evolution in the design of principles and manufacturing criteria. To date, osteoimmunology has shown that immune cells actively participate in bone pathophysiology through the release of regulatory molecules (e.g., BMP2, BMP6, VEGF, OSM, and RANKL) to exert imperative effects on osteogenesis, and the dysfunction of these cells usually leads to an imbalance between osteoclasts and osteoblasts, which results in subsequent osteoarthritis, osteolysis, and osteoporosis. Chen et al. first proposed the concept of osteoimmunomodulation (OIM, please refer to Table 1 for the abbreviations covered in this manuscript) with the interpretation of a novel favorable property of bone biomaterials to induce a beneficial immune environment for osteogenesis [14]. In the present review, we discuss the research actualities of osteogenesis and dominant bone biomaterials and their advantages and disadvantages. We will also introduce the latest achievements in the field of osteoimmunology and OIM to provide novel insights into the design and production of bone biomaterials.

Table 1. Abbreviations in this review.

Abbreviation	Full Name
OIM	osteoimmunomodulation
BMSC	bone marrow mesenchymal stem cell
RUNX	runt-related transcription factor
Osx	osterix
ALP	alkaline phosphatase
Ocn	osteocalcin
M-CSF	macrophage colony stimulating factor
NF-κB	nuclear factor kappa B

Table 1. *Cont.*

Abbreviation	Full Name
RANKL	receptor activator of nuclear factor kappa B (NF-κB) ligand
M1	classically activated macrophage
M2	alternatively activated macrophage
IL-1	interleukin 1
IL-4	interleukin 4
IL-6	interleukin 6
IL-10	interleukin 10
TNF-α	tumor necrosis factor α
TGF-β	transforming growth factor β
MMP	matrix metalloproteinases
BMP2	bone morphogenetic protein 2
IFN-γ	release interferon γ
TRAF6	TNF receptor-associated factor 6
DC	dendritic cell
COX-2	cyclooxygenase 2
Th1 cell	T helper 1 cell
Th2 cell	T helper 2 cell
Th17 cell	T helper 17 cell
Treg cell	regulatory T cells
CaSR	calcium sensing receptor
TLR	toll-like receptor
IRAK1	interleukin-1 receptor-associated kinase 1
OSM	oncostatin M
PGE2	prostaglandin E2
BMP2-CPC	BMP2-modified calcium phosphate cement
VEGF	vascular endothelial growth factor
ROS	reactive oxygen species

2. Bone and Bone Cells

Similar to many other connective tissues in the body, bone tissues are composed of three basic components: cells, fibers, and matrix. However, the most notable feature of bone is the deposition of a large amount of calcium salt in the cellular matrix, which grants it solidness to form the skeletal system in the body [15]. Four types of differentiated bone cells exist: osteoblasts, osteocytes, bone lining cells, and osteoclasts. They are responsible for the formation, composition, and degradation of bones. Notably, osteoblasts and osteoclasts are crucial for bone modeling and remodeling in the bone microenvironment, which is pivotal for maintaining bone homeostasis [16].

Modeling and remodeling occur constantly in bone tissues, which correspond to the decomposition and absorption of decrepit bone substance and the formation of new bone substance [17,18]. Osteoclasts are responsible for the removal of mineralized bone, and osteoblasts are responsible for the formation of the bone matrix and mineralization. There are three consecutive phases in the remodeling cycle: resorption, reversal, and formation. In the resorption phase, hematopoietic cells migrate to remodeling sites, where they differentiate into mature osteoclasts to digest old bones. The quiescent bone surface is covered by bone lining cells. In the reversal phase, when monocytes appear on the bone surface, they couple bone resorption to bone formation by generating an osteogenic environment at the remodeling sites. In the formation phase, mature osteoblasts predominate and constantly form new bone [19,20]. Bone modeling is one of the most important events in the skeletal system because it is responsible for altering and adjusting bone structure to meet changing mechanical needs, as well as assisting with the repair of microdamage in the bone matrix to ensure timely bone metabolism and to prevent severe pathological problems [21].

Osteoblasts are responsible for the synthesis, secretion, and mineralization of the bone matrix [22]. Bone formation starts with the maturation of osteoblasts. During preosteoblasts differentiation, they gradually transform from mesenchymal stem cells in the bone marrow (BMSCs) to mature osteoblasts [23,24]. Runx2 of the runt-related transcription factor (RUNX) family is the central transcriptional factor responsible for regulating the process of cellular differentiation in MSCs [25]. To exert transcriptional effects on downstream genes, Runx2 tends to heterodimerize with Cbfa1, granting itself a greater DNA-binding ability and stability [26]. The dominant role of Runx2 can generally be described as inducing the commitment of MSCs to the osteogenic lineage, promoting the proliferation of osteoblast progenitors, and facilitating bone mineralization by stimulating osteoblast differentiation. Runx2 has been confirmed to be solidly expressed in free MSCs, and an increasingly elevated expression of Runx2 has been observed alongside the osteogenic lineage. Specifically, Runx2 expression is the highest in immature osteoblasts, second highest in preosteoblasts, and lowest in mature osteoblasts [27,28]. This expression pattern could be considered an indication of Runx2 functioning in osteoblast differentiation and generation. In addition to Runx2, several downstream genes play vital roles in osteoblast differentiation. Osterix (Osx), identified as a novel zinc finger-containing transcription factor, is an indispensable molecule that participates in bone formation. Unlike Runx2, Osx displays a consistent expression pattern in all developing stages of bone formation, and no cortical bone or bone trabecula can be formed in the absence of Osx. This suggests that Osx is essential for osteogenesis. Moreover, it has also been shown that Osx potentially acts as a downstream target of Runx2 because Runx2/Cbfa1 expression is not affected by Osx, whereas Runx2/Cbfa1 is necessary for Osx expression in osteogenic cells [29]. Furthermore, Runx2 actively interacts with other factors to upregulate associated osteoblast differentiation markers such as alkaline phosphatase (ALP) and osteocalcin (Ocn). Runx2 functions not only in the process of osteoblast differentiation but also in bone formation-related events. Type I collagen is the main component of bone tissue, and the two chains constituting this protein are encoded by Col1a1 and Col1a2. On the one hand, it has been validated that Col1a1 is transcriptionally activated by Runx2 [30]. In contrast, Col1a1 expression has been reported to have no effect on Runx2 expression [31]. Therefore, further investigation is required to validate the association between Runx2 and collagen production. Once they grow into mature osteoblasts, the progenitors obtain the phenotype and morphology of osteoblasts, with an observed prominent structure of rough endoplasmic reticulum and active Golgi apparatus, and locate themselves on the surface of the bones.

Osteoclasts are multinucleated giant cells that incorporate several monocytes that differentiate from the monocyte-macrophage lineage of hematopoietic stem cells [32]. Proliferative monocytes/macrophages (i.e., preosteoclasts) enter the blood circulation under the chemotaxis of multiple chemical factors and fuse into multinucleated giant cells driven by various transcription factors, cytokines, and other signaling factors (e.g., such as macrophage colony stimulating factor (M-CSF) and the receptor activator of the nuclear factor kappa B (NF-κB) ligand (RANKL)), eventually growing into osteoclasts. Osteoclasts are well known for their major functions in bone resorption [33–36]. Their cytoplasm contains a well-developed endoplasmic reticulum and Golgi apparatus, as well as many actively operating mitochondria and lysosomes. Another hallmark of osteoclasts is the elevated expression of tartrate-resistant acid phosphatase, matrix metalloproteinases (MMPs), and cathepsin K in cells, which are helpful for breaking down organic matrix proteins to serve the function of bone absorption [37,38].

3. Immune Cells Regulation of Osteogenesis

In addition to MSCs, there are a large number of immune cells in bone marrow, such as B cells, T cells, monocytes, and macrophages, which account for approximately 20% of the total cells in bone marrow [39]. The immune system is a powerful and diversified defensive weapon used by higher organisms to protect themselves from foreign threats and to maintain physiological homeostasis. The primary role of the immune system is to fight infections, repair damaged tissue, and restore equilibrium in the body [40]. At the most fundamental level, the human immune system can be divided into two interrelated branches: the innate and adaptive immune systems. The innate immune system is the initial line of protection for the body and is capable of producing a non-specific immune response without prior programming when in contact with a foreign material or wounded tissue [41,42]. Unlike innate immunity, the adaptive immune system is composed of lymphocytes (B and T cells) that can recognize specific antigens. After an initial contact, the antibody is programmed to react uniquely to the antigen. This process slows down the adaptive immune system in comparison to the innate immune system, but it increases its precision and creates a crucial "immunological memory" by storing early antigens for years [43].

3.1. Macrophage Responses in Bone Regeneration

Immune cells play a significant role in bone physiology and diseases by producing regulatory chemicals that influence osteogenesis. Macrophages are among the most important immune cells [44]. They play a crucial role in the immunological and inflammatory responses induced by biomaterials in the long term. They react to the debris of dead cells as well as external infections, which prioritize phagocytosis in the immune response [45]. It is well-known that macrophages possess a high-level intrinsic plasticity and a flexible polarizable activity into M1 and M2 subtypes (Figure 1A). These two subtypes of macrophages are classically distinguished based on diverse functional features, surface markers, and inducers [46,47]. Generally, well-recognized surface markers of M1 macrophages include CCR7, CXCL9, 10, and 11, CD86, NOS2, and others, and for the M2 phenotype include CD206, CD163, CD280, Dectin-1, Arg1, and others [48,49]. Functionally, macrophages play an indispensable role in the innate immune response of the human body, which is a pivotal part of the host defense. Despite the vague boundary of identification between M1 and M2 macrophages owing to the continuous properties of certain macrophages, these two subtypes could exert distinct effects on the processes of inflammation and the immune response [50].

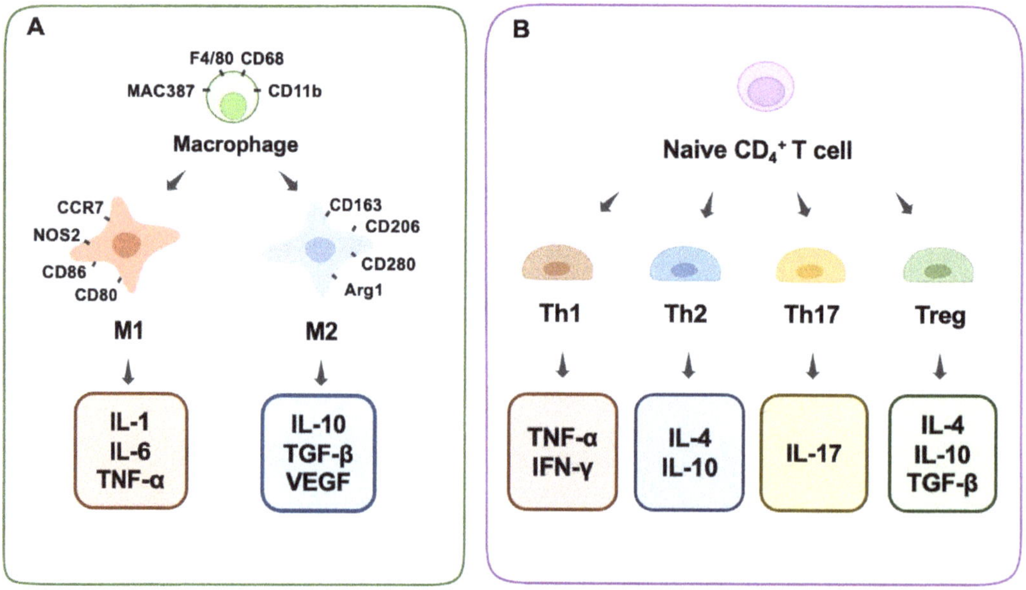

Figure 1. Macrophage phenotype and T-cell differentiation. (**A**) Macrophage phenotype, surface antibodies, and secreted cytokines. (**B**) T-cell differentiation and secreted cytokines.

Briefly, M1 macrophages are favorable for inflammatory responses and cytotoxic events in inflammation (designated as pro-inflammatory macrophages), whereas M2 macrophages tend to suppress inflammatory reactions and promote tissue repair (designated as anti-inflammatory macrophages). Classically activated inflammatory macrophages (M1) are recruited rapidly after tissue injury and participate in the early immune response, where they phagocytize pathogens and foreign infectious substances at the wound site. When M1 macrophages encounter infections, they secrete chemokines that attract other immune system cells (such as CD4+ T cells, CD8+ T cells, and dendritic cells), which transform structures from consumed cells to lymphocytes and other mature immune cells (antigen presentation) to cooperate in the phagocytosis process [51]. In addition, M1 promotes inflammation by secreting several cytokines, such as IL-1, IL-6, IL-17, IFN-γ, TNF-α, and TGF-β. In addition to serving as chemokines to boost the elimination of foreign matter during inflammation, these cytokines unintentionally harm normal tissues in several situations [52]. Due to the aggressive nature and fundamental function of the M1 phenotype in inflammation, it is impossible to ignore its important role in bone biology. Increasing evidence has shown that M1 macrophage-related inflammatory cytokines play a role in osteogenesis and bone-healing processes to some extent [53]. On the one hand, the various composition of inflammatory cytokines derived from M1 could have a diverse impact on the local osteogenesis process. Traditional recognition involves the collaboration of multiple chemokines, such as IL-17, IFN-γ, TNF-α, and TGF-β, which can induce the production of a mineralized matrix [54]. In contrast, TNF-α alone stimulates osteoclastogenesis and increases osteoclastic activity, resulting in bone resorption [55]. On the other hand, the partial enhancement of inflammation and osteogenetic reactions at a tissue level does not mean the enhancement of systemic bone formation at an organic level. In other words, the dual function of M1 macrophages in the immune response and inflammation has an undetermined role in osteogenesis and bone modeling within different situations. Generally, the robust pro-inflammation role of M1 macrophages decides the destroyed destiny of substantial bone tissues through a systematic reaction. Moreover, several recent studies hold the common view that, rather than M2 macrophages, moderately activated M1 macrophages are favorable for the osteogenic differentiation of mesenchymal stem cells (MSCs) through the mediation of OSM or bone morphogenetic protein 2 (BMP2) [56,57].

M2 macrophages, including the M2a, M2b, and M2c sub-categories, are generally considered to be responsible for tissue repair, in the form of a fibrocapsule or the formation of new bone, by alleviating the inflammation induced by the M1 phenotype [58]. M2a macrophages, produced by the activation of the cytokines IL-4 or IL-13, can suppress the secretion of several pro-inflammatory factors, such as IL-1β, IL-6, and TNF-α [58].M2b macrophages, also known as regulatory macrophages, secrete high levels of IL-10 and low levels of level of IL-12 [59]. M2c macrophages, produced by the activation of IL-10, glucocorticoids, or TGF-β, reduce the expression of various pro-inflammatory factors and enhance the ability to clear cellular debris [58]. The suppressive cytokines that M2 makes use of to resist inflammation mainly include IL-1RA, IL-10, and TGF-β, among which IL-10 and IL-1RA are known to promote osteogenesis [60]. IL-10 also inhibits osteoclast function [61]. In the repair process, M2 tends to increase multiple transforming growth factors (TGF-β1 and TGF-β3) to alleviate inflammation and facilitate fibrosis. Normally, this physiological process proceeds by the hallmark of fibrocapsule formation, which protects the normal tissue from the inflammatory microenvironment and ends in the hallmark of new bone construction [62]. It is worth mentioning that instead of a separate existence of the M1 or M2 phenotype, the macrophages continuously transform from the M1 stage to the M2 stage, and they successively perform specific duties in the whole wound-healing process. Therefore, the dynamic switching pattern highlights the significance of a proper switch from an M1 to an M2 phenotype during osteogenesis [63,64]. In other words, the appropriate proportion of the M1 and M2 stages of inflammation determines the outcome of bone regeneration. It could be concluded that any delayed switch from an M1 to an M2 phenotype would result in an inactivated M2 phenotype and poor tissue repair, and an early or excessive switch is bound to cause a deficient inflammatory reaction and delayed wound healing. Normally, the switch from the M1 to M2 phenotype could be recognized through the kinetics of gene level, surface marker expression, and protein secretion. The situation of the dominant M2 phenotype is associated with an increased expression of the M2 surface marker CD206, and a decreased expression of the M1 gene markers CCR7, IL-1β, and TNF-α, as well as an elevated level of M2-secreted proteins such as CCL18 and PDGF-BB [64].Based on a solid understanding of macrophage function, macrophage polarization and modulation could, therefore, be manipulated by designed biomaterials to fulfill specific bone-engineering events, such as osteogenesis and osteoclastogenesis [65].

3.2. T-Cell Responses in Bone Regeneration

The primary components of the adaptive immune system are lymphocytes, including B and T cells. Activated T cells can surface-express the receptor activator of nuclear factor ligand (RANKL) to stimulate osteoclast production and bone resorption. RANKL binds to the receptor activator of nuclear factor-κβ (RANK) on the surface of pro-osteoblasts, activating the RANKL/RANK signaling pathway and directly promoting osteoclast formation and differentiation through the RANKL/RANK/OPG response axis [66,67]. However, T cells can also release interferon-γ (IFN-γ) to prevent osteoclast formation, thereby interfering with TNF receptor-associated factor 6 (TRAF6), a crucial player of the RANK/RANKL signaling pathway, to block the activation of this signaling pathway and inhibit osteoclastogenesis [68]. Additionally, by interacting with bone marrow dendritic cells (DCs), CD4+ T cells are essential for the bone immune milieu by transforming into osteoclasts via the RANK/RANKL pathway [69].

Th1 cells secrete IFN-γ, IL-2, and TNF-α, which activate macrophages and promote inflammation. In contrast, Th2 cells generate IL-4, IL-10, and IL-13, which suppress macrophage function [70]. Studies have shown that some biological materials can enhance the production of cyclooxygenase-2 (COX-2) in the body after implantation, which in turn induces the release of prostaglandins (PGE2), which are chemicals that regulate inflammation [71]. PGE2 limits the growth and activity of T helper 17 cells by blocking the synthesis of IL-12p70 (Th17). In contrast, PGE2 increases Th17 cell growth and IL-17 production by increasing IL-23 production, which ultimately results in inflammation and tissue damage [71]. In addition, research has shown that tissue-derived biomaterials can potently modulate the expression of Tbx21 and Th1 canonical genes while simultaneously enhancing Th2 expression, which is important in the restoration of functioning tissues [72]. Treg cells constitute less than 10% of peripheral CD4+ T cells. They inhibit a broad spectrum of immune cells and prevent excessive immunological responses. Treg cells can inhibit osteoclast formation by producing IL-4 and IL-10 [73]. We conclude that decreasing Th1/Th17 cells while increasing Th2 cells is beneficial for tissue repair and osteogenic differentiation. Furthermore, increased Treg cells can suppress osteoclastogenesis by secreting cytokines that impede osteoclast differentiation (Figure 1B).

The body's immunological effects can be tuned in a complex and delicate balance by transforming particular CD4+ T cells (such as Th1, Th2, Treg, and Th17) into each other. Therefore, the utilization of biomaterials in regulating balance is crucial for demonstrating the efficacy of immunomodulatory treatments in wound-healing and bone-regeneration processes [72].

3.3. Other Immune Cells Responses in Bone Regeneration

Neutrophils, the initial line of defense of the innate immune system, are capable of rapidly recruiting at areas of infection or tissue damage to eliminate pathogens and clear away debris. Their overactivity, which is initiated by infection or damage, generates a large elevation of pro-inflammatory cytokines and ultimately results in tissue destruction [74]. However, a growing number of studies have indicated that neutrophils perform additional functions. They can actively coordinate the regression of inflammation and help in tissue repair by interacting with cells of the innate and adaptive immune systems to modulate the immune response [75,76]. It was shown that IL-8, which is usually regarded as the most potent neutrophil chemotactic factor [77], is secreted at the location of bone abnormalities after bone injury. Neutrophils initially arrive at the site of the defect and recruit BMSCs and macrophages. Macrophages then regulate the differentiation of BMSCs towards chondrogenesis and osteogenesis [76]. At various stages of bone regeneration, neutrophils resemble macrophages and can polarize into N1 (pro-inflammatory) or N2 (anti-inflammatory) phenotypes at different periods of the inflammatory milieu [78] and then mediate immune responses or tissue healing accordingly. In the early stage of inflammation, high levels of IL-8 produce a pro-inflammatory milieu in which neutrophils are the pro-inflammatory N1 subtype. N1 recruits other types of immune cells (such as M1 macrophages, Th1 cells, and Th17 cells), aggressively reduces inflammation, and prepares the milieu for bone regeneration. After the inflammation subsides, IL-8 levels decrease and the neutrophils recruited during this phase are the N2 subtype that express anti-inflammatory factors which facilitate bone regeneration. Kovtun et al. found that the removal of neutrophils led to hampered healing after fracture in a mouse fracture model [79].

B lymphocytes (B cells) regulate bone formation and have a significant role in the risk of bone metabolism disruption [80]. B lymphocytes' impacts on bone cells are mediated by cytokines and molecular pathways that have important effects on both immune cells and bone cell function, such as the RANKL/RANK/OPG signaling pathways [81,82]. Under normal circumstances, B cell-derived RANKL is required for B cells development; nonetheless, the overexpression of RANKL by activated B lymphocytes can have major consequences for bone metabolism [83–85]. In addition to RANKL/OPG, B cells can regulate bone homeostasis by generating several cytokines and chemokines, including TNF-α, TNF-β, IL-6, IL-10, and CCL3, which can modulate bone modeling and bone remodeling by acting directly on bone cells and modulating the immune microenvironment [86,87]. LPS-treated B lymphocytes have been shown to suppress the osteogenic function of rat bone marrow stromal cells via activating the Notch signaling system [88]. Furthermore, B cells have been shown to be able to differentiate into osteoblasts in vitro in the presence of 1,25(OH)2 vitamin D3 and ST2 stromal cells, or M-CSF and RANKL, thereby regulating bone metabolic processes [89–91].

3.4. Synergetic Regulation of Immune Cells in Bone Regeneration

Macrophages are the first biological response to allogeneic biomaterials. These extremely flexible immune sentinels govern and modulate the response to foreign and natural elements [92]. The destruction of the innate immune defense line causes macrophages to present antigen information to T cells through adaptive immunity, leading to the differentiation of T cells into Th1 and Th17 cells. IL-1, IL-6, and TNF-α are secreted by M1 leading to local inflammation. This stimulates Th1 cells to release pro-inflammatory cytokines (e.g., TNF-α, TNF-β, and IFN-γ), leading to a Th1-type inflammatory response. M2 macrophages secrete VEGF and TGF-β to help Th2 cells release cytokines such as IL-4, IL-6, IL-10, and IL-13, resulting in the generation of Th2 cells with anti-inflammatory properties that aid in tissue healing [93]. Conversely, T cells are also required for the functional polarization of M0 macrophages to the pro-inflammatory M1 or anti-inflammatory M2 type [94]. In addition, neutrophils recruit Th17 cells that can release the pro-inflammatory factor IL-17 to inflammatory areas by releasing CCL2 and CCL20 [95]. In turn, the pro-inflammatory cytokine IL-17 encourages epithelial cells to release CXC chemokines, hence enhancing neutrophil recruitment and activation [96]. In summary, when employing biomaterials to repair damaged tissues, it is critical to regulate the immune response and reveal the activation process of immune cells, notably the interplay between T cells and macrophages (Figure 2).

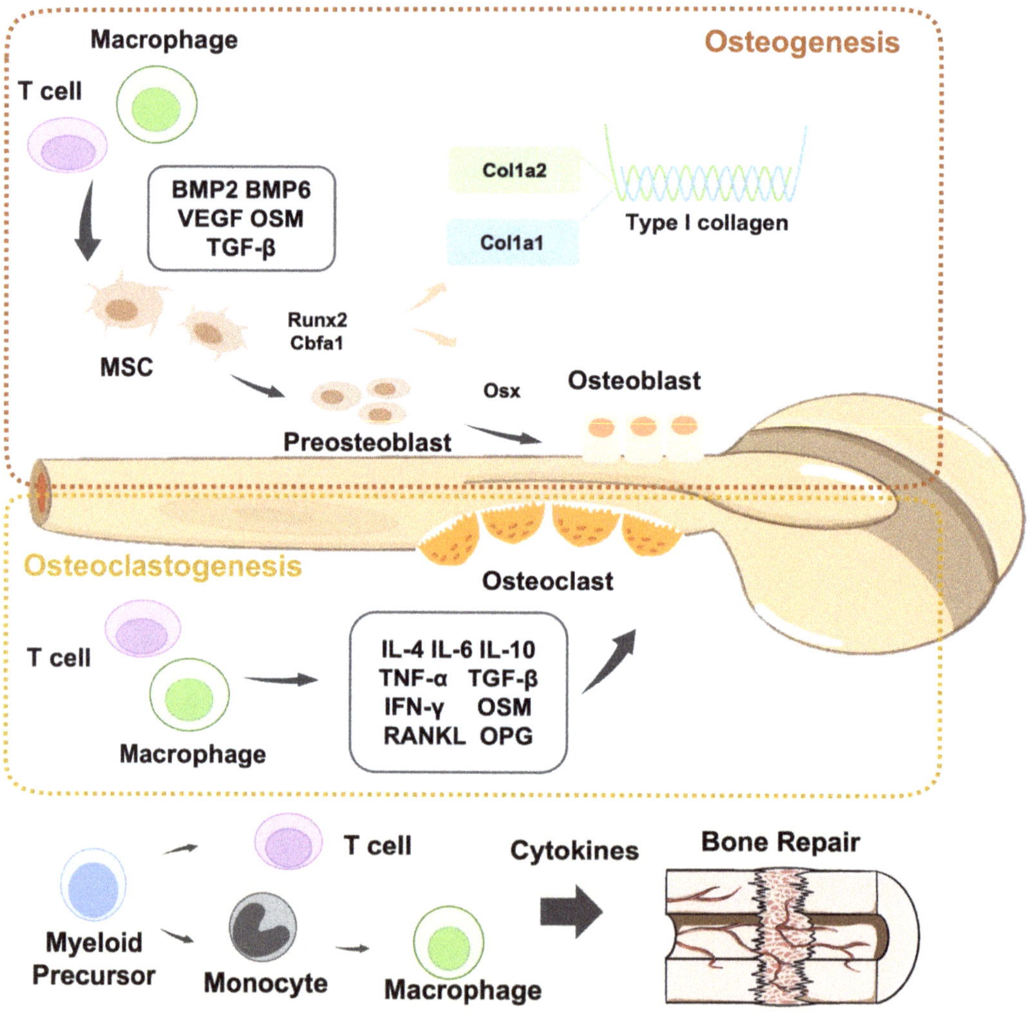

Figure 2. Schematic representation of the role of immune cells in bone modeling and remodeling. Immune cells are actively involved in osteoclastogenesis and osteogenesis.

4. Biomaterial-Mediated Bone Regeneration Immune Response

The research and development of traditional osteogenic materials largely focuses on the direct filling of bone-defect areas based on the principles of mechanical physics and chemistry; that is, they basically restore the defect in appearance, provide good mechanical support, and have a chemical composition similar to that of natural bone tissue. This approach ignores the fact that bone-defect repair is a dynamic physiological process that involves a variety of cells and cytokines. Implanting bone substitutes invariably alters the entire bone microenvironment. Osteogenic differentiation is regulated by the new bone microenvironment formed by bone substitute materials and multi-system cells, rather than by the materials acting alone. The developed bone-substitute materials may improperly regulate the microenvironment, resulting in bone regeneration failure in vivo by disregarding the importance of other system cells and their microenvironment.

Following the implantation of biomaterials into the body, the immune cells of the body will respond immediately, identify allogeneic biomaterials, and initiate host defense responses. Research has shown that bone repair can be regulated by changing the chemical composition (metal ions, proteins, and small-molecule drugs) and physical properties of biomaterials (particle size, porosity, pore size, and topology) (Figure 3).

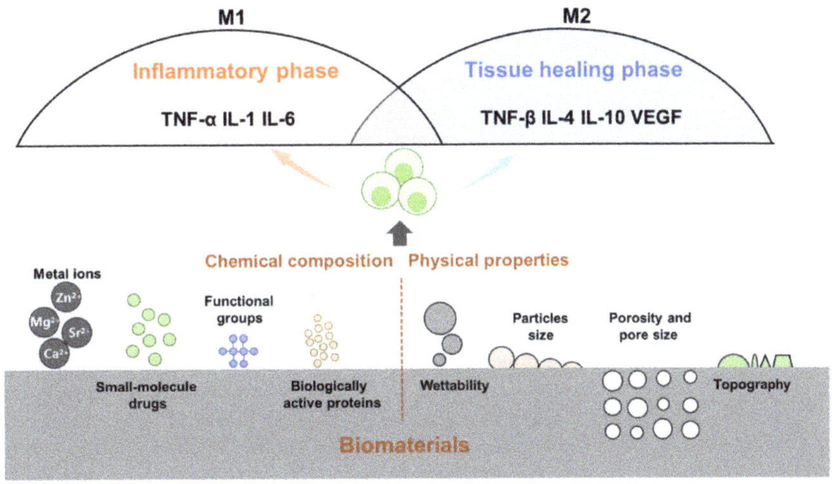

Figure 3. The physicochemical properties of bone biomaterials influence the immune response. For example, wettability, topography, particle size, porosity and pore size, release of metal ions, small molecule drugs, active proteins, and surface functional groups can modulate immune cells (e.g., macrophages) and their immune responses.

4.1. Chemical Composition of Osteogenic Biomaterials

Bone biomaterials typically degrade to varying degrees after implantation, releasing metal ions that alter the local microenvironment and impair the bone-bonding capability of the materials [97]. Metal ions can exert a crucial influence on immune and inflammatory responses through the direct regulation of macrophages, which is also considered to be an indispensable pathway [98].

Calcium (Ca) is a key component of calcium phosphate, a popular bone graft substitute that has been confirmed to connect with inflammatory signaling pathways [99]. High levels of extracellular Ca^{2+} have been reported to boost the production of Wnt5A by activating calcium-sensing receptor (CaSR) signaling, which could downregulate TNF-α and help reduce inflammatory reactions [100]. However, numerous inflammatory cytokines (e.g., IL-6, IL-1, TNF-α) can trigger bone resorption by stimulating the ligand of the receptor activator for NF-κB in osteoclasts [101]. Recently, it was found that certain cytokines such as IL-6 and IL-1 can upregulate the level of parathyroid CaSR, leading to hypocalcemia and the accumulation of phosphate, which inhibits bone resorption [102]. Hydroxyapatite is another commonly utilized bone biomaterial that promotes bone formation by stimulating osteoblast differentiation via the BMP2 and Wnt signaling pathways [103,104]. Therefore, the mutual relationship between calcium and inflammatory cytokines in the regulation of microenvironment inflammation and bone generation requires further investigation.

Magnesium (Mg), a mechanically bone-like metal ion, has a considerable biodegradability and biocompatibility in vivo and has been widely used in orthopedic implants [105]. In terms of inflammation, Mg can prevent the generation of inflammation-related cytokines by inhibiting the toll-like receptor (TLR) pathway, through which macrophages recognize foreign bodies and facilitate the innate immune reaction to deal with bone biomaterials [106]. Extracellular Mg^{2+} is reportedly an immunomodulator that modulates T-cell activation by binding to LFA-1 MIDAS [107]. It has been observed that memory T lymphocytes exhibit an Mg^{2+}-dependent dose response in the production of activation markers and cell clustering [108]. In terms of bone modulation, it has been reported that high levels of Mg^{2+} could suppress osteoclastogenesis and reduce bone resorption. The experimental validation of RAW 264.7 cells cultured on the Mg-containing surface showed that these precursor cells failed to differentiate into mature osteoclasts [109]. These promotive properties for immune repression and bone formation make Mg an excellent metal ion dopant for bone biomaterials. Bioactive ions have numerous properties, as well as complicated biological connections with in vivo molecular processes, immune responses, and cellular compositions.

Strontium (Sr), a trace element, has been proven to be necessary for bone growth, promoting osteogenesis while suppressing osteoclastogenesis, and is extensively applied in the treatment of osteoporosis [110]. Sr-doped biomaterials have been proven to considerably promote early osseointegration [111]. Sr is highly dose dependent. Low concentrations of Sr are favorable for osteogenesis, whereas excessive quantities of Sr harm the surrounding microenvironment and cause apoptosis [112,113]. It has been reported that 250–500 μM Sr^{2+} causes the best osteoinduction [114]. Shen et al. successfully fabricated Sr-doped titanium surface coatings. In large-proportion strontium-doped materials (75–100%), the expression of anti-inflammatory cytokines (e.g., IL-10) and osteogenesis-related genes (e.g., TGF-β1) increases when compared to Ti (Sr content 0%), while the expression of inflammation-favoring genes (e.g., TNF-α) decreases significantly [115]. The high-Sr samples prominently encourage macrophage polarization from M0 to M2, creating a favorable milieu for regulating OIM [116]. Research shows that Sr^{2+} can bind to CaSR, because of its comparable characteristics to Ca^{2+}, and stimulate bone formation through the MAPK/Erk 1/2 signaling pathway [117]. In addition, Sr promotes the osteogenic differentiation of mesenchymal stem cells via the Ras/MAPK signaling pathway [118].

Zinc (Zn), an essential component in maintaining the regular function of immune cells, is intimately associated with the growth and activity of macrophages [119]. Zinc homeostasis supports the differentiation of monocytes in blood into macrophages in infected tissue. Dubben et al. reported that decreased zinc levels in monocytes promoted their differentiation and increased their maturation [120]. However, excessively low concentrations of serum zinc have been shown to inhibit the growth of monocytes in the peripheral blood, and a concentration of 100 μM was suggested as the minimum concentration [121]. In addition, Brazão et al. found that zinc supplementation enormously increased the number of peritoneal macrophages, enhancing resistance to *Trypanosoma cruzi* infection [122]. In contrast, zinc determines the fate of macrophages by inducing regulated cell death in a concentration-dependent manner through different mechanisms [123]. It is worth emphasizing that both extremely high and low concentrations of zinc significantly trigger diverse forms of cell death. For example, in the macrophage RAW 264.7, a zinc oxide nanoparticle (ZnO) treatment that overloaded cells with zinc resulted in the necroptosis and apoptosis of macrophages in an Nrf2-independent manner [124]. Meanwhile, using the genetic loss of SLC39A10 to diminish the zinc level, Gao et al. found that macrophages could be induced by zinc depletion to apoptotic cell death mediated by the p-53 protein [125]. Therefore, it could be hypothesized that zinc induces programmed cell death in immune microenvironments in a concentration-dependent manner. Finally, zinc also plays an important role in the inflammatory functions of many immune cells. On one hand, zinc connects to macrophage functions by participating in TLR signaling [126]. A previous study found that a set of TLR (e.g., TLR1, TLR2, and TLR4) elicited the recruitment of macrophage phagosomes in mitochondria and enhanced their bactericidal activity [127]. TLR signaling is activated by the phosphorylation of interleukin-1 receptor-associated kinase 1 (IRAK1), whose degradation requires a certain level of zinc in vivo [128]. However, zinc deficiency is unfavorable for the regular biological processes of macrophages. It was found that long-term zinc insufficiency upregulated NLRP3 inflammasome production and activated IL-1β secretion by macrophages [129], whereas short-term zinc depletion inhibited inflammation by repressing caspase-1 activation, caspase-1 activation, and IL-1β secretion [130]. T cells are also highly vulnerable to the effects of Zn, notably on the balance of distinct T-cell subsets. The lack of Zn lowered the production of Th1 cytokines (IFN-γ, IL-2, and TNF-α), whereas it had little effect on Th2 responses (IL-4, IL-6, and IL-10), resulting in an imbalance between the Th1 and Th2 subpopulations [131]. In addition to its tight association with immune cells, Zn has a role in osteoblast and osteoclast differentiation. Multiple zinc-containing compounds have been added to bone biomaterial coatings to test their osteogenesis capacity. The upregulated abundance of differentiation markers of osteoblasts suggested that Zn may suppress the differentiation of osteoclasts while enhancing the maturation and mineralization of osteoblasts through involved immune responses [132]. Taken together, these findings suggest a multifaceted role of zinc ions in the regulation of macrophage-induced inflammation in a variety of ways, which inspired us to fully understand and better manipulate metal ions in the use of bone biomaterials. Therefore, metal ions are of great value to be fully understood and used in biomaterials. Modulation of the bone immunological microenvironment to stimulate osteogenesis by changing the concentration of different metal ions may become a significant development strategy for new bioactive bone materials.

In addition to metal ions, making use of bone material characteristics by the inclusion of biologically active proteins can also alter the ability of osteogenesis, with an emphasis on boosting M2 activation. Several signaling molecules, including oncostatin-M (OSM), prostaglandin E2 (PGE2), and BMP2, play a crucial role in the mechanism by which macrophages stimulate bone regeneration [53]. BMP2 is essential for the regulation of macrophage polarization and secretion. Based on a previous study, not only may BMP2 supplementation decrease the expression of inflammatory cytokines in M1 macrophages, it may also considerably increase the number of M2 macrophages [132]. BMP2-modified calcium phosphate cement (BMP2-CPC) causes an increase in M2 macrophages, which secrete TGF-1 and IL-10 to promote the in vitro osteogenic differentiation of MSCs [133]. In addition, it has been reported that the interaction of integrin β1 with fibronectin increases the expression of PI3 kinase signaling and promotes the phenotype of anti-inflammatory M2 macrophages, whereas blocking this mechanism induces M1 macrophages. Integrin β1 with fibrinogen on hydrophobic surfaces results in the generation of M1 macrophages, most likely through NF-κB activation [134].

Biomaterial surfaces with soluble anti-inflammatory small-molecule drugs (such as dexamethasone) can diminish the inflammatory response and fibrous encapsulation formation [135]. However, these drugs have limited applications because of their complex pharmacokinetics and decreasing concentration over time. Furthermore, the effect of anti-inflammatory agent coatings on biomaterials to promote osteogenesis needs to be considered. Glucocorticoids have been proven to improve inflammation relief and tissue repair when combined with anti-inflammatory cytokines (e.g., IL-6 and IL-10). However, glucocorticoids may decrease endogenous angiogenesis and increase the risk of infection [136]. Burgess et al. addressed this issue by administering dexamethasone and vascular endothelial growth factor (VEGF) via a hydrogel with an anti-inflammatory effect without impairing the formation of new blood vessels [137]. In addition, osthole, a coumarin-like derivative derived from traditional herbal medicine, has been found to enhance osteogenic differentiation [138]. Osthole stimulates osteoblast differentiation by activating the Wnt/β-linked protein/Bmp2 signaling pathway [139]. It has been reported that osthole can reduce inflammation by suppressing NF-κB [140] and has a negative effect on osteoblastogenesis and bone resorption induced by the nuclear factor-κB ligand (RANKL) receptor activator [141]. Moreover, sodium butyrate, a fermentation product of gut microbiota, loaded onto bone implant materials has been reported to have superior antimicrobial and osteogenic properties [142]. Sodium butyrate enhances macrophage phagocytosis by increasing reactive oxygen species (ROS) production. Sodium butyrate-loaded biomaterials increase macrophage M2 conversion and the secretion of anti-inflammatory factors, ultimately boosting bone healing [142,143]. The loading of different cytokines and biomolecules onto implant materials generates new ideas for the development of novel osteogenic materials.

Changing the characteristics of adsorbed proteins with distinct functional groups on the surface of biomaterials can also alter macrophage responsiveness. Buck et al. investigated the influence of surface chemistry on the behavior of macrophages on poly(polystyrene) surfaces with shared functional groups and comparable surface densities. The COOH group underwent adhesion with a large number of integrin-related proteins prior to macrophage attachment and improved the secretion of anti-inflammatory cytokines. After 48 h of culture, macrophages showed a pro-secretory effect on proteins associated with inflammation relief and tissue repair. In addition, unlike the NH2 and PO3H2 groups, the COOH group attenuated the LPS-stimulated inflammatory response of macrophages [144]. In addition, numerous studies have demonstrated that the COOH group is a possible anti-inflammatory surface functional component. According to Visalakshan et al., a surface with COOH groups resulted in a greater amount of LPS-stimulated macrophage IL-10 production and a lower amount of inflammatory cytokine release compared to a surface with amine or methyl groups. They also found that a hydrophilic AC surface containing COOH functional groups had a great affinity for albumin, which initiates the M2 pathway by encouraging the secretion of anti-inflammatory cytokines while blocking the production of inflammatory cytokines [144]. Conversely, a hydrophobic surface containing CH3 functional groups was more likely to adhere to IgG2, which can increase the release of inflammatory cytokines and suppress the release of anti-inflammatory cytokines, thereby activating the M1 pathway [145].

4.2. Physical Properties of Osteogenic Biomaterials

The surfaces of osteogenic biomaterials are in direct contact and react with the surrounding immune environment. Immune cells in the surrounding environment are affected by wettability, porosity, pore size, particle size, and surface microstructure.

The surface wettability of biomaterials is closely related to the adsorption of proteins, formation of blood clots, and formation of fibrin [146]. Early in the process of the implantation of biomaterials, vascular injury may lead to the extravasation of blood around the implant, triggering blood–biomaterial interactions. Blood clots play a role in the pool of cytokines that initiate wound healing. Hydrophilic polymeric surfaces may reduce protein adsorption and leukocyte activation, resulting in a lower rejection of foreign bodies [147]. By increasing the hydrophilicity of the material, implants can be more effectively integrated into the bone. A hydrophobic surface, when compared to a hydrophilic one, is generally capable of improving monocyte attachment and triggering stronger local immune responses. According to a recent study, a hydrophilic/neutral copolymer surface can significantly inhibit the adhesion of monocytes/macrophages, with the fusion of macrophages being minimal or absent, which effectively reduces the levels of vital factors, such as IL-6, IL-1β, and TNF-α [148]. According to Zischke et al., a hydrophilic surface modified by titanium (Ti) exhibited more active osteoblast differentiation, increased growth factor production, and higher osteogenic gene levels than unmodified surfaces [149].

Macrophages are approximately 20 μm in length [150]. When designing bone biomaterials, the porosity and pore size of osteogenic biomaterials must be considered, as these parameters may influence osteoblast function, macrophage polarization, and immunological responses. Micropore-size-appropriate biomaterials can promote macrophage secretion of VEGF by establishing a slightly anoxic external microenvironment, which induces the formation of microvessels and promotes bone regeneration [151]. Pores that are too small hinder the blood from transporting nutrients and oxygen, which will enhance the local inflammatory response, resulting in the formation of granulation tissue and completely blocking the micropores. The blockage prevents bone cell ingrowth from taking place, ultimately resulting in poor bone regeneration and implant failure [152,153]. Biomaterials with pore sizes of 90–120 μm have been shown to promote chondrogenesis and inhibit vascularization. Biomaterials with pore sizes up to 350 μm promote osteogenesis and vascularization [154]. Furthermore, high porosity facilitates the adhesion and growth of osteoblasts [155], resulting in the formation of a dense extracellular matrix, thereby enhancing early biological fixation [156]. In addition to their significance in bone cell activity, porosity and pore size also play an important role in the communication between the implants and the internal immune system. It has been shown that foreign body response activity diminishes as the pore size increases [156]. The foreign body reaction mediates the fibrotic response to encapsulate and remove foreign bodies from the surrounding tissue [157]. During a foreign body reaction, macrophages generate TGF-β which plays an important role in modulating the fibrotic response [157].

The size of the particles has a significant impact on the immune response. Implant particles are degraded and processed by immune cells based on their size. Macrophages can directly phagocytose particles with a small diameter (less than 0.5 μm). Although individual macrophages are no longer capable of phagocytizing structures larger than 0.5 μm, a foreign body giant cell is created as a result of the fusion of multiple macrophages in an attempt to phagocytize the particles [158,159]. When the material is excessively large (more than 100 μm), it obstructs macrophage phagocytosis and fusion [160,161], resulting in a highly inflammatory environment in which macrophages produce a substantial amount of inflammatory cytokines and ROS to destroy the substance. However, that is not to say that larger particles elicit a higher immunological response. Smaller particles in comparable-quality materials have a larger surface area, which results in an increased chemical activity and a greater immunomodulatory capacity [162,163]. Laquerriere indicated that small-diameter hydroxyapatite particles can trigger immune cells to release increased amounts of pivotal cytokines favoring inflammation [164]. Davison et al. discovered the presence of more multinucleated osteoclast-like cells surrounding calcium phosphate bioceramic crystal particles with a size of 1 μm as compared with crystals of 2–4 μm. RAW 264.7, which adhered to crystal particles with a size of 2–4 μm, was dramatically inhibited in RANKL-induced proliferation and differentiation into multinucleated osteoclast-like cells [165]. Furthermore, Li et al. compared the difference in OIM between two calcium phosphate bioceramics with submicron/micron surface topography and found that submicron biomaterials can regulate macrophage polarization to M2 by activating the PI3K/Akt pathway in vitro, thereby promoting the osteogenic differentiation of MSCs [166].

Another important property that influences immune-cell interactions is the surface topography of biomaterials [167]. There is plenty of research showing that modifying the surface topography can successfully regulate osteoblastic cell adhesion, migration, proliferation, and differentiation [168,169]. Surface roughness is an important modulator of both osteoblastogenesis and osteoclastogenesis. Polishing and sandblasting are two methods for modifying surface roughness, which can typically be portrayed on a microscale. For example, the surface of bone biomaterials has a prominent property in that it elicits modulative effects on the immune response in the host, and as a commonly used metal ion on the surface, Ti plays a regulatory role in the immune response. The surface topography of biomaterials is normally determined by the roughness of Ti, which has been shown to affect cell adhesion and spreading. When the roughness of the Ti-composed surface increases, the spreading of macrophages also advances [170]. In addition, titanium roughness also exerts significant stimulatory effects on macrophages to modulate the generation of inflammatory cytokines and chemokines [171,172]. Compared to the smooth Ti substrate, which promoted M1 polarization, Hotchkiss found that microroughened Ti surfaces encouraged the M2 macrophage phenotypic switch and increased the production of IL-4 and IL-10 [173]. According to Christo et al., materials with a 68 nm controlled surface nanotopography result in an increased synthesis of matrix metalloproteinase-9 (MMP-9) and a decreased release of pro-inflammatory cytokine secretion from primary macrophages when compared to the smooth glass control [174].

5. Definition and Research Status of OIM

Bone biomaterials modulate the local immune microenvironment and influence bone cell function, thereby regulating bone regeneration and reconstruction. Traditional bone biomaterials are primarily developed to consider whether they may lead to immune rejection and final implant failure. However, the implantation of bone biomaterials alters the microenvironment surrounding the bone. This produces a new microenvironment formed by the interaction between the bone biomaterials and multi-system cells that regulates osteogenic differentiation, rather than the materials alone. Immune cells play a central role in the local bone microenvironment as they regulate a variety of processes involved in bone regeneration (such as osteogenic differentiation, osteoclastic differentiation, fibrosis, and vascularization) by regulating the expression of chemokines and inflammatory factors. OIM is a novel concept for evaluating bone biomaterials that incorporates the new properties of biomaterials, bone cells, and immune cells to aid in the development of biomaterials. It pays more attention to the influence of the immune environment generated by the interaction with biomaterials on the behavior of bone cells rather than immune rejection. In short, biomaterials with a high OIM can elicit an appropriate inflammatory response via local immunocytes by releasing factors that drive the differentiation of BMSCs, ultimately leading to successful osteogenesis.

5.1. Evaluation Methods of OIM

Since OIM involves interactions between bone cells, immunocytes, and biomaterials, all components should be considered in the evaluation system, which can be achieved by a co-culture system. Indirect co-culture with a conditioned medium is a popular method. First, immune cells are cultured on bone biomaterials to obtain the supernatant containing cytokines related to the immune response. After mixing the supernatant with fresh medium, osteoblasts are grown in the conditioned medium to determine whether they have an influence on osteogenesis or the osteoclast reaction. This procedure is simple and reproducible. Additionally, this approach can be used when immune and bone cells originate from distinct species. However, indirect co-culture with conditioned medium cannot fully replicate the situation in vivo because bone cells actively govern the immune response rather than playing a passive role in the interaction with immune cells.

Indirect co-culture employing a Boyden chamber can more accurately mimic the in vivo environment associated with bone cell–immune cell interactions. Immune cells are seeded in the upper compartment and allowed to migrate through the pores of the membrane into the lower compartment, where bone cells and biomaterials reside [175]. A small pore size (0.4 μm) can retain cells in the upper chamber while allowing released substances to flow freely. A large pore size permits cells to migrate and can be used to evaluate the impact of activated immune cells [14]. The Boyden chamber assay saves time because it takes only a few hours for cells to pass through the porous membrane in the Boyden chamber, which is substantially less time than that required for cells to complete the cell cycle [176].

5.2. OIM-Based Development of Bone Biomaterials

OIM highlights the modulation of the immunological milieu formed by biomaterials, which plays a key role in the process of osteogenesis and the regulation of osteoclastogenesis. Bone regeneration requires the prompt modulation and transformation of the immune microenvironment in the region of bone defects from pro-inflammatory to anti-inflammatory. Macrophages are involved in a variety of biological activities, including infection, repair, and regeneration, as well as tissue homeostasis. Following biomaterial implantation, macrophages are the primary first response of the body's immune system. It is therefore important to transform M1 macrophages into M2 macrophages, which drive osteoblast development and bone production by secreting BMP-2, IL-10, and TGF-β. When developing materials, it is possible to modulate the immune response by altering the composition or structure of the materials: (i) changing the material particle size, (ii) optimizing the pore size and porosity, (iii) increasing the hydrophilicity of the materials, (iv) adding metal elements to materials, and (v) combining the use of small-molecule drugs [177]. The current focus in the design of bone biomaterials is on inhibiting M1 macrophages and increasing M2 macrophages, thus releasing cytokines with anti-inflammatory effects in order to control inflammation and ultimately promote osteogenesis. The excessive suppression of M1 macrophages, however, limits the body's ability to clear bacteria, making infection unmanageable and thus leading to the failure of osteogenic material implantation. In summary, the synergy between these strategies should be considered and any potential detrimental effects must be avoided when designing biomaterials to effectively induce bone reconstruction.

6. Conclusions

The complex process of bone healing and regeneration requires the accurate regulation of a series of molecular signals. The development of bone materials must consider the ability of the immune system to regulate the local microenvironment and facilitate bone formation. Biomaterials with a great OIM can induce an immune environment favorable for bone formation. By modifying the physical and chemical properties of biomaterials, such as their surface roughness and wettability, implants can directly influence immune cells in vivo, creating an optimal immunological microenvironment to dynamically regulate osteogenesis. Utilizing the immune system to regulate the bone-repair process precisely remains a great challenge, although we have established that immune cells are the primary force behind bone repair. Further research is required to investigate the complex signaling pathways and potential molecular targets of biomaterials, the immune system, and the skeletal system. Furthermore, numerous studies of implants have focused on the investigation of macrophages in microenvironments. However, there is still a dearth of studies exploring the role of other immune cells, including T cells, B cells, and neutrophils, in affecting the properties and functions of bone biomaterials. We suggest that more research attention should be paid to a comprehensive exploration of immune cells and their relationship with novel bone biomaterials in the future.

Author Contributions: Conceptualization, writing-original draft preparation and visualization, T.Z. and Z.C.; conceptualization, writing-review and editing, J.M. and L.O.All authors have read and agreed to the published version of the manuscript.

Funding: This work was funded by the National Natural Science Foundation of China (32000938).

Institutional Review Board Statement: Not applicable.

Informed Consent Statement: Not applicable.

Data Availability Statement: Not applicable.

Conflicts of Interest: All authors declare that there is no conflict of interest.

References

1. Schindler, O.S.; Cannon, S.R.; Briggs, T.W.; Blunn, G.W. Use of a novel bone graft substitute in peri-articular bone tumours of the knee. *Knee* **2007**, *14*, 458–464. [CrossRef] [PubMed]
2. Cook, E.A.; Cook, J.J. Bone graft substitutes and allografts for reconstruction of the foot and ankle. *Clin. Podiatr. Med. Surg.* **2009**, *26*, 589–605. [CrossRef] [PubMed]
3. Carson, J.S.; Bostrom, M.P. Synthetic bone scaffolds and fracture repair. *Injury* **2007**, *38* (Suppl. 1), S33–S37. [CrossRef] [PubMed]
4. Zhang, L.Y.; Bi, Q.; Zhao, C.; Chen, J.Y.; Cai, M.H.; Chen, X.Y. Recent Advances in Biomaterials for the Treatment of Bone Defects. *Organogenesis* **2020**, *16*, 113–125. [CrossRef] [PubMed]
5. Chen, Z.Y.; Gao, S.; Zhang, Y.W.; Zhou, R.B.; Zhou, F. Antibacterial biomaterials in bone tissue engineering. *J. Mater. Chem. B* **2021**, *9*, 2594–2612. [CrossRef] [PubMed]
6. Jin, S.; Xia, X.; Huang, J.; Yuan, C.; Zuo, Y.; Li, Y.; Li, J. Recent advances in PLGA-based biomaterials for bone tissue regeneration. *Acta Biomater.* **2021**, *127*, 56–79. [CrossRef] [PubMed]
7. Mouthuy, P.A.; Snelling, S.J.B.; Dakin, S.G.; Milković, L.; Gašparović, A.; Carr, A.J.; Žarković, N. Biocompatibility of implantable materials: An oxidative stress viewpoint. *Biomaterials* **2016**, *109*, 55–68. [CrossRef]
8. Arron, J.R.; Choi, Y. Bone versus immune system. *Nature* **2000**, *408*, 535–536. [CrossRef]
9. Takayanagi, H. Inflammatory bone destruction and osteoimmunology. *J. Periodontal Res.* **2005**, *40*, 287–293. [CrossRef] [PubMed]
10. Franz, S.; Rammelt, S.; Scharnweber, D.; Simon, J.C. Immune responses to implants—A review of the implications for the design of immunomodulatory biomaterials. *Biomaterials* **2011**, *32*, 6692–6709. [CrossRef]

11. Chen, Z.; Yi, D.; Zheng, X.; Chang, J.; Wu, C.; Xiao, Y. Nutrient element-based bioceramic coatings on titanium alloy stimulating osteogenesis by inducing beneficial osteoimmmunomodulation. *J. Mater. Chem. B* **2014**, *2*, 6030–6043. [CrossRef] [PubMed]
12. Martin, K.E.; García, A.J. Macrophage phenotypes in tissue repair and the foreign body response: Implications for biomaterial-based regenerative medicine strategies. *Acta Biomater.* **2021**, *133*, 4–16. [CrossRef] [PubMed]
13. Takayanagi, H. Osteoimmunology: Shared mechanisms and crosstalk between the immune and bone systems. *Nat. Rev. Immunol.* **2007**, *7*, 292–304. [CrossRef]
14. Chen, Z.; Klein, T.; Murray, R.Z.; Crawford, R.; Chang, J.; Wu, C.; Xiao, Y. Osteoimmunomodulation for the development of advanced bone biomaterials. *Mater. Today* **2016**, *19*, 304–321. [CrossRef]
15. Cohen, M.M., Jr. The new bone biology: Pathologic, molecular, and clinical correlates. *Am. J. Med. Genet. Part A* **2006**, *140*, 2646–2706. [CrossRef]
16. Saito, N.; Murakami, N.; Takahashi, J.; Horiuchi, H.; Ota, H.; Kato, H.; Okada, T.; Nozaki, K.; Takaoka, K. Synthetic biodegradable polymers as drug delivery systems for bone morphogenetic proteins. *Adv. Drug Deliv. Rev.* **2005**, *57*, 1037–1048. [CrossRef]
17. Geiger, M.; Li, R.H.; Friess, W. Collagen sponges for bone regeneration with rhBMP-2. *Adv. Drug Deliv. Rev.* **2003**, *55*, 1613–1629. [CrossRef] [PubMed]
18. Ludwig, S.C.; Kowalski, J.M.; Boden, S.D. Osteoinductive bone graft substitutes. *Eur. Spine J.* **2000**, *9* (Suppl. 1), S119–S125. [CrossRef]
19. Okamoto, M.; Murai, J.; Yoshikawa, H.; Tsumaki, N. Bone morphogenetic proteins in bone stimulate osteoclasts and osteoblasts during bone development. *J. Bone Miner. Res.* **2006**, *21*, 1022–1033. [CrossRef]
20. Siddiqui, J.A.; Partridge, N.C. Physiological Bone Remodeling: Systemic Regulation and Growth Factor Involvement. *Physiology* **2016**, *31*, 233–245. [CrossRef]
21. Feng, X.; McDonald, J.M. Disorders of bone remodeling. *Annu. Rev. Pathol.* **2011**, *6*, 121–145. [CrossRef] [PubMed]
22. Lecanda, F.; Avioli, L.V.; Cheng, S.L. Regulation of bone matrix protein expression and induction of differentiation of human osteoblasts and human bone marrow stromal cells by bone morphogenetic protein-2. *J. Cell. Biochem.* **1997**, *67*, 386–396. [CrossRef]
23. Liao, W.; Ning, Y.; Xu, H.J.; Zou, W.Z.; Hu, J.; Liu, X.Z.; Yang, Y.; Li, Z.H. BMSC-derived exosomes carrying microRNA-122-5p promote proliferation of osteoblasts in osteonecrosis of the femoral head. *Clin. Sci.* **2019**, *133*, 1955–1975. [CrossRef] [PubMed]
24. Xiong, Y.; Chen, L.; Yan, C.; Zhou, W.; Yu, T.; Sun, Y.; Cao, F.; Xue, H.; Hu, Y.; Chen, D.; et al. M2 Macrophagy-derived exosomal miRNA-5106 induces bone mesenchymal stem cells towards osteoblastic fate by targeting salt-inducible kinase 2 and 3. *J. Nanobiotechnol.* **2020**, *18*, 66. [CrossRef]
25. Komori, T. Roles of Runx2 in Skeletal Development. *Adv. Exp. Med. Biol.* **2017**, *962*, 83–93.
26. Kundu, M.; Javed, A.; Jeon, J.P.; Horner, A.; Shum, L.; Eckhaus, M.; Muenke, M.; Lian, J.B.; Yang, Y.; Nuckolls, G.H.; et al. Cbfbeta interacts with Runx2 and has a critical role in bone development. *Nat. Genet.* **2002**, *32*, 639–644. [CrossRef]
27. Maruyama, Z.; Yoshida, C.A.; Furuichi, T.; Amizuka, N.; Ito, M.; Fukuyama, R.; Miyazaki, T.; Kitaura, H.; Nakamura, K.; Fujita, T.; et al. Runx2 determines bone maturity and turnover rate in postnatal bone development and is involved in bone loss in estrogen deficiency. *Dev. Dyn.* **2007**, *236*, 1876–1890. [CrossRef]
28. Qin, H.; Cai, J. Effect of Runx2 silencing on autophagy and RANKL expression in osteoblasts. *Arch. Oral Biol.* **2018**, *95*, 74–78. [CrossRef]
29. Lian, J.B.; Stein, G.S. Runx2/Cbfa1: A multifunctional regulator of bone formation. *Curr. Pharm. Des.* **2003**, *9*, 2677–2685. [CrossRef]
30. Kern, B.; Shen, J.; Starbuck, M.; Karsenty, G. Cbfa1 contributes to the osteoblast-specific expression of type I collagen genes. *J. Biol. Chem.* **2001**, *276*, 7101–7107. [CrossRef]
31. Wei, J.; Shimazu, J.; Makinistoglu, M.P.; Maurizi, A.; Kajimura, D.; Zong, H.; Takarada, T.; Lezaki, T.; Pessin, J.E.; Hinoi, E.; et al. Glucose Uptake and Runx2 Synergize to Orchestrate Osteoblast Differentiation and Bone Formation. *Cell* **2015**, *161*, 1576–1591. [CrossRef] [PubMed]
32. Kylmaoja, E.; Nakamura, M.; Tuukkanen, J. Osteoclasts and Remodeling Based Bone Formation. *Curr. Stem Cell Res. Ther.* **2016**, *11*, 626–633. [CrossRef] [PubMed]
33. Imai, R.; Sato, T.; Iwamoto, Y.; Hanada, Y.; Terao, M.; Ohta, Y.; Osaki, Y.; Imai, T.; Morihana, T.; Okazaki, S.; et al. Osteoclasts Modulate Bone Erosion in Cholesteatoma via RANKL Signaling. *J. Assoc. Res. Otolaryngol.* **2019**, *20*, 449–459. [CrossRef] [PubMed]
34. Kim, J.M.; Lin, C.; Stavre, Z.; Greenblatt, M.B.; Shim, J.H. Osteoblast-Osteoclast Communication and Bone Homeostasis. *Cells* **2020**, *9*, 2073. [CrossRef]
35. Song, C.; Yang, X.; Lei, Y.; Zhang, Z.; Smith, W.; Yan, J.; Kong, L. Evaluation of efficacy on RANKL induced osteoclast from RAW264.7 cells. *J. Cell. Physiol.* **2019**, *234*, 11969–11975. [CrossRef]
36. Udagawa, N.; Koide, M.; Nakamura, M.; Nakamichi, Y.; Yamashita, T.; Uehara, S.; Kobayashi, Y.; Furuya, Y.; Yasuda, H.; Fukuda, C.; et al. Osteoclast differentiation by RANKL and OPG signaling pathways. *J. Bone Miner. Metab.* **2021**, *39*, 19–26. [CrossRef]
37. Andersen, T.L.; del Carmen Ovejero, M.; Kirkegaard, T.; Lenhard, T.; Foged, N.T.; Delaissé, J.M. A scrutiny of matrix metalloproteinases in osteoclasts: Evidence for heterogeneity and for the presence of MMPs synthesized by other cells. *Bone* **2004**, *35*, 1107–1119. [CrossRef]

38. Hou, P.; Troen, T.; Ovejero, M.C.; Kirkegaard, T.; Andersen, T.L.; Byrjalsen, I.; Ferreras, M.; Sato, T.; Shapiro, S.D.; Foged, N.T.; et al. Matrix metalloproteinase-12 (MMP-12) in osteoclasts: New lesson on the involvement of MMPs in bone resorption. *Bone* **2004**, *34*, 37–47. [CrossRef]
39. Li, C.J.; Xiao, Y.; Sun, Y.C.; He, W.Z.; Liu, L.; Huang, M.; He, C.; Huang, M.; Chen, K.X.; Hou, J.; et al. Senescent immune cells release grancalcin to promote skeletal aging. *Cell Metab.* **2021**, *33*, 1957–1973.e6. [CrossRef]
40. Yu, J.; Chen, S.; Lei, S.; Li, F.; Wang, Y.; Shu, X.; Xu, W.; Tang, X. The Effects of *Porphyromonas gingivalis* on Inflammatory and Immune Responses and Osteogenesis of Mesenchymal Stem Cells. *Stem Cells Dev.* **2021**, *30*, 1191–1201. [CrossRef]
41. Hillion, S.; Arleevskaya, M.I.; Blanco, P.; Bordron, A.; Brooks, W.H.; Cesbron, J.Y.; Kaveri, S.; Vivier, E.; Renaudineau, Y. The Innate Part of the Adaptive Immune System. *Clin. Rev. Allergy Immunol.* **2020**, *58*, 151–154. [CrossRef] [PubMed]
42. Tsukasaki, M.; Takayanagi, H. Osteoimmunology: Evolving concepts in bone-immune interactions in health and disease. *Nat. Rev. Immunol.* **2019**, *19*, 626–642. [CrossRef] [PubMed]
43. Aoshi, T.; Koyama, S.; Kobiyama, K.; Akira, S.; Ishii, K.J. Innate and adaptive immune responses to viral infection and vaccination. *Curr. Opin. Virol.* **2011**, *1*, 226–232. [CrossRef]
44. Weiskopf, D.; Weinberger, B.; Grubeck-Loebenstein, B. The aging of the immune system. *Transpl. Int.* **2009**, *22*, 1041–1050. [CrossRef]
45. Wei, Z.; Tian, P.; Liu, X.; Zhou, B. Hemocompatibility and selective cell fate of polydopamine-assisted heparinized PEO/PLLA composite coating on biodegradable AZ31 alloy. *Colloids Surf. B* **2014**, *121*, 451–460. [CrossRef] [PubMed]
46. Artyomov, M.N.; Sergushichev, A.; Schilling, J.D. Integrating immunometabolism and macrophage diversity. *Semin. Immunol.* **2016**, *28*, 417–424. [CrossRef]
47. Sica, A.; Mantovani, A. Macrophage plasticity and polarization: In vivo veritas. *J. Clin. Investig.* **2012**, *122*, 787–795. [CrossRef]
48. Blériot, C.; Chakarov, S.; Ginhoux, F. Determinants of Resident Tissue Macrophage Identity and Function. *Immunity* **2020**, *52*, 957–970. [CrossRef]
49. Cassetta, L.; Pollard, J.W. Targeting macrophages: Therapeutic approaches in cancer. *Nat. Rev. Drug Discov.* **2018**, *17*, 887–904. [CrossRef]
50. Orihuela, R.; McPherson, C.A.; Harry, G.J. Microglial M1/M2 polarization and metabolic states. *Br. J. Pharmacol.* **2016**, *173*, 649–665. [CrossRef]
51. Shapouri-Moghaddam, A.; Mohammadian, S.; Vazini, H.; Taghadosi, M.; Esmaeili, S.A.; Mardani, F.; Seifi, B.; Mohammadi, A.; Afshari, J.T.; Sahebkar, A. Macrophage plasticity, polarization, and function in health and disease. *J. Cell. Physiol.* **2018**, *233*, 6425–6440. [CrossRef] [PubMed]
52. Zhang, J.; Liu, X.; Wan, C.; Liu, Y.; Wang, Y.; Meng, C.; Zhang, Y.; Jiang, C. NLRP3 inflammasome mediates M1 macrophage polarization and IL-1β production in inflammatory root resorption. *J. Clin. Periodontol.* **2020**, *47*, 451–460. [CrossRef] [PubMed]
53. Pajarinen, J.; Lin, T.; Gibon, E.; Kohno, Y.; Maruyama, M.; Nathan, K.; Lu, L.; Yao, Z.; Goodman, S.B. Mesenchymal stem cell-macrophage crosstalk and bone healing. *Biomaterials* **2019**, *196*, 80–89. [CrossRef] [PubMed]
54. Xuan, W.; Qu, Q.; Zheng, B.; Xiong, S.; Fan, G.H. The chemotaxis of M1 and M2 macrophages is regulated by different chemokines. *J. Leukoc. Biol.* **2015**, *97*, 61–69. [CrossRef]
55. Zhao, B. Intrinsic Restriction of TNF-Mediated Inflammatory Osteoclastogenesis and Bone Resorption. *Front. Endocrinol.* **2020**, *11*, 583561. [CrossRef]
56. Cho, D.I.; Kim, M.R.; Jeong, H.Y.; Jeong, H.C.; Jeong, M.H.; Yoon, S.H.; Kim, Y.S.; Ahn, Y. Mesenchymal stem cells reciprocally regulate the M1/M2 balance in mouse bone marrow-derived macrophages. *Exp. Mol. Med.* **2014**, *46*, e70. [CrossRef]
57. Mommert, S.; Hüer, M.; Schaper-Gerhardt, K.; Gutzmer, R.; Werfel, T. Histamine up-regulates oncostatin M expression in human M1 macrophages. *Br. J. Pharmacol.* **2020**, *177*, 600–613. [CrossRef]
58. Mantovani, A.; Sozzani, S.; Locati, M.; Allavena, P.; Sica, A. Macrophage polarization: Tumor-associated macrophages as a paradigm for polarized M2 mononuclear phagocytes. *Trends Immunol.* **2002**, *23*, 549–555. [CrossRef]
59. Wang, L.X.; Zhang, S.X.; Wu, H.J.; Rong, X.L.; Guo, J. M2b macrophage polarization and its roles in diseases. *J. Leukoc. Biol.* **2019**, *106*, 345–358. [CrossRef]
60. Mahon, O.R.; Browe, D.C.; Gonzalez-Fernandez, T.; Pitacco, P.; Whelan, I.T.; Von Euw, S.; Hobbs, C.; Nicolosi, V.; Cunningham, K.T.; Mills, K.H.G.; et al. Nano-particle mediated M2 macrophage polarization enhances bone formation and MSC osteogenesis in an IL-10 dependent manner. *Biomaterials* **2020**, *239*, 119833. [CrossRef]
61. Zhang, Q.; Chen, B.; Yan, F.; Guo, J.; Zhu, X.; Ma, S.; Yang, W. Interleukin-10 inhibits bone resorption: A potential therapeutic strategy in periodontitis and other bone loss diseases. *BioMed Res. Int.* **2014**, *2014*, 284836. [CrossRef] [PubMed]
62. Witherel, C.E.; Sao, K.; Brisson, B.K.; Han, B.; Volk, S.W.; Petrie, R.J.; Han, L.; Spiller, K.L. Regulation of extracellular matrix assembly and structure by hybrid M1/M2 macrophages. *Biomaterials* **2021**, *269*, 120667. [CrossRef]
63. Alhamdi, J.R.; Peng, T.; Al-Naggar, I.M.; Hawley, K.L.; Spiller, K.L.; Kuhn, L.T. Controlled M1-to-M2 transition of aged macrophages by calcium phosphate coatings. *Biomaterials* **2019**, *196*, 90–99. [CrossRef] [PubMed]
64. Spiller, K.L.; Nassiri, S.; Witherel, C.E.; Anfang, R.R.; Ng, J.; Nakazawa, K.R.; Yu, T.; Vunjak-Novakovic, G. Sequential delivery of immunomodulatory cytokines to facilitate the M1-to-M2 transition of macrophages and enhance vascularization of bone scaffolds. *Biomaterials* **2015**, *37*, 194–207. [CrossRef] [PubMed]

65. Mia, S.; Warnecke, A.; Zhang, X.M.; Malmström, V.; Harris, R.A. An optimized protocol for human M2 macrophages using M-CSF and IL-4/IL-10/TGF-β yields a dominant immunosuppressive phenotype. *Scand. J. Immunol.* **2014**, *79*, 305–314. [CrossRef] [PubMed]
66. Walsh, M.C.; Choi, Y. Biology of the RANKL-RANK-OPG System in Immunity, Bone, and Beyond. *Front. Immunol.* **2014**, *5*, 511. [CrossRef] [PubMed]
67. Walsh, M.C.; Takegahara, N.; Kim, H.; Choi, Y. Updating osteoimmunology: Regulation of bone cells by innate and adaptive immunity. *Nat. Rev. Rheumatol.* **2018**, *14*, 146–156. [CrossRef]
68. Ota, Y.; Niiro, H.; Ota, S.; Ueki, N.; Tsuzuki, H.; Nakayama, T.; Mishima, K.; Higashioka, K.; Jabbarzadeh-Tabrizi, S.; Mitoma, H.; et al. Generation mechanism of RANKL$^+$ effector memory B cells: Relevance to the pathogenesis of rheumatoid arthritis. *Arthritis Res. Ther.* **2016**, *18*, 67. [CrossRef]
69. Boyce, B.F.; Xiu, Y.; Li, J.; Xing, L.; Yao, Z. NF-κB-Mediated Regulation of Osteoclastogenesis. *Endocrinol. Metab.* **2015**, *30*, 35–44. [CrossRef]
70. Romagnani, S. Th1/Th2 cells. *Inflamm. Bowel Dis.* **1999**, *5*, 285–294. [CrossRef]
71. Boniface, K.; Bak-Jensen, K.S.; Li, Y.; Blumenschein, W.M.; McGeachy, M.J.; McClanahan, T.K.; McKenzie, B.S.; Kastelein, R.A.; Cua, D.J.; de Waal Malefyt, R. Prostaglandin E2 regulates Th17 cell differentiation and function through cyclic AMP and EP2/EP4 receptor signaling. *J. Exp. Med.* **2009**, *206*, 535–548. [CrossRef] [PubMed]
72. Sadtler, K.; Estrellas, K.; Allen, B.W.; Wolf, M.T.; Fan, H.; Tam, A.J.; Patel, C.H.; Luber, B.S.; Wang, H.; Wagner, K.R.; et al. Developing a pro-regenerative biomaterial scaffold microenvironment requires T helper 2 cells. *Science* **2016**, *352*, 366–370. [CrossRef]
73. Bozec, A.; Zaiss, M.M. T Regulatory Cells in Bone Remodelling. *Curr. Osteoporos. Rep.* **2017**, *15*, 121–125. [CrossRef] [PubMed]
74. Fine, N.; Tasevski, N.; McCulloch, C.A.; Tenenbaum, H.C.; Glogauer, M. The Neutrophil: Constant Defender and First Responder. *Front. Immunol.* **2020**, *11*, 571085. [CrossRef] [PubMed]
75. Peiseler, M.; Kubes, P. More friend than foe: The emerging role of neutrophils in tissue repair. *J. Clin. Investig.* **2019**, *129*, 2629–2639. [CrossRef] [PubMed]
76. Cai, B.; Lin, D.; Li, Y.; Wang, L.; Xie, J.; Dai, T.; Liu, F.; Tang, M.; Tian, L.; Yuan, Y.; et al. N2-Polarized Neutrophils Guide Bone Mesenchymal Stem Cell Recruitment and Initiate Bone Regeneration: A Missing Piece of the Bone Regeneration Puzzle. *Adv. Sci.* **2021**, *8*, e2100584. [CrossRef] [PubMed]
77. Huber, A.R.; Kunkel, S.L.; Todd, R.F., 3rd; Weiss, S.J. Regulation of transendothelial neutrophil migration by endogenous interleukin-8. *Science* **1991**, *254*, 99–102. [CrossRef]
78. Das, A.; Sinha, M.; Datta, S.; Abas, M.; Chaffee, S.; Sen, C.K.; Roy, S. Monocyte and macrophage plasticity in tissue repair and regeneration. *Am. J. Pathol.* **2015**, *185*, 2596–2606. [CrossRef]
79. Kovtun, A.; Bergdolt, S.; Wiegner, R.; Radermacher, P.; Huber-Lang, M.; Ignatius, A. The crucial role of neutrophil granulocytes in bone fracture healing. *Eur. Cells Mater.* **2016**, *32*, 152–162. [CrossRef]
80. Green, A.C.; Rudolph-Stringer, V.; Chantry, A.D.; Wu, J.Y.; Purton, L.E. Mesenchymal lineage cells and their importance in B lymphocyte niches. *Bone* **2019**, *119*, 42–56. [CrossRef]
81. Okamoto, K.; Nakashima, T.; Shinohara, M.; Negishi-Koga, T.; Komatsu, N.; Terashima, A.; Sawa, S.; Nitta, T.; Takayanagi, H. Osteoimmunology: The Conceptual Framework Unifying the Immune and Skeletal Systems. *Physiol. Rev.* **2017**, *97*, 1295–1349. [CrossRef]
82. Furtado, G.E.; Letieri, R.V.; Caldo-Silva, A.; Sardão, V.A.; Teixeira, A.M.; de Barros, M.P.; Vieira, R.P.; Bachi, A.L.L. Sustaining efficient immune functions with regular physical exercise in the COVID-19 era and beyond. *Eur. J. Clin. Investig.* **2021**, *51*, e13485. [CrossRef] [PubMed]
83. Manilay, J.O.; Zouali, M. Tight relationships between B lymphocytes and the skeletal system. *Trends Mol. Med.* **2014**, *20*, 405–412. [CrossRef] [PubMed]
84. Kanematsu, M.; Sato, T.; Takai, H.; Watanabe, K.; Ikeda, K.; Yamada, Y. Prostaglandin E2 induces expression of receptor activator of nuclear factor-kappa B ligand/osteoprotegrin ligand on pre-B cells: Implications for accelerated osteoclastogenesis in estrogen deficiency. *J. Bone Miner. Res.* **2000**, *15*, 1321–1329. [CrossRef] [PubMed]
85. Horowitz, M.C.; Fretz, J.A.; Lorenzo, J.A. How B cells influence bone biology in health and disease. *Bone* **2010**, *47*, 472–479. [CrossRef] [PubMed]
86. Könnecke, I.; Serra, A.; El Khassawna, T.; Schlundt, C.; Schell, H.; Hauser, A.; Ellinghaus, A.; Volk, H.D.; Radbruch, A.; Duda, G.N.; et al. T and B cells participate in bone repair by infiltrating the fracture callus in a two-wave fashion. *Bone* **2014**, *64*, 155–165. [CrossRef] [PubMed]
87. Grčević, D.; Sironi, M.; Valentino, S.; Deban, L.; Cvija, H.; Inforzato, A.; Kovačić, N.; Katavić, V.; Kelava, T.; Kalajzić, I.; et al. The Long Pentraxin 3 Plays a Role in Bone Turnover and Repair. *Front. Immunol.* **2018**, *9*, 417. [CrossRef]
88. Pan, M.; Hong, W.; Yao, Y.; Gao, X.; Zhou, Y.; Fu, G.; Li, Y.; Guo, Q.; Rao, X.; Tang, P.; et al. Activated B Lymphocyte Inhibited the Osteoblastogenesis of Bone Mesenchymal Stem Cells by Notch Signaling. *Stem Cells Int.* **2019**, *2019*, 8150123. [CrossRef] [PubMed]
89. Manabe, N.; Kawaguchi, H.; Chikuda, H.; Miyaura, C.; Inada, M.; Nagai, R.; Nabeshima, Y.; Nakamura, K.; Sinclair, A.M.; Scheuermann, R.H.; et al. Connection between B lymphocyte and osteoclast differentiation pathways. *J. Immunol.* **2001**, *167*, 2625–2631. [CrossRef] [PubMed]

90. Sato, T.; Shibata, T.; Ikeda, K.; Watanabe, K. Generation of bone-resorbing osteoclasts from B220+ cells: Its role in accelerated osteoclastogenesis due to estrogen deficiency. *J. Bone Miner. Res.* **2001**, *16*, 2215–2221. [CrossRef]
91. Katavić, V.; Grcević, D.; Lee, S.K.; Kalinowski, J.; Jastrzebski, S.; Dougall, W.; Anderson, D.; Puddington, L.; Aguila, H.L.; Lorenzo, J.A. The surface antigen CD45R identifies a population of estrogen-regulated murine marrow cells that contain osteoclast precursors. *Bone* **2003**, *32*, 581–590. [CrossRef]
92. Rayahin, J.E.; Gemeinhart, R.A. Activation of Macrophages in Response to Biomaterials. *Results Probl. Cell Differ.* **2017**, *62*, 317–351. [PubMed]
93. Ouyang, L.; Cao, J.; Dai, Q.; Qiu, D. New insight of immuno-engineering in osteoimmunomodulation for bone regeneration. *Regen. Ther.* **2021**, *18*, 24–29. [CrossRef] [PubMed]
94. Vasandan, A.B.; Jahnavi, S.; Shashank, C.; Prasad, P.; Kumar, A.; Prasanna, S.J. Human Mesenchymal stem cells program macrophage plasticity by altering their metabolic status via a PGE(2)-dependent mechanism. *Sci. Rep.* **2016**, *6*, 38308. [CrossRef] [PubMed]
95. Pelletier, M.; Maggi, L.; Micheletti, A.; Lazzeri, E.; Tamassia, N.; Costantini, C.; Cosmi, L.; Lunardi, C.; Annunziato, F.; Romagnani, S.; et al. Evidence for a cross-talk between human neutrophils and Th17 cells. *Blood* **2010**, *115*, 335–343. [CrossRef]
96. Rosales, C. Neutrophils at the crossroads of innate and adaptive immunity. *J. Leukoc. Biol.* **2020**, *108*, 377–396. [CrossRef] [PubMed]
97. Bohner, M.; Galea, L.; Doebelin, N. Calcium phosphate bone graft substitutes: Failures and hopes. *J. Eur. Ceram. Soc.* **2012**, *32*, 2663–2671. [CrossRef]
98. Prasad, A.S. Zinc: Role in immunity, oxidative stress and chronic inflammation. *Curr. Opin. Clin. Nutr. Metab. Care* **2009**, *12*, 646–652. [CrossRef] [PubMed]
99. Hung, C.J.; Kao, C.T.; Chen, Y.J.; Shie, M.Y.; Huang, T.H. Antiosteoclastogenic activity of silicate-based materials antagonizing receptor activator for nuclear factor κB ligand-induced osteoclast differentiation of murine marcophages. *J. Endod.* **2013**, *39*, 1557–1561. [CrossRef] [PubMed]
100. MacLeod, R.J.; Hayes, M.; Pacheco, I. Wnt5a secretion stimulated by the extracellular calcium-sensing receptor inhibits defective Wnt signaling in colon cancer cells. *Am. J. Physiol. Gastrointest. Liver Physiol.* **2007**, *293*, G403–G411. [CrossRef]
101. Loi, F.; Córdova, L.A.; Pajarinen, J.; Lin, T.H.; Yao, Z.; Goodman, S.B. Inflammation, fracture and bone repair. *Bone* **2016**, *86*, 119–130. [CrossRef] [PubMed]
102. Porter, C.; Sousse, L.E.; Irick, R.; Schryver, E.; Klein, G.L. Interactions of Phosphate Metabolism with Serious Injury, including Burns. *JBMR Plus* **2017**, *1*, 59–65. [CrossRef] [PubMed]
103. Gani, M.A.; Budiatin, A.S.; Lestari, M.; Rantam, F.A.; Ardianto, C.; Khotib, J. Fabrication and Characterization of Submicron-Scale Bovine Hydroxyapatite: A Top-Down Approach for a Natural Biomaterial. *Materials* **2022**, *15*, 2324. [CrossRef] [PubMed]
104. Khotib, J.; Gani, M.A.; Budiatin, A.S.; Lestari, M.; Rahadiansyah, E.; Ardianto, C. Signaling Pathway and Transcriptional Regulation in Osteoblasts during Bone Healing: Direct Involvement of Hydroxyapatite as a Biomaterial. *Pharmaceuticals* **2021**, *14*, 615. [CrossRef]
105. Staiger, M.P.; Pietak, A.M.; Huadmai, J.; Dias, G. Magnesium and its alloys as orthopedic biomaterials: A review. *Biomaterials* **2006**, *27*, 1728–1734. [CrossRef]
106. Sugimoto, J.; Romani, A.M.; Valentin-Torres, A.M.; Luciano, A.A.; Ramirez Kitchen, C.M.; Funderburg, N.; Mesiano, S.; Bernstein, H.B. Magnesium decreases inflammatory cytokine production: A novel innate immunomodulatory mechanism. *J. Immunol.* **2012**, *188*, 6338–6346. [CrossRef]
107. Lötscher, J.; Martí, I.L.A.A.; Kirchhammer, N.; Cribioli, E.; Giordano Attianese, G.M.P.; Trefny, M.P.; Lenz, M.; Rothschild, S.I.; Strati, P.; Künzli, M.; et al. Magnesium sensing via LFA-1 regulates $CD8^+$ T cell effector function. *Cell* **2022**, *185*, 585–602.e529. [CrossRef] [PubMed]
108. Bird, L. Magnesium: Essential for T cells. *Nat. Rev. Immunol.* **2022**, *22*, 144–145. [CrossRef]
109. Roy, M.; Bose, S. Osteoclastogenesis and osteoclastic resorption of tricalcium phosphate: Effect of strontium and magnesium doping. *J. Biomed. Mater. Res. Part A* **2012**, *100*, 2450–2461. [CrossRef]
110. Shorr, E.; Carter, A.C. The usefulness of strontium as an adjuvant to calcium in the remineralization of the skeleton in man. *Bull. Hosp. Jt. Dis.* **1952**, *13*, 59–66.
111. Xu, K.; Chen, W.; Hu, Y.; Shen, X.; Xu, G.; Ran, Q.; Yu, Y.; Mu, C.; Cai, K. Influence of strontium ions incorporated into nanosheet-pore topographical titanium substrates on osteogenic differentiation of mesenchymal stem cells in vitro and on osseointegration in vivo. *J. Mater. Chem. B* **2016**, *4*, 4549–4564. [CrossRef] [PubMed]
112. Huang, Y.Z.; Wu, C.G.; Xie, H.Q.; Li, Z.Y.; Silini, A.; Parolini, O.; Wu, Y.; Deng, L.; Huang, Y.C. Strontium Promotes the Proliferation and Osteogenic Differentiation of Human Placental Decidual Basalis- and Bone Marrow-Derived MSCs in a Dose-Dependent Manner. *Stem Cells Int.* **2019**, *2019*, 4242178. [CrossRef] [PubMed]
113. Aimaiti, A.; Maimaitiyiming, A.; Boyong, X.; Aji, K.; Li, C.; Cui, L. Low-dose strontium stimulates osteogenesis but high-dose doses cause apoptosis in human adipose-derived stem cells via regulation of the ERK1/2 signaling pathway. *Stem Cell Res. Ther.* **2017**, *8*, 282. [CrossRef] [PubMed]
114. Guo, X.; Wei, S.; Lu, M.; Shao, Z.; Lu, J.; Xia, L.; Lin, K.; Zou, D. Dose-dependent Effects of Strontium Ranelate on Ovariectomy Rat Bone Marrow Mesenchymal Stem Cells and Human Umbilical Vein Endothelial Cells. *Int. J. Biol. Sci.* **2016**, *12*, 1511–1522. [CrossRef]

115. Shen, X.; Fang, K.; Ru Yie, K.H.; Zhou, Z.; Shen, Y.; Wu, S.; Zhu, Y.; Deng, Z.; Ma, P.; Ma, J.; et al. High proportion strontium-doped micro-arc oxidation coatings enhance early osseointegration of titanium in osteoporosis by anti-oxidative stress pathway. *Bioact. Mater.* **2022**, *10*, 405–419. [CrossRef]
116. Chen, Y.; Guan, M.; Ren, R.; Gao, C.; Cheng, H.; Li, Y.; Gao, B.; Wei, Y.; Fu, J.; Sun, J.; et al. Improved Immunoregulation of Ultra-Low-Dose Silver Nanoparticle-Loaded TiO_2 Nanotubes via M2 Macrophage Polarization by Regulating GLUT1 and Autophagy. *Int. J. Nanomed.* **2020**, *15*, 2011–2026. [CrossRef]
117. Chattopadhyay, N.; Quinn, S.J.; Kifor, O.; Ye, C.; Brown, E.M. The calcium-sensing receptor (CaR) is involved in strontium ranelate-induced osteoblast proliferation. *Biochem. Pharmacol.* **2007**, *74*, 438–447. [CrossRef] [PubMed]
118. Peng, S.; Zhou, G.; Luk, K.D.; Cheung, K.M.; Li, Z.; Lam, W.M.; Zhou, Z.; Lu, W.W. Strontium promotes osteogenic differentiation of mesenchymal stem cells through the Ras/MAPK signaling pathway. *Cell. Physiol. Biochem.* **2009**, *23*, 165–174. [CrossRef]
119. Prasad, A.S. Zinc in human health: Effect of zinc on immune cells. *Mol. Med.* **2008**, *14*, 353–357. [CrossRef] [PubMed]
120. Dubben, S.; Hönscheid, A.; Winkler, K.; Rink, L.; Haase, H. Cellular zinc homeostasis is a regulator in monocyte differentiation of HL-60 cells by 1 alpha,25-dihydroxyvitamin D3. *J. Leukoc. Biol.* **2010**, *87*, 833–844. [CrossRef]
121. Chang, K.L.; Hung, T.C.; Hsieh, B.S.; Chen, Y.H.; Chen, T.F.; Cheng, H.L. Zinc at pharmacologic concentrations affects cytokine expression and induces apoptosis of human peripheral blood mononuclear cells. *Nutrition* **2006**, *22*, 465–474. [CrossRef] [PubMed]
122. Brazão, V.; Caetano, L.C.; Del Vecchio Filipin, M.; Paula Alonso Toldo, M.; Caetano, L.N.; do Prado, J.C., Jr. Zinc supplementation increases resistance to experimental infection by Trypanosoma cruzi. *Vet. Parasitol.* **2008**, *154*, 32–37. [CrossRef] [PubMed]
123. Wilhelmi, V.; Fischer, U.; van Berlo, D.; Schulze-Osthoff, K.; Schins, R.P.; Albrecht, C. Evaluation of apoptosis induced by nanoparticles and fine particles in RAW 264.7 macrophages: Facts and artefacts. *Toxicol. In Vitro* **2012**, *26*, 323–334. [CrossRef] [PubMed]
124. Wilhelmi, V.; Fischer, U.; Weighardt, H.; Schulze-Osthoff, K.; Nickel, C.; Stahlmecke, B.; Kuhlbusch, T.A.; Scherbart, A.M.; Esser, C.; Schins, R.P.; et al. Zinc oxide nanoparticles induce necrosis and apoptosis in macrophages in a p47phox- and Nrf2-independent manner. *PLoS ONE* **2013**, *8*, e65704. [CrossRef]
125. Gao, H.; Zhao, L.; Wang, H.; Xie, E.; Wang, X.; Wu, Q.; Yu, Y.; He, X.; Ji, H.; Rink, L.; et al. Metal transporter Slc39a10 regulates susceptibility to inflammatory stimuli by controlling macrophage survival. *Proc. Natl. Acad. Sci. USA* **2017**, *114*, 12940–12945. [CrossRef]
126. Kapetanovic, R.; Bokil, N.J.; Achard, M.E.; Ong, C.L.; Peters, K.M.; Stocks, C.J.; Phan, M.D.; Monteleone, M.; Schroder, K.; Irvine, K.M.; et al. Salmonella employs multiple mechanisms to subvert the TLR-inducible zinc-mediated antimicrobial response of human macrophages. *FASEB J.* **2016**, *30*, 1901–1912. [CrossRef]
127. West, A.P.; Brodsky, I.E.; Rahner, C.; Woo, D.K.; Erdjument-Bromage, H.; Tempst, P.; Walsh, M.C.; Choi, Y.; Shadel, G.S.; Ghosh, S. TLR signalling augments macrophage bactericidal activity through mitochondrial ROS. *Nature* **2011**, *472*, 476–480. [CrossRef]
128. Wan, Y.; Petris, M.J.; Peck, S.C. Separation of zinc-dependent and zinc-independent events during early LPS-stimulated TLR4 signaling in macrophage cells. *FEBS Lett.* **2014**, *588*, 2928–2935. [CrossRef]
129. Summersgill, H.; England, H.; Lopez-Castejon, G.; Lawrence, C.B.; Luheshi, N.M.; Pahle, J.; Mendes, P.; Brough, D. Zinc depletion regulates the processing and secretion of IL-1β. *Cell Death Dis.* **2014**, *5*, e1040. [CrossRef]
130. Brough, D.; Pelegrin, P.; Rothwell, N.J. Pannexin-1-dependent caspase-1 activation and secretion of IL-1β is regulated by zinc. *Eur. J. Immunol.* **2009**, *39*, 352–358. [CrossRef]
131. Bonaventura, P.; Benedetti, G.; Albarède, F.; Miossec, P. Zinc and its role in immunity and inflammation. *Autoimmun. Rev.* **2015**, *14*, 277–285. [CrossRef] [PubMed]
132. Grey, Z.J.; Howie, R.N.; Durham, E.L.; Hall, S.R.; Helke, K.L.; Steed, M.B.; LaRue, A.C.; Muise-Helmericks, R.C.; Cray, J.J. Sub-clinical dose of bone morphogenetic protein-2 does not precipitate rampant, sustained inflammatory response in bone wound healing. *Wound Repair Regen.* **2019**, *27*, 335–344. [CrossRef] [PubMed]
133. Shen, H.; Shi, J.; Zhi, Y.; Yang, X.; Yuan, Y.; Si, J.; Shen, S.G.F. Improved BMP2-CPC-stimulated osteogenesis in vitro and in vivo via modulation of macrophage polarization. *Mater. Sci. Eng. C Mater. Biol. Appl.* **2021**, *118*, 111471. [CrossRef] [PubMed]
134. Lv, L.; Xie, Y.; Li, K.; Hu, T.; Lu, X.; Cao, Y.; Zheng, X. Unveiling the Mechanism of Surface Hydrophilicity-Modulated Macrophage Polarization. *Adv. Healthc. Mater.* **2018**, *7*, e1800675. [CrossRef]
135. Son, J.S.; Kim, S.G.; Oh, J.S.; Appleford, M.; Oh, S.; Ong, J.L.; Lee, K.B. Hydroxyapatite/polylactide biphasic combination scaffold loaded with dexamethasone for bone regeneration. *J. Biomed. Mater. Res. A* **2011**, *99*, 638–647. [CrossRef]
136. Koedam, J.A.; Smink, J.J.; van Buul-Offers, S.C. Glucocorticoids inhibit vascular endothelial growth factor expression in growth plate chondrocytes. *Mol. Cell. Endocrinol.* **2002**, *197*, 35–44. [CrossRef]
137. Tsianakas, A.; Varga, G.; Barczyk, K.; Bode, G.; Nippe, N.; Kran, N.; Roth, J.; Luger, T.A.; Ehrchen, J.; Sunderkoetter, C. Induction of an anti-inflammatory human monocyte subtype is a unique property of glucocorticoids, but can be modified by IL-6 and IL-10. *Immunobiology* **2012**, *217*, 329–335. [CrossRef]
138. Gao, L.N.; An, Y.; Lei, M.; Li, B.; Yang, H.; Lu, H.; Chen, F.M.; Jin, Y. The effect of the coumarin-like derivative osthole on the osteogenic properties of human periodontal ligament and jaw bone marrow mesenchymal stem cell sheets. *Biomaterials* **2013**, *34*, 9937–9951. [CrossRef]
139. Tang, D.Z.; Hou, W.; Zhou, Q.; Zhang, M.; Holz, J.; Sheu, T.J.; Li, T.F.; Cheng, S.D.; Shi, Q.; Harris, S.E.; et al. Osthole stimulates osteoblast differentiation and bone formation by activation of beta-catenin-BMP signaling. *J. Bone Miner. Res.* **2010**, *25*, 1234–1245. [CrossRef]

140. Jin, Y.; Qian, J.; Ju, X.; Bao, X.; Li, L.; Zheng, S.; Chen, X.; Xiao, Z.; Chen, X.; Zhu, W.; et al. Osthole Protects against Acute Lung Injury by Suppressing NF-κB-Dependent Inflammation. *Mediat. Inflamm.* **2018**, *2018*, 4934592. [CrossRef]
141. Ma, Y.; Wang, L.; Zheng, S.; Xu, J.; Pan, Y.; Tu, P.; Sun, J.; Guo, Y. Osthole inhibits osteoclasts formation and bone resorption by regulating NF-κB signaling and NFATc1 activations stimulated by RANKL. *J. Cell. Biochem.* **2019**, *120*, 16052–16061. [CrossRef] [PubMed]
142. Yang, C.; Ouyang, L.; Wang, W.; Chen, B.; Liu, W.; Yuan, X.; Luo, Y.; Cheng, T.; Yeung, K.W.K.; Liu, X.; et al. Sodium butyrate-modified sulfonated polyetheretherketone modulates macrophage behavior and shows enhanced antibacterial and osteogenic functions during implant-associated infections. *J. Mater. Chem. B* **2019**, *7*, 5541–5553. [CrossRef] [PubMed]
143. Li, J.; Wen, J.; Li, B.; Li, W.; Qiao, W.; Shen, J.; Jin, W.; Jiang, X.; Yeung, K.W.K.; Chu, P.K. Valence State Manipulation of Cerium Oxide Nanoparticles on a Titanium Surface for Modulating Cell Fate and Bone Formation. *Adv. Sci.* **2018**, *5*, 1700678. [CrossRef]
144. Buck, E.; Lee, S.; Stone, L.S.; Cerruti, M. Protein Adsorption on Surfaces Functionalized with COOH Groups Promotes Anti-inflammatory Macrophage Responses. *ACS Appl. Mater. Interfaces* **2021**, *13*, 7021–7036. [CrossRef]
145. Visalakshan, R.M.; MacGregor, M.N.; Sasidharan, S.; Ghazaryan, A.; Mierczynska-Vasilev, A.M.; Morsbach, S.; Mailänder, V.; Landfester, K.; Hayball, J.D.; Vasilev, K. Biomaterial Surface Hydrophobicity-Mediated Serum Protein Adsorption and Immune Responses. *ACS Appl. Mater. Interfaces* **2019**, *11*, 27615–27623. [CrossRef]
146. Chen, Z.; Bachhuka, A.; Han, S.; Wei, F.; Lu, S.; Visalakshan, R.M.; Vasilev, K.; Xiao, Y. Tuning Chemistry and Topography of Nanoengineered Surfaces to Manipulate Immune Response for Bone Regeneration Applications. *ACS Nano* **2017**, *11*, 4494–4506. [CrossRef]
147. Vishwakarma, A.; Bhise, N.S.; Evangelista, M.B.; Rouwkema, J.; Dokmeci, M.R.; Ghaemmaghami, A.M.; Vrana, N.E.; Khademhosseini, A. Engineering Immunomodulatory Biomaterials to Tune the Inflammatory Response. *Trends Biotechnol.* **2016**, *34*, 470–482. [CrossRef] [PubMed]
148. Jones, J.A.; Chang, D.T.; Meyerson, H.; Colton, E.; Kwon, I.K.; Matsuda, T.; Anderson, J.M. Proteomic analysis and quantification of cytokines and chemokines from biomaterial surface-adherent macrophages and foreign body giant cells. *J. Biomed. Mater. Res. A* **2007**, *83*, 585–596. [CrossRef] [PubMed]
149. Vlacic-Zischke, J.; Hamlet, S.M.; Friis, T.; Tonetti, M.S.; Ivanovski, S. The influence of surface microroughness and hydrophilicity of titanium on the up-regulation of TGFβ/BMP signalling in osteoblasts. *Biomaterials* **2011**, *32*, 665–671. [CrossRef]
150. Krombach, F.; Münzing, S.; Allmeling, A.M.; Gerlach, J.T.; Behr, J.; Dörger, M. Cell size of alveolar macrophages: An interspecies comparison. *Environ. Health Perspect.* **1997**, *105* (Suppl. 5), 1261–1263.
151. Fairley, M.; Unruh, D.K.; Donovan, A.; Abeysinghe, S.; Forbes, T.Z. Synthesis and characterization of homo- and heteronuclear molecular Al^{3+} and Th^{4+} species chelated by the ethylenediaminetetraacetate (edta) ligand. *Dalton Trans.* **2013**, *42*, 13706–13714. [CrossRef] [PubMed]
152. Wu, A.C.; Raggatt, L.J.; Alexander, K.A.; Pettit, A.R. Unraveling macrophage contributions to bone repair. *BoneKEy Rep.* **2013**, *2*, 373. [CrossRef] [PubMed]
153. Klinge, U.; Klosterhalfen, B.; Birkenhauer, V.; Junge, K.; Conze, J.; Schumpelick, V. Impact of polymer pore size on the interface scar formation in a rat model. *J. Surg. Res.* **2002**, *103*, 208–214. [CrossRef] [PubMed]
154. Kuboki, Y.; Jin, Q.; Kikuchi, M.; Mamood, J.; Takita, H. Geometry of artificial ECM: Sizes of pores controlling phenotype expression in BMP-induced osteogenesis and chondrogenesis. *Connect. Tissue Res.* **2002**, *43*, 529–534. [CrossRef]
155. Karageorgiou, V.; Kaplan, D. Porosity of 3D biomaterial scaffolds and osteogenesis. *Biomaterials* **2005**, *26*, 5474–5491. [CrossRef]
156. Han, Q.; Wang, C.; Chen, H.; Zhao, X.; Wang, J. Porous Tantalum and Titanium in Orthopedics: A Review. *ACS Biomater. Sci. Eng.* **2019**, *5*, 5798–5824. [CrossRef]
157. Junge, K.; Binnebösel, M.; von Trotha, K.T.; Rosch, R.; Klinge, U.; Neumann, U.P.; Lynen Jansen, P. Mesh biocompatibility: Effects of cellular inflammation and tissue remodelling. *Langenbeck's Arch. Surg.* **2012**, *397*, 255–270. [CrossRef]
158. Anderson, J.M.; Rodriguez, A.; Chang, D.T. Foreign body reaction to biomaterials. *Semin. Immunol.* **2008**, *20*, 86–100. [CrossRef]
159. Zaveri, T.D.; Lewis, J.S.; Dolgova, N.V.; Clare-Salzler, M.J.; Keselowsky, B.G. Integrin-directed modulation of macrophage responses to biomaterials. *Biomaterials* **2014**, *35*, 3504–3515. [CrossRef]
160. Ward, W.K.; Slobodzian, E.P.; Tiekotter, K.L.; Wood, M.D. The effect of microgeometry, implant thickness and polyurethane chemistry on the foreign body response to subcutaneous implants. *Biomaterials* **2002**, *23*, 4185–4192. [CrossRef]
161. Nichols, S.P.; Koh, A.; Storm, W.L.; Shin, J.H.; Schoenfisch, M.H. Biocompatible materials for continuous glucose monitoring devices. *Chem. Rev.* **2013**, *113*, 2528–2549. [CrossRef] [PubMed]
162. Oh, W.K.; Kim, S.; Choi, M.; Kim, C.; Jeong, Y.S.; Cho, B.R.; Hahn, J.S.; Jang, J. Cellular uptake, cytotoxicity, and innate immune response of silica-titania hollow nanoparticles based on size and surface functionality. *ACS Nano* **2010**, *4*, 5301–5313. [CrossRef] [PubMed]
163. Sudhakar, K.; Madhusudana Rao, K.; Subha, M.C.S.; Chowdoji Rao, K.; Sadiku, E.R. Temperature-responsive poly(N-vinylcaprolactam-co-hydroxyethyl methacrylate) nanogels for controlled release studies of curcumin. *Des. Monomers Polym.* **2015**, *18*, 705–713. [CrossRef]
164. Laquerriere, P.; Grandjean-Laquerriere, A.; Jallot, E.; Balossier, G.; Frayssinet, P.; Guenounou, M. Importance of hydroxyapatite particles characteristics on cytokines production by human monocytes in vitro. *Biomaterials* **2003**, *24*, 2739–2747. [CrossRef]

165. Davison, N.L.; Su, J.; Yuan, H.; van den Beucken, J.J.; de Bruijn, J.D.; Barrère-de Groot, F. Influence of surface microstructure and chemistry on osteoinduction and osteoclastogenesis by biphasic calcium phosphate discs. *Eur. Cells Mater.* **2015**, *29*, 314–329. [CrossRef]
166. Li, M.; Guo, X.; Qi, W.; Wu, Z.; de Bruijn, J.D.; Xiao, Y.; Bao, C.; Yuan, H. Macrophage polarization plays roles in bone formation instructed by calcium phosphate ceramics. *J. Mater. Chem. B* **2020**, *8*, 1863–1877. [CrossRef]
167. Wójciak-Stothard, B.; Madeja, Z.; Korohoda, W.; Curtis, A.; Wilkinson, C. Activation of macrophage-like cells by multiple grooved substrata. Topographical control of cell behaviour. *Cell Biol. Int.* **1995**, *19*, 485–490. [CrossRef]
168. Klymov, A.; Prodanov, L.; Lamers, E.; Jansen, J.A.; Walboomers, X.F. Understanding the role of nano-topography on the surface of a bone-implant. *Biomater. Sci.* **2013**, *1*, 135–151. [CrossRef]
169. Castro-Raucci, L.M.S.; Francischini, M.S.; Teixeira, L.N.; Ferraz, E.P.; Lopes, H.B.; de Oliveira, P.T.; Hassan, M.Q.; Rosa, A.L.; Beloti, M.M. Titanium with Nanotopography Induces Osteoblast Differentiation by Regulating Endogenous Bone Morphogenetic Protein Expression and Signaling Pathway. *J. Cell. Biochem.* **2016**, *117*, 1718–1726. [CrossRef]
170. Takebe, J.; Champagne, C.M.; Offenbacher, S.; Ishibashi, K.; Cooper, L.F. Titanium surface topography alters cell shape and modulates bone morphogenetic protein 2 expression in the J774A.1 macrophage cell line. *J. Biomed. Mater. Res. A* **2003**, *64*, 207–216. [CrossRef]
171. Refai, A.K.; Textor, M.; Brunette, D.M.; Waterfield, J.D. Effect of titanium surface topography on macrophage activation and secretion of proinflammatory cytokines and chemokines. *J. Biomed. Mater. Res. A* **2004**, *70*, 194–205. [CrossRef] [PubMed]
172. Lebre, F.; Hearnden, C.H.; Lavelle, E.C. Modulation of Immune Responses by Particulate Materials. *Adv. Mater.* **2016**, *28*, 5525–5541. [CrossRef] [PubMed]
173. Hotchkiss, K.M.; Reddy, G.B.; Hyzy, S.L.; Schwartz, Z.; Boyan, B.D.; Olivares-Navarrete, R. Titanium surface characteristics, including topography and wettability, alter macrophage activation. *Acta Biomater.* **2016**, *31*, 425–434. [CrossRef] [PubMed]
174. Christo, S.N.; Bachhuka, A.; Diener, K.R.; Mierczynska, A.; Hayball, J.D.; Vasilev, K. Te Role of Surface Nanotopography and Chemistry on Primary Neutrophil and Macrophage Cellular Responses. *Adv. Healthc. Mater.* **2016**, *5*, 956–965. [CrossRef]
175. Chen, H.C. Boyden chamber assay. *Methods Mol. Biol.* **2005**, *294*, 15–22.
176. Cary, L.A.; Chang, J.F.; Guan, J.L. Stimulation of cell migration by overexpression of focal adhesion kinase and its association with Src and Fyn. *J. Cell Sci.* **1996**, *109 Pt 7*, 1787–1794. [CrossRef]
177. Nyangoga, H.; Aguado, E.; Goyenvalle, E.; Baslé, M.F.; Chappard, D. A non-steroidal anti-inflammatory drug (ketoprofen) does not delay beta-TCP bone graft healing. *Acta Biomater.* **2010**, *6*, 3310–3317. [CrossRef]

Journal of
Functional
Biomaterials

Review

Smart Bacteria-Responsive Drug Delivery Systems in Medical Implants

Yijie Yang, Xue Jiang, Hongchang Lai * and Xiaomeng Zhang *

Shanghai Key Laboratory of Stomatology, Department of Oral and Maxillo-Facial Implantology, Shanghai Ninth People's Hospital, School of Medicine, Shanghai Jiao Tong University, Shanghai 200011, China
* Correspondence: hongchanglai@126.com (H.L.); zhangxiaomengwowo@126.com (X.Z.)

Abstract: With the rapid development of implantable biomaterials, the rising risk of bacterial infections has drawn widespread concern. Due to the high recurrence rate of bacterial infections and the issue of antibiotic resistance, the common treatments of peri-implant infections cannot meet the demand. In this context, stimuli-responsive biomaterials have attracted attention because of their great potential to spontaneously modulate the drug releasing rate. Numerous smart bacteria-responsive drug delivery systems (DDSs) have, therefore, been designed to temporally and spatially release antibacterial agents from the implants in an autonomous manner at the infected sites. In this review, we summarized recent advances in bacteria-responsive DDSs used for combating bacterial infections, mainly according to the different trigger modes, including physical stimuli-responsive, virulence-factor-responsive, host-immune-response responsive and their combinations. It is believed that the smart bacteria-responsive DDSs will become the next generation of mainstream antibacterial therapies.

Keywords: anti-bacterial; implants; drug release; stimuli-responsive

Citation: Yang, Y.; Jiang, X.; Lai, H.; Zhang, X. Smart Bacteria-Responsive Drug Delivery Systems in Medical Implants. *J. Funct. Biomater.* **2022**, *13*, 173. https://doi.org/10.3390/jfb13040173

Academic Editors: Nenad Ignjatovic and Anişoara Cîmpean

Received: 5 September 2022
Accepted: 29 September 2022
Published: 1 October 2022

Publisher's Note: MDPI stays neutral with regard to jurisdictional claims in published maps and institutional affiliations.

Copyright: © 2022 by the authors. Licensee MDPI, Basel, Switzerland. This article is an open access article distributed under the terms and conditions of the Creative Commons Attribution (CC BY) license (https://creativecommons.org/licenses/by/4.0/).

1. Introduction

Recently, the rapid development of implantable biomaterials has benefited people suffering from bone and dentition defects. However, all the surgical interventions that involve implantation of biomaterials face the risk of failure due to aseptic loosening and bacterial infections. The high predisposition for infections around post-implant sites is caused by lowered immune system efficacy and the adhesion and biofilm-forming ability of bacteria. Biofilms, in which bacteria are protected from the immune responses, dynamic environments and conventional antibiotics, are essential for the proliferation of bacteria [1]. The common treatments for peri-implant infections are limited to a combination of aggressive surgical debridements and systemic antibiotic regimens, and may eventually end up with device removal if there is no way to control the infections. Moreover, a key feature of bacterial infections is recurrence, which happens in approximately 30% of all cases [2], which indicates that repeated antibiotic treatments are necessary. However, the more frequent antibiotics are used, the higher the probability of antibiotic resistance.

Although some of the biomaterials show antibacterial properties, the majority of antibacterial activity is carried out through drug delivery systems (DDSs). Conventional DDSs load drugs mainly through physically adding large antibiotics to the matrix or covalently attaching them to the surfaces. However, the physically drug-loaded DDSs may provoke the abrupt release of drugs, which is known for its cytotoxicity. The covalently drug-loaded DDSs, in the meantime, limit antibacterial effects to the system surfaces, because of the characteristics of covalent bonds. Furthermore, the common problem with conventional DDSs is that they cannot be administered on demand [3]. The expected pattern of administration within conventional DDSs is generally sustained release of drugs. When infections occur, the local level of antibiotics may fail to reach the effective therapeutic dose according to this delivery pattern, while in the absence of infections, background leakage of antibiotics can exacerbate antibiotic resistance.

In this context, it is urgent to develop a smart bacteria-responsive DDS that can automatically release antibacterial agents from the implants when infections occur, in a more effective manner without background leakage. Namely, antibiotics should be latent in the absence of bacterial infections, but released adequately to kill bacteria immediately in response to infections. Strategies that utilize the changes specific to the bacteria-infected microenvironment as a unique key to activate drug release have attracted widespread attention in the treatments of peri-implant infections. For instance, bacterial infections can result in an acidic microenvironment (pH = 5.0–5.5) that is distinct from normal physiological conditions (pH = 7.4) [4]. Additionally, the overexpression of virulence factors, such as hyaluronidase, gelatinase and phospholipase, also makes the infected area different from the others [5,6]. Taking those features of infection sites as a stimulus for antibiotic release, a "smart" stimuli-responsive DDS can be designed to achieve more localized and controlled drug release. The greatest benefits of such smart systems are the enhanced efficacy due to higher local concentrations, minimized systemic side effects, and the ability of the released agents to diffuse into the peri-implant tissues, thereby killing bacteria both on the implant surfaces and within the surrounding environment [7–9].

Stimuli-responsive materials have been investigated in the biomedical field for several decades, including as DDSs. Here, we summarized a few smart bacteria-responsive DDSs mainly designed to prevent or solve peri-implant infections (Figure 1). The aim of this review is to summarize and analyze the design principles, autonomous reactiveness against bacteria and antibacterial effects of these systems.

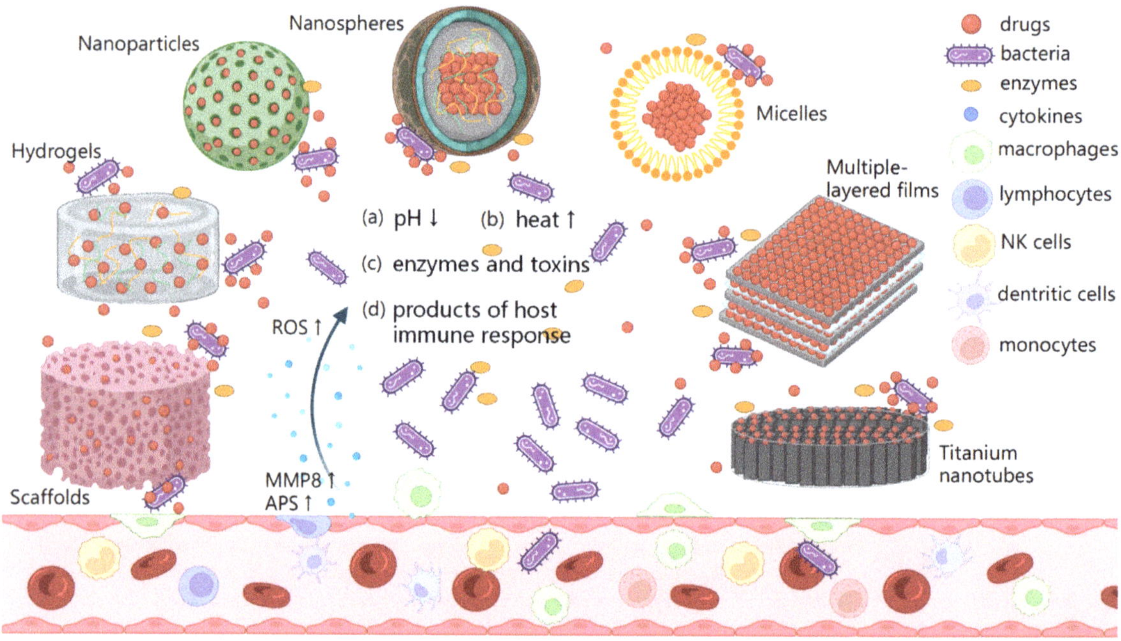

Figure 1. The schematic representation of smart bacteria-responsive drug delivery systems. Scaffolds, hydrogels, nanoparticles, nanosphere, micelles, multiple-layer films and titanium nanotubes (TNTs) loaded with drugs are triggered by the changes specific to the infection microenvironment, including the (a) pH decreasing, (b) elevated local temperature, (c) bacteria-specific enzymes and toxins and (d) products of host immune response, aiming to kill the bacteria.

2. Materials and Methods

2.1. Search Strategy and Study Selection Processes

A comprehensive search was conducted via the following medical databases: PubMed, Embase, and Web of Science, for articles published from 1 January 2012 to 18 June 2022 in English. Search terms included (antibacterial OR anti bacterial OR anti-bacterial OR antibacteria OR anti bacteria OR anti-bacteria OR antimicrobial OR anti-microbial OR antibiotics OR antibiotic OR Bacteriocidal OR Bacteriocide OR bacteriocides OR Anti-Mycobacterial OR anti mycobacterial OR Antimycobacterial OR infection OR anti-infection OR infectious) AND (implants OR implant OR prosthesis OR "Prosthetic Implants" OR "Implant, Prosthetic" OR "Implants, Prosthetic" OR "Prosthetic Implant" OR "Implants, Artificial" OR "artificial implants" OR "artificial implant" OR "Implant, Artificial" OR Prostheses OR Endoprosthesis OR endoprostheses OR nanoparticles OR nanoparticle OR nano-particles OR nano-particle) AND ("Delivery System, Drug" OR "Delivery Systems, Drug" OR "Drug Delivery System" OR "System, Drug Delivery" OR "Systems, Drug Delivery" OR "Drug Targeting" OR "Drug Targetings" OR "Targeting, Drug" OR "Targetings, Drug" OR "Drug delivering" OR "Drug release"). The electronic search showed a total number of 16,443 titles, after removing 3354 duplicates. Furthermore, relevant references were manually searched via the reference lists of the included studies and 4 studies were added to the full-text evaluation. A total of 16,294 studies with clearly irrelevant topics and abstracts or ineligible article types were excluded. The final inclusion was based on the inclusion criteria. The study flow diagram is shown in Figure 2.

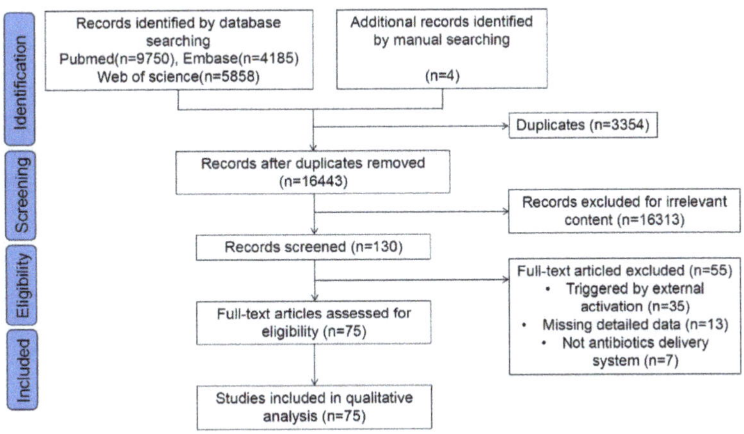

Figure 2. Search flowchart.

2.2. Inclusion and Exclusion Criteria

The inclusion criteria for the study selection were as follows:

1 Primary studies regarding autonomous bacteria-responsive DDSs.
2 Studies aiming to eliminate bacteria via releasing antibacterial drugs.
3 Studies reporting the detailed data of anti-bacterial assays in vitro or in vivo, such as bacterial inhibition rate (BIR), zones of bacterial inhibition (ZOI) and morphological characterization of bacteria (MCB).

The exclusion criteria for the study selection were as follows:

1 Studies that performed controlled drug release by additional artificial activation.
2 The DDSs were not designed for antibacterial purposes.
3 Studies missing detailed data of anti-bacterial assays.

Two researchers (Y.J.Y and X.J) independently conducted the search and screened the titles, abstracts and full text of the papers. Discrepancies were resolved via discussions amongst the researchers. An overview of the experimental details is given in Table 1.

Table 1. Features of the included studies.

Study	Drug(s)	Trigger(s)	Type	Structures	Bacteria	Outcome(s)
Li D. et al., 2022 [10]	VAN	pH switches	Nanoparticles	VAN@PEG-VAN	*S. aureus*	ZOI, MCB (in vitro), BIR (in vivo)
Fu M. et al., 2022 [11]	GTM	pH switches	Hydrogels	GTM@P(AA-co-HEMA)	*E. coli*, *S. aureus*	DLC, ZOI, BIR (in vivo)
Yang, L. et al., 2022 [12]	TOB	pH switches	Films	TOB@Protocatechualdehyde-aminoglycosides	*E. coli*, *P. aeruginosa*, *S. epidermidis*, *S. aureus*	ZOI, MCB (in vitro), BIR (in vivo)
Cámara-Torres M. et al., 2021 [13]	CFX, GTM	pH switches, ion exchange	Scaffolds	PEOT/PBT-MgAl-CFX, PEOT/PBT-Zrp-GTM	*S. epidermidis*, *P. aeruginosa*	ZOI
Guo, R. et al., 2021 [14]	Triclosan	pH switches	Micelles	PLA-PEG-PAE	*E. coli*, *S. aureus*	BIR (in vitro), MCB (in vitro)
Ramesh, S. et al., 2021 [15]	ZnONPs	pH switches	Hydrogels	ZnONPs@CS-GP	*E. coli*, *S. aureus*	BIR (in vitro), DLC
Sang, S. et al., 2021 [16]	GTM	pH switches	Films	GTM-Silk protein	*E. coli*, *S. aureus*	ZOI, DLC, BIR (in vitro), MCB (in vitro)
Yan K. et al., 2021 [17]	AgNPs	pH switches	Hydrogels	CS-AgNPs	*E. coli*, *S. aureus*	ZOI, MCB (in vitro)
Zha, J. et al., 2021 [18]	Curcumin	pH switches	Hydrogels	Curcumin@POEGMA-PEI	MRSA	BIR (in vitro)
Hassan D. et al., 2020 [19]	VAN	pH switches	Quatsomes	VAN-StBAclm	MRSA	DLC, BIR (in vitro and in vivo), MCB (in vitro)
Li, M. et al., 2020 [20]	GTM	pH switches	Films	GTM@(al-ALG/PEI)$_{10}$	*E. coli*, *S. aureus*	ZOI, DLC, BIR (in vitro), MCB (in vitro)
Chen J. et al., 2020 [21]	AMPs	pH switches	Films	TNTs-PMAA-AMP	*S. aureus*, *E. coli*, *P. aeruginosa*, MRSA	BIR (in vitro and in vivo)
Tao B. et al., 2019 [22]	GTM	pH switches	Films	TNTs-BMP2-(ADA-GTM/CS)$_{10}$	*S. aureus*, *E. coli*	BIR (in vitro), MCB (in vitro)
Jin, X. et al., 2019 [23]	GTM	pH switches	Scaffolds	Porous hydroxyapatite-GTM	*E. coli*, *S. aureus*	DLC, ZOI, BIR (in vitro and in vivo)
de Avila E.D. et al., 2019 [24]	TC	pH switches	Films	(PAA/PLL-TC)$_{10}$	*P. gingivalis*	BIR (in vitro)
Cao J. et al., 2019 [25]	CHX	pH switches	Nanoparticles	CHX@PMPC-CS	*S. aureus*	BIR (in vitro); MCB (in vitro)
Hu J. et al., 2019 [26]	TOB, ornidazole	pH switches	Hydrogels	TOB-G1-orni	*S. aureus*, *P. aeruginosa*, *C. sporogenes*, *B. fragilis*	BIR (in vitro); MCB (in vitro)
Karakeçili A. et al., 2019 [27]	VAN	pH switches	Nanoparticles	ZIF8/VAN	*S. aureus*	DLC, BIR (in vitro)

Table 1. *Cont.*

Study	Drug(s)	Trigger(s)	Type	Structures	Bacteria	Outcome(s)
Maji R. et al., 2019 [28]	VAN	pH switches	Nanoparticles	VAN@lipid–dendrimer hybrid NPs	MRSA	DLC, BIR (in vitro)
Mir M. et al., 2019 [29]	CAR	pH switches	Nanoparticles	CAR@PCL-NPs	MRSA	BIR (in vitro)
Zhou W. et al., 2018 [30]	GTM, AgNPs	pH switches	Films	CS-(AgNPs/GTM-SF)	*S. aureus*	BIR (in vitro)
Xiang Y. et al., 2018 [31]	VAN	pH switches	Quantum dots	TNTs-VAN@ZnO-FA	*S. aureus*	BIR (in vitro); MCB (in vitro)
Placente D. et al., 2018 [32]	CFX	pH switches	Nanoparticles	Lipid membrane mimetic coated nano-hydroxyapatite	*E. coli*, *P. aeruginosa*, *S. aureus*	BIR (in vitro), ZOI
Cicuéndez M. et al., 2018 [33]	LFX	pH switches	Scaffolds	MGHA-LFX	*S. aureus*	BIR (in vitro), MCB (in vitro)
Dai T. et al., 2018 [34]	AgNPs	pH switches	Hydrogels	Dex-G5-AgNPs	*E. coli*, *P. aeruginosa*, *S. aureus*, *S. epidermidis*	BIR (in vitro and in vivo), MCB (in vitro)
Dubovoy V. et al., 2018 [35]	BAC	pH swtiches	Nanoparticles	BAC-MSNs	*S. aureus*	BIR (in vitro)
Mhule D. et al., 2018 [36]	VAN	pH swtiches	Nanoparticles	VAN@NMEO	MRSA	DLC, BIR (in vitro and in vivo)
Soltani B. et al., 2018 [37]	GTM	pH swtiches	Nanoparticles	GTM@nanoscale zeolitic imidazolate frame-work-8	*E. coli*, *S. aureus*	DLC, BIR (in vitro)
Yu X. et al., 2018 [38]	VAN	pH swtiches	Microspheres	PLGA–NaHCO3–Van	*S. aureus*, *MRSA*	DLC, BIR (in vitro), ZOI
Zhang S. et al., 2018 [39]	AgNPs	pH switches	Films	AgNPs@PLL/PG	*S. aureus*	BIR (in vitro)
Zhou, W. et al., 2018 [40]	TOB	pH switches	Films	TOB@ $(CHT/HET)_2$	*S. aureus*	ZOI, BIR (in vitro), MCB (in vitro)
Wang T. et al., 2017 [41]	VAN, Ag	pH switches	Films	TNT(NH2)-VAN@Zn-BIX, TNT(NH2)-Ag@Zn-BIX	*E. coli*, *S. aureus*	ZOI, BIR (in vitro), MCB (in vitro)
Pamfil D. et al., 2017 [42]	CFX	pH switches	Hydrogels	CFX @HEMA/C-CA	*S. aureus*	ZOI
Liu Z. et al., 2017 [43]	VAN	pH switches	Nanoparticles	VAN@PVA/PLGA	*S. aureus*	ZOI, MCB (in vitro)
Dong Y. et al., 2017 [44]	AgNPs	pH switches	Films	TNTs-acetal linker-AgNPs	*E. coli*, *S. aureus*	BIR (in vitro)
Kalhapure R. S. et al., 2017 [45]	VAN	pH switches	Nanoparticles	VAN@(2-(2,4,6trimethoxyphenyl)-1,3-dioxane-5,5-diyl) bis(methylene) distearate	*S. aureus*, MRSA	DLC, BIR (in vitro and in vivo)

Table 1. *Cont.*

Study	Drug(s)	Trigger(s)	Type	Structures	Bacteria	Outcome(s)
Sang Q et al., 2017 [46]	CFX	pH switches	Nanofibers	Gelatin-sodium bicarbonate	*E. coli*, *S. aureus*	DLC, BIR (in vitro)
Onat B. et al., 2016 [47]	Triclosan	pH switches	Micelles	(Triclosan@βPDMA-b-PDPA)3	*S. aureus*, *E. coli*	BIR (in vitro), ZOI
Fullriede H. et al., 2016 [48]	CHX	pH switches	Nanoparticles	CHX@silica nanoparticles-PVP	*S. aureus*, *S. mutans*	BIR (in vitro)
Kalhapure R. S. et al., 2017 [49]	VAN	pH switches	Nanoparticles	CS@VAN-AGS	MRSA	BIR (in vitro and in vivo)
Anandhakumar S. et al., 2016 [50]	CFX	pH switches	Films	(PAH/PMAA-CFX)8	*E. coli*	ZOI
Zhang Z. et al., 2015 [51]	MNC	pH switches	Films	(DS-Mg2+-MNC)-GA	*E. coli*, *S. aureus*	BIR (in vitro)
Zhuk I. et al., 2014 [52]	GTM, TOB, polymyxin B	pH switches	Films	TA-GTM/TOB/polymyxin B (PolyB)	*S. epidermidis*, *S. aureus*	BIR (in vitro), MCB (in vitro)
Zhang Z. et al., 2014 [53]	MNC	pH switches	Films	(DS-Ca2+/MNC-Ca2+/GA-Ca2+)8	*E. coli*, *S. aureus*, MRSA, *S. epidermidis*	BIR (in vitro), MCB (in vitro)
Pichavant, L. et al., 2012 [54]	GTM	pH switches	Nanoparticles	GTM@Functionalized PEO	*S. aureus*	DLC, BIR (in vitro)
Choi H. et al., 2021 [55]	LFX	High temperatures	Films	Ti-PDEGMA-LFX	*S. aureus*	DLC, MCB (in vitro and in vivo)
Li B. et al., 2021 [56]	Glycerin, simvastatin	High temperatures	Hydrogels	TNTs-CS-glycerin-hydroxypropylmethyl	*E. coli*, *S. aureus*	BIR (in vivo)
Liang J. et al., 2019 [57]	Colistin, AMPs	Bacterial contact	Microgels	PAA-colistin, PAA-AMPs	*E. coli*, *S. epidermidis*	BIR (in vitro)
Bourgat Y. et al., 2021 [58]	CFX	Enzymes (PS)	Nanogels	Alginate-PLL-CFX	*S. aureus*	BIR (in vitro)
Timin A. et al., 2018 [59]	CFS	Enzymes (PS)	Scaffolds	PCL-CFS, PHB-CFS, (PHB-PANi)-CFS	*E. coli*	ZOI
Liao X. et al., 2021 [60]	CHX	Enzymes (PS)	Films	$(MTT/PLL\text{-}CHX)_{10}$	*S. aureus*	ZOI, DLC, BIR (in vitro and in vivo),
Yu X. et al., 2021 [61]	VAN	Enzymes (PS)	Films	$(MTT/PLL\text{-}VAN)_8$	*S. aureus*	ZOI, MCB (in vitro), BIR (in vitro and in vivo)
Zhang Y. et al., 2021 [62]	VAN	Enzymes (PS)	Films	Ti-SRP1 peptides-VAN	*S. aureus*	BIR (in vitro)
Johnson CT. et al., 2018 [63]	Lysostaphin	Enzymes (PS)	Hydrogels	PEG-4MAL-lysostaphin	*S. aureus*	BIR (in vitro and in vivo)

Table 1. *Cont.*

Study	Drug(s)	Trigger(s)	Type	Structures	Bacteria	Outcome(s)
Li Y. et al., 2020 [64]	VAN	Enzymes (HAS)	Hydrogels	VAN-HA-CS/β-glycerophosphate	*S. aureus*, *S. epidermidis*	BIR (in vitro)
Wang B. et al., 2018 [65]	GTM	Enzymes (HAS)	Films	$(MMT/HA\text{-}GTM)_{10}$	*S. aureus*, *E. coli*	BIR (in vitro and in vivo); ZOI; MCB (in vitro and in vivo)
Ji H. et al., 2016 [66]	AA, MNPs	Enzymes (HAS)	Nanosheet	AA@GMSN-HA-MNPs	*S. aureus*, *E. coli*	BIR (in vitro and in vivo), MCB (in vitro)
Shi R. et al., 2019 [67]	MNA	Enzymes (LS)	Films	PCL-dopamine-MNA	*H. pylori*	BIR (in vitro)
Filipović N. et al., 2019 [68]	SeNPs	Enzymes (LS)	Microspheres	PCL-SeNPs	*S. epidermidis*, *S. aureus*	ZOI, DLC
Yang S. et al., 2018 [69]	GTM	Enzymes (LS)	Nanoparticles	GTM@MSNs-lipid-UBI	*S. aureus*	BIR (in vitro and in vivo)
Li Y.M. et al., 2017 [70]	Triclosan, AMP, parasin I, lysozyme	Enzymes (LS)	Micelles	PEG-b-PA/PN@drugs	*S. aureus*, *E. coli*	DLC, BIR (in vitro)
Xiong M. et al., 2012 [71]	VAN	Enzymes (LS)	Nanogels	Mannosyl-PEG-polyphosphoester-VAN	*S. aureus*	DLC, BIR (in vitro and in vivo), MCB (in vivo)
Xiong M. et al., 2012 [72]	VAN	Enzymes (LS)	Nanogels	PEG-PCL-polyphosphoester-VAN	*S. aureus*	DLC, BIR (in vitro)
Qi GB. et al. 2017 [73]	AMPs	Enzymes (GS)	Nanoparticles	CS-CPC1-AMPs	*S. aureus*	BIR (in vitro and in vivo), MCB (in vitro and in vivo)
Li L.L. et al., 2014 [74]	VAN	Enzymes (GS)	Nanoparticles	SGNPs-VAN @RBC	*S. epidermidis*, *S. aureus*	DLC, BIR (in vitro), MCB
Tonkin, R. L., et al., 2022 [75]	VAN	Cytoloytic toxin	Capsosomes	$VAN@Mesosilica\text{-}PAH\text{-}(PMAA\text{-}PDA/liposome)_3$	MRSA	ZOI, survival rate
Wang T. et al., 2017 [76]	Ampicillin, CA	pH switches, enzymes (LS)	Films	VAMSC-CA/ampicillin-monopyridine functionalized β-cyclodextrin	*E. coli*, *S. aureus*, MRSA	BIR (in vitro), MCB (in vitro)
Chen M. et al., 2019 [77]	D-tyrosine, AZM	pH switches, enzymes (LS)	Micelles	DOEAz-tyrosine	*P. aeruginosa*	BIR (in vitro and in vivo), MCB (in vitro and in vivo)
Chen M. et al., 2018 [78]	VAN, CFX	pH switches, enzymes (LS)	Micelles	CFX@VAN-PECL	*P. aeruginosa*	BIR (in vitro and in vivo), MCB (in vitro and in vivo)
Qu J. et al., 2018 [79]	AMX	pH and electric field switches	Hydrogels	CP/OD-AMX	*E. coli*, *S. aureus*	BIR (in vitro)

Table 1. *Cont.*

Study	Drug(s)	Trigger(s)	Type	Structures	Bacteria	Outcome(s)
Stanton M. M. et al., 2017 [80]	CFX	pH and external magnetic guidance	Biohybrids	MSR1-CFX@MSM	*E. coli*	BIR (in vitro), MCB (in vitro)
Hu C. et al., 2020 [81]	AMIK, naproxen	pH switches, ROS	Hydrogels	(AMIK@ALG-BA)-(naproxen@HA-cholesterol)	*S. aureus*, *P. aeruginosa*	DLC, BIR (in vitro and in vivo), MCB (in vitro)
Guo J. et al., 2019 [82]	MNC, AMP	MMP-8	Hydrogels	MNC@(4-arm PEG-diacrylate)-MMP8 sensitive peptide	*P. gingivalis*	BIR (in vitro)
Polo L. et al., 2018 [83]	LFX	APS	Scaffolds	MBG-LFX-ATP	*E. coli*	BIR (in vitro)
Stavrakis A. et al., 2016 [84]	VAN, TGC	ROS	Films	Van@PEG-PPS, TGC@PEG-PPS	*S. aureus*	BIR (in vivo)

TOB: tobromycin; CFX: ciprofloxacin; GTM: gentamicin; AgNPs: silver nanoparticles; BAC: benzalkonium chloride; AMPs: antimicrobial peptides; TC: tetracycline; VAN: vancomycin; LFX: levofloxiacin; CHX: chlorhexidine; MNC: minocycline; CFS: ceftriaxone sodium; AA: ascorbic acid; MNPs: ferromagnetic nanoparticles; MNA: metronidazole; SeNPs: selenium nanoparticles; CA: cinnamaldehyde; AZM: azithromycin; AMX: amoxicillin; TGC: tigecycline; CAR: carvacrol; AMIK: amikacin; PS: protease; HAS: hyaluronidase; LS: lipase; GS: gelatinase; APS: acid phosphatase; al-ALG: aldylated sodium alginate; ROS: reactive oxygen species; PEOT: poly(ethyleneoxideterephthalate); PBT: poly(butyleneterephthalate); MgAl: magnesium aluminum layered double hydroxides; ZrP: α-zirconium phosphates; PLA: poly(lactic-co-glycolic acid); PLE: poly(β-amino ester); PEG: poly(ethylene glycol); PAE: poly(β-amino ester); GP: glycerol phosphate; CS: chitosan; POEGMA: poly(oligo(ethylene glycol) methacrylate); PEI: poly(ethyleneimine); TNTs: TiO2 nanotubes; PMPC: poly(2methacryloyloxyethyl phosphorylcholine); SGNPs: supramolecular gelatin nanoparticles; G1-orni: amine-terminated poly(amidoamine); PMAA: poly(methacrylic acid); ADA: alginate dialdehyde; PAA: poly(acrylic acid); PLL: poly-l-lysine; SF: silk fibroin; FA: folic acid; MGHA: nanocrystalline apatite uniformly embedded into a mesostructured SiO2-CaO-P2O5 glass wall; CHT: positively charged chitosan; HET: heparin miscalls; BIX: 1,4-bis(imidazol-1-ylmethyl) benzene; HEMA: 2-hydroxyethyl methacrylate; C-CA: citraconic anhydride-modified collagen; PVA: poly(vinyl alcohol); PLGA: poly(lactide-glycolide acid); βPDMA-b-PDPA: poly[3-dimethyl (methacryloyloxyethyl) ammonium propane sulfonate-b-2-(diisopropylamino)ethyl methacrylate]; PG: poly-L-glutamic; NMEO: N-(2-morpholinoethyl) oleamide; AGS: anionic gemini surfactant; PVP: poly(4-vinylpyridine); PAH: poly(allylamine hydrochloride); DS: dextran sulfate; GA: gelatin type A; TA: tannic acid; PEO: poly(ethylene oxide); PDEGMA: poly(di(ethylene glycol) methyl ether methacrylate); PCL: polycaprolactone; PHB: poly(3-hydroxybutyrate); PANi: polyaniline; MTT: montmorillonite; GMSN: graphene-mesoporous silica nanosheet; PEG-4MAL: four-arm PEG macromers functionalized with terminal maleimide groups; HA: hyaluronic acid; GNT; MMT: montmorillonite; MSNs: mesoporous silica nanoparticle;UBI: ubiquicidin; CP: chitosan-graft-polyaniline; PMA-PDA: poly(methacrylic acid) functionalized with pyridine dithioethylamine; VAMSC: vertically aligned mesoporous silica coating; PPS: poly(propylene sulfide); OD: oxidized dextran; MBG: mesoporous bioglass; PECL: poly(ethylene glycol)−poly(ε-caprolactone); PEG-b-PA: poly(ethylene glycol)-b-poly(2-((((4-acetoxybenzyl)oxy)carbonyl)amino)ethyl methacrylate); PEG-b-PN: poly(ethylene glycol)-bpoly(2-((((4-nitrobenzyl)oxy)carbonyl)amino)ethyl methacrylate); MSR1: magnetosopirrillum gryphiswalense; MSM: mesoporous silica microtube; ALG-BA: alginate-boronic acid; *S. aureus*: Staphylococcus aureus; *S. epidermidis*: Staphylococcus epidermidis; *P. aeruginosa*: Pseudomonas aeruginosa; *E. coli*: Escherichia coli; MRSA: methicillin-resistant Staphylococcus aureus; *H. pylori*: Helicobacter pylori; *P. gingivalis*: prophyromonas gingivalis; *S. mutans*: Staphylococcus mutans; BIR: bacterial inhibition rate; DLC: drug leakage concentration; ZOI: zones of bacterial inhibition; MCB: morphological characterization of bacteria.

3. Results

3.1. Physical Stimuli-Responsive Systems

There will be a few physical changes within the infected microenvironment, such as reduced pH and locally elevated temperature. These physical stimuli have already been used to activate the release of antibiotics [56,85,86].

3.1.1. pH-Responsive Systems

The most commonly used trigger is the abnormal change in local pH. As bacterial metabolism produces lactic acid and acetic acid, the local pH, dropping from 7.4 to 6.0 or lower, can be used to trigger the release of antibiotics [4]. Chemical bonds, such as the Schiff base, acetal linkage, and metal ion coordination bonds that are stable under neutral conditions but broken at lower pH, are often utilized to realize the pH-responsive release (Figure 3) [22,31,41].

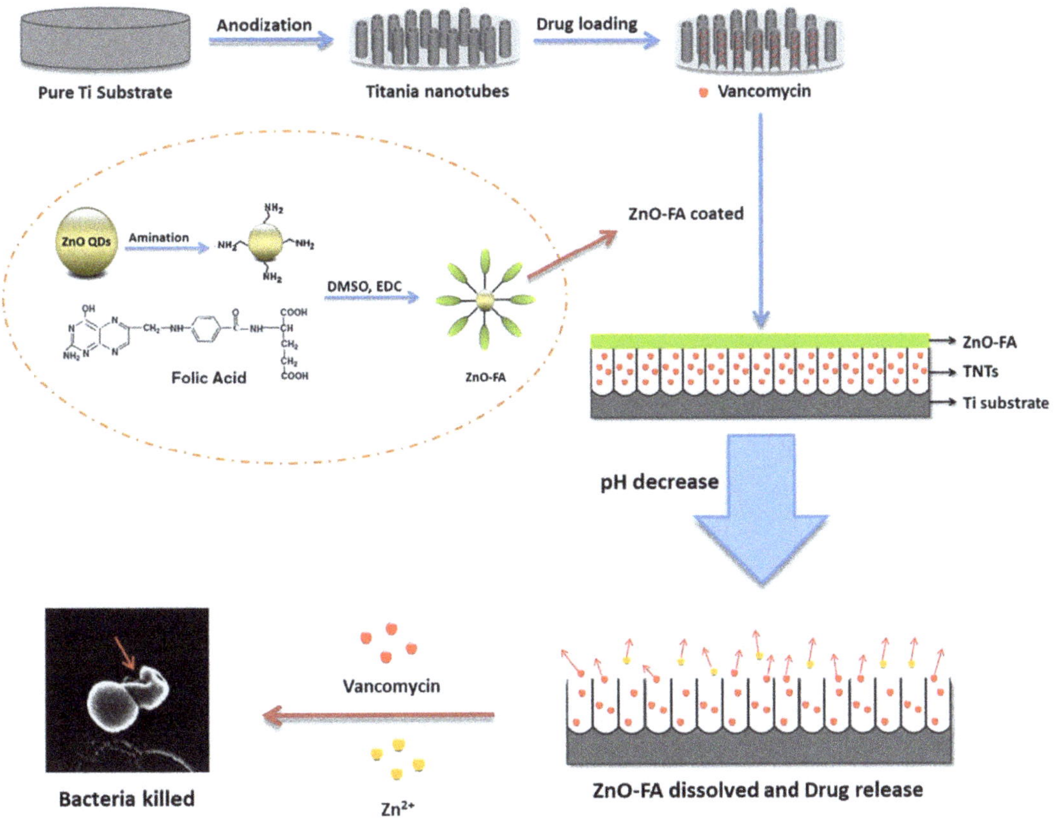

Figure 3. The schematic illustration of fabrication process of TNTs-Van@ZnO-FA system and synergistic bacteria-killing triggered by pH. Reproduced with the permission from ref. [23]. Copyright 2015 Elsevier.

In recent years, there have been studies on various kinds of pH-responsive polyelectrolyte multilayer films (PEMs), such as poly(acrylic acid) (PAA) [24] and poly(methacrylic acid) (PMAA) [21] films as antibacterial coatings. When a sudden decrease in pH disrupts the original electrostatic equilibrium between weak acidic/alkaline polyelectrolytes and incorporated antibiotics, those PEMs undergo swelling to re-balance the charge, which accomplishes the autonomous release of antibiotics. Chen et al. reported the fabrication of a smart system based on the switchable ability of PMAA as a gating element of pH-stimulated delivery of antimicrobial peptides (AMPs) [21]. The system was able to remain stable and extend the passive release of AMPs to 10 days under physiological conditions, while as the pH decreased, the PMAA collapsed to open the nanotubes and released adequate AMPs to kill the bacteria.

Besides electrostatic attraction, the Schiff reaction, which involves a dynamic covalent imine bond formation via the crosslinking of amine groups and aldehyde groups, is also a promising strategy for smart drug delivery [87]. The Schiff base is pH-responsive according to its chemical structure [88]. Researchers have already proved that the DDS, in which alginate dialdehyde (ADA) was conjugated with gentamicin (GEN) via the Schiff reaction, exhibited superior pH responsiveness and could prevent localized infections both in the early stages (6 h) and in the long term (72 h) [22].

Based on the pH response of metal ion coordination polymers (CPs) on TNTs, a novel smart DDS was designed by Wang et al., triggered by the change in the environment acidity due to *S. aureus* and *E. coli* infections [41]. TNTs were functionalized via amination by 3-aminopropyltriethoxysilane (APTES), into which drugs such as ibuprofen, vancomycin, or silver nitrate were harnessed. The researchers found that the CPs formed by 1,4-bis (imidazole-1-ylmethyl) benzene and Zn2+ or Ag+ could successfully block the drug release from TNTs in a neutral environment and could be triggered to open and release antibiotics once the environment became acidic. The release rate gradually increased as the pH value further decreased, indicating that the DDS was a controllable smart DDS for peri-implant infections.

3.1.2. Temperature-Responsive Systems

As is widely known, bacterial infections will raise the local temperature of the first place, which is also regarded as a trigger. Recently, various smart polymers that undergo a phase transition within a specific temperature range in response to an abrupt change have shown great promise in the aspect of drug delivery. The polymers are characterized by a critical solution temperature (CST), a narrow temperature range in which the hydrophobic/hydrophilic interactions between a polymer chain and aqueous medium change. These changes can lead to either chain collapse or swelling. The polymer with lower critical solution temperature (LCST) shows the solution phase below CST and becomes insoluble or forms hydrogels over CST. The polymers with LCST are mostly used for developing DDSs [89]. For example, poly(N-isopropylacrylamide) (PNIPAM) is one of the most representative smart polymers, transitioning from a two-phase to a one-phase mixture in an aqueous environment as the temperature decreases below a value of 37 °C [90].

Choi et al. reported a temperature-responsive (poly(di(ethylene glycol) methyl ether methacrylate)) (PDEGMA) brush coating that allowed the controlled release of levofloxacin [55]. The localized temperature rising of the infected site triggered the release due to the LCST behavior of the brushes. The antibacterial activity of levofloxacin, as well as the antifouling effects of PDEGMA, suppressed bacterial colonization and biofilm growth, as demonstrated in vivo tests with rats infected with *S. aureus*.

3.1.3. Contact-Responsive Systems

Another alternative involves contact killing. Polycations attached to complex biomaterial surfaces with negatively charged bacterial shells allow the polycations to penetrate the shell and kill the bacteria. Contact killing does not require time for the metabolic processes to achieve threshold levels based on other triggers, such as pH switches. However, the effective range of contact killing is more restricted, in contrast to release mechanisms that rely on diffusion [91].

Recently, a new concept called contact transfer has been introduced [57], which integrates ideas from contact killing and stimuli-responsiveness. It involves the transfer of untethered cationic antibiotics from surfaces of biomaterials to bacteria, when bacteria come close to biomaterials. Liang et al. designed anionic microgels loaded with small-molecule cationic antibiotics based on this novel concept [57]. The release of antibiotics was triggered specifically by bacterial contact, not the contact performed by macrophages or osteoblasts. Thus, the antibacterial property and biocompatibility were ensured. Researchers concluded that the negative charge and hydrophobicity of the bacterial envelope changed the local thermodynamic equilibrium that controlled antibiotic–microgel complexation, leading to antibiotics release.

Most of the DDSs mentioned above were allowed to combine passive elution of antibiotic in a physiological microenvironment with an active release in the presence of bacteria, while the background leaching was not preferred in some cases to avoid drug resistance. Moreover, all these methods have their own limitations and have proven difficult to implement in clinical trials thus far. For example, in the case of pH-responsive elements, many factors can result in changes in local pH values in the body. Further studies will be needed to assess the duration of such a release and the sensitivity of these systems in vivo.

3.2. *Virulence-Factor-Responsive Systems*

Bacteria generate various pathogenic factors in the process of adhesion, aggregation, diffusion and pathogenicity, including various enzymes and toxins, which can also be utilized to design a smart antibacterial DDS. For example, enzyme-responsive polymers, which consist of an enzyme-sensitive group such as an oligopeptide, dipeptide, or tripeptide, undergo changes when triggered by the catalytic action of enzymes, resulting in drug release. Various enzymes, such as HAS and protease, of which their concentration largely increases within the infected microenvironment, have all been explored for the controlled release of antibiotics.

3.2.1. Protease-Triggered Systems

Proteases are the general name of a class of enzymes that hydrolyze protein peptide bonds. They exist widely, mainly in human and animal digestive tracts, and can also be produced by microorganisms. Microbial proteases are mainly produced by mold and bacteria, followed by yeast and actinomycetes.

Based on the strict selectivity of protease to the substrate, investigators have designed a series of protease-responsive DDSs [58–62]. For instance, Johnson et al. engineered lysostaphin encapsulation within protease-degradable hydrogels and subsequent application to infected femurs, which led to fracture callus formation and healing [63]. The inclusion of protease-degradable peptide cross-links in lysostaphin-loaded hydrogels made it possible to deliver lysostaphin on demand in response to infections.

However, the protease-triggered release of drugs was not confined only to the presence of bacteria. The cleavage by the host proteases triggered the undesired release of antibiotics. To make the DDSs more targeted and reduce the accidental release of antibiotics, a many researchers have focused on the response to products of *S. aureus* infections, because of the specificity of their virulence factors [92]. Zhang et al. engineered a titanium coating grafted with vancomycin via a tailor-made peptide that can be cleaved by a *S. aureus*-secreted protease called serine protease-like protease (SplB), allowing the release of vancomycin specifically in the presence of *S. aureus* [62]. The bio-hydrolysis of this peptide was both sensitive and irreversible, highlighting its utility for generating a specific response to *S. aureus* infections.

3.2.2. Hyaluronidase (HAS)-Triggered Systems

Hyaluronidases (HAS) are enzymes that are capable of degrading hyaluronic acid (HA) and hyaluronate. HA constitutes an essential part of the extracellular matrix. Bacteria such as *S. aureus* and *E. coli* utilize HAS as an invasion factor to adhere to the surface of the implants [5]. A previous study has reported that HA-coated mesoporous silica nanoparticles could be degraded upon the addition of HAS [93], making HAS an available trigger for on-demand drug release. Moreover, the secretion of HAS by *S. aureus* and *E. coli* has been studied by Wang et al., where they reported that the increase in local HAS triggered the release of gentamicin from their multilayer films [65]. Similarly, Li et al. designed an intelligent vancomycin-HA-chitosan/β-glycerophophate hydrogel responsive to HAS secretion by *S. aureus* and *S. epidermidis* [64]. The hydrogel possessed antimicrobial properties both in vitro and in vivo that could be modulated by the concentration of HAS.

Interestingly, HA and hyaluronate themselves, which are generally main components of HAS-triggered systems, have been proved capable of decreasing *S. aureus* adhesion and biofilm formation [94], endowing those materials with greater antibacterial potential.

3.2.3. Lipase-Triggered Systems

Lipases are enzymes involved in the digestion of fats to fatty acids and glycerol or other alcohols. They are widely found in animals, plants and microorganisms. With regard to microorganisms, so many pathogenic bacterial species produce lipases that they have been classified as important virulence factors that exert harmful effects in combination with other bacterial enzymes, in particular the phospholipases C [95]. The capacity of lipases to break down ester bonds makes it possible for them to be used as triggers for intelligent drug delivery [96].

PCL microspheres that contain selenium nanoparticles (SeNPs) were developed as a DDS responsive to the existence of *P. aeruginosa* and lipases [68]. It was noticed that the higher lipase titer in vitro led to greater zones of bacterial inhibition. However, researchers compared the release rate of SeNPs in the lipase solution to that in the *P. aeruginosa* cell-free extract, finding that the drug concentration in the medium with bacterial extract was much more prominent. It might indicate more hidden mechanisms when the reactions occurred in vivo. In addition, Shi et al. found that metronidazole linked to dopamine-functionalized PCL nanofiber mats via ester linkage could be triggered to release in response to cholesterol esterase [67]. Furthermore, the release rate of metronidazole increased as the concentration of cholesterol esterase increased. The effective antibacterial capacity of the system indicated that it was a promising bacteria-responsive drug releasing material.

Moreover, one reason as to why bacteria are able to evade the immune system, and thus protect themselves from antibiotics, is that they can survive after phagocytosis by phagocytic cells, especially macrophages, which leads to further infection recurrence [97]. The drug delivery into macrophages is a necessary strategy in improving antibiotic therapy against intracellular infections. Lipase-activated on-demand delivery nanocarriers have been proven to be able to kill intracellular bacteria [72]. Xiong et al. reported a strategy for targeted antibiotic delivery into macrophages via mannose receptors [98,99], utilizing a mannosylated nanogel as the vancomycin carrier responsive to bacterial phospholipase (Figure 4) [71]. The nanogel contained mannosyl ligands conjugated to the shell of the poly(ethylene glycol) arm and polyphosphoester core-crosslinked nanogel. Phosphatase produced by bacteria could degrade the shell, resulting in the vancomycin release. The results suggested that mannosylated nanogels could enter macrophages via the interaction of mannosyl ligands with mannose receptors and release drugs to kill the intracellular bacteria. Similarly, Yang et al. designed a mesoporous silica nanoparticle (MSN) loaded with gentamicin that targeted both planktonic and intracellular infection [69].

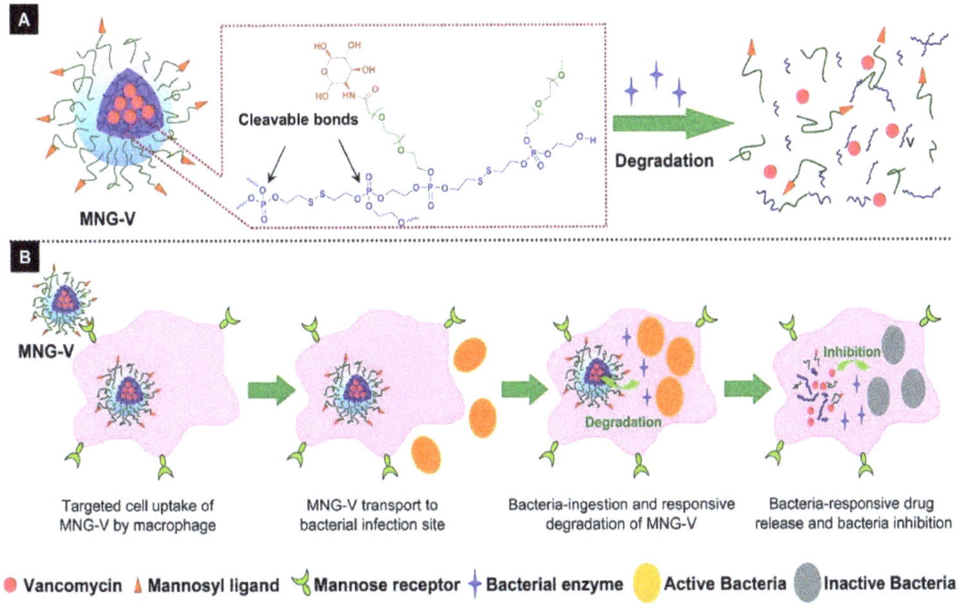

Figure 4. (**A**) Schematic illustration of a vancomycin-loaded mannosylated nanogels (MNG-V) and the bacteria-responsive drug release. (**B**) Schematic illustration of targeted uptake of MNG-V, transport, degradation, drug release and bacteria inhibition. Reproduced with the permission from ref. [52]. Copyright 2012 John Wiley and Sons.

3.2.4. Gelatinase-Triggered Systems

Gelatinases, also known as type IV collagenase, belong to the group of metalloproteinases (MMPs) and are able to cause hydrolysis of type IV collagen, leading to the breakdown of the extracellular matrix. A broad spectrum of bacterial species, including Staphylococcus, Enterococcus, and others, is known to produce gelatinases as virulence factors [100]. Based on the activity of gelatinases secreted by bacteria, a gelatin hydrolysis test has been used to distinguish the species of Bacillus, Clostridium, Proteus, Pseudomonas etc. The results revealed that pathogenic bacteria, such as *S. aureus*, were mostly gelatinase-positive [101], which indicated that gelatinase-responsive release of antibiotic agents at the infected site was achievable.

A previous study has proven that gelatinases secreted by *S. aureus* were qualified to activate the release of drugs [102]. Qi et al. further designed an "on-site transformation" system against bacterial infection composed of a chitosan backbone, a PEG-tethered gelatinase-cleavable peptide and an antibacterial peptide KLAK [73]. Cleaved by the gelatinases at the infected sites, the protecting PEG coating disappeared and the conformation changed, subsequently resulting in the release of KLAK peptide. KLAK made contact with the bacterial membranes and killed the bacteria as designed. Similarly, Li et al. [74] developed a kind of core-shell supramolecular gelatin nanoparticle that was capable of delivering vancomycin triggered by gelatinase.

While the smart DDSs are designed to be activated by bacterial enzymes, most antibiotics or AMPs are tied to the skeleton via covalent bond. With the cleavage of specific bonds by enzymes, there is a possibility that some residues are left over on antibacterial agents. The possible residues that remain on the antibiotics may raise the problem of impaired drug activity. It is worth noting that researchers have already pointed out the impact of remaining residues [58]. They demonstrated that nanogels are able to deliver on-demand ciprofloxacin triggered by trypsin, while the groups of the linker residue that remained on the ciprofloxacin negatively affected its efficacy. The antibacterial effect of the nanogels was not as good as that of ciprofloxacin alone. How to solve this impairment is an urgent problem.

Greater consideration should be given to design systems that can efficiently sense and respond to the bacteria and release antibacterial agents in response to enzymatic activities that are unique to the pathogen, while remaining stable against nonspecific cleavages by host enzymes.

3.3. Dual Responsive Systems

The concept of the smart bacteria-responsive delivery systems is ideal. As mentioned before, neither the physical changes that occur in the microenvironment nor the production of various enzymes are typically specific. The nonspecific disrupts in these systems present barriers to the success of this approach. To improve the specificity of stimuli-responsive release, researchers focus on the strategy of blending these ideas together. Such dual responsive systems have been reported previously to deliver anti-tumor drugs. Dual-responsive (pH and thermo responsive) nanoparticles from poly(NIPAAm-co-acrylic acid)-b-PCL diblock copolymers were designed to deliver paclitaxel [103]. Drug release was observed only at temperatures greater than 37 °C and at pH conditions between 4 and 6.

Likewise, dual responsive systems can also be used to control the release rate of antibacterial agents during infections. Programmable responsive antibiotics release systems have been investigated, including pH/thermal response and pH/enzymes response models. Chen et al. demonstrated that micelles are sensitive to both pH decreases and lipases [77]. The breakage of the acid-labile linkages led to the release of D-tyrosine to disintegrate the biofilm matrix, while the lipase-triggered breakdown of succinic acid linkages resulted in the release of azithromycin, killing bacteria and destructing biofilms. In addition, Wang et al. designed a vertically aligned mesoporous silica coating on the surface of stainless steel for pH and bacterial lipase-triggered antibiotics release (Figure 5) [76]. It was demonstrated that the lowering of pH triggered the opening of the cyclodextrin valve, enabling the release of the smaller cinnamaldehyde, the first step in killing bacteria. Meanwhile, lipases were shown to cause the cleavage of functionalized cyclodextrin, leading to the release of both cinnamaldehyde and AMPs. This dual release system was shown to inhibit the growth of *S. aureus*, *E. coli*, and MRSA in vitro.

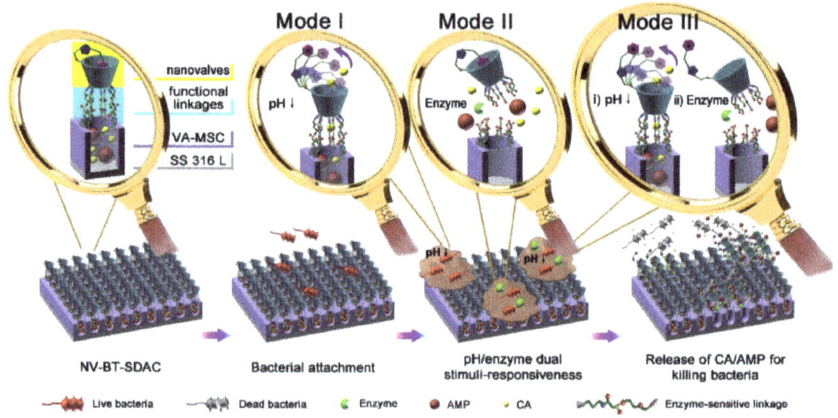

Figure 5. Schematic representation of structure and working mechanisms for NV-BT-SDACz deposited on SS316L. Reproduced with the permission from ref. [55]. Copyright 2017 American Chemical Society.

In addition to the systems that combine active releasing modes together, the trigger modes that combine active release and passive release also deserve greater attention. The combination of active "smart" release and passive controlled release can ensure the high concentration of local antibiotics when necessary. The chitosan-graft-polyaniline

(CP)/oxidized dextran (OD) hydrogels were proven to be dual responsive to electrical fields and pH [79]. Researchers used amoxicillin as the model drug, and found its release rate increased when an increase in voltage was applied or when the pH decreased. In this way, when it was clear that bacterial infection had occurred, in theory, the rapid release of antibiotics could be realized through the change in artificially applied voltage. The hydrogels presented excellent antibacterial properties and good biocompatibility both in vitro and in vivo, indicating that they are ideal candidates as smart drug delivery vehicles.

3.4. Host-Immune-Response-Responsive Systems

Apart from direct changes caused by the bacteria, hosts' immune responses to infections are also qualified to control the drug release of DDSs. When the infections occur, a large number of immunocytes accumulate nearby and secrete inflammatory factors [104]. The locally elevated levels of such inflammatory factors have aroused great concerns in researchers. Taking matrix metalloproteinases (MMPs) as an example, MMPs comprise a group of endogenous enzymes that play an important role both in physiological and pathological processes, acting on the remodeling and degradation of the extracellular matrix [105]. MMPs have been proven to be associated with the severity of periodontal destruction [106]. Guo et al. designed a degradable MMP8-responsive hydrogel that contained minocycline hydrochloride or AMPs to realize on-demand antibiotics delivery [82]. The results showed that the hydrogel had potential to be used for in situ adaptive degradation in response to peri-implantitis. Similar to MMP8, the reactive oxygen species (ROS) around infected sites also increased significantly. Stavrakis et al. reported a biodegradable coating using a branched poly(ethylene glycol)-poly(propylene sulfide) polymer [60]. The researchers noted a rapid release of antibiotics when using an oxidative environment, confirming a smart active releasing mechanism.

Bone infections will also lead to a series of host responses, including bone resorption. The concentration of acid phosphatase (APS), which takes adenosine triphosphate (ATP) as a substrate, increases significantly in bone infections due to the activation of bone resorption. Polo et al. manufactured a mesoporous bioglass that contained levofloxacin, taking ATP as the molecular gate [83]. Released levofloxacin could only be detected in the presence of APS, ensuring that on-demand release was achieved only due to the specific stimulus typical of a bone infection environment.

The most significant problem of such systems is the lack of specificity to bacterial infections. There are a variety of reasons that can lead to an inflammatory response or bone resorption in vivo, not just bacterial infections. Encapsulating antibiotics into such DDSs may result in the unintended release of antibiotics, which is contrary to what the smart DDSs are designed to do.

4. Discussion

As a promising next-generation DDS, smart bacteria-responsive DDSs, which are constructed according to the concept of self-diagnosis to self-treatment, are able to reduce the risk of antibiotic resistance for conventional passive release-based DDSs and remedy the limited delivery range of covalently drug-binding DDSs. In addition, by on-demand release of the smart DDSs, they are able to tackle the problems of external stimuli-responsive release-based antibacterial systems, namely, difficulty in monitoring bacterial growth status around the implants and precisely controlling the appropriate time for external stimuli application. In short, the temporally and spatially on-demand release of free antibacterial agents could help combat infections within a broader peri-implant tissue microenvironment, while mitigating the cytotoxicity associated with the burst release of high doses of physically entrapped antibiotics or risks for developing bacteria resistance due to inadequate or delayed antibiotic releases.

However, the reduction in drug efficiency is a noteworthy problem for both conventional and smart DDSs. During the process of design and manufacture, there are many factors that may impair the drug efficacy. With regard to PEMs, which are commonly used in both conventional and smart DDSs, more than one article reported that the manufacturing methods of PEMs affected the rate of responsive release. Zhuk et al. reported that PEMs manufactured by the spin-assisted LBL technique showed both a slower and lower release compared to dip-deposited films [52]. However, the research by Zhou et al. suggested otherwise, indicating that the antibacterial activities of dip-deposited films were limited [30]. Moreover, current experimental results have indicated that the retention/release properties of PEMs are highly dependent on the selection and matching degree of polyelectrolytes and antimicrobials. Greater caution is required when choosing manufacturing methods and corresponding antibiotics.

Although in theory, the design concept of smart bacteria-responsive DDSs is to effectively treat peri-implant infections, there are still various problems that exist in practice. DDSs triggered by pH and temperature switches, in most cases, combined passive elution and active release, while the background leaching was not preferred in some cases to avoid drug resistance. In addition, there were multiple causes for the change in physical microenvironment, which indicated that the non-specific trigger would be a major issue. This also happened in host-immune-response-responsive systems. For virulence-factor-responsive DDSs, the specificity was improved to some extent, but enzymes such as protease and lipase could be derived from the host. To conclude, although the above DDSs can theoretically be stimulated as designed, the trigger of these systems are not specific to infection. The lack of specificity is a common problem in the smart DDSs to date. Due to the complexity in vivo, researchers cannot guarantee DDS' stability under the accidental non-specific stimuli, which is undesirable. In the case of double or multiple-responsive systems, the situation becomes even more uncontrollable.

It is believed that every kind of system has its pros and cons. Inspired by smart DDSs designed particularly for MRSA [75,107] and *S. aureus* [92], finding specific substrates as triggers for various infections is considered to be promising. Just as specific virulence factors exist in MRSA-infected microenvironments, if other infected microenvironment-specific factors can be found, the problem regarding specificity can be improved and truly smart DDSs can be achieved.

In addition to improving the specificity of DDSs, it is believed that giving DDSs more functionality is one of the promising directions. First of all, smart DDSs can be designed to target intracellular infections. As is well known, intracellular bacteria are among the most dangerous causes of drug resistance. Bacteria engulfed by macrophages are able to escape from antibiotic attacks, because challenges remain in intracellular drug delivery specific to bacteria-infected cells and efficient uptake into intracellular bacteria. In this case, smart DDSs are clearly one of the best methods to deliver antibiotics directly to infected cells, such as macrophages, which may be achieved by adding specific antibodies to the surface of the system to target specific cells and utilizing materials such as nanoparticles that can be swallowed by cells to kill the intracellular bacteria. Several studies have proven its feasibility [71,72].

In addition, there are often other demands, such as anti-inflammation and bone regeneration, that must be realized when DDSs are implanted. Some of the DDSs were able to implement multiple functions at the same time. In fact, smart DDSs are also commonly used in other therapies, such as anti-inflammatory or anti-tumor therapies. Therefore, multifunctional smart DDSs are feasible, in which antibacterial properties can be achieved by the component of DDSs, such as metal nanoparticles, and the loaded drugs solve other problems. These ideas can be widely extended. For example, it is more desirable if the smart DDSs composed of hydrogels designed for oral implantation promote tissue regeneration and TNTs facilitate osseous integration. Moreover, bacterial infections are often accompanied by subsequent inflammatory responses, so smart anti-bacteria and anti-inflammation DDSs are preferred for sequential treatment [81]. The pattern of on-demand drug release can effectively circumvent drug resistance of all drugs loaded on DDSs, making it suitable for versatile drug delivery.

Overall, smart bacteria-responsive DDSs are believed to be the next generation of mainstream antibacterial therapy. The need for multimodality strategies that take into account various stages of pathogenesis [108] to improve specificity, while minimizing the negative impact on peri-implant tissues or encouraging implant-tissue integrations, is increasingly recognized. Targeted smart bacteria-responsive DDSs or versatile DDSs are believed to be promising. More attention should be paid to this area.

Author Contributions: Authorship is limited to those who have contributed substantially to the work article. The work was planned jointly. Formal analysis, Y.Y. and X.J.; investigation, Y.Y. and X.J.; data curation, Y.Y.; writing—original draft preparation, Y.Y.; writing—review and editing, Y.Y and X.Z.; visualization, Y.Y.; supervision, X.Z.; project administration, H.L.; funding acquisition, H.L. and X.Z. All authors have read and agreed to the published version of the manuscript.

Funding: This research was funded by the Cross-disciplinary Research Fund of Shanghai Ninth People's Hospital, Shanghai JiaoTong University School of Medicine, grant number JYJC202205 and the National Natural Science Foundation of China, grant number 82001098.

Data Availability Statement: Not applicable.

Acknowledgments: The authors extend their thanks to all the members of Department of Oral and Maxillo-facial Implantology, Shanghai Ninth People's Hospital for their kind support.

Conflicts of Interest: The authors declare no conflict of interest.

References

1. Feng, E.; Shen, K.; Lin, F.; Lin, W.; Zhang, T.; Zhang, Y.; Lin, F.; Yang, Y.; Lin, C. Improved osteogenic activity and inhibited bacterial biofilm formation on andrographolide-loaded titania nanotubes. *Ann. Transl. Med.* **2020**, *8*, 987. [CrossRef] [PubMed]
2. Kim, H.K.; Thammavongsa, V.; Schneewind, O.; Missiakas, D. Recurrent infections and immune evasion strategies of Staphylococcus aureus. *Curr. Opin. Microbiol.* **2012**, *15*, 92–99. [CrossRef] [PubMed]
3. Campoccia, D.; Montanaro, L.; Arciola, C.R. The significance of infection related to orthopedic devices and issues of antibiotic resistance. *Biomaterials* **2006**, *27*, 2331–2339. [CrossRef] [PubMed]
4. Radovic-Moreno, A.F.; Lu, T.K.; Puscasu, V.A.; Yoon, C.J.; Langer, R.; Farokhzad, O.C. Surface charge-switching polymeric nanoparticles for bacterial cell wall-targeted delivery of antibiotics. *ACS Nano* **2012**, *6*, 4279–4287. [CrossRef] [PubMed]
5. Fischer, C.L.; Drake, D.R.; Dawson, D.V.; Blanchette, D.R.; Brogden, K.A.; Wertz, P.W. Antibacterial activity of sphingoid bases and fatty acids against Gram-positive and Gram-negative bacteria. *Antimicrob. Agents Chemother.* **2012**, *56*, 1157–1161. [CrossRef]
6. Tang, F.; Li, L.; Chen, D. Mesoporous silica nanoparticles: Synthesis, biocompatibility and drug delivery. *Adv. Mater.* **2012**, *24*, 1504–1534. [CrossRef]
7. Zhou, Z.; Seta, J.; Markel, D.C.; Song, W.; Yurgelevic, S.M.; Yu, X.W.; Ren, W. Release of vancomycin and tobramycin from polymethylmethacrylate cements impregnated with calcium polyphosphate hydrogel. *J. Biomed. Mater. Res. Part B Appl. Biomater.* **2018**, *106*, 2827–2840. [CrossRef]
8. Jose, B.; Antoci, V., Jr.; Zeiger, A.R.; Wickstrom, E.; Hickok, N.J. Vancomycin covalently bonded to titanium beads kills Staphylococcus aureus. *Chem. Biol.* **2005**, *12*, 1041–1048. [CrossRef]
9. Gerits, E.; Kucharikova, S.; Van Dijck, P.; Erdtmann, M.; Krona, A.; Lovenklev, M.; Frohlich, M.; Dovgan, B.; Impellizzeri, F.; Braem, A.; et al. Antibacterial activity of a new broad-spectrum antibiotic covalently bound to titanium surfaces. *J. Orthop. Res. Off. Publ. Orthop. Res. Soc.* **2016**, *34*, 2191–2198. [CrossRef]
10. Li, D.; Tang, G.; Yao, H.; Zhu, Y.; Shi, C.; Fu, Q.; Yang, F.; Wang, X. Formulation of pH-responsive PEGylated nanoparticles with high drug loading capacity and programmable drug release for enhanced antibacterial activity. *Bioact. Mater.* **2022**, *16*, 47–56. [CrossRef]
11. Fu, M.; Gan, Y.; Jiang, F.; Lv, X.; Tan, N.; Zhao, X.; Yang, Y.Y.; Yuan, P.; Ding, X. Interpenetrating Polymer Network Hydrogels Formed Using Antibiotics as a Dynamic Crosslinker for Treatment of Infected Wounds. *Adv. Healthc. Mater.* **2022**, *11*, 2200902. [CrossRef] [PubMed]
12. Yang, L.; Wang, C.; Li, L.; Zhu, F.; Ren, X.; Huang, Q.; Cheng, Y.; Li, Y. Bioinspired Integration of Naturally Occurring Molecules towards Universal and Smart Antibacterial Coatings. *Adv. Funct. Mater.* **2022**, *32*, 2108749. [CrossRef]
13. Cámara-Torres, M.; Duarte, S.; Sinha, R.; Egizabal, A.; Álvarez, N.; Bastianini, M.; Sisani, M.; Scopece, P.; Scatto, M.; Bonetto, A.; et al. 3D additive manufactured composite scaffolds with antibiotic-loaded lamellar fillers for bone infection prevention and tissue regeneration. *Bioact. Mater.* **2021**, *6*, 1073–1082. [CrossRef]
14. Guo, R.; Li, K.; Tian, B.; Wang, C.; Chen, X.; Jiang, X.; He, H.; Hong, W. Elaboration on the architecture of pH-sensitive surface charge-adaptive micelles with enhanced penetration and bactericidal activity in biofilms. *J. Nanobiotechnol.* **2021**, *19*, 232. [CrossRef]

15. Ramesh, S.; Kovelakuntla, V.; Meyer, A.S. Three-dimensional printing of stimuli-responsive hydrogel with antibacterial activity. *Bioprinting* **2021**, *24*, e00106. [CrossRef]
16. Sang, S.; Guo, G.; Yu, J.; Zhang, X. Antibacterial application of gentamicin-silk protein coating with smart release function on titanium, polyethylene, and Al2O3 materials. *Mater. Sci. Eng. C-Mater. Biol. Appl.* **2021**, *124*, 112069. [CrossRef] [PubMed]
17. Yan, K.; Xu, F.; Wei, W.; Yang, C.; Wang, D.; Shi, X. Electrochemical synthesis of chitosan/silver nanoparticles multilayer hydrogel coating with pH-dependent controlled release capability and antibacterial property. *Colloids Surfaces. B Biointerfaces* **2021**, *202*, 111711. [CrossRef] [PubMed]
18. Zha, J.; Mao, X.; Hu, S.; Shang, K.; Yin, J. Acid- and Thiol-Cleavable Multifunctional Codelivery Hydrogel: Fabrication and Investigation of Antimicrobial and Anticancer Properties. *ACS Appl. Bio. Mater.* **2021**, *4*, 1515–1523. [CrossRef]
19. Hassan, D.; Omolo, C.A.; Fasiku, V.O.; Elrashedy, A.A.; Mocktar, C.; Nkambule, B.; Soliman, M.E.S.; Govender, T. Formulation of pH-Responsive Quatsomes from Quaternary Bicephalic Surfactants and Cholesterol for Enhanced Delivery of Vancomycin against Methicillin Resistant Staphylococcus aureus. *Pharmaceutics* **2020**, *12*, 1093. [CrossRef]
20. Li, M.; Wang, H.; Chen, X.; Jin, S.; Chen, W.; Meng, Y.; Liu, Y.; Guo, Y.; Jiang, W.; Xu, X.; et al. Chemical grafting of antibiotics into multilayer films through Schiff base reaction for self-defensive response to bacterial infections. *Chem. Eng. J.* **2020**, *382*, 122973. [CrossRef]
21. Chen, J.; Shi, X.; Zhu, Y.; Chen, Y.; Gao, M.; Gao, H.; Liu, L.; Wang, L.; Mao, C.; Wang, Y. On-demand storage and release of antimicrobial peptides using Pandora's box-like nanotubes gated with a bacterial infection-responsive polymer. *Theranostics* **2020**, *10*, 109–122. [CrossRef] [PubMed]
22. Tao, B.; Deng, Y.; Song, L.; Ma, W.; Qian, Y.; Lin, C.; Yuan, Z.; Lu, L.; Chen, M.; Yang, X.; et al. BMP2-loaded titania nanotubes coating with pH-responsive multilayers for bacterial infections inhibition and osteogenic activity improvement. *Colloids Surf. B Biointerfaces* **2019**, *177*, 242–252. [CrossRef] [PubMed]
23. Jin, X.; Xiong, Y.-H.; Zhang, X.-Y.; Wang, R.; Xing, Y.; Duan, S.; Chen, D.; Tian, W.; Xu, F.-J. Self-Adaptive Antibacterial Porous Implants with Sustainable Responses for Infected Bone Defect Therapy. *Adv. Funct. Mater.* **2019**, *29*, 1807915. [CrossRef]
24. de Avila, E.D.; Castro, A.G.B.; Tagit, O.; Krom, B.P.; Lowik, D.; van Well, A.A.; Bannenberg, L.J.; Vergani, C.E.; van den Beucken, J.J.J.P. Anti-bacterial efficacy via drug-delivery system from layer-by-layer coating for percutaneous dental implant components. *Appl. Surf. Sci.* **2019**, *488*, 194–204. [CrossRef]
25. Cao, J.; Zhao, Y.; Liu, Y.; Tian, S.; Zheng, C.; Liu, C.; Zhai, Y.; An, Y.; Busscher, H.J.; Shi, L.; et al. Phosphorylcholine-Based Polymer Encapsulated Chitosan Nanoparticles Enhance the Penetration of Antimicrobials in a Staphylococcal Biofilm. *ACS Macro Lett.* **2019**, *8*, 651–657. [CrossRef] [PubMed]
26. Hu, J.; Zheng, Z.; Liu, C.; Hu, Q.; Cai, X.; Xiao, J.; Cheng, Y. A pH-responsive hydrogel with potent antibacterial activity against both aerobic and anaerobic pathogens. *Biomater. Sci.* **2019**, *7*, 581–584. [CrossRef] [PubMed]
27. Karakeçili, A.; Topuz, B.; Korpayev, S.; Erdek, M. Metal-organic frameworks for on-demand pH controlled delivery of vancomycin from chitosan scaffolds. *Mater. Sci. Eng. C Mater. Biol. Appl.* **2019**, *105*, 110098. [CrossRef]
28. Maji, R.; Omolo, C.A.; Agrawal, N.; Maduray, K.; Hassan, D.; Mokhtar, C.; Mackhraj, I.; Govender, T. PH-Responsive Lipid-Dendrimer Hybrid Nanoparticles: An Approach to Target and Eliminate Intracellular Pathogens. *Mol. Pharm.* **2019**, *16*, 4594–4609. [CrossRef]
29. Mir, M.; Ahmed, N.; Permana, A.D.; Rodgers, A.M.; Donnelly, R.F.; Rehman, A.U. Enhancement in site-specific delivery of carvacrol against methicillin resistant staphylococcus aureus induced skin infections using enzyme responsive nanoparticles: A proof of concept study. *Pharmaceutics* **2019**, *11*, 606. [CrossRef]
30. Zhou, W.; Li, Y.; Yan, J.; Xiong, P.; Li, Q.; Cheng, Y.; Zheng, Y. Construction of Self-defensive Antibacterial and Osteogenic AgNPs/Gentamicin Coatings with Chitosan as Nanovalves for Controlled release. *Sci. Rep.* **2018**, *8*, 13432. [CrossRef]
31. Xiang, Y.; Liu, X.; Mao, C.; Liu, X.; Cui, Z.; Yang, X.; Yeung, K.W.K.; Zheng, Y.; Wu, S. Infection-prevention on Ti implants by controlled drug release from folic acid/ZnO quantum dots sealed titania nanotubes. *Mater. Sci. Eng. C Mater. Biol. Appl.* **2018**, *85*, 214–224. [CrossRef] [PubMed]
32. Placente, D.; Benedini, L.A.; Baldini, M.; Laiuppa, J.A.; Santillán, G.E.; Messina, P.V. Multi-drug delivery system based on lipid membrane mimetic coated nano-hydroxyapatite formulations. *Int. J. Pharm.* **2018**, *548*, 559–570. [CrossRef] [PubMed]
33. Cicuéndez, M.; Doadrio, J.C.; Hernández, A.; Portolés, M.T.; Izquierdo-Barba, I.; Vallet-Regí, M. Multifunctional pH sensitive 3D scaffolds for treatment and prevention of bone infection. *Acta Biomater.* **2018**, *65*, 450–461. [CrossRef] [PubMed]
34. Dai, T.; Wang, C.; Wang, Y.; Xu, W.; Hu, J.; Cheng, Y. A Nanocomposite Hydrogel with Potent and Broad-Spectrum Antibacterial Activity. *ACS Appl. Mater. Interfaces* **2018**, *10*, 15163–15173. [CrossRef] [PubMed]
35. Dubovoy, V.; Ganti, A.; Zhang, T.; Al-Tameemi, H.; Cerezo, J.D.; Boyd, J.M.; Asefa, T. One-Pot Hydrothermal Synthesis of Benzalkonium-Templated Mesostructured Silica Antibacterial Agents. *J. Am. Chem. Soc.* **2018**, *140*, 13534–13537. [CrossRef] [PubMed]
36. Mhule, D.; Kalhapure, R.S.; Jadhav, M.; Omolo, C.A.; Rambharose, S.; Mocktar, C.; Singh, S.; Waddad, A.Y.; Ndesendo, V.M.K.; Govender, T. Synthesis of an oleic acid based pH-responsive lipid and its application in nanodelivery of vancomycin. *Int. J. Pharm.* **2018**, *550*, 149–159. [CrossRef]
37. Soltani, B.; Nabipour, H.; Nasab, N.A. Efficient Storage of Gentamicin in Nanoscale Zeolitic Imidazolate Framework-8 Nanocarrier for pH-Responsive Drug Release. *J. Inorg. Organomet. Polym. Mater.* **2018**, *28*, 1090–1097. [CrossRef]

38. Yu, X.; Pan, Q.; Zheng, Z.; Chen, Y.; Chen, Y.; Weng, S.; Huang, L. pH-responsive and porous vancomycin-loaded PLGA microspheres: Evidence of controlled and sustained release for localized inflammation inhibition in vitro. *RSC Adv.* **2018**, *8*, 37424–37432. [CrossRef]
39. Zhang, S.; Xing, M.; Li, B. Capsule-Integrated Polypeptide Multilayer Films for Effective pH-Responsive Multiple Drug Co-Delivery. *ACS Appl. Mater. Interfaces* **2018**, *10*, 44267–44278. [CrossRef]
40. Zhou, W.; Jia, Z.; Xiong, P.; Yan, J.; Li, M.; Cheng, Y.; Zheng, Y. Novel pH-responsive tobramycin-embedded micelles in nanostructured multilayer-coatings of chitosan/heparin with efficient and sustained antibacterial properties. *Mater. Sci. Eng. C Mater. Biol. Appl.* **2018**, *90*, 693–705. [CrossRef]
41. Wang, T.; Liu, X.; Zhu, Y.; Cui, Z.D.; Yang, X.J.; Pan, H.; Yeung, K.W.K.; Wu, S. Metal Ion Coordination Polymer-Capped pH-Triggered Drug Release System on Titania Nanotubes for Enhancing Self-antibacterial Capability of Ti Implants. *ACS Biomater. Sci. Eng.* **2017**, *3*, 816–825. [CrossRef] [PubMed]
42. Pamfil, D.; Vasile, C.; Tartau, L.; Verestiuc, L.; Poiata, A. pH-Responsive 2-hydroxyethyl methacrylate/citraconic anhydride-modified collagen hydrogels as ciprofloxacin carriers for wound dressings. *J. Bioact. Compat. Polym.* **2017**, *32*, 355–381. [CrossRef]
43. Liu, Z.; Zhu, Y.; Liu, X.; Yeung, K.W.K.; Wu, S. Construction of poly (vinyl alcohol)/poly (lactide-glycolide acid)/vancomycin nanoparticles on titanium for enhancing the surface self-antibacterial activity and cytocompatibility. *Colloids Surf. B Biointerfaces* **2017**, *151*, 165–177. [CrossRef] [PubMed]
44. Dong, Y.; Ye, H.; Liu, Y.; Xu, L.; Wu, Z.; Hu, X.; Ma, J.; Pathak, J.L.; Liu, J.; Wu, G. pH dependent silver nanoparticles releasing titanium implant: A novel therapeutic approach to control peri-implant infection. *Colloids Surf. B Biointerfaces* **2017**, *158*, 127–136. [CrossRef]
45. Kalhapure, R.S.; Sikwal, D.R.; Rambharose, S.; Mocktar, C.; Singh, S.; Bester, L.; Oh, J.K.; Renukuntla, J.; Govender, T. Enhancing targeted antibiotic therapy via pH responsive solid lipid nanoparticles from an acid cleavable lipid. *Nanomed. Nanotechnol. Biol. Med.* **2017**, *13*, 2067–2077. [CrossRef]
46. Sang, Q.; Williams, G.R.; Wu, H.; Liu, K.; Li, H.; Zhu, L.-M. Electrospun gelatin/sodium bicarbonate and poly(lactide-co-epsilon-caprolactone)/sodium bicarbonate nanofibers as drug delivery systems. *Mater. Sci. Eng. C-Mater. Biol. Appl.* **2017**, *81*, 359–365. [CrossRef]
47. Onat, B.; Bütün, V.; Banerjee, S.; Erel-Goktepe, I. Bacterial anti-adhesive and pH-induced antibacterial agent releasing ultra-thin films of zwitterionic copolymer micelles. *Acta Biomater.* **2016**, *40*, 293–309. [CrossRef]
48. Fullriede, H.; Abendroth, P.; Ehlert, N.; Doll, K.; Schäske, J.; Winkel, A.; Stumpp, S.N.; Stiesch, M.; Behrens, P. PH-responsive release of chlorhexidine from modified nanoporous silica nanoparticles for dental applications. *BioNanoMaterials* **2016**, *17*, 59–72. [CrossRef]
49. Kalhapure, R.S.; Jadhav, M.; Rambharose, S.; Mocktar, C.; Singh, S.; Renukuntla, J.; Govender, T. pH-responsive chitosan nanoparticles from a novel twin-chain anionic amphiphile for controlled and targeted delivery of vancomycin. *Colloids Surf. B Biointerfaces* **2017**, *158*, 650–657. [CrossRef]
50. Anandhakumar, S.; Gokul, P.; Raichur, A.M. Stimuli-responsive weak polyelectrolyte multilayer films: A thin film platform for self triggered multi-drug delivery. *Mater. Sci. Eng. C Mater. Biol. Appl.* **2016**, *58*, 622–628. [CrossRef]
51. Zhang, Z.; Nong, J.; Zhong, Y. Antibacterial, anti-inflammatory and neuroprotective layer-by-layer coatings for neural implants. *J. Neural Eng.* **2015**, *12*, 046015. [CrossRef] [PubMed]
52. Zhuk, I.; Jariwala, F.; Attygalle, A.B.; Wu, Y.; Libera, M.R.; Sukhishvili, S.A. Self-defensive layer-by-layer films with bacteria-triggered antibiotic release. *ACS Nano* **2014**, *8*, 7733–7745. [CrossRef] [PubMed]
53. Zhang, Z.; Nix, C.A.; Ercan, U.K.; Gerstenhaber, J.A.; Joshi, S.G.; Zhong, Y. Calcium binding-mediated sustained release of minocycline from hydrophilic multilayer coatings targeting infection and inflammation. *PLoS ONE* **2014**, *9*, e84360. [CrossRef]
54. Pichavant, L.; Amador, G.; Jacqueline, C.; Brouillaud, B.; Héroguez, V.; Durrieu, M.C. pH-controlled delivery of gentamicin sulfate from orthopedic devices preventing nosocomial infections. *J. Control. Release* **2012**, *162*, 373–381. [CrossRef] [PubMed]
55. Choi, H.; Schulte, A.; Müller, M.; Park, M.; Jo, S.; Schönherr, H. Drug Release from Thermo-Responsive Polymer Brush Coatings to Control Bacterial Colonization and Biofilm Growth on Titanium Implants. *Adv. Healthc. Mater.* **2021**, *10*, e2100069. [CrossRef]
56. Li, B.; Zhang, L.; Wang, D.; Peng, F.; Zhao, X.; Liang, C.; Li, H.; Wang, H. Thermosensitive -hydrogel-coated titania nanotubes with controlled drug release and immunoregulatory characteristics for orthopedic applications. *Mater. Sci. Eng. C Mater. Biol. Appl.* **2021**, *122*, 111878. [CrossRef]
57. Liang, J.; Wang, H.; Libera, M. Biomaterial surfaces self-defensive against bacteria by contact transfer of antimicrobials. *Biomaterials* **2019**, *204*, 25–35. [CrossRef]
58. Bourgat, Y.; Mikolai, C.; Stiesch, M.; Klahn, P.; Menzel, H. Enzyme-Responsive Nanoparticles and Coatings Made from Alginate/Peptide Ciprofloxacin Conjugates as Drug Release System. *Antibiotics* **2021**, *10*, 653. [CrossRef]
59. Timin, A.S.; Muslimov, A.R.; Zyuzin, M.V.; Peltek, O.O.; Karpov, T.E.; Sergeev, I.S.; Dotsenko, A.I.; Goncharenko, A.A.; Yolshin, N.D.; Sinelnik, A.; et al. Multifunctional Scaffolds with Improved Antimicrobial Properties and Osteogenicity Based on Piezoelectric Electrospun Fibers Decorated with Bioactive Composite Microcapsules. *ACS Appl. Mater. Interfaces* **2018**, *10*, 34849–34868. [CrossRef]
60. Liao, X.; Yu, X.; Yu, H.; Huang, J.; Zhang, B.; Xiao, J. Development of an anti-infective coating on the surface of intraosseous implants responsive to enzymes and bacteria. *J. Nanobiotechnol.* **2021**, *19*, 241. [CrossRef]

61. Yu, X.; Liao, X.; Chen, H. Antibiotic-Loaded MMT/PLL-Based Coating on the Surface of Endosseous Implants to Suppress Bacterial Infections. *Int. J. Nanomed.* **2021**, *16*, 2983–2994. [CrossRef] [PubMed]
62. Zhang, Y.; Hu, K.; Xing, X.; Zhang, J.; Zhang, M.R.; Ma, X.; Shi, R.; Zhang, L. Smart Titanium Coating Composed of Antibiotic Conjugated Peptides as an Infection-Responsive Antibacterial Agent. *Macromol. Biosci.* **2021**, *21*, e2000194. [CrossRef] [PubMed]
63. Johnson, C.T.; Wroe, J.A.; Agarwal, R.; Martin, K.E.; Guldberg, R.E.; Donlan, R.M.; Westblade, L.F.; García, A.J. Hydrogel delivery of lysostaphin eliminates orthopedic implant infection by Staphylococcus aureus and supports fracture healing. *Proc. Natl. Acad. Sci. USA* **2018**, *115*, E4960–E4969. [CrossRef]
64. Li, Y.; Li, G.; Sha, X.; Li, L.; Zhang, K.; Liu, D.; Hao, Y.; Cui, X.; Wang, L.; Wang, H. An intelligent vancomycin release system for preventing surgical site infections of bone tissues. *Biomater. Sci.* **2020**, *8*, 3202–3211. [CrossRef] [PubMed]
65. Wang, B.; Liu, H.; Sun, L.; Jin, Y.; Ding, X.; Li, L.; Ji, J.; Chen, H. Construction of High Drug Loading and Enzymatic Degradable Multilayer Films for Self-Defense Drug Release and Long-Term Biofilm Inhibition. *Biomacromolecules* **2018**, *19*, 85–93. [CrossRef] [PubMed]
66. Ji, H.; Dong, K.; Yan, Z.; Ding, C.; Chen, Z.; Ren, J.; Qu, X. Bacterial Hyaluronidase Self-Triggered Prodrug Release for Chemo-Photothermal Synergistic Treatment of Bacterial Infection. *Small* **2016**, *12*, 6200–6206. [CrossRef]
67. Shi, R.; Ye, J.; Li, W.; Zhang, J.; Li, J.; Wu, C.; Xue, J.; Zhang, L. Infection-responsive electrospun nanofiber mat for antibacterial guided tissue regeneration membrane. *Mater. Sci. Eng. C Mater. Biol. Appl.* **2019**, *100*, 523–534. [CrossRef] [PubMed]
68. Filipović, N.; Veselinović, L.; Ražić, S.; Jeremić, S.; Filipič, M.; Žegura, B.; Tomić, S.; Čolić, M.; Stevanović, M. Poly (ε-caprolactone) microspheres for prolonged release of selenium nanoparticles. *Mater. Sci. Eng. C Mater. Biol. Appl.* **2019**, *96*, 776–789. [CrossRef]
69. Yang, S.; Han, X.; Yang, Y.; Qiao, H.; Yu, Z.; Liu, Y.; Wang, J.; Tang, T. Bacteria-Targeting Nanoparticles with Microenvironment-Responsive Antibiotic Release To Eliminate Intracellular Staphylococcus aureus and Associated Infection. *ACS Appl. Mater. Interfaces* **2018**, *10*, 14299–14311. [CrossRef]
70. Li, Y.-m.; Liu, S.-y. Enzyme-triggered Transition from Polymeric Vesicles to Core Cross-linked Micelles for Selective Release of Antimicrobial Agents. *Acta Polym. Sin.* **2017**, *7*, 1178–1190. [CrossRef]
71. Xiong, M.H.; Li, Y.J.; Bao, Y.; Yang, X.Z.; Hu, B.; Wang, J. Bacteria-responsive multifunctional nanogel for targeted antibiotic delivery. *Adv. Mater.* **2012**, *24*, 6175–6180. [CrossRef] [PubMed]
72. Xiong, M.H.; Bao, Y.; Yang, X.Z.; Wang, Y.C.; Sun, B.; Wang, J. Lipase-sensitive polymeric triple-layered nanogel for "on-demand" drug delivery. *J. Am. Chem. Soc.* **2012**, *134*, 4355–4362. [CrossRef] [PubMed]
73. Qi, G.B.; Zhang, D.; Liu, F.H.; Qiao, Z.Y.; Wang, H. An "On-Site Transformation" Strategy for Treatment of Bacterial Infection. *Adv. Mater.* **2017**, *29*, 1703461. [CrossRef] [PubMed]
74. Li, L.-L.; Xu, J.-H.; Qi, G.-B.; Zhao, X.; Yu, F.; Wang, H. Core-Shell Supramolecular Gelatin Nanoparticles for Adaptive and "On-Demand" Antibiotic Delivery. *ACS Nano* **2014**, *8*, 4975–4983. [CrossRef]
75. Tonkin, R.L.; Klöckner, A.; Najer, A.; Simoes da Silva, C.J.; Echalier, C.; Dionne, M.S.; Edwards, A.M.; Stevens, M.M. Bacterial Toxin-Triggered Release of Antibiotics from Capsosomes Protects a Fly Model from Lethal Methicillin-Resistant Staphylococcus aureus (MRSA) Infection. *Adv. Healthc. Mater.* **2022**, *11*, e2200036. [CrossRef]
76. Wang, T.; Wang, C.; Zhou, S.; Xu, J.H.; Jiang, W.; Tan, L.H.; Fu, J.J. Nanovalves-Based Bacteria-Triggered, Self-Defensive Antibacterial Coating: Using Combination Therapy, Dual Stimuli-Responsiveness, and Multiple Release Modes for Treatment of Implant-Associated Infections. *Chem. Mater.* **2017**, *29*, 8325–8337. [CrossRef]
77. Chen, M.; Wei, J.; Xie, S.; Tao, X.; Zhang, Z.; Ran, P.; Li, X. Bacterial biofilm destruction by size/surface charge-adaptive micelles. *Nanoscale* **2019**, *11*, 1410–1422. [CrossRef]
78. Chen, M.; Xie, S.; Wei, J.; Song, X.; Ding, Z.; Li, X. Antibacterial Micelles with Vancomycin-Mediated Targeting and pH/Lipase-Triggered Release of Antibiotics. *ACS Appl. Mater. Interfaces* **2018**, *10*, 36814–36823. [CrossRef]
79. Qu, J.; Zhao, X.; Ma, P.X.; Guo, B. Injectable antibacterial conductive hydrogels with dual response to an electric field and pH for localized "smart" drug release. *Acta Biomater.* **2018**, *72*, 55–69. [CrossRef]
80. Stanton, M.M.; Park, B.W.; Vilela, D.; Bente, K.; Faivre, D.; Sitti, M.; Sánchez, S. Magnetotactic Bacteria Powered Biohybrids Target E. coli Biofilms. *ACS Nano* **2017**, *11*, 9968–9978. [CrossRef]
81. Hu, C.; Zhang, F.; Long, L.; Kong, Q.; Luo, R.; Wang, Y. Dual-responsive injectable hydrogels encapsulating drug-loaded micelles for on-demand antimicrobial activity and accelerated wound healing. *J. Control. Release* **2020**, *324*, 204–217. [CrossRef] [PubMed]
82. Guo, J.; Sun, H.; Lei, W.; Tang, Y.; Hong, S.; Yang, H.; Tay, F.R.; Huang, C. MMP-8-Responsive Polyethylene Glycol Hydrogel for Intraoral Drug Delivery. *J. Dent. Res.* **2019**, *98*, 564–571. [CrossRef] [PubMed]
83. Polo, L.; Gómez-Cerezo, N.; García-Fernández, A.; Aznar, E.; Vivancos, J.L.; Arcos, D.; Vallet-Regí, M.; Martínez-Máñez, R. Mesoporous Bioactive Glasses Equipped with Stimuli-Responsive Molecular Gates for Controlled Delivery of Levofloxacin against Bacteria. *Chemistry* **2018**, *24*, 18944–18951. [CrossRef]
84. Stavrakis, A.I.; Zhu, S.; Hegde, V.; Loftin, A.H.; Ashbaugh, A.G.; Niska, J.A.; Miller, L.S.; Segura, T.; Bernthal, N.M. In vivo efficacy of a smart antimicrobial implant coating. *J. Bone Jt. Surg.—Am. Vol.* **2016**, *98*, 1183–1189. [CrossRef]
85. Lei, L.; Yan, R.; Chen, S.; Hao, X.; Dou, W.; Liu, H.; Guo, Z.; Kilula, D.; Seok, I. Narrow pH response multilayer films with controlled release of ibuprofen on magnesium alloy. *Mater. Sci. Eng. C Mater. Biol. Appl.* **2021**, *118*, 111414. [CrossRef] [PubMed]
86. Heffernan, J.M.; Overstreet, D.J.; Vernon, B.L.; McLemore, R.Y.; Nagy, T.; Moore, R.C.; Badha, V.S.; Childers, E.P.; Nguyen, M.B.; Gentry, D.D.; et al. In vivo evaluation of temperature-responsive antimicrobial-loaded PNIPAAm hydrogels for prevention of surgical site infection. *J. Biomed. Mater. Res. B Appl. Biomater.* **2022**, *110*, 103–114. [CrossRef]

87. McKay, C.S.; Finn, M.G. Click chemistry in complex mixtures: Bioorthogonal bioconjugation. *Chem. Biol.* **2014**, *21*, 1075–1101. [CrossRef] [PubMed]
88. Xu, J.; Liu, Y.; Hsu, S.H. Hydrogels Based on Schiff Base Linkages for Biomedical Applications. *Molecules* **2019**, *24*, 3005. [CrossRef]
89. Kuckling, D. Stimuli-Responsive Gels. *Gels* **2018**, *4*, 60. [CrossRef]
90. Alem, H.; Duwez, A.S.; Lussis, P.; Lipnik, P.; Jonas, A.M.; Demoustier-Champagne, S. Microstructure and thermo-responsive behavior of poly(N-isopropylacrylamide) brushes grafted in nano-pores of track-etched membranes. *J. Membr. Sci.* **2008**, *308*, 75–86. [CrossRef]
91. Schaer, T.P.; Stewart, S.; Hsu, B.B.; Klibanov, A.M. Hydrophobic polycationic coatings that inhibit biofilms and support bone healing during infection. *Biomaterials* **2012**, *33*, 1245–1254. [CrossRef] [PubMed]
92. Harris, L.G.; Richards, R.G. Staphylococci and implant surfaces: A review. *Injury* **2006**, *37* (Suppl. S2), S3–S14. [CrossRef] [PubMed]
93. Lai, M.; Cai, K.; Zhao, L.; Chen, X.; Hou, Y.; Yang, Z. Surface functionalization of TiO2 nanotubes with bone morphogenetic protein 2 and its synergistic effect on the differentiation of mesenchymal stem cells. *Biomacromolecules* **2011**, *12*, 1097–1105. [CrossRef] [PubMed]
94. Harris, L.G.; Richards, R.G. Staphylococcus aureus adhesion to different treated titanium surfaces. *J. Mater. Sci. Mater. Med.* **2004**, *15*, 311–314. [CrossRef]
95. Jaeger, K.E.; Ransac, S.; Dijkstra, B.W.; Colson, C.; van Heuvel, M.; Misset, O. Bacterial lipases. *FEMS Microbiol. Rev.* **1994**, *15*, 29–63. [CrossRef]
96. Chawla, J.S.; Amiji, M.M. Biodegradable poly(epsilon -caprolactone) nanoparticles for tumor-targeted delivery of tamoxifen. *Int. J. Pharm.* **2002**, *249*, 127–138. [CrossRef]
97. Proctor, R.A.; von Eiff, C.; Kahl, B.C.; Becker, K.; McNamara, P.; Herrmann, M.; Peters, G. Small colony variants: A pathogenic form of bacteria that facilitates persistent and recurrent infections. *Nat. Rev. Microbiol.* **2006**, *4*, 295–305. [CrossRef]
98. Wijagkanalan, W.; Kawakami, S.; Higuchi, Y.; Yamashita, F.; Hashida, M. Intratracheally instilled mannosylated cationic liposome/NFkappaB decoy complexes for effective prevention of LPS-induced lung inflammation. *J. Control. Release* **2011**, *149*, 42–50. [CrossRef]
99. Irache, J.M.; Salman, H.H.; Gamazo, C.; Espuelas, S. Mannose-targeted systems for the delivery of therapeutics. *Expert Opin. Drug Deliv.* **2008**, *5*, 703–724. [CrossRef]
100. Ali, L.; Goraya, M.U.; Arafat, Y.; Ajmal, M.; Chen, J.L.; Yu, D. Molecular Mechanism of Quorum-Sensing in Enterococcus faecalis: Its Role in Virulence and Therapeutic Approaches. *Int. J. Mol. Sci.* **2017**, *18*, 960. [CrossRef]
101. Nakayama, J.; Chen, S.; Oyama, N.; Nishiguchi, K.; Azab, E.A.; Tanaka, E.; Kariyama, R.; Sonomoto, K. Revised model for Enterococcus faecalis fsr quorum-sensing system: The small open reading frame fsrD encodes the gelatinase biosynthesis-activating pheromone propeptide corresponding to staphylococcal agrd. *J. Bacteriol.* **2006**, *188*, 8321–8326. [CrossRef] [PubMed]
102. Qiu, L.; Wang, C.; Lei, X.; Du, X.; Guo, Q.; Zhou, S.; Cui, P.; Hong, T.; Jiang, P.; Wang, J.; et al. Gelatinase-responsive release of an antibacterial photodynamic peptide against Staphylococcus aureus. *Biomater. Sci.* **2021**, *9*, 3433–3444. [CrossRef] [PubMed]
103. Chen, L.; Feng, W.; Zhou, X.; Yin, Z.; He, C. Thermo-and pH dual-responsive mesoporous silica nanoparticles for controlled drug release. *J. Control. Release Off. J. Control. Release Soc.* **2015**, *213*, e69–e70. [CrossRef] [PubMed]
104. Tang, Y.W.; Labow, R.S.; Santerre, J.P. Enzyme induced biodegradation of polycarbonate-polyurethanes: Dose dependence effect of cholesterol esterase. *Biomaterials* **2003**, *24*, 2003–2011. [CrossRef]
105. Leppilahti, J.M.; Hernandez-Rios, P.A.; Gamonal, J.A.; Tervahartiala, T.; Brignardello-Petersen, R.; Mantyla, P.; Sorsa, T.; Hernandez, M. Matrix metalloproteinases and myeloperoxidase in gingival crevicular fluid provide site-specific diagnostic value for chronic periodontitis. *J. Clin. Periodontol.* **2014**, *41*, 348–356. [CrossRef] [PubMed]
106. Uitto, V.J.; Overall, C.M.; McCulloch, C. Proteolytic host cell enzymes in gingival crevice fluid. *Periodontol 2000* **2003**, *31*, 77–104. [CrossRef]
107. Alkekhia, D.; LaRose, C.; Shukla, A. β-Lactamase-Responsive Hydrogel Drug Delivery Platform for Bacteria-Triggered Cargo Release. *ACS Appl. Mater. Interfaces* **2022**, *14*, 27538–27550. [CrossRef]
108. Ghimire, A.; Song, J. Anti-Periprosthetic Infection Strategies: From Implant Surface Topographical Engineering to Smart Drug-Releasing Coatings. *ACS Appl. Mater. Interfaces* **2021**, *13*, 20921–20937. [CrossRef]

Journal of
Functional
Biomaterials

Review

MOFs and MOF-Derived Materials for Antibacterial Application

Xin Zhang [1], Feng Peng [2,*] and Donghui Wang [1,3,*]

[1] School of Materials Science and Engineering, Hebei University of Technology, Tianjin 300130, China
[2] Medical Research Center, Department of Orthopedics, Guangdong Provincial People's Hospital, Guangdong Academy of Medical Sciences, Guangzhou 510080, China
[3] School of Health Sciences and Biomedical Engineering, Hebei University of Technology, Tianjin 300130, China
* Correspondence: pengfeng@gdph.org (F.P.); donghuiwang@hebut.edu.cn (D.W.); Tel.: +86-186-1693-6841 (F.P.); +86-022-6020-2214 (D.W.)

Abstract: Bacterial infections pose a serious threat to people's health. Efforts are being made to develop antibacterial agents that can inhibit bacterial growth, prevent biofilm formation, and kill bacteria. In recent years, materials based on metal organic frameworks (MOFs) have attracted significant attention for various antibacterial applications due to their high specific surface area, high enzyme-like activity, and continuous release of metal ions. This paper reviews the recent progress of MOFs as antibacterial agents, focusing on preparation methods, fundamental antibacterial mechanisms, and strategies to enhance their antibacterial effects. Finally, several prospects related to MOFs for antibacterial application are proposed, aiming to provide possible research directions in this field.

Keywords: metal organic framework (MOF); MOF-based composites; biocompatibility; antibacterial

Citation: Zhang, X.; Peng, F.; Wang, D. MOFs and MOF-Derived Materials for Antibacterial Application. *J. Funct. Biomater.* **2022**, *13*, 215. https://doi.org/10.3390/jfb13040215

Academic Editor: Chunlei Zhang

Received: 28 September 2022
Accepted: 29 October 2022
Published: 3 November 2022

Publisher's Note: MDPI stays neutral with regard to jurisdictional claims in published maps and institutional affiliations.

Copyright: © 2022 by the authors. Licensee MDPI, Basel, Switzerland. This article is an open access article distributed under the terms and conditions of the Creative Commons Attribution (CC BY) license (https://creativecommons.org/licenses/by/4.0/).

1. Introduction

A significant factor inducing human sickness and death has always been disease-causing microbes (bacteria, fungi, and viruses) [1–6]. Alexander Fleming, in 1928, discovered that penicillin was successful at preventing bacterial colonization, and researchers since then have discovered a range of drugs for antibacterial treatment [7,8]. However, drug-resistant bacterial populations have emerged as a result of long-term use of antibiotics. This poses a significant threat to human health [9–11], so there is an urgent need to find alternatives to antibiotics.

Many antibacterial materials, including metal-based (e.g., copper, silver, zinc) nanomaterials [12–14], organic materials [15–18], semiconductor photocatalytic materials [19–21], and natural biological antibacterial materials [22], have been developed and used in the field of biomedicine in recent years. However, most of them have not achieved clinical application. Metal-based nanomaterials and organic materials can kill bacteria by releasing toxic metal ions or antibacterial molecules, but the abrupt release of active substances has a short duration of efficacy and can be harmful to organisms [12–14,16,18]. Semiconductor photocatalytic materials can absorb externally stimulated light and generate photogenerated carriers to stimulate enhanced enzyme-like activity to kill bacteria. However, their limited light absorption capacity leads to low catalytic activity and an unremarkable antibacterial effect [19,20]. Natural biological antibacterial materials are harmless substances extracted from plants and animals, which can be slightly modified or even directly applied for antibacterial therapy in organisms, but the few available sources, expense, and complex process used to obtain these materials limit their wide clinical application. Furthermore, there are some prospective biomimetic technological routes to construct antibacterial drugs (e.g., the protein cages, and cell-membrane-camouflage). Ferritin has reversible self-assembly properties and has been used to construct artificial protein nanocages. Ferritin-based protein has nanocage-like structures with endowed chambers that provide sites for the encapsulation of antibacterial drugs. Furthermore, they can effectively target bacteria because of their unique positively charged nature. However, protein nanocages are known to be nanoscale, which can only encapsulate smaller molecule drugs and are limited in large molecule loading, and the existing protein encapsulation technology is not sufficient to precisely encapsulate functional substances [23]. Drugs camouflaged by various cell membranes can take advantage of the complex biological components and functions of cell membranes to successfully "trick" the body's immune system, smoothly passes immune recognition, prolongs circulation time in the blood, and enhances the efficiency of drug delivery at the lesion site. Although the coupling of cell membrane-coated technology with antibacterial drugs has showed satisfactory results in terms of screening ability from natural products, the screening efficiency still could not meet the requirements of practical applications. Additionally, its stability still needs further improvement to adapt to the complex physiological environment [24,25]. MOFs and MOF-derived materials stand out in the antimicrobial field because of their designable structure and adjustable size.

In 1995, Yaghi et al. [26] synthesized a two-dimensional coordination compound structure called a metal organic framework (MOF, please refer to Abbreviations at the end of this review) using a rigid organic ligand, BTC, and a transition metal ion, Co^{2+}. MOFs are hybrid inorganic–organic materials with pores. They are formed by bonded self-assembly of metal centers (metal ions or clusters) and organic ligands (bipedal or multi-pedal), and MOFs with desirable structures and functions can be obtained by rational design of IBUs and OBUs. As soon as this concept was proposed, it was developed at an astonishing pace in the following two decades.

Nowadays, numerous types of MOFs have been developed; the most common include IRMOF, ZIF, MIL, CPL, UIO, and PCN. IRMOF is an octahedral microporous crystalline material self-assembled from $[Zn_4O]^{6+}$ with a series of aromatic carboxylic acid ligands [27]. ZIF is synthesized by the reaction of Zn^{2+} or Co^{2+} with imidazole and exhibits a zeolite-like structure [28,29]. MIL is synthesized by transition metal ions with dicarboxylic acid ligands (succinic acid, glutaric acid, etc.) [30]. CPL consists of six metal ions coordinated to a neutral nitrogen-containing heterocycle [31]. The UIO MOFs are three-dimensional microporous materials formed by the coordination of $[Zr_6O_4(OH)_4]$ with BDC, which contains an octahedral central pore cage and eight tetrahedral corner cages [32]. PCN contains multiple cuboctahedral nanopore cages, which can form a cage–pore channel-like topology in space [33]. The above MOFs have been extensively studied and applied in the fields of adsorption and separation, energy storage, sensing, catalysis, pollutant removal, electrochemistry, and biomedicine [29,34–42]. In addition, there are many new types of MOFs (HKUST, etc.), which have also been widely studied in recent years and are gradually being applied in various fields [43,44].

In the field of biomedicine, MOFs have emerged as an ideal material for various antibacterial applications because of their preferable functions, such as controlled/stimulated decomposition, strong interaction with bacterial membranes, ROS production under irradiation, and high loading amount and controlled release of other antibacterial agents [45]. A large number of studies and reviews appeared in this field focusing on the composition, structure, and antibacterial mechanisms of different MOFs, and showed that MOFs and MOF-derived materials present a high bactericidal activity whose antibacterial rate exceeds 99% at appropriate doses (Table 1). [46–48]. However, the possible cytotoxicity limited the clinic application of MOFs and MOF-based materials, and it is still a great challenge to enhance their bactericidal activity and biocompatibility simultaneously. In this review, we summarize the preparation and antibacterial mechanisms of MOFs. Based on recent representative works on the development of MOFs, the possible mechanisms in antibacterial applications can be categorized as physical interaction, component release, CDT, PDT, PTT, SDT, and synergistic therapy. Particularly, a comprehensive review of the strategies to enhance the antibacterial effect of MOFs is presented in terms of both dynamic and thermodynamic aspects, including modulation of the size, pore size, and coordination environment of active sites, and the construction of MOF-based composites. Finally, the prospects and challenges of MOF-based antibacterial materials are discussed.

Table 1. Antibacterial activity of MOFs and MOF-derived materials.

MOF	Microbial Strain	Antibacterial Activity				Ref.
		Average Inhibition Diameter (mm)	MBC ($\mu g\ mL^{-1}$)	MIC ($\mu g\ mL^{-1}$)	Other Method	
ZIF-67	*S. cerevisiae* *P. putida* *E. coli*	15 15 15	–	– – 5		[49]
Co-SIM-1	*S. cerevisiae* *P. putida* *E. coli*	15 15 15	–	– – 5		[49]
$[(AgL)NO_3]\cdot 2H_2O$	*E. coli* *S. aureus*	13 16	–	300 297		[50]
$[(AgL)CF_3SO_3]\cdot 2H_2O$	*E. coli* *S. aureus*	15 16	–	300 307		[50]
$[(AgL)ClO_4]\cdot 2H_2O$	*E. coli* *S. aureus*	15 19	–	308 293		[50]
BioMOF-5	*S. aureus*	–	1700	4300		[51]
$\{[Zn(\mu\text{-}4\text{-}HZBA)_2]_2\cdot 4(H_2O)\}_n$	*S. aureus*	–	–	–	a half maximal effective antibacterial concentration of about 20 mg L^{-1}	[52]
ZIF-L	*E. coli* *S. aureus*	–	–	–	log reduction > 7 for *E. coli* and *S. aureus*; SEM images	[53]
CPPs	*B. subtilis* *P. vulgaris* *S. aureus* *P. aeruginosa* *S. enteritidis*	–	–	<25		[54]
Zn-PDA	*S. aureus* *B. subtilis* *A. baumannii* *K. pneumoniae* *S. entica* *E. coli*	17 16 11 9.7 9.7 8.6	–	300–308		[55]
Cu^{2+}-doped PCN-224	*S. aureus*	–	–	–	antibacterial efficacy (99.71%)	[56]
Au@ZIF-8	*S. aureus* *E. coli*	–	–	–	inhibition ratio against *S. aureus* and *E. coli* was >99.9% when the dosage of Au@ZIF-8 was 0.2 mg mL^{-1}	[57]
SPZA	*S. aureus* *E. coli*	–	–	–	The number of viable bacterial cells on SPZA is zero; SEM images	[58]
MnO_2/ZIF-8	*E. coli*	–	3.24	–	complete inactivation against *E. coli* at low concentrations (3.24 $\mu g\ mL^{-1}$)	[59]

Table 1. *Cont.*

MOF	Microbial Strain	Antibacterial Activity				Ref.
		Average Inhibition Diameter (mm)	MBC ($\mu g\ mL^{-1}$)	MIC ($\mu g\ mL^{-1}$)	Other Method	
Ag/ZnO	*E. coli* *S. aureus*	–	12.5 6.25	6.25 3.12		[60]
GO/Co-PTA	*E. coli* *S. aureus*	–	–	–	inhibited the growth of *E. coli* and *S. aureus* by up to 99%	[61]
Dual MOFs	*E. coli* *S. aureus*	–	–	–	the antibacterial efficiency against *S. aureus* and *E. coli* exceeded 99.31 and 98.68%, respectively.	[62]
ZGO-NH	*E. coli* *S. aureus*	2.59 3.82	-	-		[63]
ZIF-8@RFP	*MRSA*	-	10	-		[64]

2. Preparation Methods of MOFs

There are many methods to prepare MOFs. The same type of MOFs or even the same MOFs can be obtained by different preparation methods, which will confer different properties. In this section, we introduce several commonly used methods of synthesizing MOFs: one-pot, hydrothermal (solvent thermal), ultrasonic, electrochemical, and mechanochemical methods (Figure 1); we summarize their advantages and disadvantages for researchers to choose the appropriate preparation method based on their requirements.

2.1. One-Pot Method

One-pot synthesis is a process in which the precursors are co-mixed in a solvent and reacted under stirring. Wang et al. [65] synthesized Zr-based MOFs (Zr/UIO-66-COOH) using Zr^{4+} as the metal centers and H_2BDC as the organic ligand in the CF_3COOH solution. Huang et al. [66] synthesized Zn-based MOFs (MOCP) with a yield of 90% using $Zn(NO_3)_2$ and H_2BDC as reactants in DMF solution with TEA at room temperature.

The advantage of this method is that the cost is low, the production is large, and the experimental conditions can be easily achieved. In addition, it is a highly controllable process, since researchers can add reactants at any time during the reaction. However, synthesized MOFs usually contain impurities. Therefore, the one-pot method is not suitable for applications that require high purity.

2.2. Hydrothermal (Solvent Thermal) Method

Hydrothermal (solvent thermal) synthesis is a process in which primitive mixtures are reacted in a closed system, such as an autoclave, using water or organic solvents at a certain temperature and with autogenous pressure. Li et al. [39] synthesized MOF-5 using $Zn(NO_3)_2$ and H_2BDC as reactants in DMF by the solvent thermal method. The experimental data showed that the crystallinity of the product prepared by this method was higher compared to the one-pot method. This is because the high pressure under hydrothermal conditions increases the solubility of the precursors, consequently promoting the reaction and crystal growth. In addition, the synthesized MOF-5 had a specific surface area of up to 2900 $m^2\ g^{-1}$ and could remain thermo-stable at 500 °C in the absence of moisture.

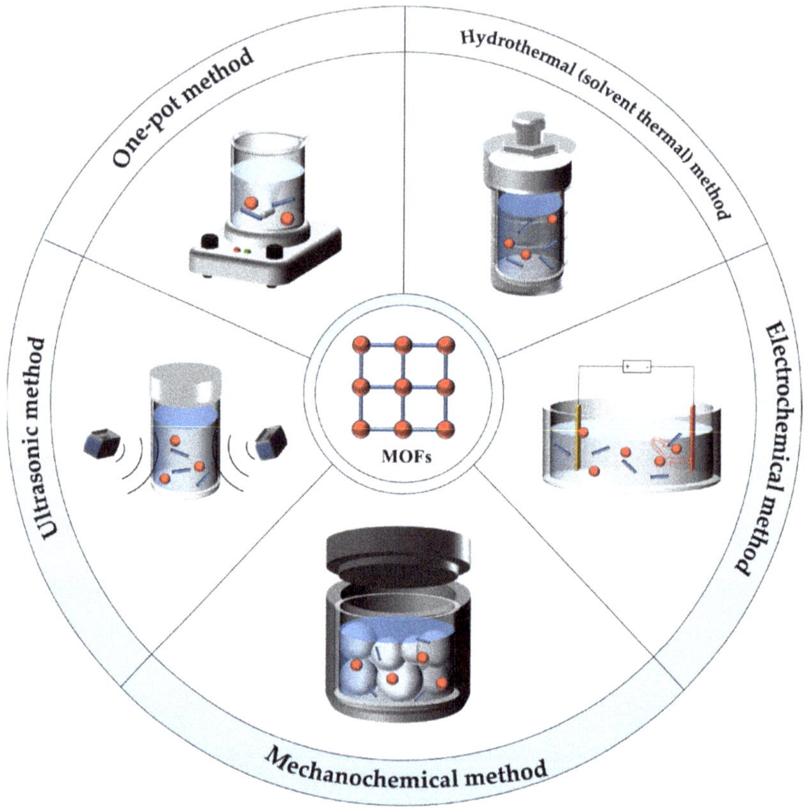

Figure 1. Schematic illustration of methods for preparing MOFs.

The hydrothermal (solvent thermal) method exhibits high specific surface area, high crystallinity, and thermal stability. However, this method is costly and cannot be well controlled, as reactants can only be added all at once. In addition, this method is not stable, since the specific pressure of the reactor, which is controlled by artificial tightening, is not a constant value, and this uncertainty will certainly affect the properties of the products.

2.3. Ultrasonic Method

Sonication is a process in which reactants are dissolved in solvent and continuously sonicated. Qiu et al. [67] first obtained $Zn_3(BTC)_2$ by sonication using an aqueous solution of $Zn(CH_3COO)_2$ and H_3BTC dissolved in ethanol for 5 min. As the sonication time increased from 5 to 90 min, the size of the product increased from 50–100 to 700–900 nm, and the yield increased from 75.3 to 85.3%. Previous studies showed that mixing $Zn(CH_3COO)_2$ with H_3BTC without ultrasound did not produce any product using the same reaction medium, temperature, and pressure. In theory, acoustic cavitation induced by the rupture of cavitation bubbles creates local hot spots with very high transient temperature (5000 K), pressure (1800 atm), and cooling rate (1010 K s^{-1}) [68,69]. The region between the microbubble and the native solution has very large temperature and pressure gradients and rapid molecular motion, leading to the generation of an excited state, breaking of bonds, free radical formation, mechanical shock, and high shear gradients, so that the ultrasonic method allows some reactions to take place that are difficult to carry out by conventional methods.

Sonication is a low-cost and environmentally friendly method. The size and yield of MOFs synthesized by the ultrasonic method can be easily controlled. Moreover, some special reactions that cannot be carried out by conventional methods can be achieved because of the acoustic cavitation effect. However, the structure and purity of MOFs formed by sonication can vary.

2.4. Electrochemical Method

Electrochemical synthesis is a process in which monomers or compounds of different species and aggregates are prepared by electro-oxidation or electro-reduction in a conductive solvent. It is a widely used method to construct MOF films on substrates. Ameloot et al. [70] first synthesized Cu-based MOF (Cu_3-$(BTC)_2$) film on copper substrate by electrochemical deposition. The anode metal plates generated metal ions during electrolysis and then self-assembled with organic ligands on the electrode surface to form MOF films. They found that increasing the voltage from 2.5 to 25 V provided a higher concentration of metal ions near the electrode surface and produced a coating with smaller crystals, while adding water to the mixture slowed down the formation of crystals and formed a coating with larger crystals.

The advantage of the electrochemical method is that the synthesis process is simple and rapid, and the thickness of MOF films can be conveniently adjusted by adjusting the voltage and current applied and the electrolyte concentration. However, this method can only construct films on conductive substrates, limiting its wide application.

2.5. Mechanochemical Method

Mechanochemical synthesis is a process in which metal salts are reacted directly with organic ligands by mechanical grinding at a specific temperature. Katsenis et al. [71] synthesized a Zn-based MOF (ZIF-8) by the mechanochemical method using ZnO and 2-methylimidazole with small amounts of acetic acid or water as catalysts. They found that the mechanochemical process led to the collapse of the ZIF-8 structure. The higher the volume of solvent added during the grinding process, the more pronounced the amorphization of the product.

The mechanochemical method is environmentally friendly because of the low solvent volatilization. In addition, the amorphization of MOFs can be controlled. However, the crystallinity of the product synthesized by this method is low and the structure is easily destroyed in the process.

3. Antibacterial Mechanisms of MOFs

MOFs have a periodic coordination network in which metal centers and organic ligands are linked to each other and can be designed to have various typologies by regulating these two components [72,73]. In the field of biomedicine, several MOFs have been extensively studied for their preferable antibacterial activity resulting from their specific physical and chemical properties (such as slow release of metal ions or organic substances and enzyme-like, photocatalytic, photothermal, and ultrasonic activity) [49,74–77]. This section summarizes in detail the possible antibacterial mechanism of MOFs, which can be subdivided into physical interaction, metal ion release, organic ligand release, antibiotic load, gas load, CDT, PDT, PTT, SDT, and synergistic therapy (Figure 2 and Table 2).

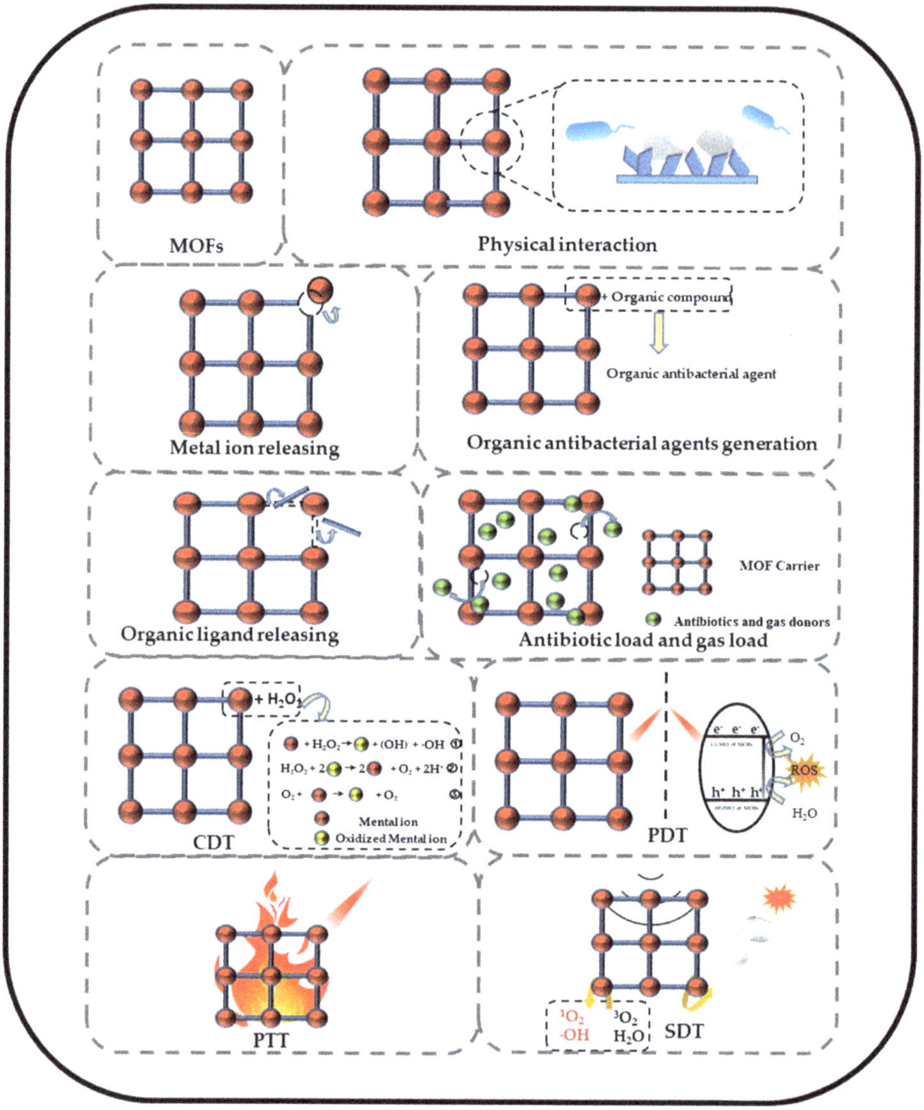

Figure 2. Schematic illustration of antibacterial mechanisms of MOFs.

Table 2. Antibacterial components and main roles of MOFs.

IBUs	OBUs	Functional Materials	Main Role of MOFs in Antibacterial Application	Ref.
Ag^+	3−(Biphenyl−4−yl)−5−(4−tertbutylphenyl)−4−phenyl-4H−1,2,4−triazole	AgTAZ	Ag^+ release	[49]
Ag^+	Tris−(4−pyridyldturyl)borane(L)	$[(AgL)NO_3]·2H_2O$	Ag^+ release	[50]
Ag^+	1−Butylimidazole	[Ag(Bim)]	Ag^+ release	[78]
Co^{2+}	Hmim	ZIF−67	Co^{2+} release	[49]
Cu^{2+}	Trimesic acid	HKUST−1	Co^{2+} release	[74]
Cu^{2+}	1,4−Benzendicarboxylic acid	Cu−SURMOF−2	Cu^{2+} release	[79]
Zn^{2+}	Azathioprine	BioMOF−5	Zn^{2+} release	[51]
Zn^{2+}	Hmim	ZIF−8	Organic compound generation	[80]

Table 2. *Cont.*

IBUs	OBUs	Functional Materials	Main Role of MOFs in Antibacterial Application	Ref.
Ni^{2+}	Hmim	Ni−Hmim	Hmin release	[77]
Zn^{2+}	4−HZBN	$[Zn(\mu-4-HZBA)_2]_2 \cdot 4(H_2O)\}_n$	HZBN release	[52]
Zr^{4+}	BA, TCPP	VAN	VAN release	[81]
Fe^{3+}	H_2BDC	VAN	VAN release	[82]
Cu^{1+}	1,3,5−Tribromobenzene + diethylamine → H_3BTTri	GSNO	NO release	[83]
Zr^{4+}	BA, TCPP	L−Arg	NO release	[84]
Hf^{4+}	2−Hydroxyterephthalic acid, acetic acid	MnCO	CO release	[85]
Zn^{2+}	Hmim	Pd(H)	H_2 release	[86]
Fe^{3+}	Acrylic acid	MIL−88B	CDT	[87]
Cu^{2+}	1,3,5−Tricarboxybenzene	HKUST−1	CDT	[88]
Zr^{4+}	H_2BDC → I2−BDP	UIO−PDT	PDT	[6]
Hf^{4+}	H_2DBP	DBP−UIO	PDT	[89]
Hf^{4+}	TCPP	Hf−TCPP NMOF	PDT	[90]
Zr^{4+}	H_2TCPP	PCN−224	PDT	[91]
Fe^{2+}/ Fe^{3+}	$K_4[Fe(CN)_6]$	PB	PTT	[92]
Mn^{2+}	IR825	Mn−IR825	PTT	[93]
Zr^{4+}	$(Fc(COOH)_2)$	Zr−Fc MOF	PTT	[94]
Fe^{3+}	GA	Fe−CPND	PTT	[95]
Zn^{2+}	Hmim	ZIF−8 → PMCS	SDT	[96]
Zn^{2+}, Mn^{2+}	Hmim	ZIF−8 → DHMS	SDT	[97]
Zr^{4+} Pt^{4+} Au^{3+}	BA	HNTM → HNTM−Pt@Au	SDT	[98]

3.1. Physical Interaction

Physical interaction refers to the process of direct interaction between materials with specific surface morphologies and bacteria in the physiological environment. Recently, some MOFs were synthesized with two-dimensional morphology that exhibited excellent antibacterial properties. Yuan et al. [53] synthesized ZIF-L with a nano-dagger surface, which is completely different from the dodecahedral structure of ZIF-8. The experimental data showed that ZIF-L exhibited excellent antibacterial activity (log reduction > 7 for *E. coli* and *S. aureus*) and sustainable bactericidal activity (more than four reapplications). They believed that the nano-dagger surface of ZIF-L could kill bacteria efficiently via physical interaction (Figure 3A). Wang et al. [54] synthesized four Cu-based MOFs (CPPs) with the same structure but different morphologies (rhombus layer, rhombus disk, rhombus lump, and bread-like morphology) by adding additives such as TEA and CH_3COOH. The rate of deprotonation of organic ligands can be controlled by the addition of additives to further control particle size and shape. Among them, the bread-like and rhombus disk CPPs exhibited weak antibacterial activity (MIC > 50 $\mu g\ mL^{-1}$), while the rhombus lump and rhombus layer CPPs showed good antibacterial activity against *B. subtilis*, *P. vulgaris*, *S. aureus*, *P. aeruginosa*, and *S. enteritidis* (MIC < 25 $\mu g\ mL^{-1}$). They believed that the unique two-dimensional sheet morphology of rhombus lump and rhombus layer CPPs effectively inhibit bacterial growth.

MOFs with two-dimensional morphology exhibit excellent antibacterial properties because their unique morphology can kill bacteria efficiently via physical interaction; however, they can also be harmful to normal cells, posing a risk to human health.

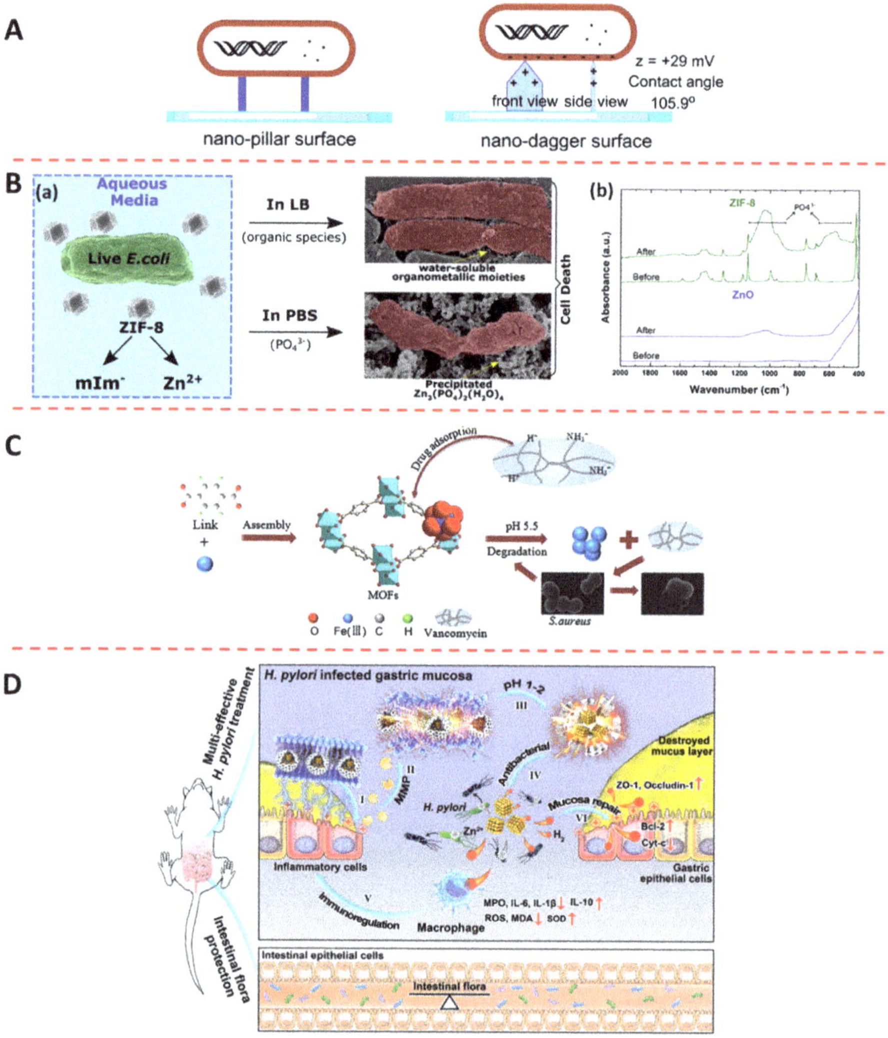

Figure 3. Various antibacterial mechanisms of MOFs: (**A**) Schematic illustration of ZIF−L with nano-dagger morphology in physical interaction with bacteria (adapted from [53]). (**B**) (**a**) Schematic illustration of ZIF−8 reacts with PBS to produce $Zn_3(PO_4)\cdot 2(H_2O)_4$ with antibacterial activity (adapted from [80]); (**b**) FTIR spectrum of the ZIF−8 and ZnO nanopowders before and after PBS incubation (adapted from [80]). (**C**) Schematic illustration of VAN@MOF−53(Fe) release Fe^{3+} and VAN to kill *S. aureus* in an acidic environment (adapted from [82]). (**D**) Schematic illustration of Pd(H)@ZIF−8 release Zn^{2+}, Pd nanoparticles, and H_2 in an acidic environment to kill H. pylori (adapted from [86]).

3.2. Metal Ion Release

Metal ion release refers to the process of dissociation and release of metal ions from the crystal structure of MOFs in the physiological environment. Various metal ions, such as silver ion (Ag^{+}), zinc ion (Zn^{2+}), iron ion ($Fe^{2/3+}$), manganese ion (Mn^{2+}), lead ion (Pb^{2+}), cobalt ion (Co^{2+}), and copper ion (Cu^{2+}), have attracted research attention because of their broad-spectrum antibacterial properties and weak toxicity to eukaryotic cells [99–103]. Researchers have synthesized a number of antibacterial MOFs by incorporating the above metal ions into the crystal structure, and they all found that the slow release of metal ions was the determining factor for the good antibacterial properties [53,54,104–106].

Liu et al. [50] synthesized three Ag-based MOFs ([(AgL)NO_3]·2H_2O, [(AgL)CF_3SO_3]·2H_2O, and [(AgL)ClO_4]·2H_2O) with different typologies using tris-(4-pyridylduryl)borane (L), Ag^+, and different coordination solvents. These three MOFs exhibited a significant bacteria-inhibiting loop with an average diameter of 13–15 mm for *E. coli* and 16−19 mm for *S. aureus*, and MIC values of 250–500 and 293–307 μg·mL^{-1} against the two bacteria, respectively. They attributed the high antibacterial activity to the sustainable and trace release of Ag^+ (0.85–1.20‰ of overall Ag^+ in the samples per day). Aguado et al. [49] synthesized two Co-based MOFs (ZIF-67 and Co-SIM-1) using 2-methylimidazole and 4-methyl-5-imidazolecarboxaldehyde, respectively, as the organic ligands. The results showed that the growth inhibition ratio for *S. cerevisiae*, *P. putida*, and *E. coli* was higher than 50% when the concentrations of ZIF-67 and Co-SIM-1 were in the range of 5–10 mg L^{-1}. Similarly, they attributed the high antibacterial activity of these MOFs to the sustainable and trace release of Co^{2+}.

To further research how released ions kill bacteria in the physiological environment, Taheri et al. [80] explored the degradation process of ZIF-8 in organisms. The experimental data showed that ZIF-8 first released Zn^{2+} and then immediately produced $Zn_3(PO_4)_2$ (a broad-spectrum antibacterial agent) when immersed in PBS, which simulated the physiological environment (Figure 3B(a,b)). So, they attributed the high antibacterial activity of ZIF-8 to the production of $Zn_3(PO_4)_2$ in the physiological environment.

Research has shown that MOFs can achieve efficient and long-term antibacterial effects through the slow release of metal ions. However, the specific metabolic behavior of released metal ions in the organism is unclear, and the antibacterial theory of the released metal ions is not unified. Some people believe that the released metal ions directly kill bacteria, while others believe that they form an antibacterial agent with organic substances to kill bacteria [80].

3.3. Organic Ligand Release

Organic ligand release refers to the process of dissociation and release of organic ligands from the crystal structure of MOFs in the physiological environment. There are many available organic antibacterial agents, including metallo-organic agents, aldehydes, phenols, acyl aniline, and heterocyclics. These antibacterial agents bind with calcium and magnesium cations of bacterial cells, which leads to the fragmentation of cellular DNA [107]. Fortunately, plenty of these organic antibacterial agents (imidazoles, benzimidazole dazoles, heterocyclic compounds, etc.) can be used in the construction of MOFs [108,109]. Restrepo et al. [52] synthesized a Zn-based MOF ({[Zn(μ-4-HZBA)$_2$]$_2$·4(H_2O)}$_n$) using 4-HZBA as the organic ligand and Zn^{2+} as the metal center. The experimental data showed that this material significantly inhibited the growth and metabolic activity of *S. aureus*, with a half maximal effective antibacterial concentration of about 20 mg L^{-1}, and it released ligands continuously over a period of days. The antibacterial effect of this material was attributed to the release of 4-HZBA; the release of Zn^{2+} not only failed to provide an antibacterial effect, but also promoted the growth and propagation of bacteria to form biofilm. The reason the Zn^{2+} from this material failed to kill bacteria is that it is one of the twenty-six essential elements in animal tissues and its concentration must be above a certain threshold (50 μM) [110] to exhibit an antibacterial effect even though it is a broad-spectrum bactericidal agent.

Research has shown that the release of organic ligands can achieve efficient and long-lasting antibacterial effects. However, the metabolic behavior of MOF-released organic substances in the physiological environment is still unclear, which raises concerns regarding biosafety.

3.4. Antibiotic Load

Antibiotics are effective at inhibiting bacterial colonization and growth, and their slow release in the physiological environment can achieve antibacterial purposes. MOFs with high specific surface area are suitable carriers for loading antibiotics, and the released antibiotics can effectively kill bacteria in the physiological environment. Chen et al. [81] synthesized a VAN-functionalized Zr-based MOF (PCN-224) for targeting and killing *S. aureus*. The functionalization endowed the VAN-PCN-224 with excellent antibacterial efficiency against *S. aureus* because of the targeting ability and antibacterial activity of VAN against Gram-positive bacteria. Lin et al. [82] encapsulated VAN into MOF-53(Fe) with strong electrostatic interactions using the one-pot method. This product had an effective drug-carrying capacity of 20 wt% and good biocompatibility. The results showed that this material exhibited a long-lasting and highly effective antibacterial effect (99.3%), because VAN and Fe^{3+} were released slowly in the inflammatory bacterial environment (Figure 3C).

Antibiotics are effective at inhibiting bacterial colonization and growth. However, the abuse of antibiotics has led to the growth of multidrug-resistant pathogens, posing a serious threat to human health.

3.5. Gas Load

Gas therapy is a treatment that produces gases to kill bacteria in the physiological environment through degradation or redox reactions. In recent years, some gas molecules, such as H_2, NO, and CO [111–114], have been reported to have antibacterial properties and to be effective at alleviating the inflammatory response [115]. MOFs with high specific surface area can be suitable carriers for loading a gas donor, and the released gas from the gas donor can effectively and safely kill bacteria in the physiological environment. GSNO is a donor of NO, which is commonly present in human blood and can decomposed into one NO equivalent and disulfide [116]. Tuttle et al. [83] encapsulated GSNO in a Cu-based MOF (CuBTTri). They found that the decomposition ratio of GSNO in the reaction catalyzed by CuBTTri was only 10% within 16 h without GSH, while in the reaction catalyzed by CuBTTri it was up to 100% with stoichiometric GSH. Therefore, this material can produce more NO and has better bactericidal effect in a bacterial infection environment with higher GSH expression. Wan et al. [84] encapsulated L-Arg as an NO donor into a Zr-based MOF (PCN-224) for gas therapy. After irradiation with a 660 nm laser at 30 mW cm^{-2}, L-Arg reacted with H_2O_2 to produce NO with a long half-life and wide diffusion range. Guan et al. [85] first synthesized an Hf-based MOF (UIO-66-OH(Hf)) using Hf^{4+} as the metal center and H_2BDC as the organic ligand. Subsequently, a PS (2I-BODIPY) was attached to UIO-66-OH(Hf) by etherification, and finally the gas donor MnCO was coordinated to the Hf cluster node. They found that the Mn^{1+} in MnCO could oxidize to Mn^{2+} and release CO when the material was exposed to light irradiation (green laser, 0.5 W/cm^2, 10 min) and oxidants containing H_2O_2 and $^1O_2$. Zhang et al. [86] encapsulated hydrogen-stored Pd nanoparticles into the inner cavity of ZIF-8 by the one-pot method, and then wrapped ascorbyl palmitate hydrogel to target *H. pylori*. The ZIF-8 shell of Pd(H)@ZIF-8 broke and subsequently released high concentrations of Pd nanoparticles and the H_2 stored in the Pd nanoparticles to kill bacteria in an acidic microenvironment induced by bacteria. It also upregulated the expression of mucosal repair proteins to repair damaged gastric mucosa (Figure 3D).

Research has shown that antibacterial gases are effective at alleviating the inflammatory response and promoting tissue restoration. However, their efficacy in sterilization is limited, so they are generally used as an adjunctive treatment in related investigations.

3.6. CDT

ROS are intermediate chemical species formed during the incomplete reduction of oxygen, mainly including H_2O_2, ·OH, ·O^{2-}, and $^1O_2$ [117]. Many natural enzymes can catalyze ROS production from low concentrations of H_2O_2, which can kill bacteria by irreversibly damaging bacterial cell walls/cell membranes/cells, DNA, proteins, polysaccharides, and nucleic acids [118]. However, the high cost, complicated purification process, poor stability, and difficult recycling of natural enzymes limit their practical application. It is urgent to find enzyme-like materials that can simulate the activities of these natural enzymes.

Recently, MOF-based catalytic systems mainly based on the simulation of GOD, POD, SOD and CAT were developed, and researchers found that the MOFs could simulate catalytic activity to produce ROS or their precursors for sterilization [119–123]. Zhang et al. [124] first found in 2016 that amorphous Fe NPs can undergo a Fenton-like reaction in the tumor microenvironment of hydrogen peroxide overexpression. The overall Fenton-like reaction involves the ionization of amorphous Fe NPs to release ferrous ions and the subsequent disproportionation of H_2O_2 to effectively produce hydroxyl radicals, which will cause irretrievable damage to DNA, lipids, and proteins [125,126]. They also referred to this treatment, which relies on Fenton-like responses, as CDT, and the overall Fenton-like response equation with Fe^{2+} as reactant is as follows:

$Fe^{2+} + H_2O_2 \rightarrow Fe^{3+} + (OH)^- + \cdot OH$ ①

$H_2O_2 + 2Fe^{3+} \rightarrow 2Fe^{2+} + O_2 + 2H^+$ ②

$O_2 + Fe^{2+} \rightarrow Fe^{3+} + O_2^-$ ③

Over the past several years, studies of CDT in MOFs have developed rapidly. This is mainly attributed to the intrinsic structural features of MOFs. In an acidic bacterial infection microenvironment, the coordination bonds between metal centers and organic ligands are easily broken, releasing metal ions and thus providing catalysts for CDT. Burachaloo et al. [87] synthesized an Fe-based MOF (MIL-88B). They found that Fe^{2+} was released when MIL-88B was fragmented and subsequently induced an intracellular Fenton-like reaction to produce high concentrations of ·OH. Hao et al. [88] synthesized a Cu-based MOF. They found that Cu^{2+} was reduced to Cu^{1+} as GSH was consumed in the bacterial infection microenvironment, and the Cu^{1+}-based MOF catalyzed the formation of ·OH from hydrogen peroxide to kill bacteria.

Research has shown that CDT is a relatively safe treatment because it does not require the introduction of antibiotic drugs, which could cause drug resistance. However, the pH of the bacterial microenvironment does not accommodate the requirements for an effective Fenton reaction (pH = 3.0–4.0) and there is a limited concentration of hydrogen peroxide in the bacterial infection microenvironment, both of which greatly limit the effectiveness of CDT.

3.7. PDT

PDT is a treatment that combines PSs and their corresponding light irradiation to catalyze ROS production through photodynamic reactions to selectively destroy target tissues in the presence of oxygen molecules. Photoexcited triplet PSs can react in two ways: (I) directly with the substrate or solvent through hydrogen or electron transfer to form free radicals, or (II) through energy transfer to oxygen molecules to form $^1O_2$.

PSs can produce ROS to kill bacteria when irradiated by a specific wavelength of light. However, the aggregation between PS nanoparticles leads to the quenching effect. To maintain the ability of PSs to produce photogenerated ROS, researchers have been working on covalently binding PSs, as OBUs, to IBUs and periodically assembling them into porous frameworks to prevent the quenching effect. In addition, the wavelength of light that PSs can absorb is also an important parameter. For biomedical applications, wavelengths are usually controlled in the "tissue transparency" window of 650–800 nm. This is because shorter wavelengths (<600 nm) have limited treatment depth and can cause skin photosensitization, while longer wavelengths (>800 nm) do not provide sufficient energy to produce $^3O_2$, which is a precursor of ROS [127]. Lu et al. [89] synthesized an Hf-based MOF (DBP-UIO) using Hf^{4+} and the free base porphyrin derivative H_2DBP under solvothermal conditions. This nanoplate-shaped material had a diameter of 100 nm and a thickness of 10 nm and exhibited photodynamic activity. They believed that the interaction of the heavy Hf^{4+} with the DBP ligand induced an enhanced inter-system crossover from monomorphic to triplet DBP, thus DBP-UIO was at least twice as efficient as H_2DBP in producing $^1O_2$ and showed excellent PDT effect. Liu et al. [90] synthesized an Hf-based MOF using Hf^{4+} as the metal center and TCPP as the organic ligand. Under irradiation by an 808 nm laser at 5 mW cm^{-2}, this MOF produced more $^1O_2$ compared to free TCPP.

They believed that the coordination of the four carboxyl groups of TCPP with metal ions broadened the Soret band of TCCP and slightly redshifted the four Q-bands. The proximity of heavy Hf^{4+} centers promoted the inter-system crossover of TCPP from the ground state to excited state, and the unique porous structure of the Hf-based MOF led to a good spatial segregation effect between TCPP ligands, avoiding agglomeration and self-quenching of the excited state.

The non-porphyrin photosensitizer BODIPY is also considered to be an ideal PDT reagent because of its low dark toxicity and high extinction coefficient in the therapeutic window. Wang et al. [6] obtained a daughter Zr-based MOF (UIO-PDT) that showed PDT effects by exchanging the I2-BDP ligand with the H_2BDC ligand of the already synthesized parent MOF via solvent-assisted ligand exchange. The mechanism of the overall solvent-assisted ligand exchange process can be explained as the exposure of the parent MOF to the solution containing a high concentration of the second ligand, then the ligand of the parent MOF being exchanged with the target ligand to obtain a daughter MOF. The experimental results showed that the daughter MOF presented a stronger ability to produce photogenerated ROS and retained the topology of the parent MOF. In addition, the UIO-PDT showed little cytotoxicity to a series of normal cells, in which the cell survival rates were all above 90%, although the concentration of UIO-PDT was adjusted up to 1.0 mg mL^{-1}.

PDT is highly selective therapy. It can generate ROS with more efficient bactericidal activity than organic antibacterial agents [128–131]. However, the production of excessive ROS also causes cellular damage and poses a threat to the health of the organism when cells are affected by microenvironmental stimulation or antioxidant system dysregulation.

3.8. PTT

PTT is a treatment in which materials generate heat after absorbing externally stimulated light and kill bacteria by high temperature in the applied environment [132,133]. The principles of photothermal conversion include the following: (I) The electron relaxation of the excited state decays back to the ground state, causing a collision of emitting chromophores with the surrounding environment and partial energy release in the form of heat. (II) Light directs the polarization of free electrons and the depolarization of accumulated charges, leading to collective electron oscillations, surface plasma jump decay, and energy dissipation in the form of heat. (III) When a semiconductor is irradiated with high-energy light, the electrons are excited to higher energy levels in the conduction band and leave holes in the VB. The electrons and holes will relax to the edge of the energy band by vibrational relaxation, which leads to the conversion of energy into heat. MOFs have attracted much attention in recent years in the field of photothermal antibacterial therapy because of their semiconductor-like behavior. In addition, some MOFs can also generate heat to kill bacteria under light irradiation because of the strong LSPR of metal ions at the active sites.

PB presents a face-centered cubic structure formed by the coordination of a carbon atom and a nitrogen atom from a carbon–nitrogen triple bond with Fe^{2+} and Fe^{3+}, respectively. PB and its analogs comprise the most important branch of intrinsic photothermal MOFs. It is also one of the oldest synthetic MOFs and widely used as a PTA, which has been approved by the Food and Drug Administration as a clinical drug for light radiation therapy [134]. Fu et al. [92] synthesized PB with controlled size by mixing solutions of $FeCl_3$ and $K_4[Fe(CN)_6]$ with citric acid as the surface capping agent. The experimental data showed that the PB had a broad absorption band at 500–900 nm with an absorption peak at 712 nm. Moreover, under irradiation by an 808 nm laser at 0.6 W cm^{-2}, the PB had a molar extinction coefficient of 1.09×10^9 M^{-1} cm^{-1}, which is slightly lower compared to Au nanorods, and its temperature increased to 43 °C in less than 3 min. However, the absorption peak was located at the edge of the near-infrared region (700–900 nm), and the photothermal conversion efficiency was only 20% at 808 nm irradiation [135].

Other intrinsic photothermal MOFs are based on the LMCT mechanism to achieve an antibacterial effect, which tends to occur when organic ligands and metal ions are in relatively low valence states. IR825 has an absorption peak at 825 nm in the near-infrared light region and is commonly used as near-infrared dye. Yang et al. [93] synthesized an Mn-based MOF using Mn^{2+} as the metal center and IR825 as the organic ligand. The temperature of this material increased rapidly to approximately 52 °C within 5 min under irradiation by an 808 nm laser at 0.6 W/cm^2, and it had a mass extinction coefficient of 78.2 L g^{-1} cm^{-1} and good photothermal cycling capability. Deng et al. [94] used $Fc(COOH)_2$, a widely used PTA, as the ligand and Zr^{4+} as the metal center to synthesize a nano-sheeted Zr-Fc MOF. The Zr-Fc MOF had broad light absorption in the range of 350–1350 nm, and its temperature reached 92 °C within 3 min under irradiation by an 808 nm laser at 0.2 W cm^{-2}, which was significantly higher than $Fc(COOH)_2$ (around 46.8 °C). The higher photothermal conversion efficiency of Zr-Fc MOF stemmed from its higher stability, so most excited photoelectrons decayed through non-radiative pathways and consequently generate heat. Liu et al. [95] synthesized an Fe-based MOF (Fe-CPND) using Fe^{3+} as the metal center and GA as the organic ligand. In an acidic environment, the GA ligand gradually dissociated and the Fe-CPND transformed from triple to double ligand, conferring a higher longitudinal relaxation rate. They also found that the preferable photothermal effect of Fe-CPND resulted from the LMCT effect triggered by the Fe–phenol structure.

The high temperature induced by PTT can kill most bacteria that are intolerant to high temperatures. However, when the temperature is too high and insufficient to completely remove the bacteria, non-local heating usually causes severe damage to healthy tissues and an inflammatory response [136].

3.9. SDT

Because of the shallow tissue penetration depth of PDT and the damage to normal cells caused by long-term exposure to external stimulating light, its clinical application is limited. SDT is a treatment in which the sound sensitizer reacts with oxygen to produce ROS to kill bacteria under ultrasonic radiation. The cavitation effect occurs during the ultrasound process, and the resulting bubbles rapidly compress in volume and burst to produce strong shock waves, local high temperature, high pressure, and hydroxyl radicals to improve the antibacterial effect. Pan et al. [96] synthesized PMCS containing porphyrin-like metal centers. PMCS can be obtained after calcination at high temperature using ZIF-8 as a template, and it is a kind of N-doped carbon material with a Zn-centered porphyrin-like structure. The unique micro/mesoporous structure of PMCS enhances the cavitation effect, resulting in high SDT efficiency. Experimental results showed that the $^1O_2$ producing efficiency of PMCS was increased by 203.6% and it could generate more ·OH than ZIF-8.

Pan et al. [97] synthesized an MOF-derived DHMS nanoparticles. DHMS can be excited by US irradiation to produce electrons and holes, which interact with O_2 and H_2O in the body to produce $^1O_2$ and ·OH, respectively. The low-valent Mn in DHMS is partially oxidized to high valence by the holes during the sonication reaction, promoting electron–hole separation and ROS production. In addition, the hollow pore structure of DHMS enhances the cavitation effect to further increase ROS generation. Yu et al. [98] synthesized a US-activated single-atom catalyst (HNTM-Pt@Au) actuated by Au nanorods. HNTM is a Zr-based MOF with a porphyrin-like structure and works as a carrier in this material. The experimental data showed that the antibacterial efficiency of HNTM-Pt@Au against methicillin-resistant *MRSA* under US was 99.93%. They summarized the ultrasonic catalysis mechanism of this material as follows: (I) Au NRs enhance ultrasonic cavitation and improve the absorption of ultrasonic energy in the system, and (II) Au NRs and single Pt atoms act as electron acceptors to promote the electron transfer generated by HNTM and improve the separation efficiency of electron–hole pairs. In addition, the products obtained by the exchange of single Au and Cu atoms by single Pt atoms in HNTM also generate a large amount of ROS, which can be attributed to the metal with a higher work function having a higher carrier separation efficiency.

Intermittent low-power ultrasound is harmless to the human body, so this non-invasive therapy has a bright future in clinical applications. However, if the ultrasound treatment time is too long or the heat is not controlled properly, it can lead to burns, inflammation, or even necrosis in the skin or deep soft tissues.

3.10. Synergistic Therapy

Taking advantage of the synergistic effect of multiple therapies is an effective way to improve the antibacterial activity. In recent years, many MOFs have achieved the effect of "1 + 1 > 2" by synergizing the antibacterial activity of their components. To synthesize MOF-derived 2D carbon nanosheets, Fan et al. [137] first synthesized MOF-derived ZnO on graphene, and then obtained TRB-ZnO@G by in situ polymerization of anchored phase change TRBs. The TRB-ZnO@G not only had nearly 100% antibacterial efficiency at low concentrations, but also exhibited rapid and safe skin wound disinfection, without damaging normal skin tissue or causing cumulative toxicity. Under NIR irradiation, TRBs absorbed NIR light and increased the local temperature, which also promoted the release of Zn^{2+} from ZnO and penetrated into bacterial cells. Moreover, the unique two-dimensional structure of 2D carbon nanosheets enhanced the physical interaction with bacterial cells. Thus, the antibacterial mechanisms working synergistically exhibited more effective antibacterial activity than any single mechanism component.

Synergistic therapy is not simply the sum of individual antibacterial mechanisms but involves the complementarity of the strengths and weaknesses of various mechanisms for mutual improvement to achieve better antibacterial effects. The synergistic antibacterial activity of MOF-based composites will be discussed in more detail in the next section.

4. Strategies to Enhance the Antibacterial Ability of MOFs

Although MOFs have shown good antibacterial effect in the experimental stage, their antibacterial ability has not yet reached clinical demand because of the complexity of the human body microenvironment. Therefore, researchers have launched extensive investigations into strategies for enhancing the antibacterial properties of MOFs in recent years. They found that these properties can be improved by modulating the size, pore size, and coordination environment of active sites of MOFs, and by synthesizing MOF-based composites. These strategies are discussed in detail in this section.

4.1. Size Modulation

The size effect influences the uptake of antibacterial MOFs by bacterial cells. Bacterial cells tend to uptake nanoparticles of small size, so the uptake of MOFs by bacterial cells can be improved by reasonably regulating the size of MOFs (Figure 4A). In addition, miniaturization could increase the interactions between MOFs and bacteria [138,139], which would enhance their antibacterial ability. Akbarzadeh et al. [55] synthesized a nanoscale Zn-based MOF (Zn-PDA) using $Zn(NO_3)_2$ and PDA. The nanoscale Zn-PDA exhibited better antibacterial properties against *S. aureus*, *S. enteritidis*, *A. baumannii*, *K. pneumoniae*, *S. entica*, and *E. coli* (average inhibition diameter, 8.6–17 mm; MIC, 300–308 μg mL^{-1}) compared with the large size Zn-PDA. They believed that the small particle size and high surface area gave the nanoscale Zn-PDA good antibacterial effect.

Recent studies have shown that nanoscale MOFs are commonly used for antibacterial application because their miniaturized size corresponds to a high specific surface area, which increases their interaction with bacteria and promotes their penetration into bacterial cell membranes. In bacterial cells, nanoscalable MOFs can interact with lipophilic acid, phosphate, or hydroxyl groups, leading to cell destruction [101,140,141] and greatly increasing the antibacterial effect. However, the strong permeability of cells to nanoscale MOFs also means that they can penetrate cell membranes and accumulate in normal cells, which would a pose risk to human health.

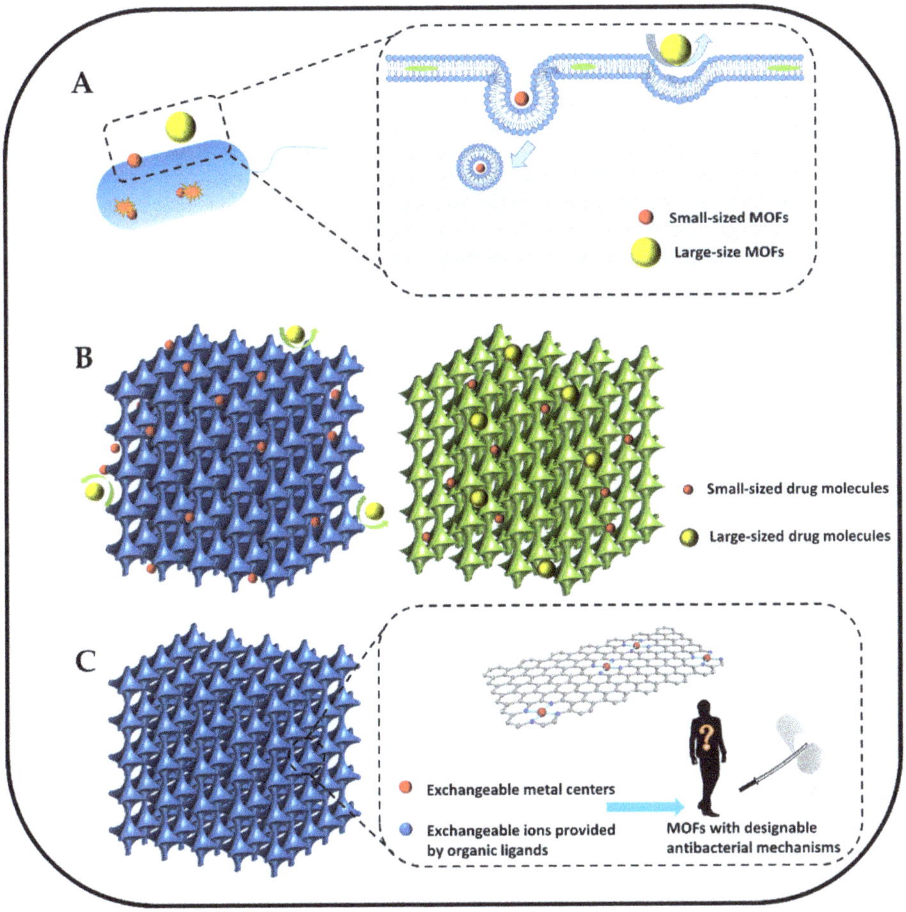

Figure 4. Schematic illustrations of strategies to enhance the antibacterial ability of MOFs: (**A**) the reduction in the size of MOFs can help the uptake of bacterial cells. (**B**) The increased pore size of MOFs can help load more large-sized drug molecules. (**C**) Alteration of the active sites of parent MOFs can confer new designable antibacterial mechanisms to daughter MOFs.

4.2. Pore Size Modulation

Rational design of MOF pore size can facilitate their use as carriers to load drug molecules (gas donors, antibiotics, etc.) for sterilization. However, drug molecules vary in size, and the pore size of the MOF should match the size of the drug molecule to ensure that the drug can be successfully loaded into the pores (Figure 4B). Pore size modulation is a strategy for obtaining MOFs with different pore sizes by changing the synthesis method, reactant ratio, and post-treatment method. Zhang et al. [142] synthesized structurally controlled ZIF-8 with a 3D mesoporous structure using polystyrene as sacrificial template. ZIF-8 precursor solution was impregnated into polystyrene particles to synthesize ZIF-8, and then polystyrene was dissolved to obtain ZIF-8 with a porous structure, with the original position of polystyrene particles becoming the pores. They believed that modulating the pore size by regulating the size of PS particles is a suitable method for designing MOFs with pore sizes matched to drug molecule sizes.

Xing et al. [143] synthesized a flower-like ZIF-8 by adding dopamine as an additional ligand to reduce the molar ratio of Hmim/Zn^{2+} during the synthesis process. The ligand-polymerized molecular sieve imidazole framework catalyzed the in situ polymerization of polydopamine, leading to microporous blocking and cross-linking of morphologically degraded ZIF nanosheets. Under the protection of polydopamine, flower-like ZIF-8, with abundant micropores, mesoporous defects, large petal spacing, high specific surface area, and high metal atom loading, was obtained by carbonization. They found that the flower-like ZIF-8 showed excellent POD activity and produced more effective ROS than normal ZIF-8 because of its wider 3D accessibility of active sites.

Yang et al. [144] synthesized mesoporous polymeric carbon nitride (PCN) and used ascorbic acid-assisted hydrothermal etching of bulk PCN to form pores and improve its crystallinity. PCN has many lone pairs of electrons on its N atoms. When PCN is placed in an inorganic weak acid solution (ascorbic acid), the acid attacks the lone pairs of electrons and the Lewis acid–base reaction occurs [145], corroding the PCN and forming pores. This simple synthesis step expands its specific surface area and provides transfer channels, which can potentially promote its photocatalytic hydrogen production (26.8 μmol h^{-1}). Deng et al. [146] synthesized ZIF-8-encapsulated Au nanoflowers (ZIF-8@Au nanoflowers), and then selectively etched them with tannic acid into a material with yolk-shell structure (Figure 5A(a)). The results showed that ZIF-8 became a very thin shell layer, and more cavities appeared between it and the Au nanoflowers, which had greater drug loading ability compared to the ZIF-8@Au nanoflowers with core–shell structure (Figure 5A(b)). Moreover, the core Au nanoflowers still maintain good photothermal effect (Figure 5A(c)).

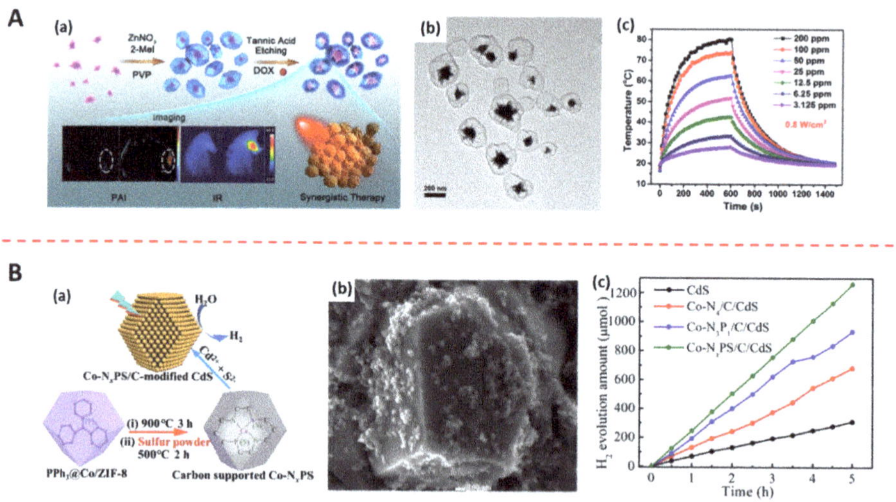

Figure 5. Methods of changing the structure of MOFs: (**A**) (**a**) Process of preparing a yolk−shell structure by selective etching of the ZIF−8 shell of Au@MOF by tannic acid (adapted from [146]); (**b**) TEM image of yolk−shell structured Au@MOF (adapted from [146]); (**c**) the temperature increasing and cooling curves of Au@MOF aqueous solution with various concentration under the irradiation of an 808 nm laser at 0.8 W cm^{-2} for 10 min (adapted from [146]). (**B**) (**a**) Process of preparing Co−N_XPS/C/CdS, a heterojunction of CdS and a MOF with an altered active site (adapted from [147]); (**b**) SEM image of Co−NxPS/C/CdS (adapted from [147]); (**c**) H_2 evolution amount graph of Co−NxPS/C/CdS and its precursors (adapted from [147]).

Studies have shown that porous materials with adjustable pore size not only are helpful for drug loading, but also could have high specific surface area via structural changes. However, other functions of porous MOFs, such as stability, adsorption dynamics, processability, and mechanical and thermal properties, which are essential for future practical applications of the materials, have been less studied.

4.3. Modulating Coordination Environment of Active Sites

Active sites are where chemical reactions occur in materials, and the coordination environment refers to the specific binding mode between ions at active sites. A metal center connected to ions provided by organic ligands constitutes the coordination environment of MOFs [148]. The ions in the coordination environment are slowly released in the physiological environment and express their special properties. Rational modulation of the metal centers in the coordination environment could re-endow MOFs with different antibacterial mechanisms. Moreover, rational modulation of the ions provided by organic ligands in the coordination environment leads to local charge inhomogeneity at the active sites, which could promote carrier separation and ultimately enhance the photodynamic effect of the material. So, modulating the coordination environment of active sites is a strategy for designing MOFs with desirable physicochemical properties by changing their metal centers and the ions provided by organic ligands (Figure 4C).

In recent years, MOFs and derivatives with altered coordination environments have attracted significant attention because of their near 100% metal dispersion [149–151]. However, methods of synthesizing MOF derivatives with same coordination environment vary slightly; some researchers use the one-pot method, in which a solution containing secondary metal ions or complexes composed of secondary metal ions and organic substances is added directly to the template precursor solution to synthesize MOFs with multiple types of active sites, then the desired coordination environment is retained, and unwanted ones are removed by calcination. Han et al. [152] loaded Fe-Phen complexes into ZIF-8 precursor solution via the one-pot method, and then pyrolyzed them under argon atmosphere. Finally, they synthesized a well-dispersed single-atom catalyst with good electrocatalytic activity. Cao et al. [153] synthesized 20 single-atom enzymes with different metal-N coordination environments (metal = V, Cr, Mn, Fe, Co, Ni, Cu, Ce, Zr, Mo, Ru, Rh, Pd, Gd, W, Re, Ir, Pt, Au, Tb). They added metal salt solution containing secondary metal ions in proportion to the primary metal ion (Zn^{2+}) and a solubilizer with antibacterial properties to the precursor solution of ZIF-8 and obtained products by sintering and centrifugation. The experimental data showed that these materials all had the catalytic behavior of OXD, POD, and simulated HPO; the Fe-based MOF with Fe-N coordination environment exhibited the highest POD activity. In addition, they found through DFT that the catalytic pathway and reactivity of ROS were closely related to the electronic structure of the metal centers, and higher ROS catalytic properties could be guaranteed if the energy barrier for ROS generation was low.

In some studies, researchers adsorbed complexes composed of secondary metal ions and organic substances onto the surface and into pores of nitrogen–carbon obtained by calcination using MOFs as templates, and then formed a new coordination environment after calcination. Wang et al. [154] synthesized a Cu-based MOF (Cu SASs/NPC). They removed the Zn^{2+} from ZIF-8 by high temperature and obtained nitrogen–carbon via acid etching, and then dipped it into a complex composed of Cu^{2+} and dicyandiamide. Finally, the complex was adsorbed on the original active sites of Zn-N_4. When dicyandiamide was burned off by high temperature calcination, the encapsulated Cu^{2+} was exposed and combined with four nitrogen atoms to form a new Cu-N_4 coordination environment. The experimental data show that Cu SASs/NPC had not only the peroxidase-like activity of ZIF-8, but also the photothermal properties of Cu^{2+}. The photothermal conversion efficiency of Cu SASs/NPC was calculated to be 82.78%, which was much stronger than the previously reported Cu-based photothermal agents (<50%), and the survival cell ratio of the Cu SASs/NPC + H_2O_2 + NIR light group against *E. coli* and *MRSA* reached 0%.

Other researchers found that co-calcining template MOFs directly with metal powder or foam also could form MOFs with new coordination environments [155]. Moreover, changing the ions provided by organic ligands in the coordination environment of active sites to ions with different electronegativity will cause uneven local charge distribution, which can accelerate the current transfer and catalyze the redox reaction of the material, producing antibacterial substances. Ji et al. [147] found that replacing the N ion in ZIF-8 with low electronegativity ions was a way to adjust the electrocatalytic properties of materials.

They synthesized a heterojunction of Co-based MOF and CdS (Co-N_XPS/C/CdS), in which P ions were introduced by adding triphenylphosphine during the MOF synthesis process, and S ion was introduced by the co-calcination of MOF and sulfur powder (Figure 5B(a,b)). The experimental data showed that the coordination environment of M-N_XPS had an inhomogeneous charge density compared to M-N_4. The inhomogeneous charge density can enhance light absorption as well as collect and store photoexcited electrons from CdS. Therefore, promoting charge carrier separation to a large extent and boosting photocatalytic hydrogen production (Figure 5B(c)).

In addition, the effects of high temperature on the structure and properties of MOFs have been studied extensively recently. It was found that high temperature does not significantly affect the morphology of MOFs, but the physical and chemical properties are slightly different. Wang et al. [156] explored changes in the crystal structure of ZIF-8 at various temperatures (500–900 °C). At temperatures less than 600 °C, tetragonal planar zinc porphyrin-like centers are formed on the nitrogen–carbon substrate, and the Zn^{2+} tends to move to the N_4 sites in the nitrogen–carbon substrate plane as the pyrolysis temperature increases. ZIF-8 synthesized at 800 °C had the best bond strength at Zn-N_4 active sites, which can maximize the adsorption of H_2O_2 and exhibit excellent peroxidase-like activity.

Studies have shown that series of MOFs with desirable properties can be designed by modulating the coordination environment of active sites, and this strategy has been widely used in recent years for the preparation of novel MOF derivatives. However, the high temperature will also change the carbon structure, bond length, and bond angle of the original MOF, causing problems when making side-by-side comparisons.

4.4. Constructing MOF-Based Composites

In MOF-based composites, the antibacterial effect can be enhanced, and the antibacterial mechanisms of individual components can be synergized, so they have better antibacterial ability than each component would have. In recent years, the synthesis of composites using MOFs with other components has also been a popular research topic (Table 3). The methods of preparing such composites can be divided into those that use MOFs or MOF precursor solution commingled with other components. MOF-based composites can be classified as MOF@metal and the oxidation products MOF@carbon, MOF@MOF, targeted MOF, and stimulus-responsive MOF.

Table 3. Composition and antibacterial mechanism of antibacterial agents for MOF-based composites.

Material 1	Material 2	Compound Mode	Main Role of MOFs in Antibacterial Application	Ref.
g-C_3N_4	Cu_3P	Heterojunction	H_2 loaded PTT PDT CDT	[136]
ZIF-8	Au NStar	Core–shell	Ag^+ release PTT	[146]
PCN-224	Cu^{2+}	stem grafting	PDT PTT	[56]
ZIF-8	Au NPs	Heterojunction	PDT	[57]
ZIF-8	Ag NPs	Heterojunction	Ag^+ release PDT	[58]
ZIF-8	MnO_2	Heterojunction	Ag^+ release PDT	[59]
ZIF-8 → ZnO	Ag NPs	Heterojunction	Zn^{2+} release Ag^+ release PDT	[60]

Table 3. *Cont.*

Material 1	Material 2	Compound Mode	Main Role of MOFs in Antibacterial Application	Ref.
Ag-BTC-S/N	CQDS	Heterojunction	Ag^+ release physical interaction CDT	[157]
Co-PTA	GO	Heterojunction	Co^{2+} release physical interaction	[61]
MIL-101	BQ	Core–shell	PDT PTT	[158]
UIO-66	PB	Core–shell	PDT PTT	[62]
multivariate MOF	Photosensitized porphyrin, BA	Core–shell	PDT	[159]
ZIF-8	$-NH_2$	stem grafting	Zn^{2+} release	[63]
ZIF-8@RFP	o-NBA	Core–shell	Zn^{2+} release Antibiotic loaded	[64]
VAN@ZIF-8	FA	Core–shell	Zn^{2+} release Antibiotic loaded	[160]

4.4.1. MOF@metal and Oxidation Products

MOF@metal and oxidation products can be obtained by introducing metal ions, metal nanoparticles, and metal oxides into MOFs. Introducing metal ions can change the regularity of the MOF structure and couple the properties of the metal ions, which is expected to improve the antibacterial properties of the composites. Han et al. [56] synthesized a Cu^{2+}-doped Zr-based MOF (PCN-224). Under irradiation by a 660 nm laser at 0.4 W cm^{-2}, the proper amount of Cu^{2+} doped in PCN-224 can trap electrons, accelerate carrier transfer, suppress electron–hole recombination, and finally catalyze ROS production. In addition, Cu^{2+}-doped PCN-224 can convert light energy into heat because of the presence of Cu^{2+}, and both of them contribute to antibacterial activity. The experimental data showed that product doped with 10% Cu^{2+} in PCN-224 had the best antibacterial efficacy (99.71%) against *S. aureus* under irradiation by 660 nm laser at 0.4 W cm^{-2} within 20 min. However, as the amount of Cu^{2+} introduced increased excessively, structure regularization effects occurred, ROS production gradually decreased, and the bactericidal effect decreased. Zhang et al. [161] constructed MOFs and mixed-metal MOFs on Ti surfaces through the co-assembly of $Au_{25}(MHA)_{18}$ and metals centers (Ti^{4+}, Zr^{4+}, Hf^{4+}, and Cu^{2+}). Results showed that the antibacterial efficiency of the Ti/Zr/Hf-based MOFs coating against both *MRSA* and *E. coli* were less than 5%. However, the antibacterial efficiency of Ti/Zr/Hf-Cu mixed-metal MOFs coating against *MRSA* was 98.86 ± 2.53%, 98.66 ± 2.0% and 98.59 ± 1.84% when Cu^{2+} was introduced in twice the amount of M^{4+}, respectively. They believed that the doping of boundary acids with Cu^{2+} results in the formation of unstable submetallic-oxygen bonds, which generally promote the release of bactericidal Cu^{2+}, leading to bacterial death.

Some metal nanoparticles can be used for antibacterial applications by themselves. However, they are thermodynamically unstable and prone to aggregation; therefore, loss of magnetic, catalytic, or rotational activity can occur [162,163]. The use of MOFs can avoid this limitation because their homogeneous pore structure can stabilize NPs and prevent their aggregation by providing spatial constraints [164–166]. Deng et al. [57] loaded Au NPs onto the ZIF-8 surface and found that the product (Au@ZIF-8) was a heterojunction of two semiconductors (Figure 6A(a,b)), which formed a Schottky junction at the Au and ZIF-8 interface, increasing the electron density of AuNPs [167] and contributing to the conversion of $^3O_2$ to $^1O_2$ (Figure 6A(c)). Moreover, AuNPs have a strong LSPR at 530 nm, increasing the visible light absorption of Au@ZIF-8. Both of them can enhance the electron transport and charge carrier separation of Au@ZIF-8 (Figure 6A(d)) The experimental data showed that the amount of ·OH produced by Au@ZIF-8 was nine times higher than that produced by ZIF-8, and the inhibition ratio against *S. aureus* and *E. coli* was >99.9% when the dosage of Au@ZIF-8 was 0.2 mg mL^{-1}. Deng et al. [146] synthesized star-shaped Au nanoflowers, and then obtained Au nanoflower@ZIF-8 by encapsulating them with ZIF-8. They found that the photothermal properties of Au nanoflower@ZIF-8 under 1064 nm (NIR-II biological window) laser irradiation were not inferior to those under 808 nm (NIR-I biological window) laser irradiation. Therefore, it can be applied to deeper tissues for antibacterial treatment because it can absorb higher wavelengths of light. Yang et al. [58] synthesized a product (SPZA) by loading ZIF-8@Ag onto sulfonated polyetheretherketone. The results showed that the sulfonated polyetheretherketone had a porous and loose structure, which allowed more ZIF-8 and Ag^+ to be loaded; the number of *E. coli* and *S. aureus* on SPZA was zero because ZIF-8 and Ag^+ were released synergistically against bacteria.

In recent years, researchers have carried out extensive studies on further inhibiting electron and hole recombination to improve PDT effectiveness. They have found that surface oxygen vacancy defects that form on the surface of oxidized MOF composites can enhance carrier transfer and effectively inhibit electron and hole recombination. Liang et al. [59] synthesized nanorods of MnO_2/ZIF-8 using the one-step method. The experimental data showed that the MnO_2/ZIF-8 hybrids exhibited complete inactivation against *E. coli* at low concentrations (3.24 μg mL^{-1}). They believed that the antibacterial mechanism of the MnO_2/ZIF-8 hybrids was mainly attributed to the production of catalytic $^1O_2$ rather than the release of Zn^{2+} or the photothermal effects under simulated solar irradiation; the increase in surface surface oxygen vacancy defects and the presence of heterojunctions could inhibit the recombination of electrons and holes to catalyze the photogeneration of ROS to kill bacteria. Cui et al. [60] synthesized Ag/ZnO hybrid cages with well-preserved polyhedral shapes and rich mesoporous structures by in situ pyrolysis of the product, obtaining by ZIF-8 impregnated with $AgNO_3$. The experimental data showed that the MIC of Ag/ZnO hybrid cages against bacteria was 6.25 μg mL^{-1}, which is comparable with commercial medicines such as streptomycin and ciprofloxacin. They believed that the LSPR effect of the encapsulated Ag nanoparticles and the band gap reduction effect caused by slight doping of Ag enhanced the absorption of visible light, resulting in a large number of light carriers. Photogenerated electrons can accumulate on Ag nanoparticles and later lead to rapid charge separation through a Schottky junction at the Ag/ZnO interface because of the high work function of the metal particles. Moreover, the electrons can react with dissolved O_2 to form $\cdot O_2^-$ in such semiconductor heterojunction systems, and the holes at the edge of the CB can participate directly in the degradation of oxide species as well as react with H_2O to form ·OH radicals for sterilization (Figure 6B). Wang et al. [136] synthesized a Z-type heterojunction of g-C_3N_4 and Cu_3P (g-C_3N_4/Cu_3P), which improved the separation efficiency of photogenerated electrons and holes, resulting in high production of H_2 and ROS. This process took advantage of the low acidity, high GSH content, and high H_2O_2 content of the bacterial infection microenvironment to achieve H_2-mediated cascade amplification of synergistic hydrogen therapy/PTT/PDT/CDT.

Research has shown that MOF@metal and oxidation products can prevent the aggregation quenching effect between metal nanoparticles. In addition, these MOF-based composites will obtain stronger antibacterial ability than individual components by the synergy of the antibacterial mechanisms of MOFs themselves and the introduced metal ions. However, if too many metal ions are introduced, it will inevitably cause metabolic toxicity in the physiological environment.

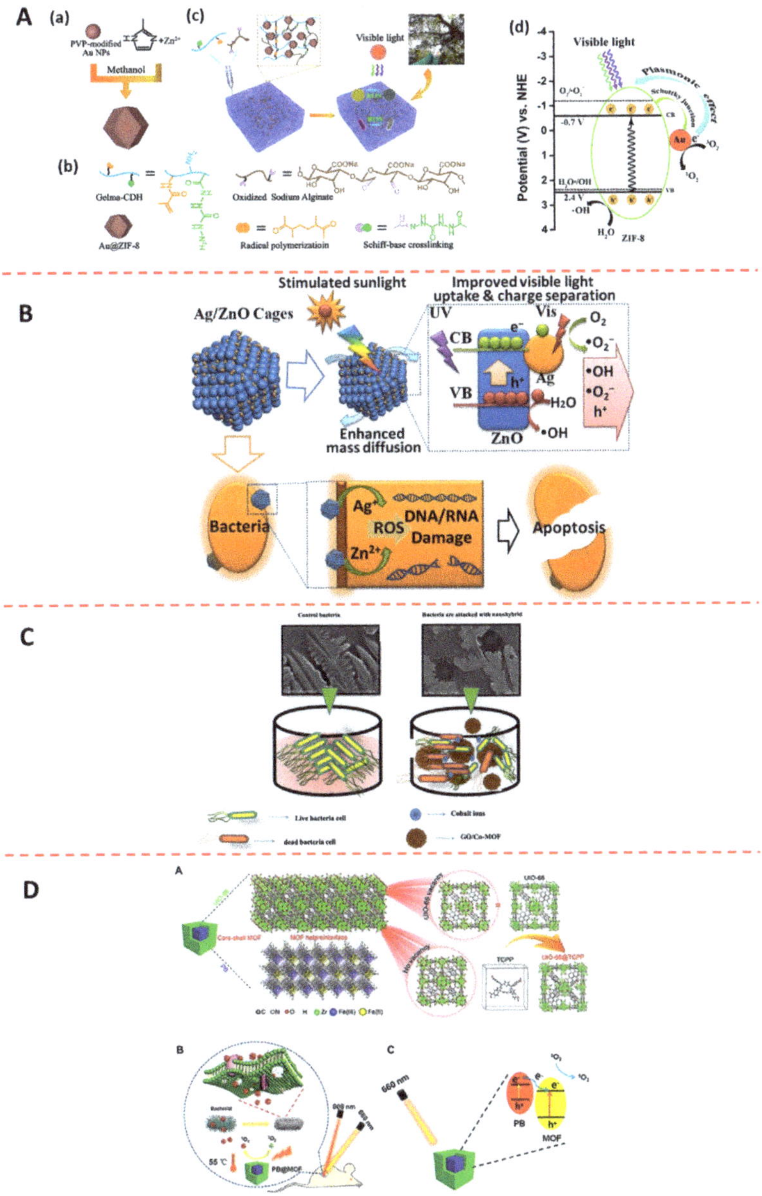

Figure 6. Enhancement of antibacterial ability of MOFs by constructing MOF-based composites: (**A**) (**a**) Process of preparing of Au@ZIF8, which possesses stronger PDT activity than ZIF−8 (adapted from [57]); (**b**) schematic illustration of infection model of Au@ZIF−8 injected into hydrogels for PDT therapy (adapted from [57]); (**c**) representative chemical structure of each component in hydrogels (adapted from [57]); (**d**) mechanism for $^1O_2$ and ·OH generation under visible−light irradiation (adapted from [57]). (**B**) Schematic illustration of a heterojunction of Ag/ZnO with efficient PDT activity synergistically releasing the respective components for antibacterial purpose (adapted from [60]). (**C**) Go/Co−MOF synergizing the physical interaction of GO and the release of Co ions to achieve efficient antibacterial effect (adapted from [61]). (**D**) Schematic illustrations of PB@UIO−66−TCPP with synergistic PTT activity of PB and PDT activity of TCPP (adapted from [62]).

4.4.2. MOF@carbon

Carbon-based materials include 0-dimensional CQDs and GQDs, one-dimensional CNTs, two-dimensional graphene and its derivatives, such as GO and RGO, and three-dimensional AC. They are well suited for antibacterial applications because of their unique physical interaction ability, photoelectric capability, and biocompatibility [168–171]. However, carbon-based nanomaterials, especially small-sized carbon nanoparticles, are prone to clustering and quenching, greatly reducing the effectiveness of their antibacterial properties. In recent years, many researchers have compounded carbon-based materials with MOFs. The porous MOFs can hinder the clustering and quenching effect of carbon-based materials and improve the stability and antibacterial effect of MOF@carbon. Travlou et al. [157] synthesized an Ag-based MOF@carbon composite antibacterial agent (Ag-BTC-S-/N-CQDs). This composite had superior bacterial inhibition ability compared to its individual components (S-/N-CQD or Ag-BTC) because of the synergistic effect of Ag^+ antibacterial release and the physical interaction of CQD with bacterial cells. Hatamie et al. [61] synthesized a Co-based MOF@carbon composite antibacterial agent (GO/Co-PTA). The experimental data showed that it inhibited the growth of *E. coli* and *S. aureus* by up to 99%, which was attributed to the synergistic antibacterial effects of Co^{2+} release and the physical interaction of GO with bacterial cells (Figure 6C).

Research has shown that MOF@carbon can prevent the aggregation quenching effect between carbon nanomaterials through the spatial segregation effect. In addition, the synergistic effects of MOFs and carbon materials can further improve the antibacterial activity of composites. However, carbon nanomaterials are expensive and cannot be put into mass production for clinical use.

4.4.3. MOF@MOF

MOF@MOF mostly refers to dual MOFs with core–shell structure, in which both the core and shell can be used as carriers to load antibacterial substances or can themselves be considered as antibacterial agents. In recent years, core and shell MOFs synthesized from PS-type and PTA-type MOFs have been extensively studied, and many MOF-based composites have been obtained with both PDT and PTT effects [172,173]. Liu et al. [158] synthesized a dual MIL-101 with core–shell structure, and then obtained a drug-loaded MOF (BQ-MIL@cat-fMIL) and used an in situ growth method to load (BQs in nuclear MIL-101 and peroxidase in shell MIL-101. They found that BQs in core MIL-101 had both photodynamic and photothermal activities because of their quantum confinement and edge effects; the photosensitivity of MIL-101 led to a widening of the gap between the triplet and ground states of the composite, which resulted in the light-excited multi-MOF reacting mainly with $^3O_2$ to produce $^1O_2$. In addition, the peroxidase in shell MIL-101 provided O_2 to the internal BQs by decomposing H_2O_2 and improved the PDT effect of the BQs (Figure 6E). The experimental data showed that the percentage of hypoxic cell apoptosis was 52.1% when this material was under irradiation by a 660 nm laser at 150 mW cm^{-2}, and the photothermal conversion efficiency of BQ-MIL@cat-fMIL was 23.3% under irradiation by an 808 nm laser at 1 W cm^{-2}, which induced a 28.7% apoptosis ratio in hypoxic cells; a 75.6% apoptosis ratio in BQ-MIL@cat-fMIL-treated hypoxic cells was achieved after applying dual light irradiation. Luo et al. [62] synthesized dual MOFs with core–shell structure using PB as the core and TCPP-doped UIO-66 as the shell (Figure 6D(a)). PB acted as the PTA and TCPP as the PS, which gave these dual MOFs the effects of PDT and PTT (Figure 6D(b)). Moreover, PB- and TCPP-doped UIO-66 are n-type semiconductors and can recombine to form an n-n heterojunction. In this heterojunction, the CB of PB is lower than that of TCPP-doped UIO-66, therefore these dual MOFs can accelerate photoelectron transfer from PB to UIO-66 and inhibit photoelectron–hole recombination under 660 nm laser irradiation (Figure 5D(c)). The experimental data showed that these dual MOFs had a high yield of $^1O_2$ under 660 nm light irradiation and high photothermal conversion efficiency up to 29.9% under 808 nm light irradiation; the antibacterial efficiency against *S. aureus* and *E. coli* exceeded 99.31 and 98.68%, respectively. In addition, the dual MOFs released trace amounts of Fe and Zr ions during the degradation process, which is beneficial to wound healing.

Research has shown that both MOFs in MOF@MOF can be used as carriers to load ideal PSs and PTAs to prevent the aggregation quenching effect, or can themselves work as PSs and PTAs to improve the antibacterial effects of the composites. In addition, the specific time and space for the release of different drug molecules in the physiological environment could be regulated by the rational design of core and shell MOFs and loaded drug molecules. However, the process of synthesizing such materials is tedious and costly.

4.4.4. Targeted and Stimulus-Responsive MOFs

Targeted and stimulus-responsive MOFs are MOFs with targeting and stimulus-responsive outer cladding, respectively. Because the antibacterial activity of PDT materials is limited by the ultra-short diffusion distance of ROS, the use of stimulus-responsive or targeted substances to modify materials to achieve precise controlled release of drugs at the correct time and place has a been popular research topic in recent years. Chen et al. [159] integrated BA ligand, which can specifically bind to bacteria and photosensitized porphyrin, into Zr- and Cu-based MOF (multivariate MOF). The results showed that the BA ligand targeted bacteria and the photosensitized porphyrin produced ROS under visible light irradiation, which synergistically improved the antibacterial effect against *MRSA* and *MDR E. coli* (10–20 times higher than without the targeting ligand) and accelerated the healing of chronic wounds infected with *MDR* bacteria (nearly 2 times faster than without the targeting ligand). Ahmad et al. [63] obtained a composite containing ZIF-8, GO, and amino groups (ZGO-NH) by grafting amino groups onto the surface of ZIF-8. The experimental data showed that when the volume of ammonium hydroxide (provider of amino groups) introduced into ZIF-8 was up to 20 mL, the modified ZIF-8 had better antibacterial effect against *E. coli* and *S. aureus* (inhibition zone diameters of 2.59 and 3.82 cm, respectively). They believed that the positively charged protonated amine groups enhanced the interaction between ZIF-8 and negatively charged bacterial cells, and they exhibited better antibacterial effect. Song et al. [64] synthesized a Zn-based MOF with an UV light-responsive rifampicin (RFP) release system (ZIF-8@RFP) by surface modification of ZIF-8 with 2-nitrobenzaldehyde (o-NBA), a pH jump reagent, then doped the RFP into the ZIF-8. The o-NBA decomposed in situ under UV irradiation at 5.25 mW cm^{-2} to produce acids and induce decreased pH in the physiological environment; ZIF-8 was degraded in the acidic environment and released Zn^{2+} and RFP continuously, which may help kill bacteria. The results showed that ZIF-8@RFP exhibited excellent antibacterial properties against MRSA under UV irradiation at 5.25 mW cm^{-2} (antibacterial ratio > 99%) at a dose of up to 10 μg mL^{-1}.

Research has shown that targeted and stimulus-responsive MOFs can precisely locate bacteria in space and exhibit stronger antibacterial ability than MOFs without the targeting and stimulus-response functions. However, most targeting and stimulus-responsive materials are harmful to organisms, which limits their clinical application.

5. Conclusions and Prospects

Over the past few years, MOFs and MOF-based composites have had a significant impact in the field of biomedicine because of their high specific surface area, atomic utilization, photodynamic activity, and photothermal activity. Moreover, they can control/stimulate the release of antibacterial components and regulate their size and morphology. Although MOFs have many advantages and are candidates for biomedical applications, they are only in the preliminary research stage, and there are still several issues and challenges that need to be addressed.

MOFs are made of metal ions and organic ligands, and it is undeniable that some of them can degrade and produce those components in certain environments. However, there is still a lack of studies on the effects of MOF degradation products on body metabolism. To address safety concerns, pathological research should be carried out. In addition, magnetic thermal therapy, microwave thermal therapy, and other treatment methods that cause less harm to normal cells, which are less studied in the field of antimicrobial therapy, may become a new development direction in the future.

Most MOFs with photodynamic and photothermal activity can only absorb light in the visible and NIR-I region, which limits their application because of the shallow depth of tissue transmission. For better effectiveness, there is an urgent need to find new MOFs with deeper tissue transmission that can absorb longer wavelengths of light.

In addition, researchers have synthesized the same MOF-derived single-atom enzymes with different amounts of active sites using different preparation methods and have found that these materials have excellent antibacterial effects, but no one has yet compared the antibacterial effects of the same single-atom enzymes with different numbers of active sites. Exploring the optimal range of the number of active sites in a material could help enrich the knowledge of antibacterial mechanisms regarding active sites and find the best balance between effectiveness and safety in MOF active sites.

Author Contributions: Conceptualization, writing—original draft preparation, and visualization, X.Z.; conceptualization, writing—review and editing, F.P. and D.W. All authors have read and agreed to the published version of the manuscript.

Funding: This study was supported by the National Natural Science Foundation of China (51901239), the GDPH Supporting Fund for Talent Program (KY0120220137 and KY012021462), and the Natural Science Foundation of Hebei Province (E2021202001).

Institutional Review Board Statement: Not applicable.

Informed Consent Statement: Not applicable.

Data Availability Statement: Not applicable.

Conflicts of Interest: All the authors declare that there is no conflict of interest.

Abbreviations

Abbreviation	Full Name
MOF	metal organic framework
IBU	inorganic building unit
OBU	organic building unit
IRMOF	isoreticular metal−organic framework
ZIF	zeolitic imidazolate framework
MIL	Matériaux de l'Institut Lavoisier
CPL	coordination pillared layer
UIO	University of Oslo
PCN	porous coordination network
HKUST	Hong Kong University of Science and Technology
ROS	reactive oxygen species
CDT	chemical dynamic therapy
PDT	photodynamic therapy
PTT	photothermal therapy
SDT	sonodynamic therapy
MIC	minimum inhibition concentration
PS	photosensitizer
PTA	photothermal agent
NPs	nanoparticles
LSPR	local surface plasma resonance
LMCT	ligand–metal charge transfer
US	ultrasound
NIR	near−infrared light
UV	ultraviolet light
DFT	density functional theory
CB	conduction band
VB	valence band
BTC	benzenetricarboxylic acid

BDC	twelve benzenedicarboxylic acid
Hmim	methylimidazole
H_3BTC	trimesic acid
BA	benzoic acid
H_2BDC	terephthalic acid
TEA	triethylamine
DMF	dimethylformamide
ZIF-L	ZIF−8 nanoblade arrays
4-HZBA	4−hydrazinobenzoic acid
VAN	vancomycin
GSNO	S−Nitrosoglutathione
GSH	glutathione
L-Arg	L−arginine (L−Arg)
2I-BODIPY	$5\lambda^4$−dipyrrolo[1,2−c:2′,1′−f][1,3,2]diazaborinine
H_2O_2	hydrogen peroxide
·OH	hydroxyl radicals
$\cdot O^{2-}$	superoxide anions
$^1O_2$	singlet oxygen
GOD	glucose oxidase
POD	peroxidase
SOD	superoxide dismutase
CAT	catalase
H_2DBP	5,15−bis(p−phenylazo)−porphyrin
TCPP	tetrakis (4−carboxyphenyl) porphyrin
BODIPY	boron dipyrromethene
I2-BDP	carboxyl-functionalized diiodo−substituted BODIPYs
H_2BDC	benzenedicarboxylate
PB	Prussian blue
$Fc(COOH)_2$	1,10−ferrocene carboxylic acid
GA	gallic acid
PMCS	porous MOF−derived carbons
DHMS	double-layer hollow manganese silicate nanoparticle
TRBs	thermoresponsive brushes
PDA	2,6−pyridinedicarboxylic acid
OXD	oxidase
HPO	halogen peroxidase
CQD	carbon quantum dots
GQD	graphene quantum dots
CNT	carbon nanotubes
GO	graphene oxide
RGO	reduced graphene oxide
AC	activated carbon
BQ	black phosphorus quantum dots
RFP	rifampicin
o-NBA	2−nitrobenzaldehyde
E. coli	*Escherichia coli*
S.aureus	*Staphylococcus aureus*
B. subtilis	*Bacillus subtilis*
P. vulgaris	*Proteus vulgaris*
P. aeruginosa	*Pseudomonas aeruginosa*
S. enteritidis	*Salmonella enteritidis*
S. cerevisiae	*Saccharomyces cerevisiae*
P. putida	*Pseudomonsa putida*
H. pylori	*Helicobacter pylori*
MRSA	*methicillin−resistant Staphylococcus aureus*
A. baumannii	*Acinetobacter baumannii*
K. pneumoniae	*Klebsiella pneumoniae*
S. entica	*Salmonella enterica*

References

1. Campoy, S.; Adrio, J.L. Antifungals. *Biochem. Pharmacol.* **2017**, *133*, 86–96. [CrossRef]
2. Costerton, J.W.; Stewart, P.S.; Greenberg, E.P. Bacterial biofilms: A common cause of persistent infections. *Science* **1999**, *284*, 1318–1322. [CrossRef]
3. Hall-Stoodley, L.; Costerton, J.W.; Stoodley, P. Bacterial biofilms: From the natural environment to infectious diseases. *Nat. Rev. Microbiol.* **2004**, *2*, 95–108. [CrossRef]
4. Khamesipour, F.; Lankarani, K.B.; Honarvar, B.; Kwenti, T.E. A systematic review of human pathogens carried by the housefly (*Musca domestica* L.). *BMC Public Health* **2018**, *18*, 1–15. [CrossRef]
5. Roemer, T.; Krysan, D.J. Antifungal Drug Development: Challenges, Unmet Clinical Needs, and New Approaches. *Csh Perspect. Med.* **2014**, *4*, a019703. [CrossRef]
6. Wang, W.; Wang, L.; Li, Z.; Xie, Z. BODIPY-containing nanoscale metal-organic frameworks for photodynamic therapy. *Chem. Commun.* **2016**, *52*, 5402–5405. [CrossRef]
7. Kovalakova, P.; Cizmas, L.; McDonald, T.J.; Marsalek, B.; Feng, M.; Sharma, V.K. Occurrence and toxicity of antibiotics in the aquatic environment: A review. *Chemosphere* **2020**, *251*, 126351. [CrossRef]
8. Lulijwa, R.; Rupia, E.J.; Alfaro, A.C. Antibiotic use in aquaculture, policies and regulation, health and environmental risks: A review of the top 15 major producers. *Rev. Aquacult.* **2020**, *12*, 640–663. [CrossRef]
9. Canica, M.; Manageiro, V.; Abriouel, H.; Moran-Gilad, J.; Franz, C.M.A.P. Antibiotic resistance in foodborne bacteria. *Trends Food Sci. Technol.* **2019**, *84*, 41–44. [CrossRef]
10. Granato, E.T.; Meiller-Legrand, T.A.; Foster, K.R. The Evolution and Ecology of Bacterial Warfare. *Curr. Biol.* **2019**, *29*, R521–R537. [CrossRef]
11. Zheng, J.; Su, C.; Zhou, J.; Xu, L.; Qian, Y.; Chen, H. Effects and mechanisms of ultraviolet, chlorination, and ozone disinfection on antibiotic resistance genes in secondary effluents of municipal wastewater treatment plants. *Chem. Eng. J.* **2017**, *317*, 309–316. [CrossRef]
12. Ingle, A.P.; Duran, N.; Rai, M. Bioactivity, mechanism of action, and cytotoxicity of copper-based nanoparticles: A review. *Appl. Microbiol. Biotechnol.* **2014**, *98*, 1001–1009. [CrossRef]
13. Sirelkhatim, A.; Mahmud, S.; Seeni, A.; Kaus, N.H.M.; Ann, L.C.; Bakhori, S.K.M.; Hasan, H.; Mohamad, D. Review on Zinc Oxide Nanoparticles: Antibacterial Activity and Toxicity Mechanism. *Nano-Micro Lett.* **2015**, *7*, 219–242. [CrossRef]
14. Zheng, K.; Setyawati, M.I.; Leong, D.T.; Xie, J. Antimicrobial silver nanomaterials. *Coord. Chem. Rev.* **2018**, *357*, 1–17. [CrossRef]
15. Sileika, T.S.; Barrett, D.G.; Zhang, R.; Lau, K.H.A.; Messersmith, P.B. Colorless Multifunctional Coatings Inspired by Polyphenols Found in Tea, Chocolate, and Wine. *Angew. Chem. Int. Edit.* **2013**, *52*, 10766–10770. [CrossRef]
16. Sun, J.; Yin, Y.; Sheng, G.-H.; Yang, Z.-B.; Zhu, H.-L. Synthesis, molecular modeling and structural characterization of vanillin derivatives as antimicrobial agents. *J. Mol. Struct.* **2013**, *1039*, 214–218. [CrossRef]
17. Wynne, J.H.; Fulmer, P.A.; McCluskey, D.M.; Mackey, N.M.; Buchanan, J.P. Synthesis and Development of a Multifunctional Self-Decontaminating Polyurethane Coating. *ACS Appl. Mater. Interfaces* **2011**, *3*, 2005–2011. [CrossRef]
18. Zhang, Z.; Xing, D.; Liang, Q.; Yong, D.; Han, X. Size controllable synthesis and antimicrobial activity of poly-N,N′-(4,5-dihydroxy-1,2-phenylene)bis(methylene) bisacrylamide microspheres. *Rsc Adv.* **2014**, *4*, 57891–57898. [CrossRef]
19. Murugesan, P.; Moses, J.A.; Anandharamakrishnan, C. Photocatalytic disinfection efficiency of 2D structure graphitic carbon nitride-based nanocomposites: A review. *J. Mater. Sci.* **2019**, *54*, 12206–12235. [CrossRef]
20. Pelaez, M.; Nolan, N.T.; Pillai, S.C.; Seery, M.K.; Falaras, P.; Kontos, A.G.; Dunlop, P.S.M.; Hamilton, J.W.J.; Byrne, J.A.; O'Shea, K.; et al. A review on the visible light active titanium dioxide photocatalysts for environmental applications. *Appl. Catal. B Environ.* **2012**, *125*, 331–349. [CrossRef]
21. Sun, Y.; Qin, H.; Yan, Z.; Zhao, C.; Ren, J.; Qu, X. Combating Biofilm Associated Infection In Vivo: Integration of Quorum Sensing Inhibition and Photodynamic Treatment based on Multidrug Delivered Hollow Carbon Nitride Sphere. *Adv. Funct. Mater.* **2019**, *29*, 1808222. [CrossRef]
22. Amos-Tautua, B.M.; Songca, S.P.; Oluwafemi, O.S. Application of Porphyrins in Antibacterial Photodynamic Therapy. *Molecules* **2019**, *24*, 2456. [CrossRef]
23. Li, Y.; Dong, L.; Mu, Z.; Liu, L.; Yang, J.; Wu, Z.; Pan, D.; Liu, L. Research Advances of Lactoferrin in Electrostatic Spinning, Nano Self-Assembly, and Immune and Gut Microbiota Regulation. *J. Agric. Food Chem.* **2022**, *70*, 10075–10089. [CrossRef]
24. Bu, Y.; Hu, Q.; Bao, T.; Xie, X.; Wang, S. Recent advances in cell membrane-coated technology for drug discovery from natural products. *Trac-Trends Anal. Chem.* **2022**, *151*, 116601. [CrossRef]
25. Wu, X.; Li, Y.; Raza, F.; Wang, X.; Zhang, S.; Rong, R.; Qiu, M.; Su, J. Red Blood Cell Membrane-Camouflaged Tedizolid Phosphate-Loaded PLGA Nanoparticles for Bacterial-Infection Therapy. *Pharmaceutics* **2021**, *13*, 99. [CrossRef]
26. Yaghi, O.M.; Li, G.; Li, H. Selective binding and removal of guests in a microporous metal–organic framework. *Nature* **1995**, *378*, 703–706. [CrossRef]
27. Eddaoudi, M.; Kim, J.; Rosi, N.; Vodak, D.; Wachter, J.; O'Keeffe, M.; Yaghi, O.M. Systematic design of pore size and functionality in isoreticular MOFs and their application in methane storage. *Science* **2002**, *295*, 469–472. [CrossRef]
28. Phan, A.; Doonan, C.J.; Uribe-Romo, F.J.; Knobler, C.B.; O'Keeffe, M.; Yaghi, O.M. Synthesis, Structure, and Carbon Dioxide Capture Properties of Zeolitic Imidazolate Frameworks. *Acc. Chem. Res.* **2010**, *43*, 58–67. [CrossRef]

29. Wang, B.; Cote, A.P.; Furukawa, H.; O'Keeffe, M.; Yaghi, O.M. Colossal cages in zeolitic imidazolate frameworks as selective carbon dioxide reservoirs. *Nature* **2008**, *453*, 207–211. [CrossRef]
30. Zhang, H.; Hu, X.; Li, T.; Zhang, Y.; Xu, H.; Sun, Y.; Gu, X.; Gu, C.; Luo, J.; Gao, B. MIL series of metal organic frameworks (MOFs) as novel adsorbents for heavy metals in water: A review. *J. Hazard. Mater.* **2022**, *429*, 128271. [CrossRef] [PubMed]
31. Kondo, M.; Okubo, T.; Asami, A.; Noro, S.i.; Yoshitomi, T.; Kitagawa, S.; Ishii, T.; Matsuzaka, H.; Seki, K. Rational synthesis of stable channel-like cavities with methane gas adsorption properties:[{Cu2 (pzdc) 2 (L)} n](pzdc= pyrazine-2, 3-dicarboxylate; L= a pillar ligand). *Angew. Chem. Int. Edit.* **1999**, *38*, 140–143. [CrossRef]
32. Yang, Q.; Wiersum, A.D.; Llewellyn, P.L.; Guillerm, V.; Serred, C.; Maurin, G. Functionalizing porous zirconium terephthalate UiO-66(Zr) for natural gas upgrading: A computational exploration. *Chem. Commun.* **2011**, *47*, 9603–9605. [CrossRef] [PubMed]
33. Ma, S.; Sun, D.; Simmons, J.M.; Collier, C.D.; Yuan, D.; Zhou, H.-C. Metal-organic framework from an anthracene derivative containing nanoscopic cages exhibiting high methane uptake. *J. Am. Chem. Soc.* **2008**, *130*, 1012–1016. [CrossRef] [PubMed]
34. Banerjee, R.; Phan, A.; Wang, B.; Knobler, C.; Furukawa, H.; O'Keeffe, M.; Yaghi, O.M. High-throughput synthesis of zeolitic imidazolate frameworks and application to CO_2 capture. *Science* **2008**, *319*, 939–943. [CrossRef]
35. Ferey, G.; Mellot-Draznieks, C.; Serre, C.; Millange, F.; Dutour, J.; Surble, S.; Margiolaki, I. A chromium terephthalate-based solid with unusually large pore volumes and surface area. *Science* **2005**, *309*, 2040–2042. [CrossRef]
36. Ferey, G.; Serre, C.; Mellot-Draznieks, C.; Millange, F.; Surble, S.; Dutour, J.; Margiolaki, I. A hybrid solid with giant pores prepared by a combination of targeted chemistry, simulation, and powder diffraction. *Angew. Chem. Int. Edit.* **2004**, *43*, 6296–6301. [CrossRef]
37. Jiao, L.; Wang, Y.; Jiang, H.-L.; Xu, Q. Metal-Organic Frameworks as Platforms for Catalytic Applications. *Adv. Mater.* **2018**, *30*, 1703663. [CrossRef]
38. Li, H.; Eddaoudi, M.; Groy, T.L.; Yaghi, O.M. Establishing Microporosity in Open MetalOrganic Frameworks: Gas Sorption Isotherms for Zn(BDC) (BDC = 1,4-Benzenedicarboxylate). *J. Am. Chem. Soc.* **1998**, *120*, 8571–8572. [CrossRef]
39. Li, H.L.; Eddaoudi, M.M.; O'Keeffe, M.; Yaghi, O.M. Design and Synthesis of an Exceptionally Stable and Highly Porous Metal-Organic Framework. *Nature* **1999**, *402*, 276–279. [CrossRef]
40. Yaghi, O.M.; Li, H. Hydrothermal Synthesis of a Metal-Organic Framework Containing Large Rectangular Channels. *J. Am. Chem. Soc.* **1995**, *117*, 10401–10402. [CrossRef]
41. Yang, J.; Yang, Y.-W. Metal-Organic Frameworks for Biomedical Applications. *Small* **2020**, *16*, 19064846. [CrossRef] [PubMed]
42. Zhang, W.; Li, R.; Zheng, H.; Bao, J.; Tang, Y.; Zhou, K. Laser-Assisted Printing of Electrodes Using Metal-Organic Frameworks for Micro-Supercapacitors. *Adv. Funct. Mater.* **2021**, *31*, 2009057. [CrossRef]
43. Cun, J.-E.; Fan, X.; Pan, Q.; Gao, W.; Luo, K.; He, B.; Pu, Y. Copper-based metal-organic frameworks for biomedical applications. *Adv. Colloid Interface Sci.* **2022**, *305*, 102686. [CrossRef] [PubMed]
44. Dong, H.; Zhang, X.; Li, H.; Hou, F.; Yang, Y.; Cui, L. Progress in Preparation of Metal Organic Frameworks HKUST-1 and Its Application. *Mater. Rev.* **2016**, *30*, 114–119, 139.
45. Li, R.; Chen, T.; Pan, X. Metal-Organic-Framework-Based Materials for Antimicrobial Applications. *ACS Nano* **2021**, *15*, 3808–3848. [CrossRef]
46. Shen, M.; Forghani, F.; Kong, X.; Liu, D.; Ye, X.; Chen, S.; Ding, T. Antibacterial applications of metal-organic frameworks and their composites. *Compr. Rev. Food. Sci. F.* **2020**, *19*, 1397–1419. [CrossRef]
47. Pettinari, C.; Pettinari, R.; Di Nicola, C.; Tombesi, A.; Scuri, S.; Marchetti, F. Antimicrobial MOFs. *Coord. Chem. Rev.* **2021**, *446*, 214121. [CrossRef]
48. Wyszogrodzka, G.; Marszalek, B.; Gil, B.; Dorozynski, P. Metal-organic frameworks: Mechanisms of antibacterial action and potential applications. *Drug Discov. Today* **2016**, *21*, 1009–1018. [CrossRef]
49. Aguado, S.; Quiros, J.; Canivet, J.; Farrusseng, D.; Boltes, K.; Rosal, R. Antimicrobial activity of cobalt imidazolate metal-organic frameworks. *Chemosphere* **2014**, *113*, 188–192. [CrossRef]
50. Liu, Y.; Xu, X.; Xia, Q.; Yuan, G.; He, Q.; Cui, Y. Multiple topological isomerism of three-connected networks in silver-based metal-organoboron frameworks. *Chem. Commun.* **2010**, *46*, 2608–2610. [CrossRef]
51. Tamames-Tabar, C.; Imbuluzqueta, E.; Guillou, N.; Serre, C.; Miller, S.R.; Elkaim, E.; Horcajada, P.; Blanco-Prieto, M.J. A Zn azelate MOF: Combining antibacterial effect. *CrystEngComm* **2015**, *17*, 456–462. [CrossRef]
52. Restrepo, J.; Serroukh, Z.; Santiago-Morales, J.; Aguado, S.; Gomez-Sal, P.; Mosquera, M.E.G.; Rosal, R. An Antibacterial Zn-MOF with Hydrazinebenzoate Linkers. *Eur. J. Inorg. Chem.* **2017**, *2017*, 574–580. [CrossRef]
53. Yuan, Y.; Zhang, Y. Enhanced biomimic bactericidal surfaces by coating with positively-charged ZIF nano-dagger arrays. *Nanomed-Nanotechnol* **2017**, *13*, 2199–2207. [CrossRef] [PubMed]
54. Wang, K.; Geng, Z.; Yin, Y.; Ma, X.; Wang, Z. Morphology effect on the luminescent property and antibacterial activity of coordination polymer particles with identical crystal structures. *CrystEngComm* **2011**, *13*, 5100–5104. [CrossRef]
55. Akbarzadeh, F.; Motaghi, M.; Chauhan, N.P.S.; Sargazi, G. A novel synthesis of new antibacterial nanostructures based on Zn-MOF compound: Design, characterization and a high performance application. *Heliyon* **2020**, *6*, e03231. [CrossRef]
56. Han, D.; Han, Y.; Li, J.; Liu, X.; Yeung, K.W.K.; Zheng, Y.; Cui, Z.; Yang, X.; Liang, Y.; Li, Z. Enhanced photocatalytic activity and photothermal effects of cu-doped metal-organic frameworks for rapid treatment of bacteria-infected wounds. *Appl. Catal. B Environ.* **2020**, *261*, 118248. [CrossRef]

57. Deng, Z.; Li, M.; Hu, Y.; He, Y.; Tao, B.; Yuan, Z.; Wang, R.; Chen, M.; Luo, Z.; Cai, K. Injectable biomimetic hydrogels encapsulating Gold/metal-organic frameworks nanocomposites for enhanced antibacterial and wound healing activity under visible light actuation. *Chem. Eng. J.* **2021**, *420*, 129668. [CrossRef]
58. Yang, X.; Chai, H.; Guo, L.; Jiang, Y.; Xu, L.; Huang, W.; Shen, Y.; Yu, L.; Liu, Y.; Liu, J. In situ preparation of porous metal-organic frameworks ZIF-8@Ag on poly-ether-ether-ketone with synergistic antibacterial activity. *Colloid Surface B* **2021**, *205*, 111920. [CrossRef]
59. Liang, Z.; Wang, H.; Zhang, K.; Ma, G.; Zhu, L.; Zhou, L.; Yan, B. Oxygen-defective MnO_2/ZIF-8 nanorods with enhanced antibacterial activity under solar light. *Chem. Eng. J.* **2022**, *428*, 131349. [CrossRef]
60. Cui, J.; Wu, D.; Li, Z.; Zhao, G.; Wang, J.; Wang, L.; Niu, B. Mesoporous Ag/ZnO hybrid cages derived from ZIF-8 for enhanced photocatalytic and antibacterial activities. *Ceram. Int.* **2021**, *47*, 15759–15770. [CrossRef]
61. Hatamie, S.; Ahadian, M.M.; Zomorod, M.S.; Torabi, S.; Babaie, A.; Hosseinzadeh, S.; Soleimani, M.; Hatami, N.; Wei, Z.-H. Antibacterial properties of nanoporous graphene oxide/cobalt metal organic framework. *Mater. Sci. Eng. C-Mater. Biol. Appl.* **2019**, *104*, 109862. [CrossRef] [PubMed]
62. Luo, Y.; Li, J.; Liu, X.; Tan, L.; Cui, Z.; Feng, X.; Yang, X.; Liang, Y.; Li, Z.; Zhu, S.; et al. Dual Metal-Organic Framework Heterointerface. *ACS Central Sci.* **2019**, *5*, 1591–1601. [CrossRef] [PubMed]
63. Ahmad, N.; Nordin, N.A.H.M.; Jaafar, J.; Ismail, A.F.; Ramli, M.K.N.B. Significant improvement in antibacterial property of ZIF-8 decorated graphene oxide by post-synthetic modification process. *J. Environ. Chem. Eng.* **2021**, *9*, 105887. [CrossRef]
64. Song, Z.; Wu, Y.; Cao, Q.; Wang, H.; Wang, X.; Han, H. pH-Responsive, Light-Triggered on-Demand Antibiotic Release from Functional Metal-Organic Framework for Bacterial Infection Combination Therapy. *Adv. Funct. Mater.* **2018**, *28*, 1800011. [CrossRef]
65. Wang, P.; Wang, S.; Zhang, W.; Li, X.; Gu, Z.; Li, W.; Zhao, S.; Fu, Y. Preparation of MOF catalysts and simultaneously modulated metal nodes and ligands via a one-pot method for optimizing cycloaddition reactions. *New J. Chem.* **2020**, *44*, 9611–9615. [CrossRef]
66. Huang, L. Synthesis, morphology control, and properties of porous metal–organic coordination polymers. *Microporous Mesoporous Mater.* **2003**, *58*, 105–114. [CrossRef]
67. Qiu, L.-G.; Li, Z.-Q.; Wu, Y.; Wang, W.; Xu, T.; Jiang, X. Facile synthesis of nanocrystals of a microporous metal–organic framework by an ultrasonic method and selective sensing of organoamines. *Chem. Commun.* **2008**, 3642–3644. [CrossRef]
68. Petrier, C.; Luche, J.; Luche, J. *Synthetic Organic Sonochemistry*; Plenum Press: New York, NY, USA, 1998; pp. 53–56.
69. Suslick, K.S.; Nyborg, W.L. ULTRASOUND: Its Chemical, Physical and Biological Effects. *J. Acoust. Soc. Am.* **1989**, *87*, 919–920. [CrossRef]
70. Ameloot, R.; Stappers, L.; Fransaer, J.; Alaerts, L.; Sels, B.F.; De Vos, D.E. Patterned Growth of Metal-Organic Framework Coatings by Electrochemical Synthesis. *Chem. Mater.* **2009**, *21*, 2580–2582. [CrossRef]
71. Katsenis, A.D.; Puskaric, A.; Strukil, V.; Mottillo, C.; Julien, P.A.; Uzarevic, K.; Minh-Hao, P.; Trong-On, D.; Kimber, S.A.J.; Lazic, P.; et al. In situ X-ray diffraction monitoring of a mechanochemical reaction reveals a unique topology metal-organic framework. *Nat. Commun.* **2015**, *6*, 6662. [CrossRef]
72. Freund, R.; Canossa, S.; Cohen, S.M.; Yan, W.; Deng, H.; Guillerm, V.; Eddaoudi, M.; Madden, D.G.; Fairen-Jimenez, D.; Lyu, H.; et al. 25 Years of Reticular Chemistry. *Angew. Chem. Int. Edit.* **2021**, *60*, 23946–23974. [CrossRef] [PubMed]
73. Jiang, H.; Alezi, D.; Eddaoudi, M. A reticular chemistry guide for the design of periodic solids. *Nat. Rev. Mater.* **2021**, *6*, 466–487. [CrossRef]
74. Chiericatti, C.; Carlos Basilico, J.; Zapata Basilico, M.L.; Manuel Zamaro, J. Novel application of HKUST-1 metal-organic framework as antifungal: Biological tests and physicochemical characterizations. *Microporous Mesoporous Mater.* **2012**, *162*, 60–63. [CrossRef]
75. Iram, S.; Imran, M.; Kanwal, F.; Iqbal, Z.; Deeba, F.; Iqbal, Q.J. Bismuth(III) based Metal Organic Frameworks: Luminescence, Gas Adsorption, and Antibacterial Studies. *Z. Anorg. Allg. Chem.* **2019**, *645*, 50–56. [CrossRef]
76. Li, P.; Li, J.; Feng, X.; Li, J.; Hao, Y.; Zhang, J.; Wang, H.; Yin, A.; Zhou, J.; Ma, X.; et al. Metal-organic frameworks with photocatalytic bactericidal activity for integrated air cleaning. *Nat. Commun.* **2019**, *10*, 2177. [CrossRef]
77. Raju, P.; Ramalingam, T.; Nooruddin, T.; Natarajan, S. In vitro assessment of antimicrobial, antibiofilm and larvicidal activities of bioactive nickel metal organic framework. *J. Drug Deliv. Sci. Tec.* **2020**, *56*, 101560. [CrossRef]
78. Lu, X.; Ye, J.; Sun, Y.; Bogale, R.F.; Zhao, L.; Tian, P.; Ning, G. Ligand effects on the structural dimensionality and antibacterial activities of silver-based coordination polymers. *Dalton Trans.* **2014**, *43*, 10104–10113. [CrossRef]
79. Sancet, M.P.A.; Hanke, M.; Wang, Z.; Bauer, S.; Azucena, C.; Arslan, H.K.; Heinle, M.; Gliemann, H.; Woell, C.; Rosenhahn, A. Surface anchored metal-organic frameworks as stimulus responsive antifouling coatings. *Biointerphases* **2013**, *8*, 29. [CrossRef]
80. Taheri, M.; Ashok, D.; Sen, T.; Enge, T.G.; Verma, N.K.; Tricoli, A.; Lowe, A.; Nisbet, D.R.; Tsuzuki, T. Stability of ZIF-8 nanopowders in bacterial culture media and its implication for antibacterial properties. *Chem. Eng. J.* **2021**, *413*, 127511. [CrossRef]
81. Chen, L.-J.; Liu, Y.-Y.; Zhao, X.; Yan, X.-P. Vancomycin-Functionalized Porphyrinic Metal-Organic Framework PCN-224 with Enhanced Antibacterial Activity against Staphylococcus Aureus. *Chem. Asian J.* **2021**, *16*, 2022–2026. [CrossRef]
82. Lin, S.; Liu, X.; Tan, L.; Cui, Z.; Yang, X.; Yeung, K.W.K.; Pan, H.; Wu, S. Porous Iron-Carboxylate Metal-Organic Framework: A Novel Bioplatform with Sustained Antibacterial Efficacy and Nontoxicity. *ACS Appl. Mater. Interfaces* **2017**, *9*, 19248–19257. [CrossRef] [PubMed]

83. Tuttle, R.R.; Rubin, H.N.; Rithner, C.D.; Finke, R.G.; Reynolds, M.M. Copper ion vs copper metal organic framework catalyzed NO release from bioavailable S-Nitrosoglutathione en route to biomedical applications: Direct H-1 NMR monitoring in water allowing identification of the distinct, true reaction stoichiometries and thiol dependencies. *J. Inorg. Biochem.* **2019**, *199*, 110760. [CrossRef] [PubMed]
84. Wan, S.-S.; Zeng, J.-Y.; Cheng, H.; Zhang, X.-Z. ROS-induced NO generation for gas therapy and sensitizing photodynamic therapy of tumor. *Biomaterials* **2018**, *185*, 51–62. [CrossRef] [PubMed]
85. Guan, Q.; Zhou, L.-L.; Li, Y.-A.; Dong, Y.-B. A nanoscale metal-organic framework for combined photodynamic and starvation therapy in treating breast tumors. *Chem. Commun.* **2019**, *55*, 14898–14901. [CrossRef]
86. Zhang, W.; Zhou, Y.; Fan, Y.; Cao, R.; Xu, Y.; Weng, Z.; Ye, J.; He, C.; Zhu, Y.; Wang, X. Metal-Organic-Framework-Based Hydrogen-Release Platform for Multieffective Helicobacter Pylori Targeting Therapy and Intestinal Flora Protective Capabilities. *Adv. Mater.* **2022**, *34*, 2105738. [CrossRef]
87. Ranji-Burachaloo, H.; Karimi, F.; Xie, K.; Fu, Q.; Gurr, P.A.; Dunstan, D.E.; Qiao, G.G. MOF-Mediated Destruction of Cancer Using the Cell's Own Hydrogen Peroxide. *ACS Appl. Mater. Interfaces* **2017**, *9*, 33599–33608. [CrossRef]
88. Hao, Y.-N.; Qu, C.-C.; Shu, Y.; Wang, J.-H.; Chen, W. Construction of Novel Nanocomposites (Cu-MOF/GOD@HA) for Chemodynamic Therapy. *Nanomaterials* **2021**, *11*, 1843. [CrossRef]
89. Lu, K.; He, C.; Lin, W. Nanoscale Metal-Organic Framework for Highly Effective Photodynamic Therapy of Resistant Head and Neck Cancer. *J. Am. Chem. Soc.* **2014**, *136*, 16712–16715. [CrossRef]
90. Liu, J.; Yang, Y.; Zhu, W.; Yi, X.; Dong, Z.; Xu, X.; Chen, M.; Yang, K.; Lu, G.; Jiang, L.; et al. Nanoscale metal-organic frameworks for combined photodynamic & radiation therapy in cancer treatment. *Biomaterials* **2016**, *97*, 1–9. [CrossRef]
91. Park, J.; Jiang, Q.; Feng, D.; Zhou, H.-C. Controlled Generation of Singlet Oxygen in Living Cells with Tunable Ratios of the Photochromic Switch in Metal-Organic Frameworks. *Angew. Chem. Int. Edit.* **2016**, *55*, 7188–7193. [CrossRef]
92. Fu, G.; Liu, W.; Feng, S.; Yue, X. Prussian blue nanoparticles operate as a new generation of photothermal ablation agents for cancer therapy. *Chem. Commun.* **2012**, *48*, 11567–11569. [CrossRef] [PubMed]
93. Yang, Y.; Liu, J.; Liang, C.; Feng, L.; Fu, T.; Dong, Z.; Chao, Y.; Li, Y.; Lu, G.; Chen, M.; et al. Nanoscale Metal-Organic Particles with Rapid Clearance for Magnetic Resonance Imaging-Guided Photothermal Therapy. *ACS Nano* **2016**, *10*, 2774–2781. [CrossRef]
94. Deng, Z.; Fang, C.; Ma, X.; Li, X.; Zeng, Y.-J.; Peng, X. One Stone Two Birds: Zr-Fc Metal-Organic Framework Nanosheet for Synergistic Photothermal and Chemodynamic Cancer Therapy. *ACS Appl. Mater. Interfaces* **2020**, *12*, 20321–20330. [CrossRef] [PubMed]
95. Liu, F.; He, X.; Chen, H.; Zhang, J.; Zhang, H.; Wang, Z. Gram-scale synthesis of coordination polymer nanodots with renal clearance properties for cancer theranostic applications. *Nat. Commun.* **2015**, *6*, 8003. [CrossRef] [PubMed]
96. Pan, X.; Bai, L.; Wang, H.; Wu, Q.; Wang, H.; Liu, S.; Xu, B.; Shi, X.; Liu, H. Metal-Organic-Framework-Derived Carbon Nanostructure Augmented Sonodynamic Cancer Therapy. *Adv. Mater.* **2018**, *30*, 1800180. [CrossRef]
97. Pan, X.; Wang, W.; Huang, Z.; Liu, S.; Guo, J.; Zhang, F.; Yuan, H.; Li, X.; Liu, F.; Liu, H. MOF-Derived Double-Layer Hollow Nanoparticles with Oxygen Generation Ability for Multimodal Imaging-Guided Sonodynamic Therapy. *Angew. Chem. Int. Edit.* **2020**, *59*, 13557–13561. [CrossRef]
98. Yu, Y.; Tan, L.; Li, Z.; Liu, X.; Zheng, Y.; Feng, X.; Liang, Y.; Cui, Z.; Zhu, S.; Wu, S. Single-atom catalysis for efficient sonodynamic therapy of methicillin-resistant Staphylococcus aureus-infected osteomyelitis. *ACS Nano* **2021**, *15*, 10628–10639. [CrossRef]
99. Bouhidel, Z.; Cherouana, A.; Durand, P.; Doudouh, A.; Morini, F.; Guillot, B.; Dahaoui, S. Synthesis, spectroscopic characterization, crystal structure, Hirshfeld surface analysis and antimicrobial activities of two triazole Schiff bases and their silver complexes. *Inorg. Chim. Acta* **2018**, *482*, 34–47. [CrossRef]
100. Liu, A.; Wang, C.-C.; Wang, C.-z.; Fu, H.-f.; Peng, W.; Cao, Y.-L.; Chu, H.-Y.; Du, A.-F. Selective adsorption activities toward organic dyes and antibacterial performance of silver-based coordination polymers. *J. Colloid Interface Sci.* **2018**, *512*, 730–739. [CrossRef]
101. Lu, X.; Ye, J.; Zhang, D.; Xie, R.; Bogale, R.F.; Sun, Y.; Zhao, L.; Zhao, Q.; Ning, G. Silver carboxylate metal-organic frameworks with highly antibacterial activity and biocompatibility. *J. Inorg. Biochem.* **2014**, *138*, 114–121. [CrossRef]
102. Pulido, M.D.; Parrish, A.R. Metal-induced apoptosis: Mechanisms. *Mutat. Res.* **2003**, *533*, 227–241. [CrossRef] [PubMed]
103. Seyedpour, S.F.; Rahimpour, A.; Najafpour, G. Facile in-situ assembly of silver-based MOFs to surface functionalization of TFC membrane: A novel approach toward long-lasting biofouling mitigation. *J. Membr. Sci.* **2019**, *573*, 257–269. [CrossRef]
104. Jo, J.H.; Kim, H.-C.; Huh, S.; Kim, Y.; Lee, D.N. Antibacterial activities of Cu-MOFs containing glutarates and bipyridyl ligands. *Dalton Trans.* **2019**, *48*, 8084–8093. [CrossRef] [PubMed]
105. Yuan, Y.; Wu, H.; Lu, H.; Zheng, Y.; Ying, J.Y.; Zhang, Y. ZIF nano-dagger coated gauze for antibiotic-free wound dressing. *Chem. Commun.* **2019**, *55*, 699–702. [CrossRef]
106. Zhuang, W.; Yuan, D.; Li, J.-R.; Luo, Z.; Zhou, H.-C.; Bashir, S.; Liu, J. Highly Potent Bactericidal Activity of Porous Metal-Organic Frameworks. *Adv. Healthc. Mater.* **2012**, *1*, 225–238. [CrossRef] [PubMed]
107. Lin, T.; Wang, J.; Yin, X.; Wei, X. Modification of Bentonite and Its Application in Antimicrobial Material. *Bentonite Modif.* **2020**, *5*, 54–61.
108. Xu, Y.-T.; Ye, Z.-M.; Ye, J.-W.; Cao, L.-M.; Huang, R.-K.; Wu, J.-X.; Zhou, D.-D.; Zhang, X.-F.; He, C.-T.; Zhang, J.-P.; et al. Non-3d Metal Modulation of a Cobalt Imidazolate Framework for Excellent Electrocatalytic Oxygen Evolution in Neutral Media. *Angew. Chem. Int. Edit.* **2019**, *58*, 139–143. [CrossRef]

109. Zhang, W.; Wang, Y.; Zheng, H.; Li, R.; Tang, Y.; Li, B.; Zhu, C.; You, L.; Gao, M.-R.; Liu, Z.; et al. Embedding Ultrafine Metal Oxide Nanoparticles in Monolayered Metal-Organic Framework Nanosheets Enables Efficient Electrocatalytic Oxygen Evolution. *ACS Nano* **2020**, *14*, 1971–1981. [CrossRef]
110. Wang, X.; Liu, S.; Li, M.; Yu, P.; Chu, X.; Li, L.; Tan, G.; Wang, Y.; Chen, X.; Zhang, Y.; et al. The synergistic antibacterial activity and mechanism of multicomponent metal ions-containing aqueous solutions against Staphylococcus aureus. *J. Inorg. Biochem.* **2016**, *163*, 214–220. [CrossRef]
111. Carpenter, A.W.; Schoenfisch, M.H. Nitric oxide release: Part II. Therapeutic applications. *Chem. Soc. Rev.* **2012**, *41*, 3742–3752. [CrossRef]
112. Ma, W.; Chen, X.; Fu, L.; Zhu, J.; Fan, M.; Chen, J.; Yang, C.; Yang, G.; Wu, L.; Mao, G.; et al. Ultra-efficient Antibacterial System Based on Photodynamic Therapy and CO Gas Therapy for Synergistic Antibacterial and Ablation Biofilms. *ACS Appl. Mater. Interfaces* **2020**, *12*, 22479–22491. [CrossRef] [PubMed]
113. Schairer, D.O.; Chouake, J.S.; Nosanchuk, J.D.; Friedman, A.J. The potential of nitric oxide releasing therapies as antimicrobial agents. *Virulence* **2012**, *3*, 271–279. [CrossRef]
114. Yu, S.; Li, G.; Zhao, P.; Cheng, Q.; He, Q.; Ma, D.; Xue, W. NIR-Laser-Controlled Hydrogen-Releasing PdH Nanohydride for Synergistic Hydrogen-Photothermal Antibacterial and Wound-Healing Therapies. *Adv. Funct. Mater.* **2019**, *29*, 1905697. [CrossRef]
115. Fan, W.; Lu, N.; Huang, P.; Liu, Y.; Yang, Z.; Wang, S.; Yu, G.; Liu, Y.; Hu, J.; He, Q.; et al. Glucose-Responsive Sequential Generation of Hydrogen Peroxide and Nitric Oxide for Synergistic Cancer Starving-Like/Gas Therapy. *Angew. Chem. Int. Edit.* **2017**, *56*, 1229–1233. [CrossRef] [PubMed]
116. Tsikas, D.; Schmidt, M.; Boehmer, A.; Zoerner, A.A.; Gutzki, F.-M.; Jordan, J. UPLC-MS/MS measurement of S-nitrosoglutathione (GSNO) in human plasma solves the S-nitrosothiol concentration enigma. *J. Chromatogr. B* **2013**, *927*, 147–157. [CrossRef]
117. Yang, B.; Chen, Y.; Shi, J. Reactive Oxygen Species (ROS)-Based Nanomedicine. *Chem. Rev.* **2019**, *119*, 4881–4985. [CrossRef] [PubMed]
118. Jiang, D.; Ni, D.; Rosenkrans, Z.T.; Huang, P.; Yan, X.; Cai, W. Nanozyme: New horizons for responsive biomedical applications. *Chem. Soc. Rev.* **2019**, *48*, 3683–3704. [CrossRef]
119. Chen, Q.; Zhang, X.; Li, S.; Tan, J.; Xu, C.; Huang, Y. MOF-derived Co_3O_4@Co-Fe oxide double-shelled nanocages as multi-functional specific peroxidase-like nanozyme catalysts for chemo/biosensing and dye degradation. *Chem. Eng. J.* **2020**, *395*, 125130. [CrossRef]
120. Liu, X.; Yan, Z.; Zhang, Y.; Liu, Z.; Sun, Y.; Ren, J.; Qu, X. Two-Dimensional Metal-Organic Framework/Enzyme Hybrid Nanocatalyst as a Benign and m Self-Activated Cascade Reagent for in Vivo Wound Healing. *ACS Nano* **2019**, *13*, 5222–5230. [CrossRef]
121. Xu, W.; Jiao, L.; Yan, H.; Wu, Y.; Chen, L.; Gu, W.; Du, D.; Lin, Y.; Zhu, C. Glucose Oxidase-Integrated Metal-Organic Framework Hybrids as Biomimetic Cascade Nanozymes for Ultrasensitive Glucose Biosensing. *ACS Appl. Mater. Interfaces* **2019**, *11*, 22096–22101. [CrossRef]
122. Yin, S.-Y.; Song, G.; Yang, Y.; Zhao, Y.; Wang, P.; Zhu, L.-M.; Yin, X.; Zhang, X.-B. Persistent Regulation of Tumor Microenvironment via Circulating Catalysis of $MnFe_2O_4$@Metal-Organic Frameworks for Enhanced Photodynamic Therapy. *Adv. Funct. Mater.* **2019**, *29*, 1901417. [CrossRef]
123. Zheng, H.-Q.; Liu, C.-Y.; Zeng, X.-Y.; Chen, J.; Lu, J.; Lin, R.-G.; Cao, R.; Lin, Z.-J.; Su, J.-W. MOF-808: A Metal-Organic Framework with Intrinsic Peroxidase-Like Catalytic Activity at Neutral pH for Colorimetric Biosensing. *Inorg. Chem.* **2018**, *57*, 9096–9104. [CrossRef] [PubMed]
124. Cai, W.; Wang, J.; Chu, C.; Chen, W.; Wu, C.; Liu, G. Metal Organic Framework-Based Stimuli-Responsive Systems for Drug Delivery. *Adv. Sci.* **2019**, *6*, 1801526. [CrossRef] [PubMed]
125. Pan, X.; Wang, H.; Wang, S.; Sun, X.; Wang, L.; Wang, W.; Shen, H.; Liu, H. Sonodynamic therapy (SDT): A novel strategy for cancer nanotheranostics. *Sci. China-Life Sci.* **2018**, *61*, 415–426. [CrossRef] [PubMed]
126. Qian, X.; Zhang, J.; Gu, Z.; Chen, Y. Nanocatalysts-augmented Fenton chemical reaction for nanocatalytic tumor therapy. *Biomaterials* **2019**, *211*, 1–13. [CrossRef]
127. Lismont, M.; Dreesen, L.; Wuttke, S. Metal-Organic Framework Nanoparticles in Photodynamic Therapy: Current Status and Perspectives. *Adv. Funct. Mater.* **2017**, *27*, 1606314. [CrossRef]
128. Hou, H.; Huang, X.; Wei, G.; Xu, F.; Wang, Y.; Zhou, S. Fenton Reaction-Assisted Photodynamic Therapy for Cancer with Multifunctional Magnetic Nanoparticles. *ACS Appl. Mater. Interfaces* **2019**, *11*, 29579–29592. [CrossRef]
129. Huang, Y.; Skripka, A.; Labrador-Paez, L.; Sanz-Rodriguez, F.; Haro-Gonzalez, P.; Jaque, D.; Rosei, F.; Vetrone, F. Upconverting nanocomposites with combined photothermal and photodynamic effects. *Nanoscale* **2018**, *10*, 791–799. [CrossRef]
130. Jia, T.; Xu, J.; Dong, S.; He, F.; Zhong, C.; Yang, G.; Bi, H.; Xu, M.; Hu, Y.; Yang, D.; et al. Mesoporous cerium oxide-coated upconversion nanoparticles for tumor-responsive chemo-photodynamic therapy and bioimaging. *Chem. Sci.* **2019**, *10*, 8618–8633. [CrossRef]
131. Pan, C.; Ou, M.; Cheng, Q.; Zhou, Y.; Yu, Y.; Li, Z.; Zhang, F.; Xia, D.; Mei, L.; Ji, X. Z-Scheme Heterojunction Functionalized Pyrite Nanosheets for Modulating Tumor Microenvironment and Strengthening Photo/Chemodynamic Therapeutic Effects. *Adv. Funct. Mater.* **2020**, *30*, 1906466. [CrossRef]

132. Chen, Y.; Gao, Y.; Chen, Y.; Liu, L.; Mo, A.; Peng, Q. Nanomaterials-based photothermal therapy and its potentials in antibacterial treatment. *J. Control. Release* **2020**, *328*, 251–262. [CrossRef] [PubMed]
133. Chen, S.; Lu, J.; You, T.; Sun, D. Metal-organic frameworks for improving wound healing. *Coord. Chem. Rev.* **2021**, *439*, 213929. [CrossRef]
134. Cai, X.; Gao, W.; Ma, M.; Wu, M.; Zhang, L.; Zheng, Y.; Chen, H.; Shi, J. A Prussian Blue-Based Core-Shell Hollow-Structured Mesoporous Nanoparticle as a Smart Theranostic Agent with Ultrahigh pH-Responsive Longitudinal Relaxivity. *Adv. Mater.* **2015**, *27*, 6382–6389. [CrossRef] [PubMed]
135. Yu, Z.; Hu, W.; Zhao, H.; Miao, X.; Guan, Y.; Cai, W.; Zeng, Z.; Fan, Q.; Tan, T.T.Y. Generating New Cross Relaxation Pathways by Coating Prussian Blue on NaNdF4 for Enhanced Photothermal Agents. *Angew. Chem.* **2019**, *58*, 8624–8628. [CrossRef] [PubMed]
136. Wang, Q.; Ji, Y.; Shi, J.; Wang, L. NIR-Driven Water Splitting H-2 Production Nanoplatform for H-2-Mediated Cascade-Amplifying Synergetic Cancer Therapy. *ACS Appl. Mater. Interfaces* **2020**, *12*, 23677–23688. [CrossRef]
137. Fan, X.; Yang, F.; Huang, J.; Yang, Y.; Nie, C.; Zhao, W.; Ma, L.; Cheng, C.; Zhao, C.; Haag, R. Metal-Organic-Framework-Derived 2D Carbon Nanosheets for Localized Multiple Bacterial Eradication and Augmented Anti-infective Therapy. *Nano Lett.* **2019**, *19*, 5885–5896. [CrossRef]
138. Horcajada, P.; Chalati, T.; Serre, C.; Gillet, B.; Sebrie, C.; Baati, T.; Eubank, J.F.; Heurtaux, D.; Clayette, P.; Kreuz, C.; et al. Porous metal-organic-framework nanoscale carriers as a potential platform for drug delivery and imaging. *Nat. Mater.* **2010**, *9*, 172–178. [CrossRef]
139. Liu, Y.; Hou, W.; Xia, L.; Cui, C.; Wan, S.; Jiang, Y.; Yang, Y.; Wu, Q.; Qiu, L.; Tan, W. ZrMOF nanoparticles as quenchers to conjugate DNA aptamers for target-induced bioimaging and photodynamic therapy. *Chem. Sci.* **2018**, *9*, 7505–7509. [CrossRef]
140. Medici, S.; Peana, M.; Crisponi, G.; Nurchi, V.M.; Lachowicz, J.I.; Remelli, M.; Zoroddu, M.A. Silver coordination compounds: A new horizon in medicine. *Coord. Chem. Rev.* **2016**, *327*, 349–359. [CrossRef]
141. Yamanaka, M.; Hara, K.; Kudo, J. Bactericidal actions of a silver ion solution on Escherichia coli, studied by energy-filtering transmission electron microscopy and proteomic analysis. *Appl. Environ. Microbiol.* **2005**, *71*, 7589–7593. [CrossRef]
142. Zhang, X.; Han, X.; Jiang, Z.; Xu, J.; Chen, L.; Xue, Y.; Nie, A.; Xie, Z.; Kuang, Q.; Zheng, L. Atomically dispersed hierarchically ordered porous Fe-N-C electrocatalyst for high performance electrocatalytic oxygen reduction in Zn-Air battery. *Nano Energy* **2020**, *71*, 104547. [CrossRef]
143. Xing, Y.; Wang, L.; Wang, L.; Huang, J.; Wang, S.; Xie, X.; Zhu, J.; Ding, T.; Cai, K.; Zhang, J. Flower-Like Nanozymes with Large Accessibility of Single Atom Catalysis Sites for ROS Generation Boosted Tumor Therapy. *Adv. Funct. Mater.* **2022**, *32*, 2111171. [CrossRef]
144. Yang, X.; Jiang, D.; Zhang, X.; Gu, L.; Yuan, Y. Ascorbic acid-assisted hydrothermal route to create mesopores in polymeric carbon nitride for increased photocatalytic hydrogen generation. *Int. J. Hydrog. Energy* **2021**, *46*, 38310–38318. [CrossRef]
145. Jensen, W.B. The Lewis acid-base definitions: A status report. *Chem. Rev.* **1978**, *78*, 1–22. [CrossRef]
146. Deng, X.; Liang, S.; Cai, X.; Huang, S.; Cheng, Z.; Shi, Y.; Pang, M.; Ma, P.A.; Lin, J. Yolk-Shell Structured Au Nanostar@Metal-Organic Framework for Synergistic Chemo-photothermal Therapy in the Second Near-Infrared Window. *Nano Lett.* **2019**, *19*, 6772–6780. [CrossRef]
147. Ji, Q.; Xu, J.; Wang, C.; Wang, L. Controlling the coordination environment of Co atoms derived from Co/ZIF-8 for boosting photocatalytic H-2 evolution of CdS. *J. Colloid Interface Sci.* **2021**, *596*, 139–147. [CrossRef]
148. Yang, M.; Zhang, J.; Wei, Y.; Zhang, J.; Tao, C. Recent advances in metal-organic framework-based materials for anti-staphylococcus aureus infection. *Nano Res.* **2022**, *15*, 6220–6242. [CrossRef]
149. Fu, Q.; Saltsburg, H.; Flytzani-Stephanopoulos, M. Active nonmetallic Au and Pt species on ceria-based water-gas shift catalysts. *Science* **2003**, *301*, 935–938. [CrossRef]
150. Qiao, B.; Wang, A.; Yang, X.; Allard, L.F.; Jiang, Z.; Cui, Y.; Liu, J.; Li, J.; Zhang, T. Single-atom catalysis of CO oxidation using Pt-1/FeOx. *Nat. Chem.* **2011**, *3*, 634–641. [CrossRef]
151. Yin, P.; Yao, T.; Wu, Y.; Zheng, L.; Lin, Y.; Liu, W.; Ju, H.; Zhu, J.; Hong, X.; Deng, Z.; et al. Single Cobalt Atoms with Precise N-Coordination as Superior Oxygen Reduction Reaction Catalysts. *Angew. Chem. Int. Edit.* **2016**, *55*, 10800–10805. [CrossRef]
152. Han, J.; Meng, X.; Lu, L.; Bian, J.; Li, Z.; Sun, C. Single-Atom Fe-N-x-C as an Efficient Electrocatalyst for Zinc-Air Batteries. *Adv. Funct. Mater.* **2019**, *29*, 1800872. [CrossRef]
153. Cao, S.; Zhao, Z.; Zheng, Y.; Wu, Z.; Ma, T.; Zhu, B.; Yang, C.; Xiang, X.; Ma, L.; Han, X.; et al. A Library of ROS-Catalytic Metalloenzyme Mimics with Atomic Metal Centers. *Adv. Mater.* **2022**, *34*, 2200255. [CrossRef] [PubMed]
154. Wang, X.; Shi, Q.; Zha, Z.; Zhu, D.; Zheng, L.; Shi, L.; Wei, X.; Lian, L.; Wu, K.; Cheng, L. Copper single-atom catalysts with photothermal performance and enhanced nanozyme activity for bacteria-infected wound therapy. *Bioact. Mater.* **2021**, *6*, 4389–4401. [CrossRef] [PubMed]
155. Sun, L.; Li, C.; Yan, Y.; Yu, Y.; Zhao, H.; Zhou, Z.; Wang, F.; Feng, Y. Engineering DNA/Fe-N-C single-atom nanozymes interface for colorimetric biosensing of cancer cells. *Anal. Chim. Acta* **2021**, *1180*, 338856. [CrossRef] [PubMed]
156. Wang, Q.; Ina, T.; Chen, W.-T.; Shang, L.; Sun, F.; Wei, S.; Sun-Waterhouse, D.; Telfer, S.G.; Zhang, T.; Waterhouse, G.I.N. Evolution of Zn(II) single atom catalyst sites during the pyrolysis-induced transformation of ZIF-8 to N-doped carbons. *Sci. Bull.* **2020**, *65*, 1743–1751. [CrossRef]

157. Travlou, N.A.; Algarra, M.; Alcoholado, C.; Cifuentes-Rueda, M.; Labella, A.M.; Lazaro-Martinez, J.M.; Rodriguez-Castellon, E.; Bandosz, T.J. Carbon Quantum Dot Surface-Chemistry-Dependent Ag Release Governs the High Antibacterial Activity of Ag-Metal-Organic Framework Composites. *ACS Appl. Bio Mater.* **2018**, *1*, 693–707. [CrossRef]
158. Liu, J.; Liu, T.; Du, P.; Zhang, L.; Lei, J. Metal-Organic Framework (MOF) Hybrid as a Tandem Catalyst for Enhanced Therapy against Hypoxic Tumor Cells. *Angew. Chem. Int. Edit.* **2019**, *58*, 7808–7812. [CrossRef]
159. Chen, M.; Zhang, J.; Qi, J.; Dong, R.; Liu, H.; Wu, D.; Shao, H.; Jiang, X. Boronic Acid-Decorated Multivariate Photosensitive Metal-Organic Frameworks for Combating Multi-Drug-Resistant Bacteria. *ACS Nano* **2022**, *16*, 7732–7744. [CrossRef]
160. Chowdhuri, A.R.; Das, B.; Kumar, A.; Tripathy, S.; Roy, S.; Sahu, S.K. One-pot synthesis of multifunctional nanoscale metal-organic frameworks as an effective antibacterial agent against multidrug-resistant Staphylococcus aureus. *Nanotechnology* **2017**, *28*, 095102. [CrossRef]
161. Zhang, C.; Chu, G.; Ruan, Z.; Tang, N.; Song, C.; Li, Q.; Zhou, W.; Jin, J.; Haick, H.; Chen, Y.; et al. Biomimetic Self-Assembling Metal–Organic Architectures with Non-Iridescent Structural Coloration for Synergetic Antibacterial and Osteogenic Activity of Implants. *ACS Nano* **2022**, *16*, 16584–16597. [CrossRef]
162. Guo, L.; Xu, Y.; Ferhan, A.R.; Chen, G.; Kim, D.-H. Oriented Gold Nanoparticle Aggregation for Colorimetric Sensors with Surprisingly High Analytical Figures of Merit. *J. Am. Chem. Soc.* **2013**, *135*, 12338–12345. [CrossRef] [PubMed]
163. Sadhukha, T.; Wiedmann, T.S.; Panyam, J. Enhancing therapeutic efficacy through designed aggregation of nanoparticles. *Biomaterials* **2014**, *35*, 7860–7869. [CrossRef] [PubMed]
164. White, R.J.; Luque, R.; Budarin, V.L.; Clark, J.H.; Macquarrie, D.J. Supported metal nanoparticles on porous materials. Methods and applications. *Chem. Soc. Rev.* **2009**, *38*, 481–494. [CrossRef] [PubMed]
165. Aghayi-Anaraki, M.; Safarifard, V. Fe_3O_4@MOF Magnetic Nanocomposites: Synthesis and Applications. *Eur. J. Inorg. Chem.* **2020**, *2020*, 1916–1937. [CrossRef]
166. Li, S.; Huo, F. Metal-organic framework composites: From fundamentals to applications. *Nanoscale* **2015**, *7*, 7482–7501. [CrossRef]
167. Yang, Q.; Xu, Q.; Jiang, H.-L. Metal-organic frameworks meet metal nanoparticles: Synergistic effect for enhanced catalysis. *Chem. Soc. Rev.* **2017**, *46*, 4774–4808. [CrossRef]
168. Long, R.; Mao, K.; Gong, M.; Zhou, S.; Hu, J.; Zhi, M.; You, Y.; Bai, S.; Jiang, J.; Zhang, Q.; et al. Tunable Oxygen Activation for Catalytic Organic Oxidation: Schottky Junction versus Plasmonic Effects. *Angew. Chem. Int. Edit.* **2014**, *53*, 3205–3209. [CrossRef]
169. Han, W.; Wu, Z.; Li, Y.; Wang, Y. Graphene family nanomaterials (GFNs)-promising materials for antimicrobial coating and film: A review. *Chem. Eng. J.* **2019**, *358*, 1022–1037. [CrossRef]
170. Lakshmi, S.D.; Avti, P.K.; Gurumurthy, H. Activated carbon nanoparticles from biowaste as new generation antimicrobial agents: A review. *Nano Struct. Nano Objects* **2018**, *16*, 306–321. [CrossRef]
171. Shi, L.; Chen, J.; Teng, L.; Wang, L.; Zhu, G.; Liu, S.; Luo, Z.; Shi, X.; Wang, Y.; Ren, L. The Antibacterial Applications of Graphene and Its Derivatives. *Small* **2016**, *12*, 4165–4184. [CrossRef]
172. Xin, Q.; Shah, H.; Nawaz, A.; Xie, W.; Akram, M.Z.; Batool, A.; Tian, L.; Jan, S.U.; Boddula, R.; Guo, B.; et al. Antibacterial Carbon-Based Nanomaterials. *Adv. Mater.* **2019**, *31*, 1804838. [CrossRef] [PubMed]
173. Zheng, Q.; Liu, X.; Zheng, Y.; Yeung, K.W.K.; Cui, Z.; Liang, Y.; Li, Z.; Zhu, S.; Wang, X.; Wu, S. The recent progress on metal-organic frameworks for phototherapy. *Chem. Soc. Rev.* **2021**, *50*, 5086–5125. [CrossRef] [PubMed]

Journal of *Functional Biomaterials*

MDPI

Article

AMPK/mTOR Pathway Is Involved in Autophagy Induced by Magnesium-Incorporated TiO_2 Surface to Promote BMSC Osteogenic Differentiation

Guifang Wang [1,2,†], Jiaxin Luo [1,2,†], Yuqin Qiao [3,*], Dongdong Zhang [3], Yulan Liu [1,2], Wenjie Zhang [1,2], Xuanyong Liu [3] and Xinquan Jiang [1,2,*]

1 Department of Prosthodontics, Shanghai Ninth People's Hospital, Shanghai Jiao Tong University School of Medicine, 639 Zhizaoju Road, Shanghai 200011, China

2 College of Stomatology, Shanghai Jiao Tong University, National Center for Stomatology, National Clinical Research Center for Oral Diseases, Shanghai Key Laboratory of Stomatology, Shanghai Engineering Research Center of Advanced Dental Technology and Materials, 639 Zhizaoju Road, Shanghai 200011, China

3 State Key Laboratory of High Performance Ceramics and Superfine Microstructure, Shanghai Institute of Ceramics, Chinese Academy of Sciences, No. 1295 Dingxi Road, Shanghai 200050, China

* Correspondence: qiaoyq@mail.sic.ac.cn (Y.Q.); xinquanj@aliyun.com (X.J.)

† These authors contributed equally to this work.

Abstract: Magnesium has been extensively utilized to modify titanium implant surfaces based on its important function in promoting osteogenic differentiation. Autophagy has been proven to play a vital role in bone metabolism. Whether there is an association between autophagy and magnesium in promoting osteogenic differentiation remains unclear. In the present study, we focused on investigating the role of magnesium ions in early osteogenic activity and the underlying mechanism related to autophagy. Different concentrations of magnesium were embedded in micro-structured titanium surface layers using the micro-arc oxidation (MAO) technique. The incorporation of magnesium benefited cell adhesion, spreading, and viability; attenuated intracellular ATP concentrations and p-mTOR levels; and upregulated p-AMPK levels. This indicates the vital role of the ATP-related AMPK/mTOR signaling pathway in the autophagy process associated with osteogenic differentiation of bone marrow mesenchymal stem cells (BMSCs) induced by magnesium modification on titanium surfaces. The enhanced osteogenic differentiation and improved cellular autophagy activity of BMSCs in their extraction medium further confirmed the function of magnesium ions. The results of the present study advance our understanding of the mechanism by which magnesium regulates BMSC osteogenic differentiation through autophagy regulation. Moreover, endowing implants with the ability to activate autophagy may be a promising strategy for enhancing osseointegration in the translational medicine field in the future.

Keywords: magnesium; surface modification; titanium; osteogenic activity; autophagy

Citation: Wang, G.; Luo, J.; Qiao, Y.; Zhang, D.; Liu, Y.; Zhang, W.; Liu, X.; Jiang, X. AMPK/mTOR Pathway Is Involved in Autophagy Induced by Magnesium-Incorporated TiO_2 Surface to Promote BMSC Osteogenic Differentiation. *J. Funct. Biomater.* **2022**, *13*, 221. https://doi.org/10.3390/jfb13040221

Academic Editor: Hervé Petite

Received: 3 September 2022
Accepted: 30 October 2022
Published: 5 November 2022

Publisher's Note: MDPI stays neutral with regard to jurisdictional claims in published maps and institutional affiliations.

Copyright: © 2022 by the authors. Licensee MDPI, Basel, Switzerland. This article is an open access article distributed under the terms and conditions of the Creative Commons Attribution (CC BY) license (https://creativecommons.org/licenses/by/4.0/).

1. Introduction

Titanium and its alloys have been widely used as dental implants due to their excellent biocompatibility [1]. To improve the surface bioactivity and strengthen the binding ability of implants and bone tissues, various methods have been utilized for the surface modification of titanium. Magnesium is the fourth most abundant cation in the human body and the second largest cation in cells. It can be incorporated into titanium substrates using various methods to improve surface bioactivity and has been widely used to promote bone growth and regeneration. However, the role of magnesium in bone mineralization is controversial. It has been reported that appropriate concentrations of magnesium can promote the mineralization of BMSCs [2,3], while excessive magnesium can also impair their osteogenesis [4]. The entire osteogenesis process involves early osteogenic differentiation and late extracellular mineralization. During the osteogenesis process, especially the early osteogenic differentiation stage induced by magnesium ions, cell signaling triggers gene expression downstream. However, the underlying mechanisms of the facilitation effect of magnesium in regulating cell signal transduction are not fully understood.

Autophagy is an intracellular degradation process in eukaryotic cells that can transfer cytoplasmic components, including damaged macromolecules and organelles, to lysosomes for degradation and recycling [5]. It plays a very important role in the growth, mature differentiation, and homeostasis maintenance of cells and organisms [6]. Autophagy is considered to be highly involved in bone metabolism [7]. Increasing evidence suggests that an appropriate level of autophagy enables bone cells to survive hypoxic, nutrition-deficient, or even hypertonic environments. The induction of cellular autophagy plays a vital role in preosteoblast differentiation as well as osteoblast–osteocyte transitions [8–10]. Recent studies found that modulating magnesium transporters could accelerate osteogenic differentiation, partly via the activation of autophagy, giving clues about the correlation between magnesium, autophagy, and osteogenic differentiation [11,12]. Studies have also discovered that high concentrations of magnesium attenuated the osteogenic differentiation and mineralization ability of bone-related cells by modulating basal cellular autophagy [13,14]. Incorporating an appropriate concentration of magnesium on titanium surfaces is believed to be beneficial for improving its bioactivity as well as osteogenic differentiation. However, whether there is an association between the facilitation effect on early-stage osteogenic differentiation induced by magnesium and the modulation of autophagy remains unclear.

In the present study, we utilized the MAO technique to prepare magnesium-incorporated porous dioxide films on titanium surfaces and further investigated the role of magnesium in modulating autophagy and its possible signaling pathway in enhanced early osteogenic activity.

2. Materials and Methods

2.1. Fabrication and Modification of Specimens

Commercially pure titanium plates (Cp Ti, TA1, purity 99.85%) were ground with 400# abrasive paper and then ultrasonically washed with acetone, ethyl alcohol, and distilled water prior to plasma electrolyte oxidation (PEO) treatment. PEO was conducted in electrolytes composed of 0.1 mol/L calcium acetate monohydrate (CA, $C_4H_6O_4Ca{\cdot}H_2O$), 0.05 mol/L glycerophosphate disodium salt pentahydrate (GP, $C_3H_7Na_2O_6P{\cdot}5H_2O$) and various amounts (0.01, 0.02, and 0.03 mol/L) of magnesium acetate tetrahydrate ($Mg(CH_3COO)_2{\cdot}4H_2O$) to prepare Mg-free (M0) and Mg-incorporated (M1, M2, and M3) coatings. These coatings were formed at a current density of 16.5 A/dm^2 and frequency of 800 Hz, with a duty cycle of 10%.

2.2. Surface Characterization of Specimens

The surface morphologies and elemental compositions of Mg-free and Mg-incorporated coatings were examined and measured by scanning electron microscopy (Hitachi, Tokyo, Japan) and energy-dispersive X-ray spectrometry (EDS) (IXRF, Austin, TX, USA) attached to an electron probe X-ray microanalysis system (JEOL, Akishima, Japan), respectively. Their phase composition was analyzed by X-ray diffraction (XRD) (Bruker, Karlsruhe, Germany). The pore size and pore distribution in three random fields of each sample were measured by ImageJ (National Institutes of Health, Bethesda, MD, USA, V1.8.0).

The surface wettability of all coatings was assessed using a contact angle instrument (Solon, Shanghai, China). For this process, 2 μL of ultrapure water was vertically dropped on the sample surface, and the contact angles of the droplets were analyzed. Three samples per group were measured, and the average was taken as the final value.

For Mg ion release, each titanium substrate (1 × 1 cm) was immersed in 10 mL deionized water for 3 days, then moved to fresh 10 mL deionized water for the next 3 days. The accumulated extract was collected, and the Mg ion concentration was detected by inductively coupled plasma optical emission spectrometry (ICP-OES) (Agilent, Palo Alto, CA, USA).

2.3. Cell Culture

Four-week-old SD rats were provided by the SPF Experimental Animal Center, Ninth People's Hospital, affiliated with the Shanghai Jiao Tong University School of Medicine. All animal protocols were approved by the Animal Care and Experiment Committee of the Ninth People's Hospital (SH9H-2020-A619-1). BMSCs were isolated and cultured according to our previously published procedures [15]. SD rats were first anesthetized with chloral hydrate; then, both femurs were obtained. The marrow cavity was rinsed twice with high-glucose DMEM (Shanghai BasalMedia Technologies, Shanghai, China) containing 10% fetal bovine serum (Biological Industries, Kibbutz Beit Haemek, Israel). The cells were centrifuged at 1800 rpm for 10 min and subsequently cultured with 10 mL fresh high-glucose DMEM in a humidified atmosphere of 95% air and 5% CO_2. The culture medium was changed 48 h later, and cells at passages 2–3 and 80–90% confluency were used for subsequent studies.

2.4. Cell Adhesion and Spreading

Cells were seeded on titanium surfaces for 24 h, then were fixed with 4% paraformaldehyde. For permeabilization, cells were treated with 0.2% Triton X-100 (Yeasen Biotechnology, Shanghai, China) in PBS for 20 min at room temperature. Then, actin microfilaments were stained with phalloidin–iFluor 488 solution (Yeasen Biotechnology, Shanghai, China) at a concentration of 100 nM for 90 min, and nuclei were stained with DAPI at a concentration of 1 μg/mL for 10 min. The number of nuclei in five random fields of each sample was counted by ImageJ to represent cell adhesion ability. The extension area of cells in three random fields of each sample was measured by ImageJ to represent cell spreading.

2.5. Cell Morphology and Metabolic Activity Assay

BMSCs were cultured on the substrates for 3 days. Then, the cells were fixed with 2.5% glutaraldehyde. After being dehydrated by increasing concentrations of ethanol (30, 50, 75, 90, 95, and 100%) and dried by hexamethyldisilazane, they were sputter-coated with gold and observed by SEM. The metabolic activity of rat BMSCs was determined by MTT assay. After 4 and 7 days of culture, the cell metabolic activity was measured by comparing the optical density (OD) at 540 nm with the absorbance at 630 nm. The MTT assay was performed in triplicate, and each sample was analyzed twice.

2.6. Live/Dead Assay

BMSCs were cultured on substrates for 3 days; then, a live/dead staining assay was conducted using a Calcein–AM/PI double staining kit according to the manufacturer's protocol (Yeasen Biotechnology, Shanghai, China). Briefly, cells were washed with assay buffer three times; then, calcein AM (2 μM) and propidium iodide (1.5 mM) solutions were added to each well, and cells were incubated at 37 °C for another 30 min. The specimens were subsequently observed under a confocal microscope (Leica, Wetzlar, Germany).

2.7. Luciferase Assay

ATP concentration was measured by using a luciferase assay according to the manufacturer's protocol (Beyotime Biotechnology, Shanghai, China). After cells were cultured on titanium surfaces for 2 and 4 days, they were lysed and centrifuged at 12,000 rpm for 10 min. The collected supernatant was mixed with ATP detection solution; then, the RLU value was determined by a luminometer (Tecan, Männedorf, Switzerland). The ATP concentration was calculated according to the standard ATP concentration curve. A Micro BCA protein assay kit (Thermo Fisher Scientific, Waltham, MA, USA) was used to detect the concentration of total protein. The ratio of ATP to total protein (mg) represented the relative intracellular ATP level. The luciferase assay was performed in triplicate, and each sample was analyzed twice.

2.8. Reverse-Transcription Real-Time Polymerase Chain Reaction Assay

Cells were cultured on substrates to examine the direct effects of magnesium. To determine the effect of dissolved magnesium ions on cell bioactivity, extracts were collected using the following procedure. Briefly, titanium samples were immersed in 1 mL DMEM for 3 days, then moved to fresh 1 mL DMEM for another 3 days. Then, the collected extracts were supplemented with 10% fetal bovine serum. After 7 days of culture on substrates or in extracts, total RNA was extracted using TRIzol reagent. Complementary DNA was synthesized by using a PrimeScript RT reagent kit. The primer sequences were as follows: ALP: forward: GTCCCACAAGACCCCACAAT, reverse: CAACGGCAGAGCCAGGAAT; OCN: forward: GCCCTGACTGCATTCTGCCTCT, reverse: TCACCACCTTACTGCCCTCCTG; Runx2: forward: ATCCAGCCACCTTCACTTACACC, reverse: GGGACCATTGGGAACTGATAG; BMP2: forward: TGAACACAGCTGGTCTCAGG, reverse: TGACGCTTTTCTCGTTTGTG. The expression of these genes was quantified by real-time polymerase chain reaction (PCR) with SYBR Premix Ex TaqII (TaKaRa, Osaka, Japan). The relative expression levels for these genes were normalized to that of the housekeeping gene, GAPDH (forward: GGCAAGTTCAACGGCACAGT, reverse: GCCAGTAGACTCCACGACAT). The real-time PCR assay was performed twice, and each sample was analyzed in triplicate.

2.9. Immunofluorescence Microscopy

After 4 days of incubation, the expression of p-mTOR, p-AMPK, OCN, and LC3B of BMSCs cultured on titanium surfaces was identified by polyclonal rabbit antibodies (all from Beyotime Biotechnology, Shanghai, China) against p-mTOR (1:200), p-AMPK (1:200), OCN (1:200), and LC3B (1:200). Alexa Fluor 594 AffiniPure donkey anti-rabbit IgG (1:200; Yeasen Biotechnology, Shanghai, China) was further conjugated on the primary antibody. Actin microfilaments were stained by phalloidin–iFluor 488 solution (Yeasen Biotechnology, Shanghai, China). Nuclei were stained by DAPI solution (Beyotime Biotechnology, Shanghai, China) for 10 min. The samples were subsequently photographed by a confocal microscope (Leica, Germany).

2.10. Adenovirus Transfection

Cells were first seeded on the titanium surfaces to reach a confluency of about 70–80%; then, adenoviral mCherry-GFP-LC3B (Beyotime Biotechnology, Shanghai, China) with a multiplicity of infection (MOI) ratio of 20 was used for transfection for 24 h. Subsequently, cells were placed in a fresh culture medium and cultured for another 24 h for transfection confirmation. After another day of culture, specimens were fixed with 4% paraformaldehyde, and the fluorescence expression was observed under a confocal microscope (Leica, Germany).

2.11. Western Blotting Analysis

The Western blotting analysis was performed using standard techniques. Cells were cultured in different extracts for 3 days; then, protein samples were collected. Equal amounts of protein were placed on SDS-PAGE gel for electrophoretic separation. Proteins were then transferred to a pre-activated PVDF membrane. The membrane was incubated with polyclonal rabbit antibodies against OCN (1:1000), LC3B (1:1000), p-AMPK (1:1000), and p-mTOR (1:500) and monoclonal mouse antibody against β-actin (Wuhan Servicebio Technology, Wuhan, China) at 4 °C overnight, and then incubated with goat anti-rabbit peroxidase-conjugated secondary antibodies (Yeasen Biotechnology, Shanghai, China) at room temperature for 1 h. The protein band was scanned by the UVItec ALLIANCE 4.7 gel imaging system. The grey value of the typical protein band was calculated by ImageJ. The band intensity of each targeted protein normalized to β-actin represented its relative expression.

2.12. Statistical Analysis

The experimental data, expressed as mean ± standard deviation, were analyzed by SPSS 19.0 statistical analysis software (SPSS, Chicago, IL, USA). Statistical differences ($p < 0.05$) among various groups were compared using one-way ANOVA and SNK analysis.

3. Results

3.1. Surface Characterization

Figure 1A shows the surface morphology of dioxide titanium surfaces. After MAO treatment, a typical micro-porous network with a crater-like structure was observed on sample M0. The surface morphology seemed to have no obvious change after magnesium was doped into the coating. Figure 1B,C shows the pore distribution and average pore size of the four groups, with pores of around 3–70 μm and an average pore size of 15–20 μm. Most of the pores ranged in size from 5 to 25 μm. There were no significant differences in pore distribution and pore size among the four groups. The wettability of these samples was examined by water contact angle analysis. The contact angles were 57.62 ± 4.83, 63.27 ± 4.01, 52.05 ± 2.50, and 46.66 ± 2.74 for sample groups M0, M1, M2, and M3, respectively (Figure 1D). These four groups all showed a certain surface hydrophilicity, while the contact angle was slightly decreased in M3. The detection of magnesium ion release from these coatings is illustrated in Figure 1E. Magnesium ions were successfully detected in the extracts of M1, M2, and M3 at low concentrations ranging from 0.08 to 0.63 mg/L. The magnesium ion concentration increased with increased magnesium content on the titanium surface.

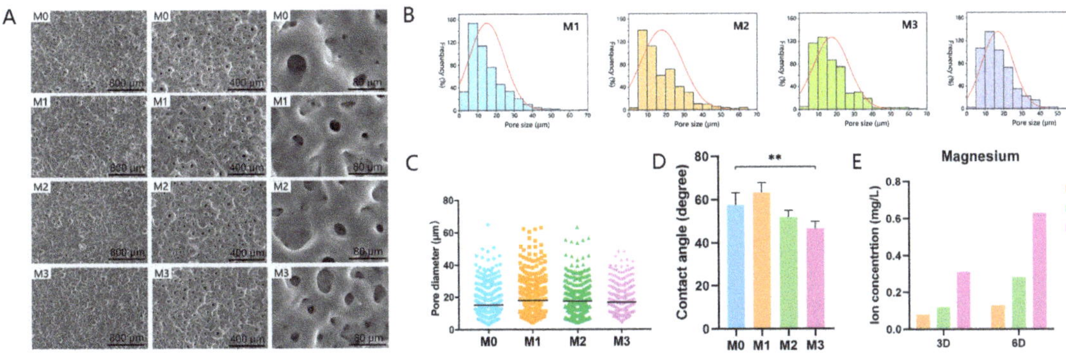

Figure 1. (**A**) Surface morphology observation of four samples at different magnifications. (**B**) Frequency distribution of pore size. (**C**) Pore size and (**D**) wettability of four substrates. (**E**) Accumulated Mg ion released from different specimens (** $p < 0.01$).

X-ray diffraction (XRD) analysis was performed on various samples to determine the crystallinity of the titanium dioxide layer. As shown in Figure 2, typical peaks of titanium and anatase TiO_2 phase were detected on the four titanium surfaces. In general, there were no significant alteration in surface morphology, wettability and phase composition of these coatings after magnesium was incorporated.

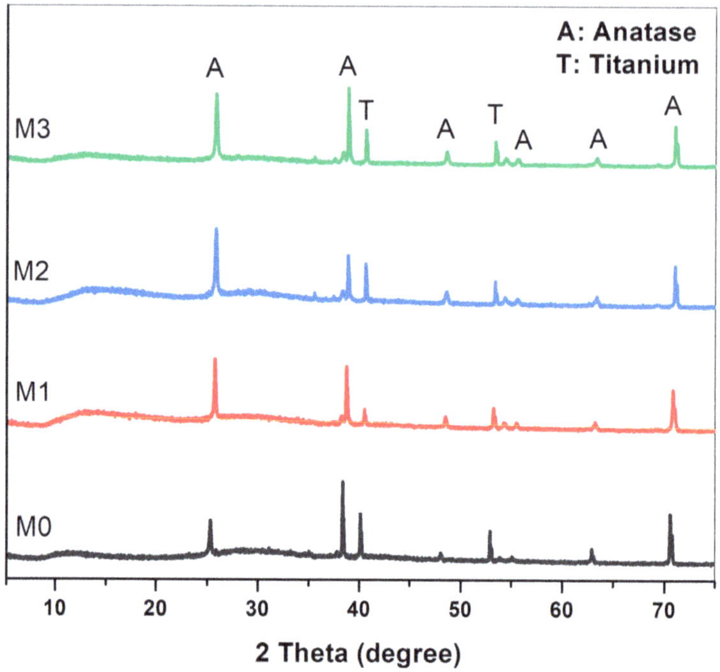

Figure 2. XRD patterns of four titanium substrates.

EDS analysis was performed to examine the chemical composition of titanium surfaces. As illustrated in Table 1, oxygen and titanium were the main elements on the four titanium substrates. Both Pi and Ca were detected in the four groups, with small differences among them. No magnesium was detected in group M0, while it was detected in the magnesium-containing electrolytic treatment groups. The magnesium percentage was 0.87 ± 0.16, 1.72 ± 0.04, and 2.51 ± 0.04% for groups M1, M2, and M3, respectively.

Table 1. Chemical composition of Mg-free and Mg-containing coatings detected by energy-dispersive X-ray spectra (EDS).

Groups	Element (wt.%)				
	O	P	Ca	Ti	Mg
M0	47.81 ± 0.06	11.99 ± 0.19	9.50 ± 0.31	30.71 ± 0.12	——
M1	48.30 ± 0.27	11.15 ± 0.08	7.27 ± 0.21	32.41 ± 0.34	0.87 ± 0.16
M2	48.51 ± 0.50	9.81 ± 0.18	8.35 ± 0.49	31.61 ± 0.13	1.72 ± 0.04
M3	47.61 ± 0.69	9.45 ± 0.34	7.57 ± 0.32	32.86 ± 0.21	2.51 ± 0.04

3.2. Bioactivity and Biocompatibility

The results of cell adhesion and spreading are shown in Figure 3A,B. After cells were incubated on the samples for 24 h, more attached cells were observed in the magnesium-incorporated groups than in the control group. The statistical analysis showed that there were no significant differences among the magnesium-incorporated groups. As shown in Figure 3C,D, the coverage area of attached cells was significantly larger in groups M1, M2, and M3 than in the group M0. There were no significant differences among groups M1, M2, and M3.

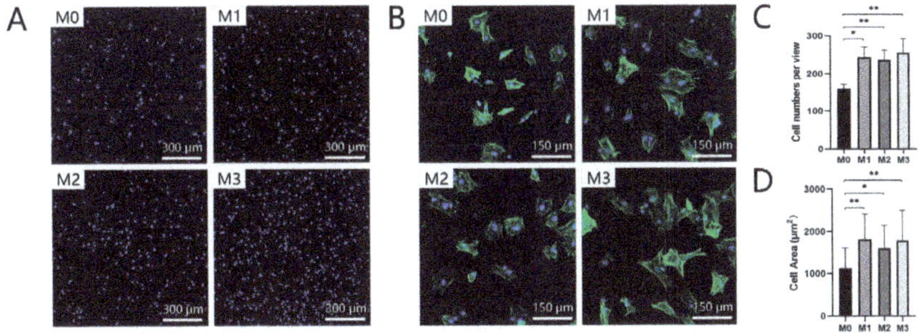

Figure 3. (**A**,**C**) Cell adhesion and (**B**,**D**) cell spreading assays were performed after BMSCs were cultured on coatings for 24 h (** $p < 0.01$, * $p < 0.05$).

The results of cell morphology observation are shown in Figure 4. After 3 days of culture, cells were observed to grow well on the four titanium specimens. There were numerous pseudopodia extending from the cell surface attached to the micro-porous titanium surfaces.

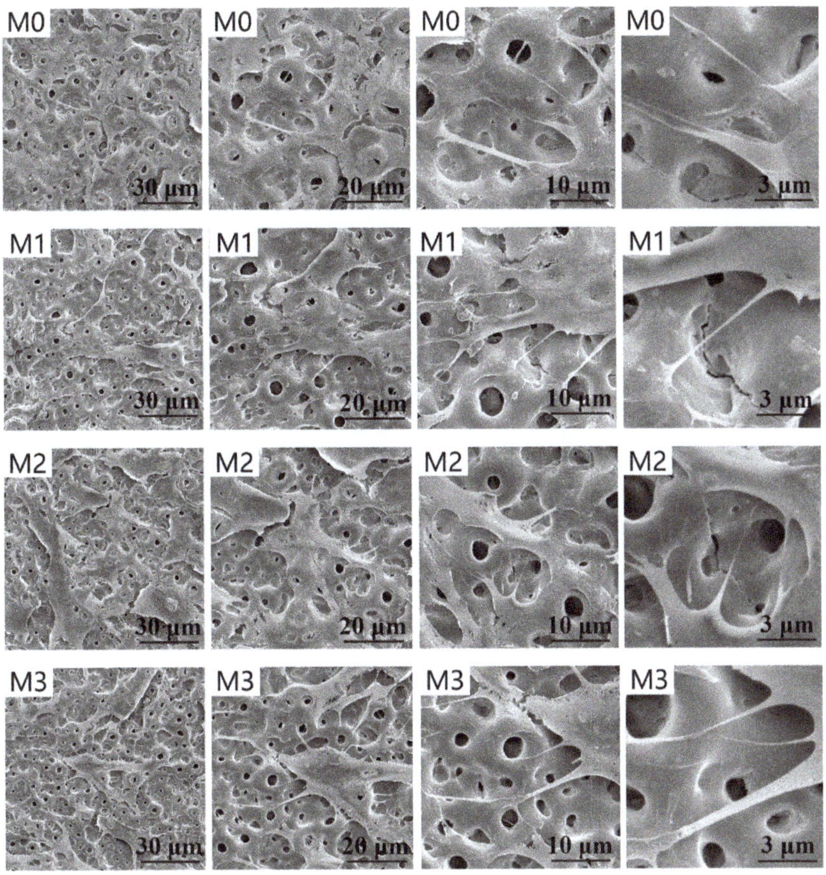

Figure 4. SEM observation of morphology of cells cultured on Mg-free and Mg-incorporated coatings for 3 days at different magnifications.

The live/dead assay results are presented in Figure 5A. The majority of cells were alive, and few dead cells were observed on the titanium surfaces, suggesting good biocompatibility for cell growth. The results of the MTT assay, shown in Figure 5B, represent total cell metabolic activity. After 4 days of culture, no significant differences were observed between Mg-free and Mg-incorporated titanium oxide surfaces. After 7 days of culture, total cell metabolic activity was higher on the M3 titanium surface than on the other three (M0, M1, and M2). The addition of magnesium showed good biocompatibility for cell viability, while a high magnesium content improved cell metabolic activity.

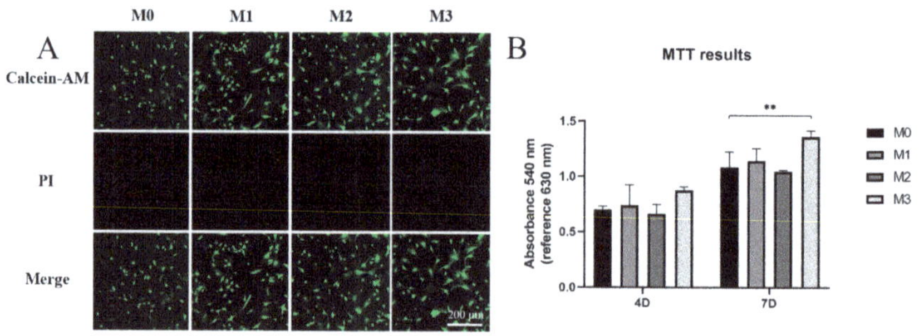

Figure 5. (**A**) Live/dead assay of cells cultured on titanium surfaces for 3 days. (**B**) MTT assay of cell viability (** $p < 0.01$).

3.3. Osteogenic Differentiation

Cells were cultured on different coatings for 7 days. A real-time PCR assay was performed to examine the mRNA expression of osteogenic-related genes, including ALP, OCN, Runx2, and BMP2. As illustrated in Figure 6A, the expression of ALP was upregulated in the three magnesium-incorporated groups, while the mRNA expression of ALP on the M3 coating was significantly higher compared to the other groups. The expressions of OCN, Runx2, and BMP2 were significantly improved on the Mg-incorporated coatings compared to the control group. M3 exhibited the highest expression of ALP, OCN, Runx2, and BMP2 mRNA among the four groups. As shown in Figure 6B, the intensity of immunofluorescence staining of OCN protein was the lowest in the control group and was more pronounced in the three magnesium-incorporated groups. Among the four groups, M3 exhibited the strongest OCN protein expression.

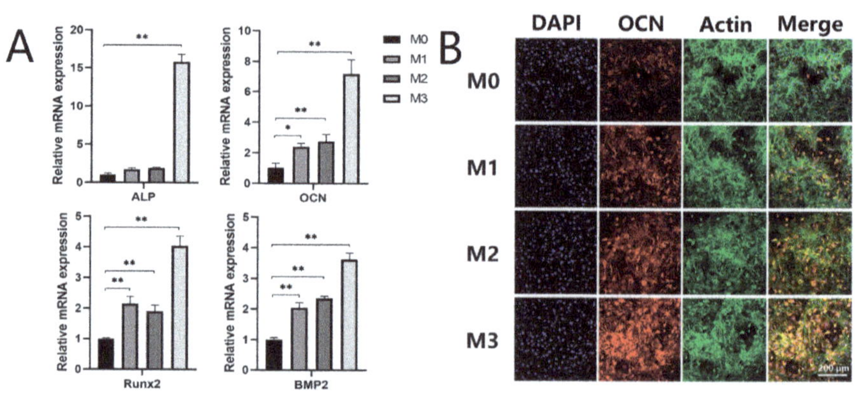

Figure 6. (**A**) Expression of osteogenic-related ALP, OCN, Runx2, and BMP2 genes. (**B**) Immunofluorescence staining for OCN protein (** $p < 0.01$, * $p < 0.05$).

3.4. Autophagy Activation and Its Possible Mechanism

Previous investigations demonstrated that autophagy plays an important role in the process of osteogenic differentiation of stem cells. In this study, we examined the protein level of autophagic marker LC3 to assess cellular autophagic activity. After 4 days of culture, the LC3 protein expressions of BMSCs cultured on different titanium surfaces were determined by immunofluorescence staining; and a typical image is shown in Figure 7. The distribution of LC3 protein was dispersed in group M0, and only a few LC3 protein puncta could be observed, while the Mg-incorporated titanium surfaces induced LC3 protein expression and accumulation. Group M3 exhibited the strongest fluorescent intensity of LC3 protein, and its distribution was spot-like.

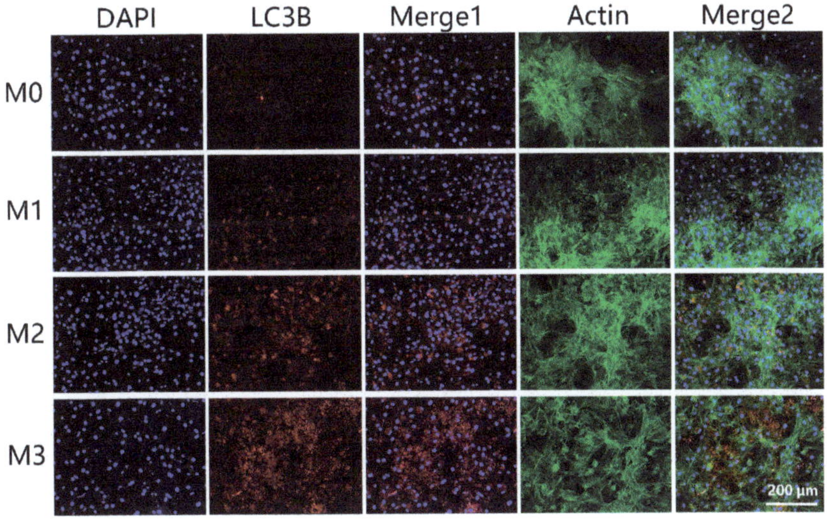

Figure 7. Immunofluorescence staining for LC3 protein.

To further elucidate the effects of incorporated magnesium on the autophagy flux of cells cultured on titanium dioxide surfaces, BMSCs were transfected with mCherry-GFP-LC3B adenovirus. Typical images in Figure 8 show that co-localization of mCherry-LC3B and GFP-LC3B dots was higher in groups M1, M2, and M3 compared to M0, indicating a higher level of autophagy. The M3 coating showed the strongest co-localization of mCherry-LC3B and GFP-LC3B dots.

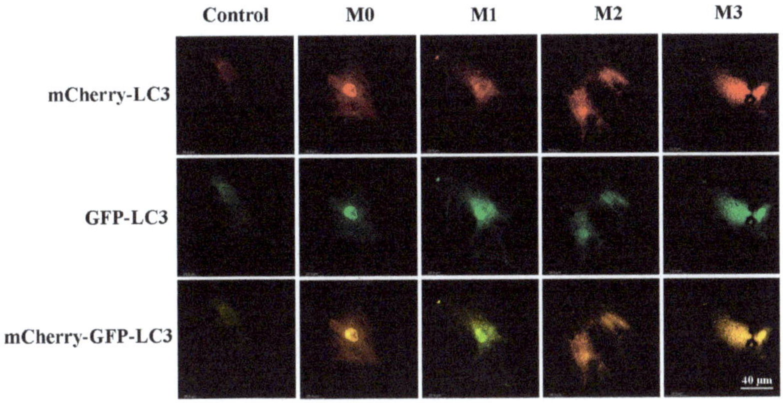

Figure 8. Observation of autophagy flux in cells transfected with mCherry-GFP-LC3B adenovirus.

In the regulation of osteogenic differentiation and autophagy, evidence has shown a critical role for the AMPK axis. AMPK senses low energy levels in cells. The sudden fall of cellular energy is known to activate AMPK, resulting in the phosphorylation of downstream targets and the inhibition of mTOR, which can result in the regulation of autophagy. To determine the potential mechanism of autophagy induction, we first examined the intracellular level of ATP using a luciferase assay. The statistical analysis is shown in Figure 9A. After 2 and 4 days of incubation, the intracellular ATP levels of cells cultured on magnesium-incorporated coatings were decreased compared to the control group. The level of p-AMPK expression was further investigated by immunofluorescence staining. As shown in Figure 9B, phosphorylation of AMPK protein was found to be strengthened in groups M1, M2, and M3 compared to the control. M3 exhibited the highest p-AMPK protein expression level.

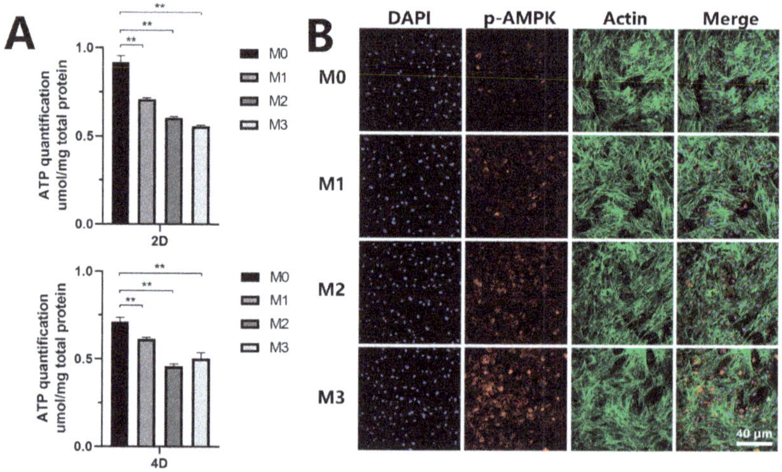

Figure 9. (**A**) Intracellular ATP level. (**B**) Immunofluorescence staining for p-AMPK protein (** $p < 0.01$).

The detection of the p-mTOR protein is shown in Figure 10. The fluorescence intensity of p-mTOR protein was obviously less pronounced in groups M1, M2, and M3 compared to M0. The M3 coating showed the lowest p-mTOR protein expression among the four specimens. The incorporated magnesium downregulated cellular energy levels in the process of enhancing osteogenic differentiation, then activated AMPK and inhibited mTOR, subsequently initiating autophagy. Taken together, these data point out that the AMPK/mTOR signaling pathway plays a vital role in autophagy-induced osteogenic differentiation regulated by magnesium.

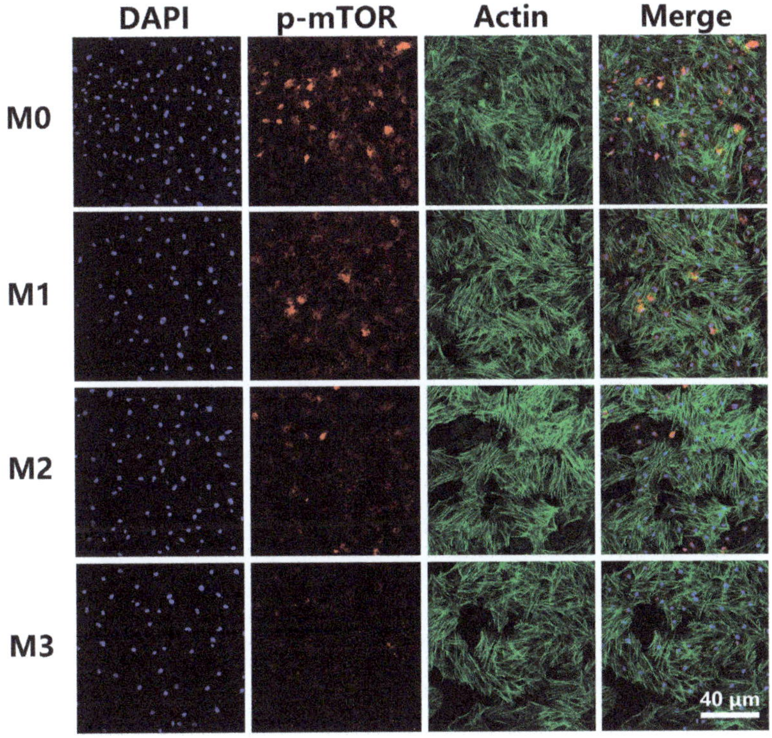

Figure 10. Immunofluorescence staining for p-mTOR protein.

3.5. The Effect of Extracts on Osteogenic Differentiation and Autophagy

BMSCs were cultured in extracts of different coatings for 7 days. The results of relative mRNA expression of osteogenic-related genes (ALP, OCN, Runx2, and BMP2) are shown in Figure 11A. ALP expression was significantly upregulated in the three magnesium-containing titanium extracts compared to the M0 group, while M2 exhibited the highest expression level of ALP mRNA. OCN expression was significantly increased in groups M1, M2, and M3 compared to the control group, and the highest level was observed in M3. Runx2 enhancement was observed in group M2, while there was no significant change observed in M1 and M3 compared to the control group. The expression of BMP2 was slightly improved in M2 and M3 compared to M0, although the difference was not significant. After cells were incubated in different extracts for 4 days, the protein expression of OCN, LC3, p-AMPK, and p-mTOR was examined by Western blotting analysis (Figure 11B,C). OCN and LC3 II protein expression was higher in the magnesium-incorporating groups, M1, M2, and M3. M3 exhibited the deepest gray band and the highest gray value ratio among the four groups. The expression of p-mTOR showed an opposite trend to OCN and LC3 II. The expression of LC3 I was downregulated in M1 and upregulated in M2 and M3, compared to M0. The gray band of p-AMPK seemed deeper in M1 compared to the control group, and the gray value ratio of p-AMPK to β-actin was slightly higher in M1 compared to the control. The gray value ratio of p-AMPK to β-actin in M2 and M3 seemed to have no obvious enhancement.

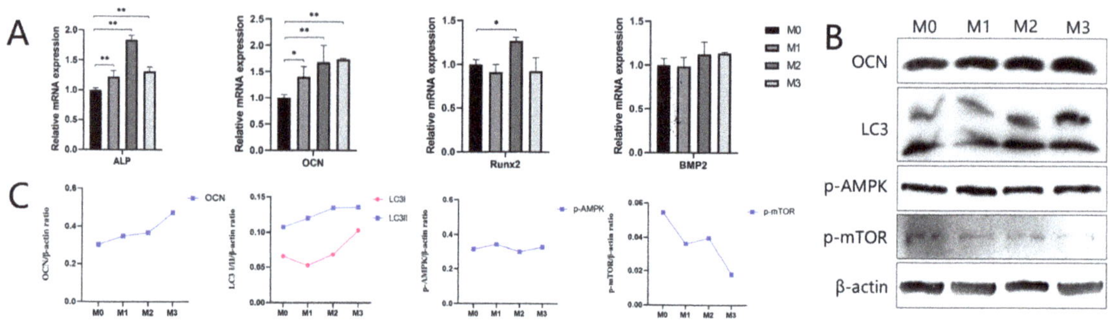

Figure 11. (**A**) Expression of osteogenic-related genes, ALP, OCN, Runx2, and BMP2. (**B**) Western blotting for OCN- and autophagy-related proteins. (** $p < 0.01$, * $p < 0.05$) (**C**) The gay value ratio of OCN- and autophagy-related proteins to β-actin.

4. Discussion

Magnesium is an essential inorganic component in bone tissue that plays a crucial role in skeletal development [16]. Mg deficiency has been linked to atherosclerosis, endothelial dysfunction, inflammation, insulin resistance, and dyslipidemia [17–19], while excess magnesium impairs the mineralization of BMSCs and thus has adverse biological effects on bone [20,21]. Therefore, local delivery of magnesium at appropriate concentrations might be a better approach for clinical application. MAO has been widely used to fabricate micro-structured ceramic oxide layers composed of several kinds of bioactive metal ions on titanium substrates [22]. In the present study, we fabricated a magnesium-containing dioxide titanium surface via the MAO technique and focused on its influence on BMSC osteogenic differentiation and its possible underlying mechanism related to autophagy regulation.

The surface topography, chemical composition, and wettability of titanium have been reported to interactively regulate and control local cytokine production and osteogenic differentiation, thus finally determining in vivo bone-implant contact. By keeping the electrolytic parameters consistent, the surface morphology, including average pore size and pore distribution, and surface crystallization remained the same among the four groups. During the MAO process, magnesium ions participated in the plasma–chemical reaction occurring at the metal–electrolyte interface and then were embedded in the porous coating formed on the titanium substrates. By adjusting the magnesium ion concentration in the electrolyte, different amounts of magnesium were embedded in the coatings. The increasing concentration of magnesium released from different coatings further confirmed that magnesium was incorporated into the surface layer of the material and thus could be released in a content-dependent manner. A small amount of incorporated magnesium ion did not change the surface wettability, while a larger magnesium content improved surface hydrophilicity, probably owing to the cumulative –OH groups on the surface of the crystals [23]. Although increased surface wettability has been well recognized for its vital function in promoting cell response, a score of degrees is usually considered as little difference. In a comparison study, an appreciable improvement in hydrophilicity from ~108° to 47° did not affect the initial single-cell detachment force and process of detachment at the titanium implant surfaces in vitro [24]. In another study, compared to a contact angle of ~65°, a contact angle of about 27° showed no obvious stimulation effect on cell integrin expression [25]. Moreover, magnesium incorporated in the scaffold was proved to be able to induce desirable protein adsorption and cell membrane integrin receptor expression, thus enhancing initial cell adhesion [26]. Based on previous studies and our data, we believe that the major difference among the four coatings is the different amounts of magnesium incorporated.

The results of ALP, OCN, Runx2, and BMP2 gene expression and OCN protein expression demonstrated the increased osteogenic differentiation ability of the magnesium-modified coatings. The higher the magnesium content, the better the enhancement effect that could be observed. The evidence described above indicates the critical role of magnesium in cell responses, including cell compatibility, adhesion, attachment and early-stage osteogenic differentiation, which suggests a potentially beneficial role in improving titanium-bone bonding.

Autophagy is an evolutionarily conserved cellular pathway that plays a crucial role in bone remodeling [27] by mediating cellular metabolism. MSCs with a high level of autophagy have increased osteogenic differentiation potential [28]. Recently, titanium surfaces with autophagy-modulating ability have attracted attention in orthopedic and dental implant research. Titanium-based rough surfaces were confirmed to promote osteogenic differentiation via the autophagy-dependent PI3/Akt signal transduction pathway [29]. Nanotopography on titanium was recognized as promoting osteogenic differentiation via autophagy-mediated signaling between YAP and β-catenin [30]. A combination of micro-nano topography exhibited an ability to create an anti-inflammation microenvironment, thus promoting cell proliferation and osteogenic differentiation by inducing cellular autophagy activity [31]. Beyond the surface topography, the chemical composition was also utilized to modulate autophagy activity on titanium surfaces in order to obtain better osteogenic differentiation [32,33]. Autophagy is greatly influenced by magnesium, but the effect is quite complex. Studies have reported that dietary magnesium deficiency reduces the number of autolysosomes and autophagosomes [34]. Magnesium incorporated in biomaterials triggers autophagy to mediate apoptosis in osteosarcoma cells [35,36]. Other studies have reported that magnesium negatively regulates cellular autophagy levels, resulting in an inhibitory effect of extracellular matrix calcification and osteogenic differentiation of ADTC cells, as well as human BMSCs [13,14,37]. This phenomenon may be explained by the fact that magnesium plays a dual role in bone metabolism while it changes the crystalline morphology of HA and inhibits collagen calcification [38]. Consistent with numerous previous studies [26,39,40], the present study demonstrated that an appropriate amount of magnesium incorporation was beneficial for BMSC proliferation as well as early-stage osteogenic differentiation. The upregulated expression of LC3 protein and enhanced autophagy influx indicated that cellular autophagy was activated in the process of osteogenic differentiation induced by magnesium incorporation.

For a better understanding of the role of magnesium in BMSC osteogenic differentiation and autophagy activation, extracts collected from different titanium surfaces were prepared and utilized in cell cultures. Along with improved magnesium content on the titanium surface, the magnesium concentration in the extracts was increased. Its promotional effect on osteogenic differentiation and activation of cellular autophagy and its possible signaling pathway was further confirmed, suggesting a positive role of dissolved magnesium in the local environment. However, the beneficial effect was obviously attenuated when cells were cultured in the extractions, indicating that the major stimulation effect in autophagy-induced osteogenic differentiation may be attributed to the magnesium present on the titanium dioxide surface.

An exploration of the observed effects on autophagy, osteogenic differentiation, and magnesium stimulated a further investigation of the possible mechanism. Evidence has shown that there is a critical role of the AMPK axis in regulating osteogenic differentiation and autophagy. AMPK has been reported to control the osteogenic differentiation of human MSCs through early mTOR inhibition-mediated autophagy [41,42]. The AMPK activator has been demonstrated to have a role in inducing osteogenic differentiation and mineralization of osteoblastic cell lines and bone marrow progenitor cells [43–46]. A recent study reported that magnesium released from biomaterials could activate the AMPK/mTOR signaling pathway, thus upregulating autophagy in osteosarcoma cells [35]. We further explored the level of AMPK phosphorylation. Though applied to different cell lines and in different amounts, it was found that magnesium could upregulate the level of p-AMPK in a dose-dependent manner. In addition, phosphorylation levels of molecular mTOR were altered in an opposite trend to AMPK variation. Known as an energy sensor, AMPK is an evolutionarily conserved serine/threonine protein kinase that is precisely regulated by the ratios of AMP/ATP and ADP/ATP in the cells. In our study, we detected intracellular ATP levels and found that the incorporation of magnesium in the titanium dioxide surface resulted in a significant reduction in ATP levels. The cellular contents of ADP and AMP increase in response to energy stress. An increase in the intracellular AMP to ATP ratio, as a consequence of decreased intracellular ATP, may be responsible for the activation of AMPK, as well as the subsequent induction of cellular autophagy. Serving as a cofactor of ATP, magnesium plays an important role in regulating glucose metabolism [47], resulting in a metabolic shift from oxidative phosphorylation to glycolysis [48]. Compared to oxidative phosphorylation, glycolysis gives rise to faster but lower ATP production [49], which is consistent with our results showing that there was a significantly decreased amount of ATP with the presence of magnesium. In conclusion, the ATP-related AMPK/mTOR pathway is a pivotal regulator involved in early-stage osteogenic differentiation and autophagy induced by magnesium incorporation.

5. Conclusions

Using the MAO technique, we fabricated micro-structured titanium dioxide coatings with different contents of magnesium. The incorporation of magnesium did not obviously change the surface morphology, porosity, or surface wettability. We found that incorporating magnesium benefited cellular responses, including cell adhesion, spreading, and viability. Osteogenic differentiation and autophagy activity were more pronounced on titanium oxide coatings incorporated with magnesium. Consistently, intracellular ATP and p-mTOR levels were decreased, whereas p-AMPK levels were upregulated after magnesium modification. The findings of the present study demonstrate that the AMPK/mTOR signaling pathway is involved in the process of autophagy associated with the osteogenic differentiation of BMSCs induced by magnesium incorporation. These results advance our understanding of the link between autophagy induction and osteogenic differentiation regulated by magnesium-modified biomaterials and reveal a potential mechanism of magnesium-mediated BMSC osteogenic differentiation and autophagy.

Author Contributions: G.W.: conceptualization, methodology, writing—original draft preparation, funding acquisition. J.L.: validation, data curation. D.Z.: methodology, investigation. Y.L.: validation, visualization. W.Z.: conceptualization, formal analysis. X.L.: resources. Y.Q.: resources, writing—review and editing, funding acquisition. X.J.: resources, funding acquisition. All authors have read and agreed to the published version of the manuscript.

Funding: This work was supported by the National Natural Science Foundation of China (82001011, 31870944, 81921002 and 82130027).

Institutional Review Board Statement: Not applicable.

Informed Consent Statement: Not applicable.

Data Availability Statement: Not applicable.

Conflicts of Interest: The authors declare no conflict of interest.

References

1. Wang, G.; Li, J.; Lv, K.; Zhang, W.; Ding, X.; Yang, G.; Liu, X.; Jiang, X. Surface thermal oxidation on titanium implants to enhance osteogenic activity and in vivo osseointegration. *Sci. Rep.* **2016**, *6*, 31769. [CrossRef] [PubMed]
2. Belluci, M.M.; de Molon, R.S.; Rossa, C., Jr.; Tetradis, S.; Giro, G.; Cerri, P.S.; Marcantonio, E., Jr.; Orrico, S.R.P. Severe magnesium deficiency compromises systemic bone mineral density and aggravates inflammatory bone resorption. *J. Nutr. Biochem.* **2020**, *77*, 108301. [CrossRef] [PubMed]
3. Zhang, X.; Chen, Q.; Mao, X. Magnesium Enhances Osteogenesis of BMSCs by Tuning Osteoimmunomodulation. *Biomed. Res. Int.* **2019**, *2019*, 7908205. [CrossRef] [PubMed]
4. Hénaut, L.; Massy, Z.A. Magnesium as a Calcification Inhibitor. *Adv. Chronic Kidney Dis.* **2018**, *25*, 281–290. [CrossRef] [PubMed]
5. Wang, X.; Wu, R.; Liu, Y.; Zhao, Y.; Bi, Z.; Yao, Y.; Liu, Q.; Shi, H.; Wang, F.; Wang, Y. m(6)A mRNA methylation controls autophagy and adipogenesis by targeting Atg5 and Atg7. *Autophagy* **2020**, *16*, 1221–1235. [CrossRef]

6. Ma, Y.; Qi, M.; An, Y.; Zhang, L.; Yang, R.; Doro, D.H.; Liu, W.; Jin, Y. Autophagy controls mesenchymal stem cell properties and senescence during bone aging. *Aging Cell* **2018**, *17*, e12709. [CrossRef] [PubMed]
7. Lin, N.Y.; Chen, C.W.; Kagwiria, R.; Liang, R.; Beyer, C.; Distler, A.; Luther, J.; Engelke, K.; Schett, G.; Distler, J.H. Inactivation of autophagy ameliorates glucocorticoid-induced and ovariectomy-induced bone loss. *Ann. Rheum. Dis.* **2016**, *75*, 1203–1210. [CrossRef]
8. Nollet, M.; Santucci-Darmanin, S.; Breuil, V.; Al-Sahlanee, R.; Cros, C.; Topi, M.; Momier, D.; Samson, M.; Pagnotta, S.; Cailleteau, L.; et al. Autophagy in osteoblasts is involved in mineralization and bone homeostasis. *Autophagy* **2014**, *10*, 1965–1977. [CrossRef]
9. Zhu, L.; Chen, J.; Zhang, J.; Guo, C.; Fan, W.; Wang, Y.M.; Yan, Z. Parathyroid Hormone (PTH) Induces Autophagy to Protect Osteocyte Cell Survival from Dexamethasone Damage. *Med. Sci. Monit.* **2017**, *23*, 4034–4040. [CrossRef]
10. Zheng, L.; Wang, W.; Ni, J.; Mao, X.; Song, D.; Liu, T.; Wei, J.; Zhou, H. Role of autophagy in tumor necrosis factor-α-induced apoptosis of osteoblast cells. *J. Investig. Med.* **2017**, *65*, 1014–1020. [CrossRef]
11. Castiglioni, S.; Romeo, V.; Locatelli, L.; Cazzaniga, A.; Maier, J.A.M. TRPM7 and MagT1 in the osteogenic differentiation of human mesenchymal stem cells in vitro. *Sci. Rep.* **2018**, *8*, 16195. [CrossRef] [PubMed]
12. Castiglioni, S.; Romeo, V.; Locatelli, L.; Zocchi, M.; Zecchini, S.; Maier, J.A.M. The simultaneous downregulation of TRPM7 and MagT1 in human mesenchymal stem cells in vitro: Effects on growth and osteogenic differentiation. *Biochem. Biophys. Res. Commun.* **2019**, *513*, 159–165. [CrossRef] [PubMed]
13. Yue, J.; Jin, S.; Gu, S.; Sun, R.; Liang, Q. High concentration magnesium inhibits extracellular matrix calcification and protects articular cartilage via Erk/autophagy pathway. *J. Cell Physiol.* **2019**, *234*, 23190–23201. [CrossRef] [PubMed]
14. Li, Y.; Wang, J.; Yue, J.; Wang, Y.; Yang, C.; Cui, Q. High magnesium prevents matrix vesicle-mediated mineralization in human bone marrow-derived mesenchymal stem cells via mitochondrial pathway and autophagy. *Cell Biol. Int.* **2018**, *42*, 205–215. [CrossRef]
15. Wang, G.; Li, J.; Zhang, W.; Xu, L.; Pan, H.; Wen, J.; Wu, Q.; She, W.; Jiao, T.; Liu, X.; et al. Magnesium ion implantation on a micro/nanostructured titanium surface promotes its bioactivity and osteogenic differentiation function. *Int. J. Nanomed.* **2014**, *9*, 2387–2398.
16. Lin, S.; Yang, G.; Jiang, F.; Zhou, M.; Yin, S.; Tang, Y.; Tang, T.; Zhang, Z.; Zhang, W.; Jiang, X. A Magnesium-Enriched 3D Culture System that Mimics the Bone Development Microenvironment for Vascularized Bone Regeneration. *Adv. Sci.* **2019**, *6*, 1900209. [CrossRef] [PubMed]
17. Kostov, K.; Halacheva, L. Role of Magnesium Deficiency in Promoting Atherosclerosis, Endothelial Dysfunction, and Arterial Stiffening as Risk Factors for Hypertension. *Int. J. Mol. Sci.* **2018**, *19*, 1724. [CrossRef]
18. López-Baltanás, R.; Encarnación Rodríguez-Ortiz, M.; Canalejo, A.; Díaz-Tocados, J.M.; Herencia, C.; Leiva-Cepas, F.; Torres-Peña, J.D.; Ortíz-Morales, A.; Muñoz-Castañeda, J.R.; Rodríguez, M.; et al. Magnesium supplementation reduces inflammation in rats with induced chronic kidney disease. *Eur. J. Clin. Investig.* **2021**, *51*, e13561. [CrossRef]
19. Rezazadeh, H.; Sharifi, M.R.; Sharifi, M.; Soltani, N. Magnesium sulfate improves insulin resistance in high fat diet induced diabetic parents and their offspring. *Eur. J. Pharmacol.* **2021**, *909*, 174418. [CrossRef]
20. Tsao, Y.-T.; Shih, Y.-Y.; Liu, Y.-A.; Liu, Y.-S.; Lee, O.K. Knockdown of SLC41A1 magnesium transporter promotes mineralization and attenuates magnesium inhibition during osteogenesis of mesenchymal stromal cells. *Stem Cell Res. Ther.* **2017**, *8*, 39. [CrossRef]
21. Leidi, M.; Dellera, F.; Mariotti, M.; Maier, J.A.M. High magnesium inhibits human osteoblast differentiation in vitro. *Magnes. Res.* **2011**, *24*, 1–6. [CrossRef] [PubMed]
22. Tan, L.; Hu, Y.; Hou, Y.; Chen, M.; Xue, C.; Chen, M.; Sun, Y.; Mu, C.; Luo, Z.; Cai, K. Osteogenic differentiation of mesenchymal stem cells by silica/calcium micro-galvanic effects on the titanium surface. *J. Mater. Chem. B* **2020**, *8*, 2286–2295. [CrossRef] [PubMed]
23. Wang, J.; Li, D.; Liu, Q.; Yin, X.; Zhang, Y.; Jing, X.; Zhang, M. Fabrication of hydrophobic surface with hierarchical structure on Mg alloy and its corrosion resistance. *Electrochim. Acta* **2010**, *55*, 6897–6906. [CrossRef]
24. Becker, M.; Schmied, F.; Kadem, L.F.; Freitag-Wolf, S.; Naujokat, H.; Mehl, C.; Kern, M.; Harder, S. Single-cell adhesion of human osteoblasts on plasma-conditioned titanium implant surfaces in vitro. *J. Mech. Behav. Biomed. Mater.* **2020**, *109*, 103841. [CrossRef] [PubMed]
25. Park, J.H.; Wasilewski, C.E.; Almodovar, N.; Olivares-Navarrete, R.; Boyan, B.D.; Tannenbaum, R.; Schwartz, Z. The responses to surface wettability gradients induced by chitosan nanofilms on microtextured titanium mediated by specific integrin receptors. *Biomaterials* **2012**, *33*, 7386–7393. [CrossRef]
26. Zhang, J.; Ma, X.; Lin, D.; Shi, H.; Yuan, Y.; Tang, W.; Zhou, H.; Guo, H.; Qian, J.; Liu, C. Magnesium modification of a calcium phosphate cement alters bone marrow stromal cell behavior via an integrin-mediated mechanism. *Biomaterials* **2015**, *53*, 251–264. [CrossRef]
27. Bo, T.; Yan, F.; Guo, J.; Lin, X.; Zhang, H.; Guan, Q.; Wang, H.; Fang, L.; Gao, L.; Zhao, J.; et al. Characterization of a Relatively Malignant Form of Osteopetrosis Caused by a Novel Mutation in the PLEKHM1 Gene. *J. Bone Miner. Res.* **2016**, *31*, 1979–1987. [CrossRef]
28. Chen, M.; Hu, Y.; Hou, Y.; Li, M.; Chen, M.; Mu, C.; Tao, B.; Zhu, W.; Luo, Z.; Cai, K. Differentiation regulation of mesenchymal stem cells via autophagy induced by structurally-different silica based nanobiomaterials. *J. Mater. Chem. B* **2019**, *7*, 2657–2666. [CrossRef]

29. Kaluderovic, M.R.; Mojic, M.; Schreckenbach, J.P.; Maksimovic-Ivanic, D.; Graf, H.L.; Mijatovic, S. A Key Role of Autophagy in Osteoblast Differentiation on Titanium-Based Dental Implants. *Cells Tissues Organs* **2014**, *200*, 265–277. [CrossRef]
30. Li, L.; Yang, S.; Xu, L.; Li, Y.; Fu, Y.; Zhang, H.; Song, J. Nanotopography on titanium promotes osteogenesis via autophagy-mediated signaling between YAP and beta-catenin. *Acta Biomater.* **2019**, *96*, 674–685. [CrossRef]
31. Zhang, T.; Jiang, M.; Yin, X.; Yao, P.; Sun, H. The role of autophagy in the process of osseointegration around titanium implants with micro-nano topography promoted by osteoimmunity. *Sci. Rep.* **2021**, *11*, 18418. [CrossRef] [PubMed]
32. Wang, H.; Ma, Y.; Li, J.; Zhou, C.; Xu, A.; Xu, Y.; He, F. Modulating autophagy by strontium-doped micro/nano rough titanium surface for promotion of osteogenesis and inhibition of osteoclastogenesis. *Colloids Surf. B Biointerfaces* **2022**, *210*, 112246. [CrossRef] [PubMed]
33. Zhang, X.; Cui, J.; Cheng, L.; Lin, K. Enhancement of osteoporotic bone regeneration by strontium-substituted 45S5 bioglass via time-dependent modulation of autophagy and the Akt/mTOR signaling pathway. *J. Mater. Chem. B* **2021**, *9*, 3489–3501. [CrossRef]
34. Bai, R.; Miao, M.Z.; Li, H.; Wang, Y.; Hou, R.; He, K.; Wu, X.; Jin, H.; Zeng, C.; Cui, Y.; et al. Increased Wnt/β-catenin signaling contributes to autophagy inhibition resulting from a dietary magnesium deficiency in injury-induced osteoarthritis. *Arthritis Res Ther* **2022**, *24*, 165. [CrossRef]
35. Wei, X.; Tang, Z.; Wu, H.; Zuo, X.; Dong, H.; Tan, L.; Wang, W.; Liu, Y.; Wu, Z.; Shi, L.; et al. Biofunctional magnesium-coated Ti6Al4V scaffolds promote autophagy-dependent apoptosis in osteosarcoma by activating the AMPK/mTOR/ULK1 signaling pathway. *Mater. Today Bio.* **2021**, *12*, 100147. [CrossRef] [PubMed]
36. Li, T.; Yu, Y.; Shi, H.; Cao, Y.; Liu, X.; Hao, Z.; Ren, Y.; Qin, G.; Huang, Y.; Wang, B. Magnesium in Combinatorial With Valproic Acid Suppressed the Proliferation and Migration of Human Bladder Cancer Cells. *Front Oncol.* **2020**, *10*, 589112. [CrossRef]
37. Zhang, L.; Yang, C.; Li, J.; Zhu, Y.; Zhang, X. High extracellular magnesium inhibits mineralized matrix deposition and modulates intracellular calcium signaling in human bone marrow-derived mesenchymal stem cells. *Biochem. Biophys. Res. Commun.* **2014**, *450*, 1390–1395. [CrossRef]
38. Zhang, J.; Tang, L.; Qi, H.; Zhao, Q.; Liu, Y.; Zhang, Y. Dual Function of Magnesium in Bone Biomineralization. *Adv. Healthc Mater.* **2019**, *8*, e1901030. [CrossRef]
39. Okuzu, Y.; Fujibayashi, S.; Yamaguchi, S.; Yamamoto, K.; Shimizu, T.; Sono, T.; Goto, K.; Otsuki, B.; Matsushita, T.; Kokubo, T.; et al. Strontium and magnesium ions released from bioactive titanium metal promote early bone bonding in a rabbit implant model. *Acta Biomater.* **2017**, *63*, 383–392. [CrossRef] [PubMed]
40. Li, S.; Pu, X.; Chen, X.; Liao, X.; Huang, Z.; Yin, G. A novel bi-phase Sr-doped magnesium phosphate/calcium silicate composite scaffold and its osteogenesis promoting effect. *Ceram. Int.* **2018**, *44*, 16237–16245. [CrossRef]
41. Pantovic, A.; Krstic, A.; Janjetovic, K.; Kocic, J.; Harhaji-Trajkovic, L.; Bugarski, D.; Trajkovic, V. Coordinated time-dependent modulation of AMPK/Akt/mTOR signaling and autophagy controls osteogenic differentiation of human mesenchymal stem cells. *Bone* **2013**, *52*, 524–531. [CrossRef] [PubMed]
42. Chen, W.R.; Yang, J.Q.; Liu, F.; Shen, X.Q.; Zhou, Y.J. Melatonin attenuates vascular calcification by activating autophagy via an AMPK/mTOR/ULK1 signaling pathway. *Exp. Cell Res.* **2020**, *389*, 111883. [CrossRef] [PubMed]
43. Cheng, Y.; Huang, L.; Wang, Y.; Huo, Q.; Shao, Y.; Bao, H.; Li, Z.; Liu, Y.; Li, X. Strontium promotes osteogenic differentiation by activating autophagy via the the AMPK/mTOR signaling pathway in MC3T3E1 cells. *Int. J. Mol. Med.* **2019**, *44*, 652–660. [PubMed]
44. Vidoni, C.; Ferraresi, A.; Secomandi, E.; Vallino, L.; Gardin, C.; Zavan, B.; Mortellaro, C.; Isidoro, C. Autophagy drives osteogenic differentiation of human gingival mesenchymal stem cells. *Cell Commun. Signal.* **2019**, *17*, 98. [CrossRef]
45. Sun, X.; Zheng, W.; Qian, C.; Wu, Q.; Hao, Y.; Lu, G. Focal adhesion kinase promotes BMP2-induced osteogenic differentiation of human urinary stem cells via AMPK and Wnt signaling pathways. *J. Cell Physiol.* **2020**, *235*, 4954–4964. [CrossRef] [PubMed]
46. Zhao, X.; Pathak, J.L.; Huang, W.; Zhu, C.; Li, Y.; Guan, H.; Zeng, S.; Ge, L.; Shu, Y. Metformin enhances osteogenic differentiation of stem cells from human exfoliated deciduous teeth through AMPK pathway. *J. Tissue Eng. Regen. Med.* **2020**, *14*, 1869–1879. [CrossRef]
47. Oost, L.J.; Kurstjens, S.; Ma, C.; Hoenderop, J.G.J.; Tack, C.J.; de Baaij, J.H.F. Magnesium increases insulin-dependent glucose uptake in adipocytes. *Front Endocrinol.* **2022**, *13*, 986616. [CrossRef] [PubMed]
48. Lin, S.; Yin, S.; Shi, J.; Yang, G.; Wen, X.; Zhang, W.; Zhou, M.; Jiang, X. Orchestration of energy metabolism and osteogenesis by Mg(2+) facilitates low-dose BMP-2-driven regeneration. *Bioact. Mater.* **2022**, *18*, 116–127. [CrossRef] [PubMed]
49. Yellen, G. Fueling thought: Management of glycolysis and oxidative phosphorylation in neuronal metabolism. *J. Cell Biol.* **2018**, *217*, 2235–2246. [CrossRef] [PubMed]

Journal of *Functional Biomaterials*

MDPI

Article

Why Is Tantalum Less Susceptible to Bacterial Infection?

Xin Chen [1,2,†], Yikang Bi [3,4,†], Moran Huang [5,†], Huiliang Cao [6,*] and Hui Qin [1,*]

1 Department of Orthopedic Surgery, Shanghai Sixth People's Hospital Affiliated to Shanghai Jiao Tong University School of Medicine, Shanghai 200233, China
2 Department of Laboratory Medicine, The Affiliated Hospital of Xuzhou Medical University, Xuzhou 221002, China
3 Department of Orthopedics, The Eighth People's Hospital, Jiang Su University, Shanghai 200235, China
4 Department of Orthopedics, Xuhui Branch of Shanghai Sixth People's Hospital Affiliated to Shanghai Jiao Tong University School of Medicine, Shanghai 200235, China
5 Department of Orthopedic Surgery, Shanghai General Hospital, Shanghai Jiao Tong University School of Medicine, Shanghai 201620, China
6 Interfacial Electrochemistry and Biomaterials, Lab of Low-Dimensional Materials Chemistry, Key Laboratory for Ultrafine Materials of Ministry of Education, Shanghai Engineering Research Center of Hierarchical Nanomaterials, School of Materials Science and Engineering, East China University of Science & Technology, Shanghai 200237, China
* Correspondence: huiliang.cao@ecust.edu.cn (H.C.); qh197826@163.com (H.Q.)
† These authors contributed equally to this work.

Abstract: Periprosthetic infection is one of the trickiest clinical problems, which often leads to disastrous consequences. The emergence of tantalum and its derivatives provides novel ideas and effective methods to solve this problem and has attracted great attention. However, tantalum was reported to have different anti-infective effects in vivo and in vitro, and the inherent antibacterial capability of tantalum is still controversial, which may restrict its development as an antibacterial material to some extent. In this study, the polished tantalum was selected as the experimental object, the implant-related tibia osteomyelitis model was first established to observe whether it has an anti-infective effect in vivo compared to titanium, and the early studies found that the tantalum had a lower infectious state in the implant-related tibia osteomyelitis model in vivo than titanium. However, further in vitro studies found that the polished tantalum was not superior to the titanium against bacterial adhesion and antibacterial efficacy. In addition, we focus on the state of interaction between cells, bacteria and materials to restore the internal environment as realistically as possible. We found that the adhesion of fibroblasts to tantalum was faster and better than that of titanium. Moreover, what is more, interesting is that, in the early period, bacteria were more likely to adhere to cells that had already attached to the surface of tantalum than to the bare surface of it, and over time, the cells eventually fell off the biomaterials and took away more bacteria in tantalum, making it possible for tantalum to reduce the probability of infection in the body through this mechanism. Moreover, these results also explained the phenomenon of the "race for the surface" from a completely different perspective. This study provides a new idea for further exploring the relationship between bacteria and host tissue cells on the implant surface and a meaningful clue for optimizing the preparation of antibacterial implants in the future.

Keywords: biomaterials; tantalum; race for the surface; implant-associated infection

Citation: Chen, X.; Bi, Y.; Huang, M.; Cao, H.; Qin, H. Why Is Tantalum Less Susceptible to Bacterial Infection? *J. Funct. Biomater.* **2022**, *13*, 264. https://doi.org/10.3390/jfb13040264

Academic Editor: Masaaki Nakai

Received: 2 October 2022
Accepted: 19 November 2022
Published: 22 November 2022

Publisher's Note: MDPI stays neutral with regard to jurisdictional claims in published maps and institutional affiliations.

Copyright: © 2022 by the authors. Licensee MDPI, Basel, Switzerland. This article is an open access article distributed under the terms and conditions of the Creative Commons Attribution (CC BY) license (https://creativecommons.org/licenses/by/4.0/).

1. Introduction

As an indispensable treatment for the restoration of function in modern medicine, biomaterial implants often fail because of implant-associated infection (IAI) [1]. Orthopedic IAI is a serious complication occurring in 1–2% after closed and in up to 30% after open fractures [2]. There are about 55,000 deaths from IAI annually in the USA, and the average cost of treating plant infections is up to $100,000 [1,3], which brings great pain and financial burden on patients. In addition, owing to the increase in antibiotic resistance and the difficulties in developing new antibiotics, hopeful strategies are urgently needed to change this situation [4,5]. In recent years, the antibacterial materials based on inorganic metal nanoparticles and their derivatives have been widely studied to prevent and treat biomaterial-related infections [6–9]. However, due to the inherent characteristics of most antibacterial nanomaterials, they not only have toxic effects on bacteria but also negatively affect the functions of normal human cells and tissues, thus limiting the application and clinical transformation of these new materials as implant coatings [10–14].

In recent years, clinically, titanium (Ti) and its derivatives have been widely used as implant materials in orthopedic surgery. Until now, the development of new titanium alloys has greatly optimized the mechanical properties of implants [15–19].Therefore, the focus of the urgent solution should be to select potential biomaterials to improve the effective antibacterial activity of implants and minimize their biological toxicity [20–23]. Tantalum (Ta) is considered a bioinert metal, which has attracted much attention because of its excellent chemical stability, biocompatibility, osteogenic activity, and corrosion resistance, and it has been widely applied in orthopedics and dentistry [24–28]. Moreover, Ta can effectively improve the capability of adhesion, proliferation, and differentiation of bone marrow mesenchymal stem cells and osteoblasts and promote osteogenic-related gene expression [29,30]. Especially porous Ta, as one of the excellent tantalum derivatives, has achieved gratifying results in clinical studies because of its appropriate elastic modulus and superior ability to induce bone regeneration [31–33]. In addition, tantalum and its derivatives are also reported to have effective antibacterial properties, but the inherent antibacterial activity of tantalum is still controversial. Different forms of tantalum, including metal ions and solid state, have been reported to have significant antibacterial activity [34,35]. Schildhauer et al. described that *S. aureus* had lower adhesion to pure tantalum than other common metal implant materials in vitro bacterial adhesion test [35]. At the same time, the structure of porous tantalum materials creates a suitable environment for bacterial colonization [36]. Moreover, in the in vitro experiments, porous tantalum did not show significant antibacterial properties [37]. However, it was demonstrated with satisfactory bony integration even if long-standing infection at the implantation site in the in vivo experiments [38]. Therefore, it is essential to further explore the inherent antibacterial properties of tantalum [26].

In this study, we first carried out in vivo tests and observed that the infection degree of the tantalum group was lighter than that of the titanium group, which prompted the authors to further continue in vitro tests to explore the possible mechanism behind these results. However, some interesting phenomena have been found in the author's experiments; compared with titanium, fibroblasts not only adhered to tantalum more effectively. In the early period, bacteria also tend to adhere to the surface of cells rather than tantalum in the co-culture environment of cells and bacteria. The experimental results provide an idea for a further understanding of the relationship between host tissue and bacteria in the internal implant environment and can also be said to further explore the theory of "the race for the surface" from another point of view.

2. Materials and Methods

2.1. Sample Preparation and Characterization

Commercial pure titanium (Grade IV, ASTM) and tantalum (Ta2, YST 751-2011) were supplied by BAOTI Group (Baoji, Shaanxi, China)., China. In the in vitro tests, pure titanium and pure tantalum were prepared in 10 mm × 10 mm × 1 mm size and carefully polished to a mirror finish with abrasive paper and ultrasonically washed using ethanol and ultrapure water. The surface topography of Ta and Ti was evaluated using a field-emission scanning electron microscope (FE-SEM; S-4800, Hitachi, Japan). The depth profiles and chemical state of elements were examined by X-ray photoelectron spectroscopy. (XPS; PHI 5802, Physical Electronics Inc., Eden Prairie, MN, USA). The roughness of the surface in

Ta and Ti has been evaluated by atomic force microscopy (AFM) before, as described in our previous paper [39]. Moreover, the contact-angle instrument (SL200B; Shanghai Solon Information Technology Co., Ltd., Shanghai, China) was used to measure the water contact angles of each sample. In addition, in the in vivo tests, the same materials were used to fabricate polished Kirschner wires with a length of 10 mm and a diameter of 1.2 mm.

2.2. Bacteria Preparation and Characterization

Freeze-dried *Staphylococcus aureus* (*S. a*; ATCC 43300) were get from the American Type Culture Collection (Manassas, VA, USA). Referring to the methodology of the author's previous research [40], the Bacterial concentration used in the in vitro tests was 1×10^6 colony forming units (CFUs)/mL in Trypticase Soy Broth (TSB; BD Biosciences, Franklin Lakes, NJ, USA), which was 1×10^5 CFUs/mL in PBS in the in vivo tests.

2.3. Implant-Related Tibia Osteomyelitis Model in Rats

The experimental protocol was approved by the Animal Care and Experiment Committee of Shanghai Sixth People's Hospital, Affiliated with Shanghai Jiao Tong University School of Medicine (No: DWLL2018-0339). A total of 35 male Sprague Dawley rats (8 weeks) with an average weight of 215 g (180–255 g) were divided into three groups (Table 1). 0.6% pentobarbital sodium (0.9 mL/100 g body wt) was injected intraperitoneally into all the rats. The rats were treated as follows (Table 1). After shaving and disinfecting, a 0.5 cm long longitudinal incision was made on the tibial anteromedial side of the hind leg. A hole in the cancellous bone of the proximal metaphysis was drilled between the tibial tuberosity and the tibial plateau with a 1.0 mm-diameter Kirschner wire to enter the medullary cavity while ensuring the integrity of the surrounding periosteum. Then, 10 μL PBS or PBS containing *S. aureus* with a concentration of 1×10^5 CFUs/mL was injected into the medullary cavity. According to the treatment, three groups of rats were divided into six groups according to their left and right legs. After bacterial inoculation, a polished metal Ta or Ti Kirschner wire (length:10 mm, diameter: 0.8 mm) was inserted into the medullary cavity. The subcutaneous tissue was irrigated carefully with a povidone-iodine solution, and the fascia and skin were closed in layers. After the operation, the animals were returned to different cages and allowed to move freely. The animals were allowed to bear weight and monitored daily. Buprenorphine was used as an analgesic for 2 days, but no antibiotics were used.

Table 1. Details of animal experiments.

Group	Number (n)	Left/Right	Implant	Inoculation	Abbreviation
I	10	right leg	no	10 μL PBS	Blank
		left leg	no	*S. a* 10^3 CFU/10 μL	*S. a*
II	10	right leg	Ti wire	10 μL PBS	Ti
		left leg	Ta wire	10 μL PBS	Ta
III	15	right leg	Ti wire	*S. a* 10^3 CFU/10 μL	Ti + *S. a*
		left leg	Ta wire	*S. a* 10^3 CFU/10 μL	Ta + *S. a*

2.3.1. Radiographic Evaluation

Two weeks after the operation, the animals were photographed with high-resolution lateral X-ray films under general anesthesia using chloral hydrate. Three independent observers unaware of the study group evaluated three regions of interest (ROI): proximal epi-/metaphyseal area, diaphyseal region, and distal epi-/metaphyseal area. Radiographic assessment was based on the system used by Lucke et al. [41].

2.3.2. Imaging Assay

At 14 days post-surgery, each rat (n = 5) was injected with 18.13–18.87 MBq (490–510 μCi) of 18F-FDG (Atomic Firm Sinovac Pharmaceutical co., LTD., Beijing, China) in 1.0 mL of saline via tail vein. The animals fasted for at least 8 h before the tracer injection. PET/CT scanning was performed at 40 min post tracer injection with GE Discovery VCT (General Electric Medical Systems, Milwaukee, WI, USA). The scanning conditions were as follows: CT scan, 120 kV and 80 mA, 64 slices, thickness 3.75 mm. PET scans were obtained in 3D, with 2.5 min/bed scanning time, and the animals were anesthetized by weight-adopted intraperitoneal injection of 0.6% pentobarbital sodium (0.9 mL/100 g body wt). Ordered subset expectation maximization (OSEM) was used to reconstruct the image iteratively. Attenuation correction was used in CT. 18F-FDG uptake was reported as the maximum standardized uptake value (SUVmax), which was calculated as the radioactivity of the region of interest (ROI) divided SUVmax for Ti+ *S. a* group, Ta+ *S. a* group were acquired, then the SUVmax differences between groups were compared.

2.3.3. Microbiological Evaluation

After 14 days, 5 rats from III were sacrificed, and bilateral tibia was retrieved aseptically, followed by soft tissue removal and Kirschner wire explanted. To quantify bacteria adhesion, the explanted Kirschner wires were sonicated and vortexed to remove adhered bacteria in 4 mL PBS. The spread plate method was used to count the adhered bacteria. Then tibiae were selected randomly, frozen quickly, and ground into powder under aseptic conditions [42]. One tibial powder was agitated for 3 min in 2 mL PBS. After 10,000$\times$ *g* centrifugation for 15 s, the supernatant was continuously diluted (10 times), and the CFU/tibia was analyzed by the spread plate method.

2.3.4. Histopathologic Evaluation

Masson's trichrome staining was used to evaluate the morphological change on the tibia, and Giemsa staining was chosen to identify the residual bacteria. Five rats from III were sacrificed, and Kirschner wires were explanted. The proximal tibia obtained from each group was decalcified in EDTA for 2 weeks, followed by dehydration and embedded in paraffin. Each specimen was cut to a 5 cm sagittal section and dyed by Masson's trichrome staining and Giemsa staining, respectively.

2.4. *In Vitro Antibacterial Assay*

A total of 1 mL of prepared *S. aureus* suspension (1×10^6 CFU/mL) was added into the 24-well plate containing Ti and Ta plates and cultured for different periods (3, 6, 12, and 24 h) at 37 °C. At each time point, the spread plate method was used to calculate the number of planktonic bacteria and to analyze the bacteria adhered to the surface of the samples using SEM after ultrasonic vibration [43].

The viable planktonic bacteria were calculated in the culture medium at each time point. The bacteria loosely adhered to the samples were carefully cleaned with PBS, and the adhered bacteria were separated ultrasonically for 5 min to 3 mL PBS under the frequency of 50 Hz in an ultrasonic bath (B3500S-MT, Shanghai Branson Ultrasonic Co., Shanghai, China) [44]. The solution was diluted 10 folds and plated on sheep blood Agar in triplicate, followed by incubation overnight at 37 °C. The account of CFUs was calculated according to the National Standard of China GB/T 4789.2 protocol. The samples were rinsed with PBS, fixed with glutaraldehyde solution for 4 h, then dehydrated for 10 min in gradient ethanol series, and finally dehydrated in anhydrous ethanol (twice), freeze-dried, coated with gold, and examined by SEM.

2.5. *Cell-Surface Interactions*

Human gingival fibroblasts (HGF-1) (Stem Cell Bank, Chinese Academy of Sciences, Shanghai, China) were used in the cell adhesion experiment; the cells were inoculated in 24-well plates containing samples with the concentration of 1×10^5 cells/well and incubated in the Dulbecco's modified eagle medium (DMEM; Gibco Invitrogen, Inc., Carlsbad, CA, USA) with 10% fetal bovine serum (HyClone, South Logan, UT, USA), 100 U/mL penicillin and 100 μg/mL streptomycin at 37 °C. For morphological observations, the samples were seeded with HCF-1 in 24-well plates at the concentration of 1×10^5 cells per well. In different cultured periods (3, 6, 12 and 24 h), the cells were fixed for 10 min with paraformaldehyde, followed by permeabilized for 5 min by Triton X-100 (Amresco, WA, USA). Then, Rhodamine-phalloidin (Sigma, CA, USA) and DAPI were used to stain the cells for 30 min and 10 min, respectively. Finally, the cells were observed under a fluorescence microscope. The cell's coverage area was measured by Scion image software. There were three different samples in each group, and each sample was selected randomly from five different horizons. For cell counting, the samples were carefully cleaned with PBS twice and digested using Trypsin-EDTA (TE, Gibco Invitrogen, Inc., Carlsbad, CA, USA). Cell number was counted by Cell Counting Instrument (AMQAX 1000, Thermo Fisher Scientific, Waltham, MA, USA).

The protein concentration was measured by a Bio-Rad protein analysis kit. Equivalent to polyvinylidene fluoride (PVDF) membrane (PAL). The membrane was incubated with rabbit antibody FAK, phosphorylated FAK (Tyr397) and GAPDH (Cst,1:1000 dilution) overnight at 4 °C. Then, the goat anti-rabbit antibody labeled with horseradish peroxidase was used to detect the first anti-60 min in TBST. The second binding antibody was displayed by enhanced chemiluminescence (ECL). Protein bands were quantified by TotalLabQuant of amersham in England. In the western blot experiment, after incubation for different periods (3, 6, 12, and 24 h), the protein extraction reagent was used to lyse the cells according to the protocols [45]. A Bio-Rad protein analysis kit was used to measure the protein concentration. The proteins were separated and transferred to a polyvinylidene fluoride membrane. The membranes were probed with primary antibodies overnight at 4°C, followed by incubation with horseradish peroxidase-labeled secondary antibody for 60 min and washed in TBST three times for 10 min. The proteins were measured using the Total Lab Quant (Amersham, UK).

2.6. Co-Culture Assay

To detect bacterial adhesion in the co-culture environment of bacteria and cells. 100 μL *S. aureus* suspension with the concentration of 1×10^6 CFUs/mL and 1×10^4 cells were co-cultured to each of the samples in a 24-well plate with a modified culture medium and statically incubated for different periods (3, 6, 12 and 24 h) at 37 °C. The modified culture medium included 98% regular growth medium and 2% TSB. To observe the interaction of cells and bacteria with materials in the co-culture environment, the cells and bacteria were fixed and dyed using rhodamine-phalloidin and DAPI following the above protocols, followed by stained using SYTO 9 (Invitrogen, Carlsbad, CA, USA) for 15 min and examined under a fluorescence microscope. In addition, the samples were fixed with glutaraldehyde solution for 4 h, followed by dehydrated successively with a gradient ethanol series for 10 min and dehydrated as mentioned above, and observed using SEM. To quantify bacteria adhesion for 3, 6, 12, and 24 h, the samples were carefully washed with PBS three times and sonicated and vortexed to remove adhered bacteria in 4 mL PBS. The spread plate method was used to count the adhered bacteria.

2.7. Statistical Analysis

The experiments were repeated in triplicate, and the data were presented as means ± standard deviations. The differences were analyzed using the one-way ANOVA and Student-Newman-Keuls post hoc tests and considered to be significant or highly significant if "p" values < 0.05 or 0.01, respectively.

3. Results

3.1. Sample Characterization

The surface topography of the samples is exhibited in Figure 1. SEM images of the polished Ta and Ti are illustrated in Figure 1a. The surface structure of both samples shows a flat and smooth topography. As shown in Figure 1b, characteristic peaks of Ta 4f, Ta 4d5, Ta 4s, and Ta 4p in the Ta sample were examined by XPS. The surface roughness of

Ti and Ta examined by AFM was 42.95 ± 5.13 nm and 42.87 ± 5.13 nm, respectively, in our previous paper [39]. Furthermore, there was no significant difference between Ta and Ti in the surface roughness. In addition, no significant difference was observed between polished Ti and Ta, considering the surface wettability (Figure 1c).

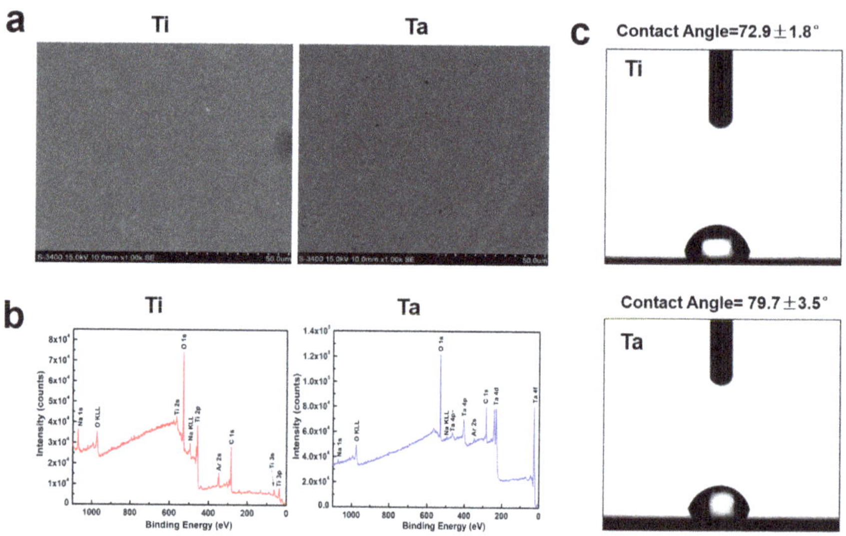

Figure 1. Sample characterization. (**a**) The surface topography of Ti and Ta was examined by FE-SEM. (**b**) XPS full spectra of Ti and Ta samples. (**c**) The water contact angle values for the polished Ti and Ta groups.

3.2. *In Vivo Antibacterial Property*

3.2.1. Radiographical Assessment

As shown in Figure 2a, the radiographic signs of obvious osteolysis, periosteal reaction and slight soft tissue swelling were observed using X-ray after 2 weeks. There were no obvious signs of osteomyelitis in Blank, *S. a*, Ti and Ta groups. Ta + *S., a* group, was demonstrated with similar radiographic signs but less than Ti + *S., a* group, by X-rays. There was no deformity but swelling soft tissue in group Ta + *S. a*. In both groups, however, there were no significant advances in new bone formation, periosteal elevation and continuous deformities.

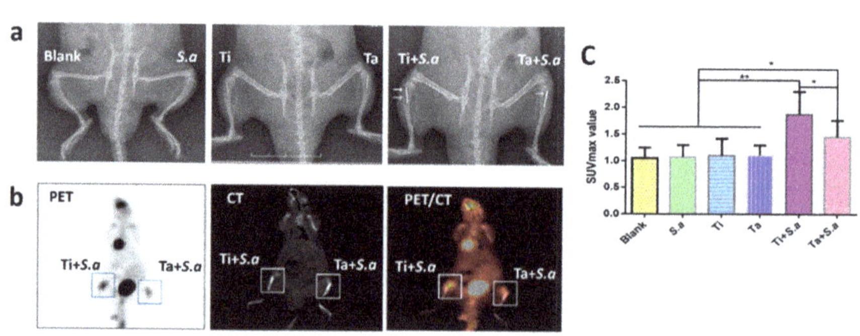

Figure 2. Imaging results and assessment. (**a**) X-rays of the tibia in lateral view. (**b**) PET/CT imaging of Ti + *S. a* and Ta+ *S.* groups. (**c**) SUVmax value in different groups. (* "*p*" < 0.05, ** "*p*" < 0.01).

3.2.2. 18F-FDG PET/CT Imaging Evaluation

PET/CT imaging showed that the uptake of 18F-FDG in Ti + *S. a* site was significantly higher than that in Ta+ *S. a* site with the mean SUVmax ratios of 1.92 (SD 0.37) and 1.43 (SD 0.34), respectively ("p" < 0.022) (Figure 2b,c).

3.2.3. CFU of Tibia and Kirschner Wires

There were no bacteria cultured from the bone powder in Blank, *S. a*, Ti and Ta groups and no bacteria can be cultured from Ti and Ta groups (Figure 3a). The CFU of the tibia and Kirschner wires from group Ti+ *S. a* were more than that of group Ta+ *S. a*. (Figure 3a–c).

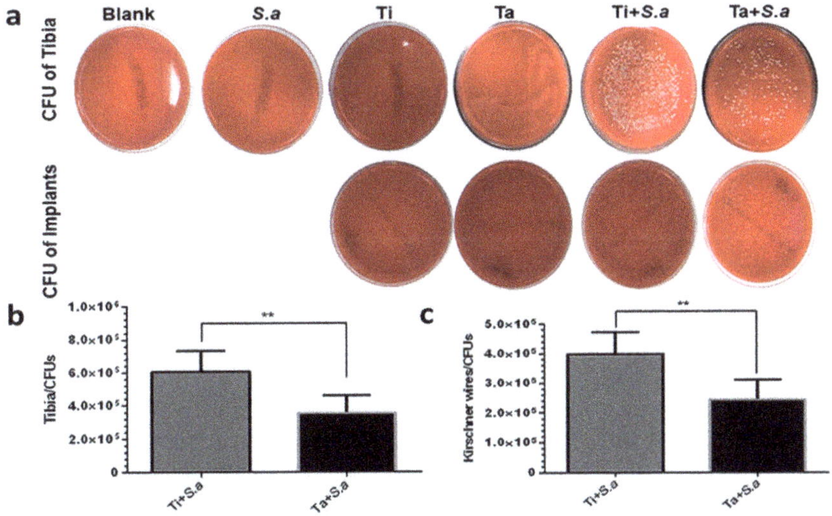

Figure 3. The antimicrobial results of different groups in implant-related tibia osteomyelitis model. (**a–c**) quantitative results of tibia powders and Kirschner wires were obtained from different groups. (** "p" < 0.01).

3.2.4. Histological Evaluation

Surrounding tissues stained with Masson's trichrome after extraction of Ti and Ta Kirschner wires (Figure 4a) showed signs of bone infection. There were plenty of fibrous tissue and inflammatory cells around without bone formation. Giemsa staining section showed that more bacteria were observed in the fibrous tissue junction and intramedullary tissue of the Ti + *S. a* group than Ta+ *S. a* group (Figure 4b). These results indicated that the severity of osteomyelitis in group Ta+ *S. a* was milder than that in group Ti + *S. a*.

3.3. *In Vitro Anti-Biofilm Property*

SEM imaging showed that bacteria were observed uniformly adhering to the surface of Ta and Ti without obvious agglomeration (Figure 5a). There was no significant difference in bacteria adhesion of polished Ta and Ti for 3, 6, 12, and 24 h (Figure 5b). CFU was no significant reduction when comparing Ti with Ta for 3, 6, 12 and 24 h, since planktonic bacteria in both materials with plate counts at > 1×10^8 CFUs/mL for 24 h (p = 0.969) (Figure 5c), There was no significant difference in the antimicrobial activity when comparing Ti with Ta.

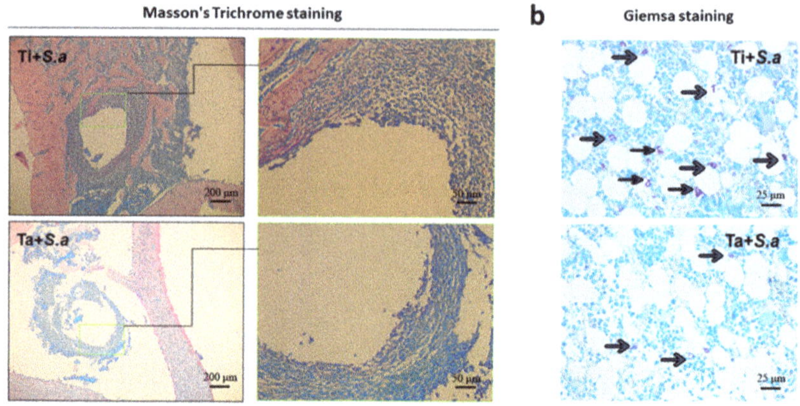

Figure 4. Histological evaluation of soft tissue of implant-related tibia osteomyelitis model. (**a**) Masson's Trichrome staining of Ti + *S. a* and Ta+ *S. a* groups. (**b**) Giemsa staining of Ti + *S. a* and Ta+ *S. a* groups and scale bar = 25 μm. (the black arrow represents bacteria).

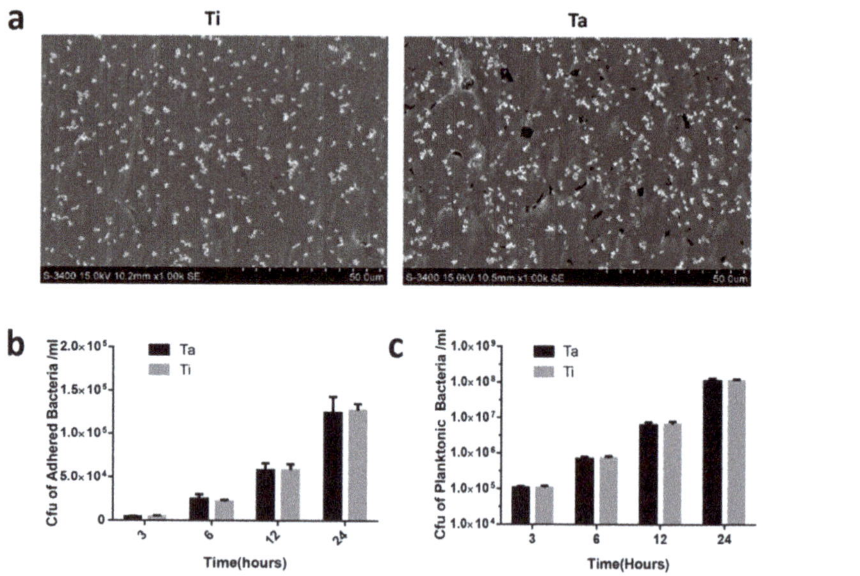

Figure 5. In vitro antibacterial assay. (**a**) SEM morphology of bacteria on Ta and Ti surface. (**b**,**c**) CFU of adhered and planktonic bacteria on Ta and Ti group at different time points.

3.4. Cell-Surface Interactions

As indicated in Figure 6a, HGF-1 numbers on Ta samples were significantly more than those on Ti after 3, 6, and 12 h incubation, but no difference for 24 h. Moreover, the above statement was also demonstrated by the results of data statistics. ("p"< 0.05) (Figure 6b). The area spread by cells on the metal ta was larger than those on Ti after incubation for 3, 6 and 12 h and also no difference for 24 h. The cell coverage area on the surface of Ta exhibited larger than that on the Ti surface at 3, 6, and 12 h (Figure 6c), The cytoskeleton and cell morphology showed more extended and multipolar spindle shape compared with those on the Ti surface at 3, 6 and 12 h. (Figure 6a–c). In addition, Western blot confirmed that the phosphorylated FAK protein level on the Ta surface was higher than that on the Ti surface at 3, 6 and 12 h, but no difference at 24 h. (Figure 6d,e).

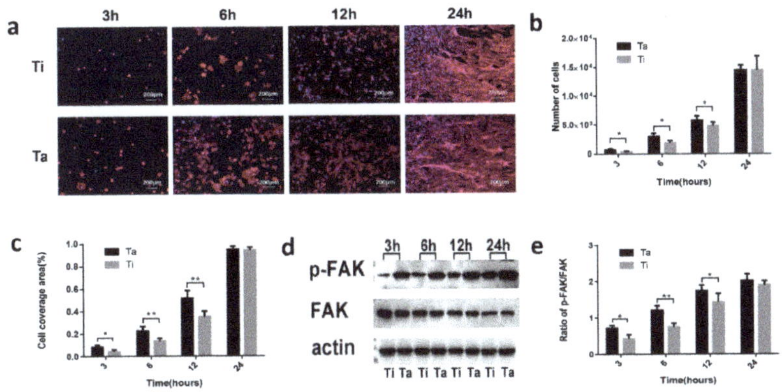

Figure 6. Cell-surface interactions and expressions of adhesion-related proteins. (**a**) Fluorescence images of cell-sample co-culture. Red represents actin, blue represents nuclei, and the scale bar = 200 μm. (**b**,**c**) The number of cells and cell coverage area on Ta and Ti samples at different time points. (**d**,**e**) Expressions of FAK, phospho-FAK proteins on Ta and Ti. (* "p" < 0.05, ** "p" < 0.01).

3.5. The Race between Bacterial and Mammalian Cells

In observation of immunofluorescence staining shown in Figure 7a, bacteria and cells were closely attached, rather than scattered respectively, on the surface of tantalum at the early stage of co-culture (within 3–6 h), moreover, as shown in Figure 7b, the results of the SEM images further indicated that the cells earlier adhered to Ta surface, then bacteria adhered to the surface of the cell rather than Ta bare surface, bacteria agglomerated on the surface of cells rather than uniformly adhered to materials shown in Figure 7a,b. After 12 h, the cells began to necrosis. After washing, they were easy to fall off. As shown in Figure 7c, at 3 and 6 h of the co-culture system, more bacteria adhered to the tantalum surface than that titanium, but there was no statistical significance. At 12 h, more bacteria were adhering to the surface of titanium than that tantalum ("p" < 0.05). After 24 h, there was no significant difference in the count of bacteria.

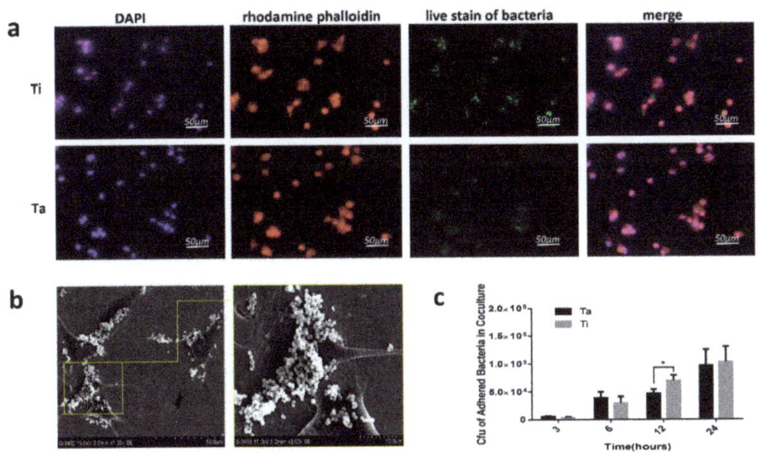

Figure 7. Cells and bacteria co-culture experiments. (**a**) Fluorescence images of cell-bacteria-sample co-culture. Red represents actin, blue represents nuclei, green represents live bacteria, and the scale bar = 50 μm. (**b**) Bacteria and cells on the Ta sample were examined by FE-SEM. (**c**) CFU adhered bacteria on Ta and Ti samples in co-culture. (* "p" < 0.05).

4. Discussion

Titanium and its derivatives have been widely used in orthopedics and other fields because of their superior stability, and a variety of modification methods have been selected to improve the osseointegration and antibacterial properties of titanium-based implants [7,46–48]. However, many of the antibacterial mechanisms play a role in killing bacteria through toxicity, including reactive oxygen species (ROS) production and the destruction of DNA [49]. While these internal implant candidates are sterilized, they will inevitably hurt the normal host tissue around the implantation site, which in turn may produce negative effects such as foreign body reactions to prevent the implants from remaining in the host and failing [50]. Therefore, the preparation of an antibacterial surface must not only achieve an effective antibacterial effect but also maintain the normal functional state of tissues and cells and achieve dynamic balance. On the other hand, the selection of the modification mode, which has superior antibacterial properties and can promote the adhesion of host cells, is of great significance in improving the ability of bone integration in implants.

Tantalum has excellent corrosion resistance, and as a component of the implant, it is beneficial to reduce the local inflammatory reaction of the implant [51]. The presence of tantalum metal gives the implant good cell biocompatibility and plays an excellent role in promoting bone regeneration [52,53]. In this study, polished tantalum was selected as the experimental group, and polished titanium as the control group (Figure 1) to exclude the influence of the surface structure of materials on the results and explore the inherent biological properties of Ta. The results also proved that pure titanium could effectively promote cell adhesion, cytoskeleton extension, and diffusion (Figure 6). The authors compared the early adhesion of fibroblasts to Ta and Ti; the results showed that tantalum with good histocompatibility had better adhesion to cells (within 12 h, especially 3 to 6 h) by rhodamine-phalloidin staining and cell counting on the material surface, however, there was no difference at 24 h because of the proliferation of fibroblasts, with enough cells covering the surface of the material (Figure 6). The attachment of cells in a complex biological environment largely depends on the specific interaction between cell surface receptors and extracellular matrix (ECM) proteins. [54–56]. The results of Western blot for signaling proteins showed that phosphorylation of FAK was stimulated by Ta surfaces at an early stage, which is similar to the results of Zhu et al. [57], except that the object of the latter is Ta -modified micro-nanostructure titanium. Therefore, the early adhesion of fibroblasts to tantalum is not only reflected in the number of cells but also in the spreading area.

In terms of the antibacterial efficiency of tantalum, tantalum exhibited excellent antibacterial properties in the form of metal ions, and the mechanisms include DNA denaturation and destruction of the signal pathway [34,58]. It can exist in the state of tantalum oxide and nitride, and both of them have been found to have certain antibacterial activity [59,60]. In addition, Zhang et al. observed that *F. nucleatum* (*F. n.*) and *Porphyromonas gingivalis* (*P. g.*) also adhere less to the tantalum coating [61]. However, Harrison et al. observed that Ta was not demonstrated to possess inherent antibacterial activity compared with Ti [37]. In vivo studies and clinical studies reported a decline in infection rates in cases of revision hip arthroplasty with porous tantalum [62,63]. while in the rabbit prosthesis model, porous tantalum cannot prevent nail infection [64]. In this study, at the early experimental stage, the differences in antimicrobial properties between Ta and Ti were found in a rat implant-related infection model (Figure 2). It is worth mentioning that the X-ray findings in vivo are not very obvious because this study mainly focuses on the early manifestations of osteomyelitis models, mainly local low-density images in the bone marrow cavity. There was no obvious periosteal reaction, bone capsule formation, and dead bone formation, and it was to evaluate the severity of early osteomyelitis more objectively. Therefore, PET/CT was used to evaluate the severity of osteomyelitis in the Ta-Kirschner wires group, which showed that the uptake of 18F-FDG was lower in Ta- the Kirschner wires group. The results of histological examination and microbiological cultures also showed that the number of bone destruction, inflammatory cell infiltration, and bacteria in the Ta group were less than those in the Ti group (Figures 3 and 4). To further explore the potential mechanisms behind the inconsistent infection degree of tantalum and titanium in vivo, the authors designed and carried out in vitro experiments. In the subsequent experiments, it was observed that pure tantalum had no significant inhibitory effect on *S. aureus* in vitro compared with titanium, indicating that tantalum does not exhibit intrinsic antimicrobial properties (Figure 5). Conditional pathogens on implant surfaces, operating rooms, surgical equipment, surgeons, patients themselves, contaminated disinfectants, and others can reach the implant surface through direct contact, hematogenous dissemination, or endogenous displacement. The characteristics of implant materials and implant site are important factors affecting the occurrence of infection [65]. The study has confirmed that bacterial colonization and adhesion are the initial factors of orthopedic implant infection. After implantation, proteins (such as fibrinogen, bole in, compliment, fibronectin, etc.) cells (such as fibroblasts, neutrophils, mesenchymal stem cells, etc.) from blood or tissue fluid quickly adhere to the implant surface to form a temporary surface matrix, at the same time, Local coagulation reactions and complement systems are activated, then immune cells are activated [66]. Therefore, in this study, the authors established an in vitro model of co-culture with bacteria and cells to simulate the microenvironment in vivo to some extent (Figure 7). Time points including 3, 6, 12, and 24 h were selected to simulate the early adhesion in vivo and the characteristics of the interaction discussed between bacteria and cells on the tantalum surface. To simulate peri-operative infection, in the co-culture experiment, bacteria were permitted to adhere for 2 h before cell adhesion. In co-culture experiments, cells are planned to be stained for different periods (3, 6, 12, and 24 h). However, it was found that after the co-culture of bacteria and cells for more than 6 h, cells began to damage or even die. Cells will be lost in the process of washing and cannot be evaluated objectively. Surprisingly, early-stage live bacteria staining demonstrated that the bacteria were extremely close to the cells. Moreover, SEM observation more intuitively revealed that bacteria were more likely to adhere to the cell surface and reunite in the co-cultured instead of the surface of Ta (Figure 7a,b). The study has confirmed that bacterial colonization and adhesion are the initial factors of orthopedic implant infection. Various factors, such as polarity, van der Waals force, and hydrophobicity, together regulate the initial bacterial adhesion [67]. This means that the early adhesion of bacteria or cells to the implant surface has a great impact on the occurrence of implant infection. More bacteria adhered to the Ta because more cells existed within hours. As the experiment continued, at 12 h, the cells began to exfoliate, and cell exfoliation also took away the bacteria attached to them. The bacteria on the surface of cells were taken away after washing. Considering that there is a powerful immune system in the normal body environment, including a variety of immune cells that will destruct bacteria [68], which cannot be reflected in this co-culture model, therefore, at 24 h, the cells were almost exhausted and could not be compared, and bacteria have completely taken over the position.

Previously, the theory of "the race for the surface" has been put forward and embraced by some researchers [69–72]; this theory properly explains the phenomenon of competitive colonization of bacteria and cells on the implant surface. Moreover, the mainstream view was that host cell attachment helps to prevent infection (reduce the incidence of infection) [69,71,73]. On the basis of this theory, foreign bodies could trigger the competition between host cells and bacteria to colonize the implant surface. [40–42]. In other words, when the host cells win the competition, the cells will occupy the surface, thus inhibiting the adhesion and colonization of bacteria. On the contrary, bacteria will invade the surface of the implant to form biofilms, and the host cells will be suppressed by virulent bacterial substances, which eventually lead to infection. The results of this competition greatly determine the infection severity and inhibit organizational integration. Phenomena that bacteria adhere to the cell surface preferentially rather than directly on the material surface were observed in these experiments but have not been paid special attention to. The authors hold the opinions that these phenomena are extremely surprising and interesting and have remarkable significance. In the early stage, tantalum has the superior capability to promote cell adhesion, making cells preferentially adhere to the surface of Ta, and then bacteria are more likely to adhere to the cells that have already adhered to the surface of Ta rather than the exposed surface of Ta. Over time, the cells die and fall off, causing the bacteria attached to their surfaces to fall off as well. As a result, the chance of bacteria directly adhering to Ta is greatly reduced, which is helpful in reducing the production of biofilm and the occurrence of implant infection. In addition, bacteria adhered to the cell surface are more likely to be involved by antibiotics and immune cells regardless of whether they fall off or not [74]. Considering the different antibacterial properties of tantalum in vivo and in vitro, some scholars have also given relevant explanations for the phenomenon that tantalum has a certain capability to resist infection in the body. Schildhauer et al. reported that porous tantalum effectively activates the immune system represented by leukocytes, creates a microenvironment conducive to killing bacteria, and may enhance the defense ability of local hosts [75]. In addition, Yang et al. observed that the Ta nanomembrane can significantly improve the ability of neutrophils to phagocytize bacteria and promote macrophages to release pro-inflammatory cytokines. Thus, the host can exterminate bacteria by regulating the immune environment [76]. Combined with in vitro experimental results, the bacteria peeled off with the cells exfoliation will be attacked by many immune cells activated in vivo microenvironment, and tantalum possesses dual effects on eliminating bacteria in addition to its immune initiating ability, to achieve the role of preventing implant infection.

This study has some limitations, this study found the above surprising phenomenon through in vitro immunostaining and SEM observation, but more in vitro, ex vivo, and in vivo experiments are needed to further explore this phenomenon. Moreover, the authors chose the most common pathogen of osteomyelitis (*S. aureus*) as the research object. In future work, the authors will further study the interaction between Gram-negative bacteria and cells on the surface of Ta. In addition, the authors will continue to explore and improve the mechanism behind this phenomenon in the following work. This article is more inclined to put forward the concept, appealing for more attention to focus on the appearance.

5. Conclusions

The original purpose of the fabrication and development of antibacterial implant materials is to seek materials with both excellent antibacterial properties and host tissue protection. The results of this study show that tantalum has good biological compatibility and contributes to cell adhesion and proliferation. In addition, although tantalum does not have the inherent antibacterial ability, the discovery that bacteria tend to adhere to the cell surface rather than the tantalum surface further explains the reason for the difference between tantalum antibacterial results in vivo and in vitro, which provides a certain basis for the fabrication of tantalum-based antibacterial implants in the future. In addition, this finding also brings a new perspective to the manufacture of orthopedic antibacterial implant materials. On the one hand, more focus needs to be paid to the interaction between the host, bacteria, and implants, instead of directly targeting bacteria, on the other hand, biomaterials with good cell adhesion ability deserve more attention, because they may exert their ability to resist bacterial infection through the way found in this study.

Author Contributions: Conceptualization, H.Q. and H.C.; methodology, X.C., Y.B. and M.H.; software, Y.B. and M.H.; validation, H.Q., H.C. and X.C.; resources, H.Q. and H.C.; data curation, X.C., Y.B. and M.H.; writing—original draft preparation, X.C.; writing—review and editing, H.Q. and H.C.; supervision, Y.B. and M.H.; funding acquisition, H.Q. and H.C. All authors have read and agreed to the published version of the manuscript.

Funding: This work was jointly supported by grants from the National Natural Science Foundation of China (31771022, 32271399, and 31870945), and the Natural Science Foundation of Shanghai (21ZR1415700).

Institutional Review Board Statement: The animal study protocol was approved by the Institutional Review Board of Shanghai Sixth People's Hospital (protocol code: DWLL2018-0339 and date of approval: 3 December 2018).

Informed Consent Statement: Not applicable.

Data Availability Statement: Not applicable.

Conflicts of Interest: The authors declare no conflict of interest.

References

1. Grainger, D.W.; van der Mei, H.C.; Jutte, P.C.; van den Dungen, J.J.; Schultz, M.J.; van der Laan, B.F.; Zaat, S.A.; Busscher, H.J. Critical factors in the translation of improved antimicrobial strategies for medical implants and devices. *Biomaterials* **2013**, *34*, 9237–9243. [CrossRef] [PubMed]
2. Trampuz, A.; Widmer, A.F. Infections associated with orthopedic implants. *Curr. Opin. Infect. Dis.* **2006**, *19*, 349–356. [CrossRef] [PubMed]
3. van Oosten, M.; Schäfer, T.; Gazendam, J.A.; Ohlsen, K.; Tsompanidou, E.; de Goffau, M.C.; Harmsen, H.J.; Crane, L.M.; Lim, E.; Francis, K.P.; et al. Real-time in vivo imaging of invasive- and biomaterial-associated bacterial infections using fluorescently labelled vancomycin. *Nat. Commun.* **2013**, *4*, 2584. [CrossRef] [PubMed]
4. Theuretzbacher, U.; Outterson, K.; Engel, A.; Karlén, A. The global preclinical antibacterial pipeline. *Nat. Rev. Microbiol.* **2020**, *18*, 275–285. [CrossRef] [PubMed]
5. Baker, S.J.; Payne, D.J.; Rappuoli, R.; De Gregorio, E. Technologies to address antimicrobial resistance. *Proc. Natl. Acad. Sci. USA* **2018**, *115*, 12887–12895. [CrossRef] [PubMed]
6. Nazarov, D.; Ezhov, I.; Yudintceva, N.; Shevtsov, M.; Rudakova, A.; Kalganov, V.; Tolmachev, V.; Zharova, Y.; Lutakov, O.; Kraeva, L.; et al. Antibacterial and Osteogenic Properties of Ag Nanoparticles and Ag/TiO(2) Nanostructures Prepared by Atomic Layer Deposition. *J. Funct. Biomater.* **2022**, *13*, 62. [CrossRef]
7. Zhang, C.; Li, X.; Xiao, D.; Zhao, Q.; Chen, S.; Yang, F.; Liu, J.; Duan, K. Cu(2+) Release from Polylactic Acid Coating on Titanium Reduces Bone Implant-Related Infection. *J. Funct. Biomater.* **2022**, *13*, 78. [CrossRef]
8. Siddiqui, H.; Qureshi, M.S.; Haque, F.Z. Biosynthesis of Flower-Shaped CuO Nanostructures and Their Photocatalytic and Antibacterial Activities. *Nano-Micro Lett.* **2020**, *12*, 29. [CrossRef]
9. Jia, B.; Zhang, Z.; Zhuang, Y.; Yang, H.; Han, Y.; Wu, Q.; Jia, X.; Yin, Y.; Qu, X.; Zheng, Y.; et al. High-strength biodegradable zinc alloy implants with antibacterial and osteogenic properties for the treatment of MRSA-induced rat osteomyelitis. *Biomaterials* **2022**, *287*, 121663. [CrossRef]
10. Croes, M.; Bakhshandeh, S.; van Hengel, I.A.J.; Lietaert, K.; van Kessel, K.P.M.; Pouran, B.; van der Wal, B.C.H.; Vogely, H.C.; Van Hecke, W.; Fluit, A.C.; et al. Antibacterial and immunogenic behavior of silver coatings on additively manufactured porous titanium. *Acta Biomater.* **2018**, *81*, 315–327. [CrossRef]
11. Huang, M.; Ye, K.; Hu, T.; Liu, K.; You, M.; Wang, L.; Qin, H. Silver Nanoparticles Attenuate the Antimicrobial Activity of the Innate Immune System by Inhibiting Neutrophil-Mediated Phagocytosis and Reactive Oxygen Species Production. *Int. J. Nanomed.* **2021**, *16*, 1345–1360. [CrossRef] [PubMed]
12. Zhou, J.; Xu, N.S.; Wang, Z.L. Dissolving Behavior and Stability of ZnO Wires in Biofluids: A Study on Biodegradability and Biocompatibility of ZnO Nanostructures. *Adv. Mater.* **2006**, *18*, 2432–2435. [CrossRef]
13. Cady, N.C.; Behnke, J.; Strickland, A.D. Copper-Based Nanostructured Coatings on Natural Cellulose: Nanocomposites Exhibiting Rapid and Efficient Inhibition of a Multi-Drug Resistant Wound Pathogen, A. baumannii, and Mammalian Cell Biocompatibility In Vitro. *Adv. Funct. Mater.* **2011**, *21*, 2506–2514. [CrossRef]
14. Cao, H.; Qin, H.; Li, Y.; Jandt, K.D. The Action-Networks of Nanosilver: Bridging the Gap between Material and Biology. *Adv. Health Mater.* **2021**, *10*, e2100619. [CrossRef] [PubMed]
15. Celesti, C.; Gervasi, T.; Cicero, N.; Giofrè, S.V.; Espro, C.; Piperopoulos, E.; Gabriele, B.; Mancuso, R.; Lo Vecchio, G.; Iannazzo, D. Titanium Surface Modification for Implantable Medical Devices with Anti-Bacterial Adhesion Properties. *Materials* **2022**, *15*, 3283. [CrossRef] [PubMed]
16. Yuan, Y.; Luo, R.; Ren, J.; Zhang, L.; Jiang, Y.; He, Z. Design of a new Ti-Mo-Cu alloy with excellent mechanical and antibacterial properties as implant materials. *Mater. Lett.* **2022**, *306*, 130875. [CrossRef]
17. Zhang, S.; Yu, Y.; Wang, H.; Ren, L.; Yang, K. Study on mechanical behavior of Cu-bearing antibacterial titanium alloy implant. *J. Mech. Behav. Biomed. Mater.* **2022**, *125*, 104926. [CrossRef]
18. Jang, T.-S.; Jung, H.-D.; Kim, S.; Moon, B.-S.; Baek, J.; Park, C.; Song, J.; Kim, H.-E. Multiscale porous titanium surfaces via a two-step etching process for improved mechanical and biological performance. *Biomed. Mater.* **2017**, *12*, 025008. [CrossRef]
19. Lee, H.; Jung, H.-D.; Kang, M.-H.; Song, J.; Kim, H.-E.; Jang, T.-S. Effect of HF/HNO3-treatment on the porous structure and cell penetrability of titanium (Ti) scaffold. *Mater. Des.* **2018**, *145*, 65–73. [CrossRef]
20. Zhao, F.; Zhao, Y.; Liu, Y.; Chang, X.; Chen, C.; Zhao, Y. Cellular uptake, intracellular trafficking, and cytotoxicity of nanomaterials. *Small* **2011**, *7*, 1322–1337. [CrossRef]

21. Cao, H.; Dauben, T.J.; Helbing, C.; Jia, Z.; Zhang, Y.; Huang, M.; Müller, L.; Gu, S.; Zhang, X.; Qin, H.; et al. The antimicrobial effect of calcium-doped titanium is activated by fibrinogen adsorption. *Mater. Horiz.* **2022**, *9*, 1962–1968. [CrossRef] [PubMed]
22. Cao, H.; Qiao, S.; Qin, H.; Jandt, K.D. Antibacterial Designs for Implantable Medical Devices: Evolutions and Challenges. *J. Funct. Biomater.* **2022**, *13*, 86. [CrossRef] [PubMed]
23. Kim, S.; Park, C.; Cheon, K.-H.; Jung, H.-D.; Song, J.; Kim, H.-E.; Jang, T.-S. Antibacterial and bioactive properties of stabilized silver on titanium with a nanostructured surface for dental applications. *Appl. Surf. Sci.* **2018**, *451*, 232–240. [CrossRef]
24. Qian, H.; Lei, T.; Lei, P.; Hu, Y. Additively Manufactured Tantalum Implants for Repairing Bone Defects: A Systematic Review. *Tissue Eng. Part B Rev.* **2021**, *27*, 166–180. [CrossRef] [PubMed]
25. Wang, X.; Ning, B.; Pei, X. Tantalum and its derivatives in orthopedic and dental implants: Osteogenesis and antibacterial properties. *Colloids Surf. B Biointerfaces* **2021**, *208*, 112055. [CrossRef]
26. Cheon, K.-H.; Park, C.; Kang, M.-H.; Kang, I.-G.; Lee, M.-K.; Lee, H.; Kim, H.-E.; Jung, H.-D.; Jang, T.-S. Construction of tantalum/poly(ether imide) coatings on magnesium implants with both corrosion protection and osseointegration properties. *Bioact. Mater.* **2021**, *6*, 1189–1200. [CrossRef]
27. Park, C.; Seong, Y.-J.; Kang, I.-G.; Song, E.-H.; Lee, H.; Kim, J.; Jung, H.-D.; Kim, H.-E.; Jang, T.-S. Enhanced Osseointegration Ability of Poly(lactic acid) via Tantalum Sputtering-Based Plasma Immersion Ion Implantation. *ACS Appl. Mater. Interfaces* **2019**, *11*, 10492–10504. [CrossRef]
28. Lee, M.-K.; Lee, H.; Park, C.; Kang, I.-G.; Kim, J.; Kim, H.-E.; Jung, H.-D.; Jang, T.-S. Accelerated biodegradation of iron-based implants via tantalum-implanted surface nanostructures. *Bioact. Mater.* **2022**, *9*, 239–250. [CrossRef]
29. Huang, G.; Pan, S.T.; Qiu, J.X. The Clinical Application of Porous Tantalum and Its New Development for Bone Tissue Engineering. *Materials* **2021**, *14*, 2647. [CrossRef]
30. Wang, N.; Li, H.; Wang, J.; Chen, S.; Ma, Y.; Zhang, Z. Study on the anticorrosion, biocompatibility, and osteoinductivity of tantalum decorated with tantalum oxide nanotube array films. *ACS Appl. Mater. Interfaces* **2012**, *4*, 4516–4523. [CrossRef]
31. Minagar, S.; Berndt, C.C.; Wang, J.; Ivanova, E.; Wen, C. A review of the application of anodization for the fabrication of nanotubes on metal implant surfaces. *Acta Biomater.* **2012**, *8*, 2875–2888. [CrossRef] [PubMed]
32. Konan, S.; Duncan, C.P.; Masri, B.A.; Garbuz, D.S. Porous tantalum uncemented acetabular components in revision total hip arthroplasty: A minimum ten-year clinical, radiological and quality of life outcome study. *Bone Jt. J.* **2016**, *98-b*, 767–771. [CrossRef] [PubMed]
33. Garbuz, D.S.; Hu, Y.; Kim, W.Y.; Duan, K.; Masri, B.A.; Oxland, T.R.; Burt, H.; Wang, R.; Duncan, C.P. Enhanced gap filling and osteoconduction associated with alendronate-calcium phosphate-coated porous tantalum. *J. Bone Jt. Surg. Am.* **2008**, *90*, 1090–1100. [CrossRef] [PubMed]
34. Lemire, J.A.; Harrison, J.J.; Turner, R.J. Antimicrobial activity of metals: Mechanisms, molecular targets and applications. *Nat. Rev. Microbiol.* **2013**, *11*, 371–384. [CrossRef] [PubMed]
35. Schildhauer, T.A.; Robie, B.; Muhr, G.; Köller, M. Bacterial adherence to tantalum versus commonly used orthopedic metallic implant materials. *J. Orthop. Trauma* **2006**, *20*, 476–484. [CrossRef] [PubMed]
36. Subramani, K.; Jung, R.E.; Molenberg, A.; Hammerle, C.H. Biofilm on dental implants: A review of the literature. *Int. J. Oral Maxillofac. Implant.* **2009**, *24*, 616–626.
37. Harrison, P.L.; Harrison, T.; Stockley, I.; Smith, T.J. Does tantalum exhibit any intrinsic antimicrobial or antibiofilm properties? *Bone Jt. J.* **2017**, *99-b*, 1153–1156. [CrossRef] [PubMed]
38. Wahl, P.; Sprecher, C.M.; Brüning, C.; Meier, C.; Milz, S.; Gautier, E.; Fintan Moriarty, T. Successful bony integration of a porous tantalum implant despite longlasting and ongoing infection: Histologic workup of an explanted shoulder prosthesis. *J. Biomed. Mater. Res. B Appl. Biomater.* **2018**, *106*, 2924–2931. [CrossRef] [PubMed]
39. Lu, M.; Zhuang, X.; Tang, K.; Wu, P.; Guo, X.; Yin, L.; Cao, H.; Zou, D. Intrinsic Surface Effects of Tantalum and Titanium on Integrin $\alpha5\beta1$/ ERK1/2 Pathway-Mediated Osteogenic Differentiation in Rat Bone Mesenchymal Stromal Cells. *Cell. Physiol. Biochem.* **2018**, *51*, 589–609. [CrossRef]
40. Qin, H.; Cao, H.; Zhao, Y.; Jin, G.; Cheng, M.; Wang, J.; Jiang, Y.; An, Z.; Zhang, X.; Liu, X. Antimicrobial and osteogenic properties of silver-ion-implanted stainless steel. *ACS Appl Mater. Interfaces* **2015**, *7*, 10785–10794. [CrossRef]
41. Lucke, M.; Schmidmaier, G.; Sadoni, S.; Wildemann, B.; Schiller, R.; Haas, N.P.; Raschke, M. Gentamicin coating of metallic implants reduces implant-related osteomyelitis in rats. *Bone* **2003**, *32*, 521–531. [CrossRef]
42. Chen, X.; Tsukayama, D.T.; Kidder, L.S.; Bourgeault, C.A.; Schmidt, A.H.; Lew, W.D. Characterization of a chronic infection in an internally-stabilized segmental defect in the rat femur. *J. Orthop. Res.* **2005**, *23*, 816–823. [CrossRef] [PubMed]
43. Qin, H.; Cao, H.; Zhao, Y.; Zhu, C.; Cheng, T.; Wang, Q.; Peng, X.; Cheng, M.; Wang, J.; Jin, G.; et al. In vitro and in vivo anti-biofilm effects of silver nanoparticles immobilized on titanium. *Biomaterials* **2014**, *35*, 9114–9125. [CrossRef] [PubMed]
44. Tan, H.; Peng, Z.; Li, Q.; Xu, X.; Guo, S.; Tang, T. The use of quaternised chitosan-loaded PMMA to inhibit biofilm formation and downregulate the virulence-associated gene expression of antibiotic-resistant staphylococcus. *Biomaterials* **2012**, *33*, 365–377. [CrossRef]
45. Wang, X.; Lu, T.; Wen, J.; Xu, L.; Zeng, D.; Wu, Q.; Cao, L.; Lin, S.; Liu, X.; Jiang, X. Selective responses of human gingival fibroblasts and bacteria on carbon fiber reinforced polyetheretherketone with multilevel nanostructured TiO_2. *Biomaterials* **2016**, *83*, 207–218. [CrossRef]

46. Shimabukuro, M.; Hiji, A.; Manaka, T.; Nozaki, K.; Chen, P.; Ashida, M.; Tsutsumi, Y.; Nagai, A.; Hanawa, T. Time-Transient Effects of Silver and Copper in the Porous Titanium Dioxide Layer on Antibacterial Properties. *J. Funct. Biomater.* **2020**, *11*, 44. [CrossRef]
47. Vidakis, N.; Petousis, M.; Mountakis, N.; Korlos, A.; Papadakis, V.; Moutsopoulou, A. Trilateral Multi-Functional Polyamide 12 Nanocomposites with Binary Inclusions for Medical Grade Material Extrusion 3D Printing: The Effect of Titanium Nitride in Mechanical Reinforcement and Copper/Cuprous Oxide as Antibacterial Agents. *J. Funct. Biomater.* **2022**, *13*, 115. [CrossRef]
48. Kim, J.; Lee, H.; Jang, T.-S.; Kim, D.; Yoon, C.-B.; Han, G.; Kim, H.-E.; Jung, H.-D. Characterization of Titanium Surface Modification Strategies for Osseointegration Enhancement. *Metals* **2021**, *11*, 618. [CrossRef]
49. Hajipour, M.J.; Fromm, K.M.; Ashkarran, A.A.; Jimenez de Aberasturi, D.; de Larramendi, I.R.; Rojo, T.; Serpooshan, V.; Parak, W.J.; Mahmoudi, M. Antibacterial properties of nanoparticles. *Trends Biotechnol.* **2012**, *30*, 499–511. [CrossRef]
50. Campoccia, D.; Montanaro, L.; Arciola, C.R. A review of the biomaterials technologies for infection-resistant surfaces. *Biomaterials* **2013**, *34*, 8533–8554. [CrossRef]
51. Zhang, L.; Haddouti, E.M.; Beckert, H.; Biehl, R.; Pariyar, S.; Rüwald, J.M.; Li, X.; Jaenisch, M.; Burger, C.; Wirtz, D.C.; et al. Investigation of Cytotoxicity, Oxidative Stress, and Inflammatory Responses of Tantalum Nanoparticles in THP-1-Derived Macrophages. *Mediat. Inflamm.* **2020**, *2020*, 3824593. [CrossRef] [PubMed]
52. Hu, X.; Mei, S.; Wang, F.; Qian, J.; Xie, D.; Zhao, J.; Yang, L.; Wu, Z.; Wei, J. Implantable PEKK/tantalum microparticles composite with improved surface performances for regulating cell behaviors, promoting bone formation and osseointegration. *Bioact. Mater.* **2021**, *6*, 928–940. [CrossRef] [PubMed]
53. Hwang, C.; Park, S.; Kang, I.G.; Kim, H.E.; Han, C.M. Tantalum-coated polylactic acid fibrous membranes for guided bone regeneration. *Mater. Sci. Eng. C Mater. Biol. Appl.* **2020**, *115*, 111112. [CrossRef] [PubMed]
54. Dewez, J.L.; Doren, A.; Schneider, Y.J.; Rouxhet, P.G. Competitive adsorption of proteins: Key of the relationship between substratum surface properties and adhesion of epithelial cells. *Biomaterials* **1999**, *20*, 547–559. [CrossRef]
55. Snyder, K.L.; Holmes, H.R.; McCarthy, C.; Rajachar, R.M. Bioactive vapor deposited calcium-phosphate silica sol-gel particles for directing osteoblast behavior. *J. Biomed. Mater. Res. A* **2016**, *104*, 2135–2148. [CrossRef]
56. Saldarriaga Fernández, I.C.; Busscher, H.J.; Metzger, S.W.; Grainger, D.W.; van der Mei, H.C. Competitive time- and density-dependent adhesion of staphylococci and osteoblasts on crosslinked poly(ethylene glycol)-based polymer coatings in co-culture flow chambers. *Biomaterials* **2011**, *32*, 979–984. [CrossRef]
57. Zhu, Y.; Gu, Y.; Qiao, S.; Zhou, L.; Shi, J.; Lai, H. Bacterial and mammalian cells adhesion to tantalum-decorated micro-/nano-structured titanium. *J. Biomed. Mater. Res. A* **2017**, *105*, 871–878. [CrossRef]
58. Kolmas, J.; Groszyk, E.; Kwiatkowska-Różycka, D. Substituted hydroxyapatites with antibacterial properties. *Biomed. Res. Int.* **2014**, *2014*, 178123. [CrossRef]
59. Chang, Y.-Y.; Huang, H.-L.; Chen, H.-J.; Lai, C.-H.; Wen, C.-Y. Antibacterial properties and cytocompatibility of tantalum oxide coatings. *Surf. Coat. Technol.* **2014**, *259*, 193–198. [CrossRef]
60. Cristea, D.; Cunha, L.; Gabor, C.; Ghiuta, I.; Croitoru, C.; Marin, A.; Velicu, L.; Besleaga, A.; Vasile, B. Tantalum Oxynitride Thin Films: Assessment of the Photocatalytic Efficiency and Antimicrobial Capacity. *Nanomaterials* **2019**, *9*, 476. [CrossRef]
61. Zhang, X.M.; Li, Y.; Gu, Y.X.; Zhang, C.N.; Lai, H.C.; Shi, J.Y. Ta-Coated Titanium Surface With Superior Bacteriostasis And Osseointegration. *Int. J. Nanomed.* **2019**, *14*, 8693–8706. [CrossRef] [PubMed]
62. Jafari, S.M.; Bender, B.; Coyle, C.; Parvizi, J.; Sharkey, P.F.; Hozack, W.J. Do tantalum and titanium cups show similar results in revision hip arthroplasty? *Clin. Orthop. Relat. Res.* **2010**, *468*, 459–465. [CrossRef] [PubMed]
63. Tokarski, A.T.; Novack, T.A.; Parvizi, J. Is tantalum protective against infection in revision total hip arthroplasty? *Bone Jt. J.* **2015**, *97-b*, 45–49. [CrossRef] [PubMed]
64. Chou, T.G.; Petti, C.A.; Szakacs, J.; Bloebaum, R.D. Evaluating antimicrobials and implant materials for infection prevention around transcutaneous osseointegrated implants in a rabbit model. *J. Biomed. Mater. Res. A* **2010**, *92*, 942–952. [CrossRef]
65. Hickok, N.J.; Shapiro, I.M.; Chen, A.F. The Impact of Incorporating Antimicrobials into Implant Surfaces. *J. Dent. Res.* **2018**, *97*, 14–22. [CrossRef]
66. Franz, S.; Rammelt, S.; Scharnweber, D.; Simon, J.C. Immune responses to implants—A review of the implications for the design of immunomodulatory biomaterials. *Biomaterials* **2011**, *32*, 6692–6709. [CrossRef]
67. Ridenhour, B.J.; Metzger, G.A.; France, M.; Gliniewicz, K.; Millstein, J.; Forney, L.J.; Top, E.M. Persistence of antibiotic resistance plasmids in bacterial biofilms. *Evol. Appl.* **2017**, *10*, 640–647. [CrossRef]
68. Diacovich, L.; Gorvel, J.P. Bacterial manipulation of innate immunity to promote infection. *Nat. Rev. Microbiol.* **2010**, *8*, 117–128. [CrossRef]
69. Gristina, A.G. Biomaterial-centered infection: Microbial adhesion versus tissue integration. *Science* **1987**, *237*, 1588–1595. [CrossRef]
70. Subbiahdoss, G.; Kuijer, R.; Grijpma, D.W.; van der Mei, H.C.; Busscher, H.J. Microbial biofilm growth vs. tissue integration: "The race for the surface" experimentally studied. *Acta Biomater.* **2009**, *5*, 1399–1404. [CrossRef]
71. Pham, V.T.; Truong, V.K.; Orlowska, A.; Ghanaati, S.; Barbeck, M.; Booms, P.; Fulcher, A.J.; Bhadra, C.M.; Buividas, R.; Baulin, V.; et al. "Race for the Surface": Eukaryotic Cells Can Win. *ACS Appl. Mater. Interfaces* **2016**, *8*, 22025–22031. [CrossRef] [PubMed]

72. Pérez-Tanoira, R.; Han, X.; Soininen, A.; Aarnisalo, A.A.; Tiainen, V.M.; Eklund, K.K.; Esteban, J.; Kinnari, T.J. Competitive colonization of prosthetic surfaces by staphylococcus aureus and human cells. *J. Biomed. Mater. Res. A* **2017**, *105*, 62–72. [CrossRef] [PubMed]
73. Busscher, H.J.; van der Mei, H.C.; Subbiahdoss, G.; Jutte, P.C.; van den Dungen, J.J.; Zaat, S.A.; Schultz, M.J.; Grainger, D.W. Biomaterial-associated infection: Locating the finish line in the race for the surface. *Sci. Transl. Med.* **2012**, *4*, 153rv110. [CrossRef] [PubMed]
74. Ribeiro, M.; Monteiro, F.J.; Ferraz, M.P. Infection of orthopedic implants with emphasis on bacterial adhesion process and techniques used in studying bacterial-material interactions. *Biomatter* **2012**, *2*, 176–194. [CrossRef] [PubMed]
75. Schildhauer, T.A.; Peter, E.; Muhr, G.; Köller, M. Activation of human leukocytes on tantalum trabecular metal in comparison to commonly used orthopedic metal implant materials. *J. Biomed. Mater. Res. A* **2009**, *88*, 332–341. [CrossRef] [PubMed]
76. Yang, C.; Li, J.; Zhu, C.; Zhang, Q.; Yu, J.; Wang, J.; Wang, Q.; Tang, J.; Zhou, H.; Shen, H. Advanced antibacterial activity of biocompatible tantalum nanofilm via enhanced local innate immunity. *Acta Biomater.* **2019**, *89*, 403–418. [CrossRef]

Journal of
Functional
Biomaterials

MDPI

Article

Enhanced Antibacterial Effect on Zirconia Implant Abutment by Silver Linear-Beam Ion Implantation

Yang Yang [1,2,†], Mingyue Liu [2,3,†], Zhen Yang [1,2], Wei-Shao Lin [4], Li Chen [1,2,*] and Jianguo Tan [1,2]

1 Department of Prosthodontics, Peking University School, Hospital of Stomatology, Beijing 100081, China
2 National Center of Stomatology, National Clinical Research Center for Oral Diseases, National Engineering Research Center of Oral Biomaterials and Digital Medical Devices, Beijing Key Laboratory of Digital Stomatology, Research Center of Engineering and Technology for Computerized Dentistry Ministry of Health, NMPA Key Laboratory for Dental Materials, No. 22, Zhongguancun South Avenue, Haidian District, Beijing 100081, China
3 First Clinical Division, Peking University School, Hospital of Stomatology, Beijing 100081, China
4 Department of Prosthodontics, Indiana University School of Dentistry, Indianapolis, IN 46202, USA
* Correspondence: chenli@pkuss.bjmu.edu.cn
† These authors contributed equally to this work.

Abstract: Peri-implant lesions, such as peri-implant mucositis and peri-implantitis, are bacterial-derived diseases that happen around dental implants, compromising the long-term stability and esthetics of implant restoration. Here, we report a surface-modification method on zirconia implant abutment using silver linear-beam ion implantation to reduce the bacterial growth around the implant site, thereby decreasing the prevalence of peri-implant lesions. The surface characteristics of zirconia after ion implantation was evaluated using energy dispersive spectroscopy, X-ray photoelectron spectroscopy, and a contact-angle device. The antibacterial properties of implanted zirconia were evaluated using *Streptococcus mutans* and *Porphyromonas gingivalis*. The biocompatibility of the material surface was evaluated using human gingival fibroblasts. Our study shows that the zirconia surface was successfully modified with silver nanoparticles by using the ion-implantation method. The surface modification remained stable, and the silver-ion elution was below 1 ppm after one-month of storage. The modified surface can effectively eliminate bacterial growth, while the normal gingiva's cell growth is not interfered with. The results of the study demonstrate that a silver-ion-implanted zirconia surface possesses good antibacterial properties and good biocompatibility. The surface modification using silver-ion implantation is a promising method for future usage.

Keywords: zirconia; silver; ion implantation; peri-implant lesions; implant interface

Citation: Yang, Y.; Liu, M.; Yang, Z.; Lin, W.-S.; Chen, L.; Tan, J. Enhanced Antibacterial Effect on Zirconia Implant Abutment by Silver Linear-Beam Ion Implantation. *J. Funct. Biomater.* **2023**, *14*, 46. https://doi.org/10.3390/jfb14010046

Academic Editor: Daniele Botticelli

Received: 26 November 2022
Revised: 30 December 2022
Accepted: 11 January 2023
Published: 13 January 2023

Copyright: © 2023 by the authors. Licensee MDPI, Basel, Switzerland. This article is an open access article distributed under the terms and conditions of the Creative Commons Attribution (CC BY) license (https://creativecommons.org/licenses/by/4.0/).

1. Introduction

Dental implantation is now widely applied as a reliable method to restore missing human teeth. Successful implants rely not only on solid integration with the surrounding bone tissues (osseointegration), but also on soft-tissue integration. A robust soft-tissue integration can act as an intact seal, preventing oral bacterial invasion, as well as providing good esthetics [1]. When the soft-tissue seal is interrupted or destructed by oral bacteria, the implant stability is destroyed. There are two types of peri-implant lesions that happen frequently around implant-surrounding tissues: peri-implant mucositis and peri-implantitis [2]. Peri-implant mucositis is a localized infection in the surrounding soft tissues, while peri-implantitis endangers the integrated bones underneath. The prevalence of these two peri-implant lesions is growing at a high speed. A study in 2019 reported that 34% of patients experienced peri-implantitis within 2 years of implant placement [3]. The prevalence of peri-implant mucositis is even higher, since it is believed that most cases of peri-implantitis are derived from peri-implant mucositis [4]. Oral bacteria are one of the major causes of peri-implant lesions, since more and more studies confirm that biofilm accumulation leads to the occurrence of peri-implant mucositis and eventually evolves into peri-implantitis [5]. *Porphyromonas gingivalis*, an anaerobic bacterium and an opportunistic pathogen, has been identified as a pathogen around peri-implant. It has been recognized as a late colonizer that adheres to the extracellular matrix formed by early colonizers such as *Streptococcus mutans* [6]. These bacteria synergistically contribute to the emergence of peri-implant lesions [7,8].

The management of peri-implant lesions is difficult and costly. The commonly used non-surgical approaches include scaling, sand-blasting, laser treatment, and localized/general antibiotic administration; the surgical approaches include periodontal flap surgery, guided tissue regeneration, and other regeneration therapies [9]. If these interventions fail, the implant might eventually be removed. An alternative perspective of peri-implant lesions management is to prevent them from emerging [10]. Since peri-implant lesions are recognized as bacterial-derived diseases, plaque control around the implant site can effectively reduce its prevalence. On the other hand, the managing of peri-implant mucositis can reduce the prevalence of peri-implantitis [11], as peri-implant mucositis is recognized as a precursor of peri-implantitis. Over the years, the surface antibacterial-modification method of implant-abutment materials has been an interesting topic to prevent peri-implant lesions. Numerous studies have proved that the surface modification on the implant-abutment surface has the potential to reduce bacterial load around the implant site, thereby facilitating the formation of soft-tissue seals and reducing the emergence of peri-implant lesions [12]. Different surface-modification methods of titanium abutments and implants have been investigated by researchers; the use of either organic or inorganic antibacterial agents on titanium surfaces showed effective bacterial-inhibition efficacy [13,14]. Compared with the organic antibacterial agents, the inorganic antibacterial agents show better stability and greater antibacterial ability. Inorganic antibacterial agents such as silver, copper, zinc, and their oxides possess a broader antibacterial spectrum [15]. Among all inorganic antibacterial agents, silver has been extensively utilized due to its excellent antimicrobial potency. Although widely used in wound-dressing materials, the cell toxicity of silver remains under debate nowadays.

Recent studies revealed that silver poses little biocidal threat to normal cells when at a nano-sized state [16,17]. The silver nanoparticles (AgNPs) cause bacterial death mainly by direct contact with bacterial cell walls [18]. The electrolyte of AgNPs into silver ions can also inhibit bacterial growth without interfering with normal living cells [19]. There is a study that prepared AgNPs with positive or negative charges, and it proved useful in the antibacterial aspect [20]. Among the many ways to fabricate AgNPs, ion-implantation technology showed effectiveness to incorporate AgNPs into solid surfaces. Ion-implantation technology has been widely used in semiconductor device fabrication, metal finishing, as well as materials-science research. There are mainly two types of ion implantation: linear-beam ion implantation and plasma-induced ion implantation [21]. Compared with plasma-induced ion implantation, linear-beam ion implantation can allow to perform an accurate ion implantation through a linear beam to a single point or a flat surface. The size of the implanted nanoparticle is well-controlled through implantation voltage and time [22]. Studies have reported successful silver-nanoparticle implantations on titanium implant surfaces [23,24]. The implanted AgNPs on the implant surface enhanced its anti-microbial efficacy and had little toxicity to normal human cells. More studies have showed that silver-ion implantation is a viable way to modify an object's surface without changing its original nature [25–28].

With the development of dental materials, zirconia has gradually become a proper substitute for titanium in implant dentistry. With its ivory color, zirconia implant abutments are believed to be more advantageous than titanium abutments, especially in the esthetic zone [29]. Meanwhile, the zirconia surface attracts less bacteria due to its low surface free energy [30]. However, peri-implant lesions still happen around zirconia implants or abutments [31,32], suggesting that plaque control is still necessary on the zirconia surface. Numerous studies have reported that different methods of surface modification on zirconia surfaces can effectively enhance their antimicrobial properties with good biocompatibilities [33–35]. To date, no investigations have laid grounds on silver-ion implantation on zirconia to eliminate bacterial growth, thereby preventing peri-implant lesions. Therefore, this study is designed to explore the method of silver-ion implantation on zirconia surfaces. After the silver-ion implantation, the topography change was evaluated by performing the EDS and XPS analyses and a water-contact-angle device. The antibacterial efficacy was evaluated using *Streptococcus mutans* and *Porphyromonas gingivalis*, and the biocompatibility was evaluated using human gingival fibroblasts. This study provides a viable surface-modification method on zirconia abutments to prevent peri-implant lesions.

2. Materials and Methods

2.1. Specimen Preparation

Twenty-four yttrium-stabilized zirconia disks (Wieland, Bamberg, Germany) with 15 mm diameter and 2 mm thickness were designed and milled by computer-aided design and computer-aided manufacturing (CAD/CAM) process. An 800-grit SiC abrasive paper was used to polish the disks until a unified roughness height of 0.1 μm was achieved. Then, the zirconia disks were ultrasonically cleaned using absolute ethanol and deionized water, each for 20 min. They were naturally dried and stored at room temperature before silver-ion implantation.

Prior to silver-ion implantation, the zirconia disks were randomly divided into four groups, with six specimens in each group. One group was left untreated and was used as the negative control. The other three groups were implanted using ion bombardment at 30 keV and a nominal dose of 1×10^{14} ions/cm^2, 1×10^{15} ions/cm^2, and 1×10^{16} ions/cm^2, respectively. The implantation energy and nominal dose were determined by preliminary experiments. After ion implantation, the specimens were cooled down to room temperature (25 °C). All specimens were stored at room temperature before use. Before various tests, the disks were sterilized with 75% ethanol for 40 min and washed 3 times using 0.1 M PBS buffer (Solarbio, Beijing, China).

2.2. Surface Characteristics

2.2.1. Surface Chemical Composition

The energy dispersive spectroscopy analysis (EDS) was performed to determine surface element composition. The X-ray photoelectron spectroscopy (XPS) (ESCALAB 250; ThermoFisher Scientific, Waltham, MA, USA) examination was performed to further determine the surface chemical composition. In all XPS tests, the survey spectra within the range of 0–600 eV were collected and calibrated using C1s peak at 284.6 eV.

2.2.2. Surface Wettability

The surface wettability was evaluated by measuring the contact angle of 1 μL deionized water droplet using a surface-wettability survey device (OCA15Pro, Dataphysics, Filderstadt, Germany). For each sample, five locations were randomly chosen for the measurement.

2.3. Silver-Ion-Elution Test

The zirconia disks with different silver doping were soaked in 10 mL deionized water at 37 °C for 1 day up to 1 month. After the incubation period, the leaching liquid was collected and subjected to inductively coupled plasma mass spectrometry (ICP-MS, iCAP Qc, ThermoFisher Scientific, Waltham, MA, USA). Silver-ion concentration in the liquid was recorded and recognized as a silver release from zirconia disks.

2.4. Ethics Statement

This study was approved by the Institutional Review Board of the Peking University School of Stomatology (PKUSSIRB-201943034). The unstimulated human saliva was collected from 5 healthy volunteers, and human gingival fibroblasts (HGFs) were grown from the biopsies obtained from another 10 healthy periodontal human volunteers during periodontal surgery. All participants signed the written informed consent before the procedure.

2.5. *Bacterial Response*

2.5.1. Saliva Coating of Zirconia Disks

Saliva was collected as previously described [36]. Briefly, the whole saliva from healthy donors was collected by unstimulated method and stored on ice. To clarify the saliva samples, they were centrifuged at 3000× *g* for 20 min at 4 °C. After clarification, distilled water was used to dilute the supernatant with a 3:1 ratio *v*/*v*. The 25% saliva was then flited with 0.22 μm PES membrane (Millex GP, Millpore, MA, USA) and stored at −80 °C until use. Prior to the usage, the saliva was immediately thawed at 37 °C and centrifuged again at 1430× *g* for 5 min. All zirconia disks, including the untreated zirconia, were immersed in the saliva for 2 h at 37 °C.

2.5.2. Bacterial Culture

The antimicrobial activity of sterilized samples was tested against gram-positive *Streptococcus mutans* (*S. mutans*, UA159) bacteria and gram-negative *Porphyromonas gingivalis* (*P. gingivalis*, ATCC 33277) bacteria, which were provided by the Institute of Microorganisms, Chinese Academy of Sciences. The bacteria strains were maintained using brain–heart infusion (BHI) agar plate (Difco, MI, USA). The incubation condition for *S. mutans* was the standard cell condition (5% CO_2, 95% humidified air, at 37 °C), and the incubation condition for *P. gingivalis* was the standard anaerobic condition (80% N_2, 10% H_2, 10% CO_2, at 37 °C). After the exponential growth phase in the liquid medium, the bacterial cells were collected, centrifuged at 3000× *g* for 15 min, and washed two times with 0.1M PBS buffer. Bacterial suspensions with different final concentrations were shaken for 30 s (Vortex 2, IKA, Königswinter, Germany) to obtain single cells or pairs, then seeded onto the sample disks for further experiments. Each experiment was run in triplicate and repeated on three separate occasions.

2.5.3. Spot-Assay Analysis

The antibacterial efficiency of silver-implanted zirconia was evaluated using a spot assay described by Suppi and Kasemets [37]. Briefly, the final concentration of bacteria suspensions was adjusted to 1×10^5 CFU/mL before added onto saliva-coated zirconia disks. After 4 h of contact with disks, 10 μL of bacterial suspension was pipetted as a 'spot' onto the BHI agar plates to assess the viability of the bacteria cells. The agar plates were cultivated in standard cultivation conditions and the numbers of bacterial colonies were recorded.

2.5.4. LIVE/DEAD Staining Assay

To evaluate the viability of the bacteria after seeding on silver-implanted zirconia, the LIVE/DEAD BacLight Bacterial Viability Kit (L-7012, Invitrogen, Carlsbad, CA, USA) was used. Briefly, staining components A (SYTO 9) and B (propidium iodide) were premixed and diluted with PBS at a ratio of 1.5:1.5:1000 *v*/*v*. The bacteria were seeded to the zirconia surface and cultivated for 24 h, then washed three times with PBS to remove non-adhered cells. A 300 μL mixed staining dilution was added to the zirconia surface. After the staining process, zirconia disks were observed using confocal laser scanning microscopy (CLSM; LSM710, Zeiss, Jena, Germany) at 100-fold magnification. The live cells were stained as green fluorescent and dead cells were stained as red fluorescent.

2.5.5. MTT Colorimetric Assay

The MTT assay is based on the cleavage of MTT (3-(4,5-Dimethylthiazol-2-yl)-2,5-diphenyltetrazolium bromide) into a blue formazan by living cell enzymes. As previously described [38], the amount of formazan formed is positively correlated to the total viable cell counts. To prepare the MTT solution, 5 mg/mL MTT (Biosynth, Itasca, IL, USA) was dissolved in PBS and purified. The bacterial suspension with a final concentration of 1 × 10^8 CFU/mL was added to saliva-coated zirconia disks and cultivated for 24 h. After removing the culture medium, the disks were washed three times with sterile PBS to remove non-attached bacterial cells. To perform the assay, 5 μL of MTT solution was added into 500 μL BHI broth culture medium. The mixed solution was added to the disks and incubated 3 h in the dark in standard cultivation conditions. After discarding the MTT solution, 500 μL DMSO (ThermoFisher Scientific, Waltham, MA, USA) was added to dissolve formazan crystals formed by viable cells. The optical density of the solution was determined at 570 nm using a microplate reader (ELX808, BioTek, Winooski, VT, USA). Solutions without seeding bacteria were used as blank controls.

2.5.6. Crystal Violet Assay

The total amount of biofilm formed on the zirconia disks was evaluated by using crystal violet (CV) assay. The bacterial suspension with a final concentration of 1 × 10^8 CFU/mL was added to saliva-coated zirconia disks and cultivated for 48 h to form a biofilm. After discarding the growth medium, the samples were washed 3 times using PBS and fixed for 20 min at 37 °C using 2.5% glutaraldehyde. The crystal violet solution (Sigma-Aldrich, St. Louis, MO, USA) was used to stain the fixed biofilm by 10 min incubation at room temperature. The unbound dye was washed with gently running deionized water, and the bound dye was extracted using absolute ethanol. The amount of biofilm was measured at an optical density of 570 nm using a microplate reader. The background staining was corrected by subtracting the mean value for CV bound to negative controls.

2.6. *Cellular Response*

2.6.1. Cell Culture

Primary HGFs were grown from tissue explants. Briefly, healthy gingival tissue obtained from patients who underwent periodontal surgery (crown-lengthening surgery) was cut into pieces (~1 mm^3) and placed in 35 mm cell-culture dishes containing Dulbecco's modified Eagle's medium (DMEM; Gibco, Gaithersburg, MD, USA) supplemented with 10% fetal bovine serum (FBS; Gibco, Gaithersburg, MD, USA). Fibroblasts were obtained by trypsinization of the primary outgrowth of cells and were maintained and routinely passaged in 10 cm dishes in DMEM with 10% FBS and 1% antibiotic–antimycotic solution at 37 °C under a humidified atmosphere of 5% CO_2 in 95% air. Cells in the third to sixth passage were used for the experiments.

2.6.2. CCK-8 Assay

The cell proliferation on sample disks was evaluated by a quantitative colorimetric cell-counting kit-8 assay (CCK-8; Dojindo, Kyushu, Japan). Briefly, the HGF cells were seeded onto the disks at a density of 10^5 cells. After incubation for 1 day, 3 days, and 7 days, the zirconia disks were washed with PBS three times. The CCK-8 diluted with cell-culture medium (1:10 ratio, v/v) was added to each well, followed by a 2 h incubation time at 37 °C. The supernatant was then transferred to 96-well plates and subjected to the optical density test using a microplate reader at 450 nm wavelength.

2.7. *Statistical Analysis*

The experimental data were obtained by three repeated experiments performed in triplicate. One-way analysis of variance (ANOVA) was used for comparisons among different groups. The level of significance was set at 0.05. The confidence level was set as 95%. Data analyses were performed using SPSS statistics software (ver. 25.0; SPSS Inc., Chicago, IL, USA).

3. Results

3.1. Surface Characteristics

3.1.1. Surface Chemical Composition

The EDS results of the untreated zirconia and the silver-implanted zirconia at a nominal dose of 1×10^{16} ions/cm^2 show different chemical elements on the surface. As shown in Figure 1 and Table 1, the untreated zirconia surface was composed solely of oxygen and zirconium, while several peaks of silver can be observed on the silver-implanted zirconia surface. The nominal dose of 1×10^{16} ions/cm^2 will lead to a 0.62% silver atomic percentage increase on the zirconia surface.

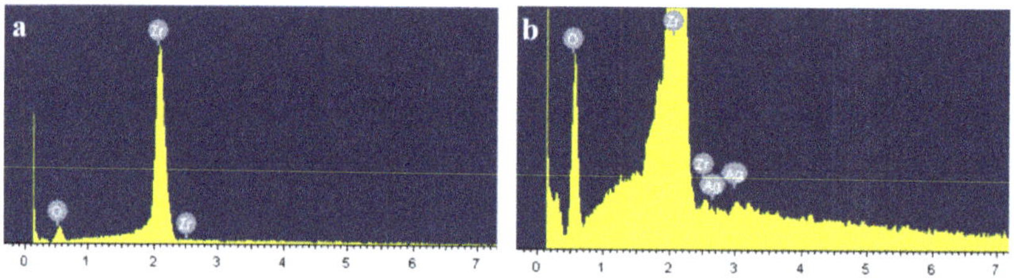

Figure 1. EDS results of (**a**) untreated surface and (**b**) silver-implanted surface at a nominal dose of 1×10^{16} ions/cm^2.

Table 1. Element percentage change on zirconia disks after silver-ion implantation.

Elements	Untreated Zirconia		Silver-Implanted Zirconia (1×10^{16} ions/cm^2)	
	Weight Percentage (%)	Atomic Percentage (%)	Weight Percentage (%)	Atomic Percentage (%)
O	20.3	59.3	22.5	62.4
Zr	79.7	40.7	76.0	37.0
Ag	-	-	1.5	0.6

The XPS results (Figure 2) further reveal that surface composition changes after silver-ion implantation. Generally, the surface composition remained unchanged after the silver-ion implantation. However, as shown in Figure 3, a Ag3d spectrum can be observed on the high-resolution spectra of silver-implanted zirconia surface. The Ag3d doublet at 374.05 eV (Ag $3d_{3/2}$) and 368.04 eV (Ag $3d_{5/2}$) corresponds to metallic silver. These results confirm the successful implantation of silver nanoparticles to the zirconia surface.

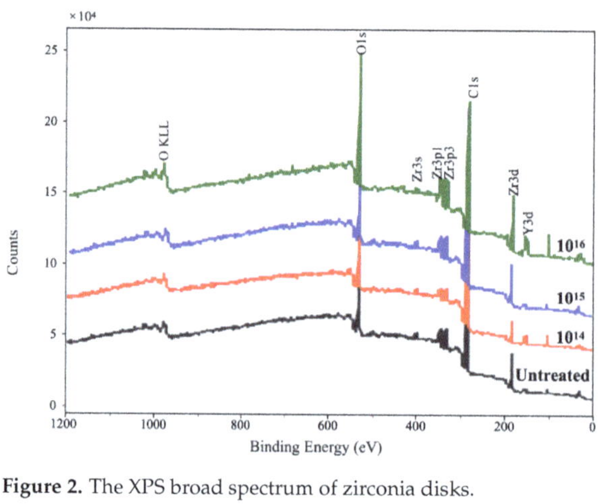

Figure 2. The XPS broad spectrum of zirconia disks.

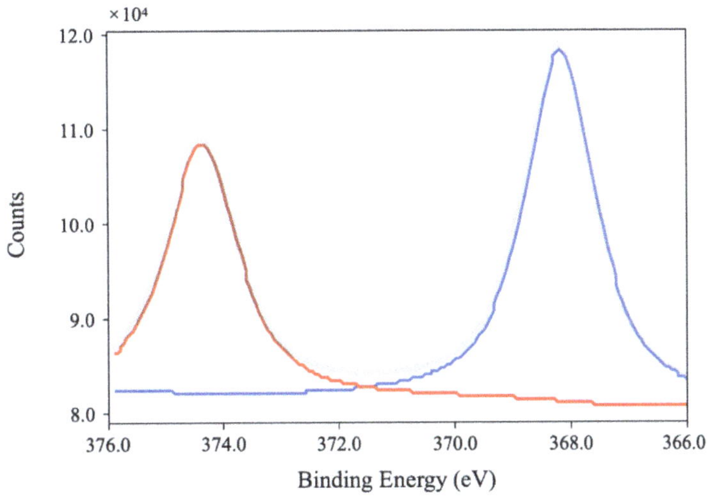

Figure 3. The Ag3d high-resolution image of silver-implanted zirconia disk (nominal dose of 1×10^{16} ions/cm^2).

3.1.2. Surface Wettability

The results of surface wettability are shown in Table 2. Before silver implantation, the surface-water contact angle is $62.27 \pm 2.52°$, and the surface wettability remains after silver implantation. The silver-ion implantation does not change the overall surface wettability of zirconia disks. Since the surface free energy is related to the surface wettability [39], we presumed that surface free energy after the ion implantation remained as well.

Table 2. Surface-water contact-angle results of zirconia disks (Mean ± SD).

Group	Untreated	1×10^{14}/cm^2	1×10^{15}/cm^2	1×10^{16}/cm^2
Surface-water contact angle (°)	62.27 ± 2.40	63.18 ± 2.47	60.51 ± 2.68	62.83 ± 3.73
Image				

3.1.3. Silver-Ion Elution

For all liquid samples stored for 1, 3, 5, 7, 10, 14, 17, 21, 25, and 30 days, the silver-ion concentration is less than 0.1 ppb, which is barely detectable by ICP-MS. This result confirms that silver-implanted zirconia disks are stable, with little silver-ion elution to the surrounding environment.

3.2. Bacterial Response

3.2.1. Bacterial Viability on the Silver-Implanted Zirconia Surface

To first evaluate the bactericidal ability of silver-implanted zirconia against red-complex pathogen *P. gingivalis*, a spot analysis was performed. As shown in Figure 4, a significant decrease in bacterial load can be observed on the silver-implanted samples compared with the untreated disks. As the implant dose increases, there are fewer viable bacteria after coming into contact with the zirconia disks. This preliminary result indicates that silver-implanted zirconia shows a good bactericidal ability against gram-negative bacteria *P. gingivalis*.

Figure 4. Spot-analysis result of *P. gingivalis*.

To further evaluate the bactericidal ability of silver-implanted zirconia, the LIVE/DEAD staining assay was performed. Figure 5 shows *P. gingivalis*'s viability on different zirconia surfaces. There are mainly viable (green fluorescent, stained by SYTO 9) bacterial cells on untreated zirconia surfaces. As the silver implant dose becomes higher, the number of viable bacteria cells decreases, and the number of inviable bacteria cells increases. The inviable bacterial cells were stained red by propidium iodide. For the gram-positive *S. mutans* bacteria, the same trend can be observed. As shown in Figure 6, the viability of bacterial cells decreases with the increase in silver implantation dose, which indicates an improved bactericidal effect with the increase in silver implantation dose. Taken together, an enhanced antibacterial effect can be observed with the accumulation of implanted silver nanoparticles on the zirconia surface.

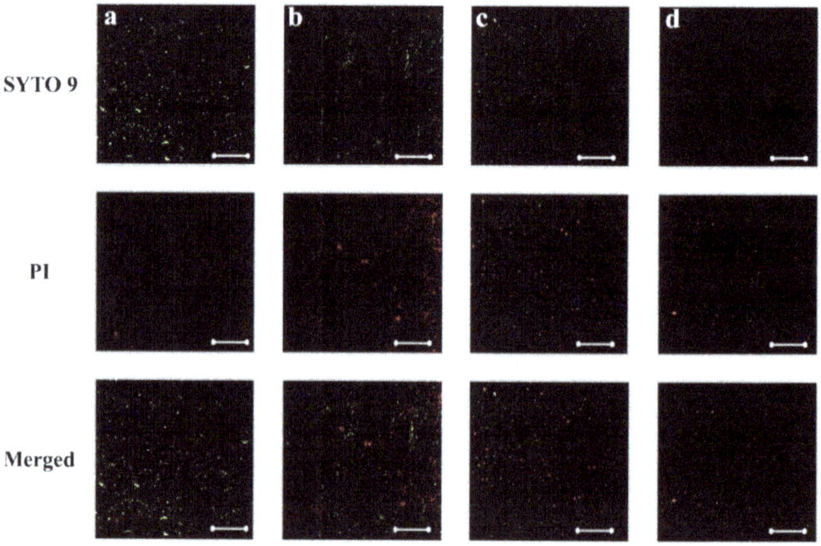

Figure 5. *P. gingivalis* viability on (**a**) untreated, (**b**) silver-implanted zirconia disks with a nominal dose of 10^{14}, (**c**) 10^{15}, and (**d**) 10^{16} ions/cm^2, respectively. (Scale bar is 50 μm).

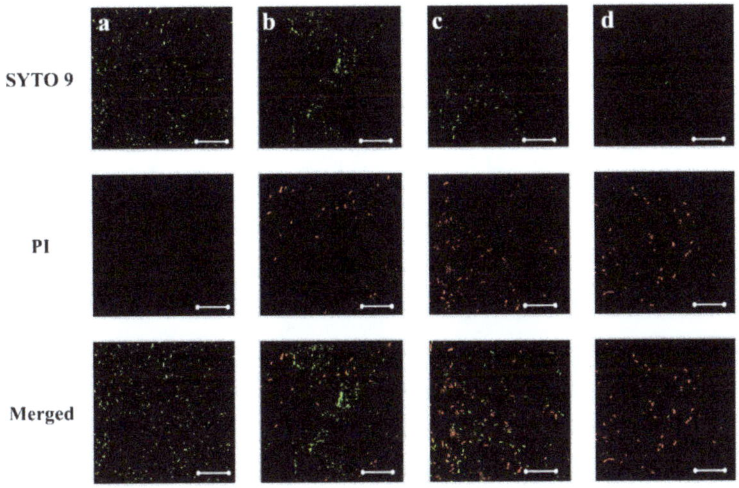

Figure 6. *S. mutans* viability on (**a**) untreated, (**b**) silver-implanted zirconia disks with a nominal dose of 10^{14}, (**c**) 10^{15}, and (**d**) 10^{16} ions/cm^2, respectively. (Scale bar is 50 μm).

3.2.2. Bacterial Adhesion on Silver-Implanted Zirconia Surface

The MTT colorimetric assay was utilized to evaluate the bacterial adhesion to untreated zirconia disks or silver-implanted zirconia disks. As shown in Figure 7, the adhesion of *P. gingivalis* was inhibited on the silver-implanted zirconia surface after a one-day cultivation. The group of 10^{16}/cm^2 showed the least-adhered bacterial load on the surface among all groups. As the cultivation time extends, the same trend can be observed (fewer bacteria are adhering to the zirconia surface as the implantation dose increases).

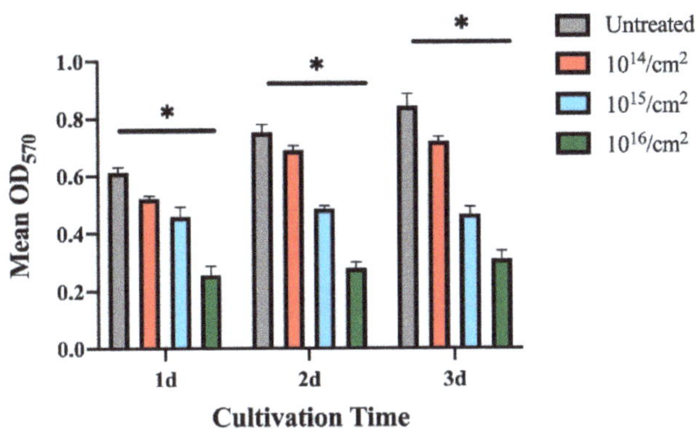

Figure 7. MTT results for *P. gingivalis*. (Data shown: mean and standard deviations; * indicates significant differences among different experimental groups, $p < 0.05$).

Since *S. mutans* grows faster than *P. gingivalis*, the MTT assays were performed 12 h after seeding to the sample surfaces. The results in Figure 8 show that *S. mutans*'s adhesion was also interfered with on the silver-implanted zirconia surfaces. This adhesion-inhibition effect also shows a dose-dependent pattern, with fewer bacteria adhering to high implant-dose groups. Collectively, these results show that silver-implanted zirconia attenuates bacterial adhesion for both gram-positive and gram-negative pathogens. An enhanced inhibition can be seen as the implantation dose becomes higher.

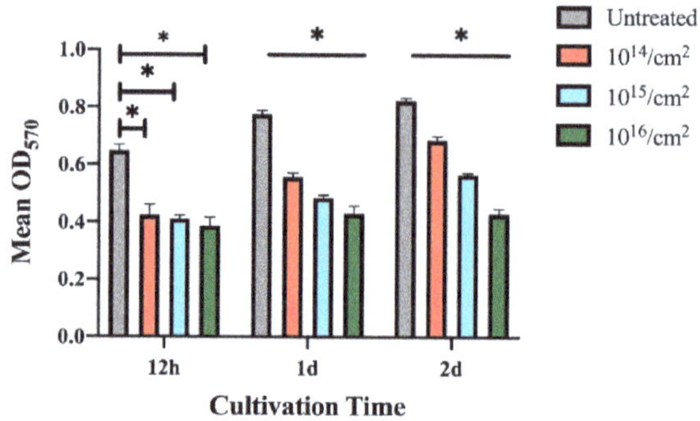

Figure 8. MTT results for *S. mutans*. (Data shown: mean and standard deviations; * indicates significant differences among different experimental groups, $p < 0.05$).

3.2.3. Bacterial-Biofilm Formation on the Silver-Implanted Zirconia Surface

The biofilm formation on silver-implanted zirconia surfaces is shown in Figures 9 and 10. For *P. gingivalis*, a total reduction in biofilm mass can be observed after a two-day cultivation. The zirconia disks with nominal doses of $10^{15}/cm^2$ and $10^{16}/cm^2$ accumulate fewer biofilms than the $10^{14}/cm^2$ group. Indeed, after a four-day cultivation, there are no significant differences in the biofilms formed between the $10^{14}/cm^2$ group and the untreated group. However, the $10^{15}/cm^2$ and $10^{16}/cm^2$ groups still show a biofilm-inhibition effect after a 4-day cultivation. This result indicates that, after a certain cultivation time, the group with an implant dose of $10^{14}/cm^2$ might eventually lose its anti-biofilm formation ability.

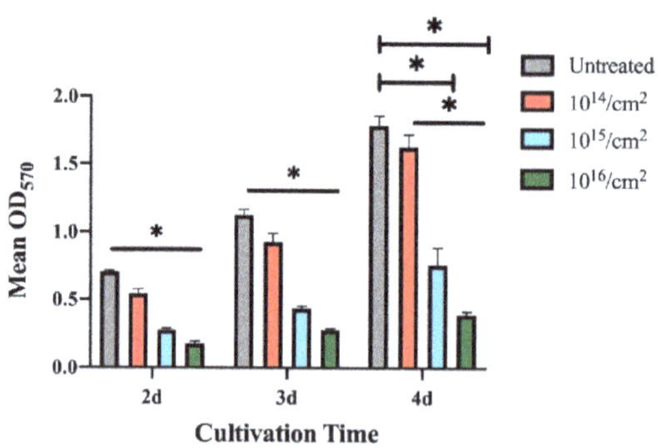

Figure 9. Biofilm formation on zirconia disks of *P. gingivalis*. (Data shown: mean and standard deviations; * indicates significant differences among different experimental groups, $p < 0.05$).

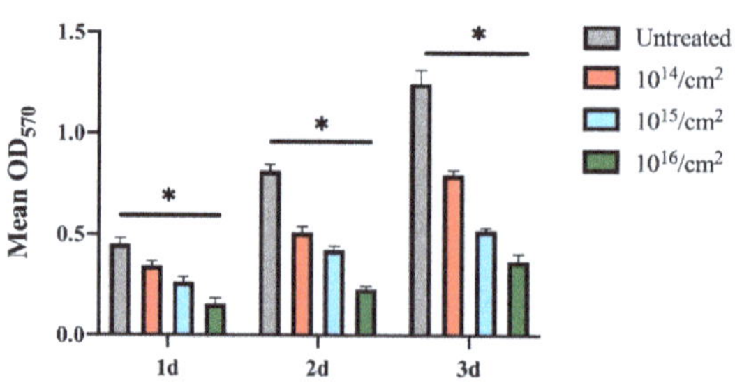

Figure 10. Biofilm formation on zirconia disks of *S. mutans*. (Data shown: mean and standard deviations; * indicates significant differences among different experimental groups, $p < 0.05$).

The biofilm formed by *S. mutans* on the zirconia disks shows a more unified trend within 3 days of seeding bacteria to the sample surface. As shown in Figure 10, the $10^{15}/cm^2$ and $10^{16}/cm^2$ groups also show a good anti-biofilm-formation ability. The crystal violet results reveal that the decrease in biofilm formation can be observed on silver-implanted zirconia for both gram-positive and gram-negative bacteria.

3.3. Cellular Response by CCK-8 Assay

To evaluate the cytotoxicity of silver-implanted zirconia, the CCK-8 colorimetric assay was performed. As shown in Figure 11, there is no significant difference between the untreated group and the silver-implanted groups. All groups show a similar growth rate, and this trend is retained after 7 days of incubation. The CCK-8 results show silver-implanted zirconia barely has any cytotoxicity to human gingival fibroblast, which is a crucial cell during soft-tissue wound healing after implant surgery and abutment installation.

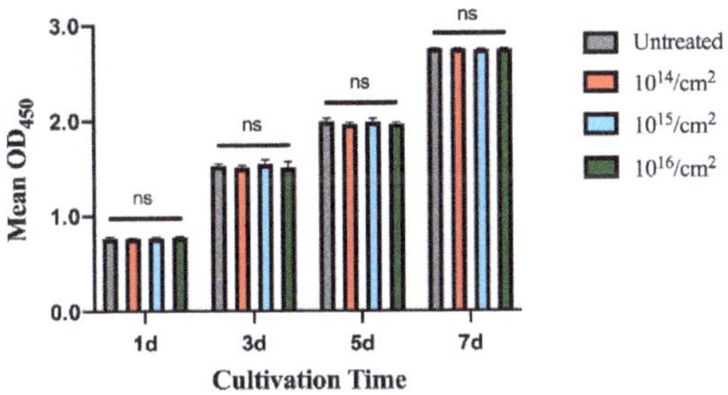

Figure 11. The CCK-8 results for human gingival fibroblast on zirconia disks. (Data shown: mean and standard deviations; ns: not significant).

4. Discussion

Implant-associated infection has been widely researched due to its high prevalence. Until now, most researchers and clinicians believed it was bacterially derived, thus lots of researchers have focused on the surface modification of implant materials to improve their antimicrobial efficacy. The surface modification of implant materials should show a great antimicrobial effect without releasing antibiotic agents at a fast rate, while being non-hazardous to the surrounding living human cells. Although many antimicrobial coatings have been presented, most of them are prone to degradation in the oral environment quickly. The method of silver nanoparticles embedded into zirconia by ion implantation is presented in this article. Using EDX, silver can be detected on the surface of zirconia. In the XPS results, the Ag3d doublet at 374.05 eV ($Ag3d_{3/2}$) and 368.04 eV ($Ag3d_{5/2}$) corresponds to metallic silver, indicating the nanoparticles are metallic silver. The ICP-MS test shows that silver nanoparticles remain stable and have little ion release (the leaching of silver ions is under the detection limit of 0.1 ppb even after 1 month of immersion). The low leaching property indicates silver ions will not accumulate in the tissues around implants; otherwise, it would result in cytotoxicity or other side effects.

This study investigates the antimicrobial ability of silver-nanoparticle-implanted zirconia, and the results show it can significantly enhance the antimicrobial ability in vitro and cause little harm to live human gingival fibroblasts. In antimicrobial assays, both the 1×10^{15} ions/cm^2 and the 1×10^{16} ions/cm^2 nominal-dose groups show positive antimicrobial properties on *P. gingivalis* and *S. mutans* in all assays, while the 1×10^{16} ions/cm^2 group shows the strongest effects on bacterial inhibition. The LIVE/DEAD staining assay results demonstrated a decrease in live cells on the silver-implanted samples, and a decrease in the total number of cells was also observed. The crystal violet staining test shows the anti-biofilm activity of silver nanoparticles, which was in agreement with previously conducted studies [40,41]. The study did not choose the Zone of Inhibition (ZOI) assay for the reason that the antimicrobial effect of silver nanoparticles does not come from the leaching of silver ions from the material's surface. It is believed that silver nanoparticles firmly attached to the zirconia surface can interact with microorganisms by direct contact. Furthermore, although the size of silver nanoparticles is important since it has an influence on cytotoxicity and antibacterial effects [20,42], it is hard to evaluate after the ion-implantation process, as the nanoparticles are embedded in the zirconia surface.

Although many research studies have investigated the antimicrobial activity of silver nanoparticles, the mechanism of the antibacterial activities and potential hazards remain unclear. However, possible theories have been made, as suggested by Marambio-Jones and Hoek et al. [43]: (a) uptake of free silver ions followed by disruption of ATP production and DNA replication, (b) reactive oxygen species (ROS) generation by silver nanoparticles and silver ions, and (c) direct damage to the cell membrane by silver nanoparticles. From the result of this study, it is believed that silver nanoparticles interact with microorganisms in the latter two ways, since silver ions are not released in a remarkable manner. The silver nanoparticles can produce high levels of ROS and free-radical species such as hydrogen peroxide, superoxide anion, hydroxyl radical, hypochlorous acid, and singlet oxygen [44,45]. The excessive ROS cannot be eliminated by antioxidant systems and will inhibit cell respiration and growth. Increased ROS levels will lead to an apoptosis-like response, which will cause DNA damage and, eventually, cell death. On the other hand, silver nanoparticles can anchor to the bacterial cell wall and, consequently, infiltrate it. This action will cause physical changes in the bacterial membrane, e.g., membrane damage, which can lead to cellular-content leakage and bacterial death [46,47].

Soft-tissue connection is crucial to implant dentistry. Ideal dental implant bio-materials require effective soft-tissue integration for long-term stability after implantation. A sound soft-tissue seal on the implant-abutment site can not only inhibit unwanted soft-tissue recession and marginal bone loss, but also resist bacterial invasion. Following implant surgery, HGFs are involved in the formation of connective tissues around the implant site. The CCK-8 proliferation assay indicates that silver nanoparticles show no obvious damage to HGFs' proliferation. An effective HGF attachment to implant abutment plays a critical role in soft-tissue connection, and a strong attachment starts with a undisturbed initial proliferation. At the same time, the antimicrobial effect of silver-implanted surfaces can further enhance soft-tissue integration. The silver-implanted zirconia surface can provide an antibacterial effect without obvious cytotoxicity; however, as many cells are involved in the soft-tissue wound-healing process, more experiments should be carried out in the future to evaluate their response to the silver-implanted zirconia.

Taken together, the results suggest that nanostructured surfaces with silver nanoparticles incorporated in zirconia may provide a better choice for dental implant abutments than traditional ones. Silver nanoparticles show a promising bactericidal ability that will be effective in the oral environment. However, only two pathogens were evaluated in this study. More pathogens, such as *Staphylococcus aureus* and other red-complex pathogens, should be studied further. Moreover, mono-culture in vitro studies are sometimes at odds with clinical reality. To avoid this, further in vivo studies to mimic clinical situations are required. Furthermore, the long-term release of silver ions and their safety for human beings need to be carefully evaluated. The release of silver ions in a different medium (such as saliva or blood serum) should also be evaluated in the future. The possible mechanical property changes (such as surface hardness or the Young's modulus) after silver-ion implantation should also be investigated before its clinical usage.

5. Conclusions

In this work, we present an effective surface-modification technique for zirconia dental implant materials. Silver was implanted into zirconia with a nanostructured surface. Within the limitations of this in vitro study, the results suggest that silver-implanted zirconia has a strong antimicrobial effect against oral microorganisms, including *P. gingivalis* and *S. mutans*. It also shows no harm to human gingival fibroblast proliferation. Silver-implanted zirconia shows promise for both preventing peri-implant lesions and forming peri-implant soft-tissue attachments in clinical application.

Author Contributions: Conceptualization, L.C., W.-S.L. and J.T.; methodology, Y.Y.; formal analysis, Z.Y. and M.L.; investigation, Y.Y.; data curation, Y.Y.; writing—original draft preparation, Y.Y and M.L.; writing—review and editing, Z.Y., M.L., W.-S.L. and L.C.; funding acquisition, Y.Y., L.C. and J.T. All authors have read and agreed to the published version of the manuscript.

Funding: This research was funded by the National Natural Science Foundation of China, grant number 82201018; the Beijing Natural Science Foundation, grant number 7212138; the Natural Science Foundation of Tibet Autonomous Region, grant number XZ2021ZR-ZY07(Z); the Program for New Clinical Techniques and Therapies of Peking University School and Hospital of Stomatology, grant number PKUSSNCT-20A02; and the National Clinical Key Discipline Construction Project of Peking University School and Hospital of Stomatology, grant number PKUSSNKP-202108.

Data Availability Statement: Research data are available upon request to the authors.

Conflicts of Interest: The authors declare no conflict of interest.

References

1. Luchinskaya, D.; Du, R.; Owens, D.M.; Tarnow, D.; Bittner, N. Various surface treatments to implant provisional restorations and their effect on epithelial cell adhesion: A comparative in vitro study. *Implant Dent.* **2017**, *26*, 12–23. [CrossRef] [PubMed]
2. Berglundh, T.; Armitage, G.; Araujo, M.G.; Avila-Ortiz, G.; Blanco, J.; Camargo, P.M.; Chen, S.; Cochran, D.; Derks, J.; Figuero, E.; et al. Peri-implant diseases and conditions: Consensus report of workgroup 4 of the 2017 World Workshop on the Classification of Periodontal and Peri-Implant Diseases and Conditions. *J. Clin. Periodontol.* **2018**, *45*, S286–S291. [CrossRef] [PubMed]
3. Kordbacheh Changi, K.; Finkelstein, J.; Papapanou, P.N. Peri-implantitis prevalence, incidence rate, and risk factors: A study of electronic health records at a U.S. dental school. *Clin. Oral Implant. Res.* **2019**, *30*, 306–314. [CrossRef] [PubMed]
4. Heitz-Mayfield, L.J.A.; Salvi, G.E. Peri-implant mucositis. *J. Clin. Periodontol.* **2018**, *45*, S237–S245. [CrossRef]
5. Salvi, G.E.; Cosgarea, R.; Sculean, A. Prevalence and mechanisms of peri-implant diseases. *J. Dent. Res.* **2016**, *96*, 31–37. [CrossRef]
6. Jordan, R.P.C.; Marsh, L.; Ayre, W.N.; Jones, Q.; Parkes, M.; Austin, B.; Sloan, A.J.; Waddington, R.J. An assessment of early colonisation of implant-abutment metal surfaces by single species and co-cultured bacterial periodontal pathogens. *J. Dent.* **2016**, *53*, 64–72. [CrossRef]
7. Jang, W.; Kim, H.S.; Alam, K.; Ji, M.K.; Cho, H.S.; Lim, H.P. Direct-deposited graphene oxide on dental implants for antimicrobial activities and osteogenesis. *Int. J. Nanomed.* **2021**, *16*, 5745–5754. [CrossRef]
8. Alharbi, M.S.; Alshehri, F.A. High molecular weight hyaluronic acid reduces the expression of virulence genes fima, mfa1, haga, rgpa, and kgp in the oral pathogen Porphyromonas gingivalis. *Pharmaceutics* **2022**, *14*, 1628. [CrossRef]
9. Berglundh, T.; Jepsen, S.; Stadlinger, B.; Terheyden, H. Peri-implantitis and its prevention. *Clin. Oral Implant. Res.* **2019**, *30*, 150–155. [CrossRef]
10. RÖSing, C.K.; Fiorini, T.; Haas, A.N.; Muniz, F.W.M.G.; Oppermann, R.V.; Susin, C. The impact of maintenance on peri-implant health. *Braz. Oral Res.* **2019**, *33*, e074. [CrossRef]
11. Jepsen, S.; Berglundh, T.; Genco, R.; Aass, A.M.; Demirel, K.; Derks, J.; Figuero, E.; Giovannoli, J.L.; Goldstein, M.; Lambert, F.; et al. Primary prevention of peri-implantitis: Managing peri-implant mucositis. *J. Clin. Periodontol.* **2015**, *42*, S152–S157. [CrossRef] [PubMed]
12. Han, A.; Tsoi, J.K.H.; Rodrigues, F.P.; Leprince, J.G.; Palin, W.M. Bacterial adhesion mechanisms on dental implant surfaces and the influencing factors. *Int. J. Adhes. Adhes.* **2016**, *69*, 58–71. [CrossRef]
13. Kulkarni Aranya, A.; Pushalkar, S.; Zhao, M.; LeGeros, R.Z.; Zhang, Y.; Saxena, D. Antibacterial and bioactive coatings on titanium implant surfaces. *J. Biomed. Mater. Res. A* **2017**, *105*, 2218–2227. [CrossRef]
14. Jiao, Y.; Tay, F.R.; Niu, L.-N.; Chen, J.-H. Advancing antimicrobial strategies for managing oral biofilm infections. *Int. J. Oral Sci.* **2019**, *11*, 28. [CrossRef] [PubMed]
15. Chouirfa, H.; Bouloussa, H.; Migonney, V.; Falentin-Daudre, C. Review of titanium surface modification techniques and coatings for antibacterial applications. *Acta Biomater.* **2019**, *83*, 37–54. [CrossRef] [PubMed]
16. Abram, S.L.; Fromm, K.M. Handling (nano) silver as antimicrobial agent: Therapeutic window, dissolution dynamics, detection methods and molecular interactions. *Chemistry* **2020**, *26*, 10948–10971. [CrossRef]
17. Liao, C.; Li, Y.; Tjong, S.C. Bactericidal and cytotoxic properties of silver nanoparticles. *Int. J. Mol. Sci.* **2019**, *20*, 449. [CrossRef]
18. Tang, S.; Zheng, J. Antibacterial activity of silver nanoparticles: Structural effects. *Adv. Healthc. Mater.* **2018**, *7*, e1701503. [CrossRef]
19. Durán, N.; Durán, M.; De Jesus, M.B.; Seabra, A.B.; Fávaro, W.J.; Nakazato, G. Silver nanoparticles: A new view on mechanistic aspects on antimicrobial activity. *Nanomedicine* **2016**, *12*, 789–799. [CrossRef]
20. Sharonova, A.; Loza, K.; Surmeneva, M.; Surmenev, R.; Prymak, O.; Epple, M. Synthesis of positively and negatively charged silver nanoparticles and their deposition on the surface of titanium. *IOP Conf. Ser. Mater. Sci. Eng.* **2016**, *116*, 012009. [CrossRef]

21. Donkelaar, J.V.; Yang, C.; Alves, A.D.C.; Mccallum, J.C.; Hougaard, C.; Johnson, B.C.; Hudson, F.E.; Dzurak, A.S.; Morello, A.; Spemann, D. Single atom devices by ion implantation. *J. Phys. Condens. Matter* **2015**, *27*, 154204. [CrossRef] [PubMed]
22. Zhou, X.; Chen, X.; Mao, T.C.; Li, X.; Shi, X.H.; Fan, D.L.; Zhang, Y.M. Carbon ion implantation: A good method to enhance the biocompatibility of silicone rubber. *Plast. Reconstr. Surg.* **2016**, *137*, 690e–699e. [CrossRef] [PubMed]
23. Zheng, Y.; Li, J.; Liu, X.; Sun, J. Antimicrobial and osteogenic effect of Ag-implanted titanium with a nanostructured surface. *Int. J. Nanomed.* **2012**, *7*, 875–884.
24. Lampé, I.; Beke, D.; Biri, S.; Csarnovics, I.; Csik, A.; Dombrádi, Z.; Hajdu, P.; Hegedűs, V.; Rácz, R.; Varga, I.; et al. Investigation of silver nanoparticles on titanium surface created by ion implantation technology. *Int. J. Nanomed.* **2019**, *14*, 4709–4721. [CrossRef] [PubMed]
25. Cao, H.; Liu, X.; Meng, F.; Chu, P.K. Biological actions of silver nanoparticles embedded in titanium controlled by micro-galvanic effects. *Biomaterials* **2011**, *32*, 693–705. [CrossRef]
26. Wickens, D.J.; West, G.; Kelly, P.J.; Verran, J.; Whitehead, K.A. Antimicrobial activity of nanocomposite zirconium nitride/silver coatings to combat external bone fixation pin infections. *Int. J. Artif. Organs* **2012**, *35*, 817–825. [CrossRef]
27. Wang, X.; Lu, T.; Wen, J.; Xu, L.; Zeng, D.; Wu, Q.; Cao, L.; Lin, S.; Liu, X.; Jiang, X. Selective responses of human gingival fibroblasts and bacteria on carbon fiber reinforced polyetheretherketone with multilevel nanostructured TiO_2. *Biomaterials* **2016**, *83*, 207–218. [CrossRef] [PubMed]
28. Slate, A.J.; Wickens, D.J.; Mohtadi, M.E.; Dempseyhibbert, N.; West, G.; Banks, C.E.; Whitehead, K.A. Antimicrobial activity of Ti-ZrN/Ag coatings for use in biomaterial applications. *Sci. Rep.* **2018**, *8*, 1497. [CrossRef]
29. Hashim, D.; Cionca, N.; Courvoisier, D.S.; Mombelli, A. A systematic review of the clinical survival of zirconia implants. *Clin. Oral Investig.* **2016**, *20*, 1403–1417. [CrossRef]
30. Roehling, S.; Astasovfrauenhoffer, M.; Hausergerspach, I.; Braissant, O.; Woelfler, H.; Waltimo, T.; Kniha, H.; Gahlert, M. In vitro biofilm formation on titanium and zirconia implant surfaces. *J. Periodontol.* **2017**, *88*, 298–307. [CrossRef]
31. Schwarz, F.; John, G.; Hegewald, A.; Becker, J. Non-surgical treatment of peri-implant mucositis and peri-implantitis at zirconia implants: A prospective case series. *J. Clin. Periodontol.* **2015**, *42*, 783–788. [CrossRef] [PubMed]
32. Fretwurst, T.; Müller, J.; Larsson, L.; Bronsert, P.; Hazard, D.; Castilho, R.M.; Kohal, R.; Nelson, K.; Iglhaut, G. Immunohistological composition of peri-implantitis affected tissue around ceramic implants-A pilot study. *J. Periodontol.* **2021**, *92*, 571–579. [CrossRef] [PubMed]
33. Yang, Y.; Zhou, J.; Liu, X.; Zheng, M.; Yang, J.; Tan, J. Ultraviolet light-treated zirconia with different roughness affects function of human gingival fibroblasts in vitro: The potential surface modification developed from implant to abutment. *J. Biomed. Mater. Res. B Appl. Biomater.* **2015**, *103*, 116–124. [CrossRef] [PubMed]
34. Liu, M.; Zhou, J.; Yang, Y.; Zheng, M.; Yang, J.; Tan, J. Surface modification of zirconia with polydopamine to enhance fibroblast response and decrease bacterial activity in vitro: A potential technique for soft tissue engineering applications. *Colloids Surf. B Biointerfaces* **2015**, *136*, 74–83. [CrossRef]
35. Yang, Y.; Zheng, M.; Yang, Y.; Li, J.; Su, Y.-F.; Li, H.-P.; Tan, J.-G. Inhibition of bacterial growth on zirconia abutment with a helium cold atmospheric plasma jet treatment. *Clin. Oral Investig.* **2020**, *24*, 1465–1477. [CrossRef]
36. Kuboniwa, M.; Amano, A.; Hashino, E.; Yamamoto, Y.; Inaba, H.; Hamada, N.; Nakayama, K.; Tribble, G.D.; Lamont, R.J.; Shizukuishi, S. Distinct roles of long/short fimbriae and gingipains in homotypic biofilm development by Porphyromonas gingivalis. *BMC Microbiol.* **2009**, *9*, 105. [CrossRef] [PubMed]
37. Suppi, S.; Kasemets, K.; Ivask, A.; Kunnis-Beres, K.; Sihtmae, M.; Kurvet, I.; Aruoja, V.; Kahru, A. A novel method for comparison of biocidal properties of nanomaterials to bacteria, yeasts and algae. *J. Hazard. Mater.* **2015**, *286*, 75–84. [CrossRef]
38. Sylvester, P.W. Optimization of the tetrazolium dye (MTT) colorimetric assay for cellular growth and viability. *Methods Mol. Biol.* **2011**, *716*, 157–168.
39. Nakamura, M.; Hori, N.; Ando, H.; Namba, S.; Toyama, T.; Nishimiya, N.; Yamashita, K. Surface free energy predominates in cell adhesion to hydroxyapatite through wettability. *Mater. Sci. Eng. C Mater. Biol. Appl.* **2016**, *62*, 283–292. [CrossRef]
40. Franci, G.; Falanga, A.; Galdiero, S.; Palomba, L.; Rai, M.; Morelli, G.; Galdiero, M. Silver nanoparticles as potential antibacterial agents. *Molecules* **2015**, *20*, 8856–8874. [CrossRef]
41. van Hengel, I.A.J.; Riool, M.; Fratila-Apachitei, L.E.; Witte-Bouma, J.; Farrell, E.; Zadpoor, A.A.; Zaat, S.A.J.; Apachitei, I. Selective laser melting porous metallic implants with immobilized silver nanoparticles kill and prevent biofilm formation by methicillin-resistant Staphylococcus aureus. *Biomaterials* **2017**, *140*, 1–15. [CrossRef] [PubMed]
42. Yin, I.X.; Zhang, J.; Zhao, I.S.; Mei, M.L.; Li, Q.; Chu, C.H. The antibacterial mechanism of silver nanoparticles and its application in dentistry. *Int. J. Nanomed.* **2020**, *15*, 2555–2562. [CrossRef] [PubMed]
43. Marambiojones, C.; Hoek, E.M.V. A review of the antibacterial effects of silver nanomaterials and potential implications for human health and the environment. *J. Nanopart. Res.* **2010**, *12*, 1531–1551. [CrossRef]
44. Gomaa, E.Z. Silver nanoparticles as an antimicrobial agent: A case study on Staphylococcus aureus and Escherichia coli as models for Gram-positive and Gram-negative bacteria. *J. Gen. Appl. Microbiol.* **2017**, *63*, 36–43. [CrossRef]
45. Zhao, R.; Lv, M.; Li, Y.; Sun, M.; Kong, W.; Wang, L.; Song, S.; Fan, C.; Jia, L.; Qiu, S.; et al. Stable nanocomposite based on PEGylated and silver nanoparticles loaded graphene oxide for long-term antibacterial activity. *ACS Appl. Mater. Interfaces* **2017**, *9*, 15328–15341. [CrossRef]

46. Khalandi, B.; Asadi, N.; Milani, M.; Davaran, S.; Abadi, A.J.; Abasi, E.; Akbarzadeh, A. A review on potential role of silver nanoparticles and possible mechanisms of their actions on bacteria. *Drug Res.* **2017**, *11*, 70–76. [CrossRef]
47. Seong, M.; Dong, G.L. Silver nanoparticles against Salmonella enterica Serotype Typhimurium: Role of inner membrane dysfunction. *Curr. Microbiol.* **2017**, *74*, 661–670. [CrossRef]

Disclaimer/Publisher's Note: The statements, opinions and data contained in all publications are solely those of the individual author(s) and contributor(s) and not of MDPI and/or the editor(s). MDPI and/or the editor(s) disclaim responsibility for any injury to people or property resulting from any ideas, methods, instructions or products referred to in the content.

Journal of *Functional Biomaterials*

MDPI

Article

Submicron-Grooved Films Modulate the Directional Alignment and Biological Function of Schwann Cells

Zhen Zhang [1,2], Yuanliang Lv [1], Javad Harati [1,2], Jianan Song [3], Ping Du [1], Peiyan Ou [1,2], Jiaqi Liang [1], Huaiyu Wang [1,2,*] and Peng-Yuan Wang [3,4,*]

[1] Center for Human Tissues and Organs Degeneration, Shenzhen Institute of Advanced Technology, Chinese Academy of Sciences, Shenzhen 518055, China
[2] University of Chinese Academy of Sciences, Beijing 100049, China
[3] Oujiang Laboratory, Key Laboratory of Alzheimer's Disease of Zhejiang Province, Institute of Aging, Wenzhou Medical University, Wenzhou 325000, China
[4] Shenzhen Key Laboratory of Biomimetic Materials and Cellular Immunomodulation, Shenzhen Institute of Advanced Technology, Chinese Academy of Sciences, Shenzhen 518055, China
* Correspondence: hy.wang1@siat.ac.cn (H.W.); py.wang@ojlab.ac.cn (P.-Y.W.)

Abstract: Topographical cues on material surfaces are crucial for guiding the behavior of nerve cells and facilitating the repair of peripheral nerve defects. Previously, micron-grooved surfaces have shown great potential in controlling nerve cell alignment for studying the behavior and functions of those cells and peripheral nerve regeneration. However, the effects of smaller-sized topographical cues, such as those in the submicron- and nano-scales, on Schwann cell behavior remain poorly understood. In this study, four different submicron-grooved polystyrene films (800/400, 800/100, 400/400, and 400/100) were fabricated to study the behavior, gene expression, and membrane potential of Schwann cells. The results showed that all submicron-grooved films could guide the cell alignment and cytoskeleton in a groove depth-dependent manner. Cell proliferation and cell cycle assays revealed that there was no significant difference between the submicron groove samples and the flat control. However, the submicron grooves can direct the migration of cells and upregulate the expression of critical genes in axon regeneration and myelination (e.g., MBP and Smad6). Finally, the membrane potential of the Schwann cells was significantly altered on the grooved sample. In conclusion, this study sheds light on the role of submicron-grooved patterns in regulating the behavior and function of Schwann cells, which provides unique insights for the development of implants for peripheral nerve regeneration.

Keywords: topographical cues; Schwann cells; submicron grooves; cell alignment; peripheral nerve

Citation: Zhang, Z.; Lv, Y.; Harati, J.; Song, J.; Du, P.; Ou, P.; Liang, J.; Wang, H.; Wang, P.-Y. Submicron-Grooved Films Modulate the Directional Alignment and Biological Function of Schwann Cells. *J. Funct. Biomater.* **2023**, *14*, 238. https://doi.org/10.3390/jfb14050238

Academic Editor: Lidy Fratila-Apachitei

Received: 8 March 2023
Revised: 12 April 2023
Accepted: 21 April 2023
Published: 23 April 2023

Copyright: © 2023 by the authors. Licensee MDPI, Basel, Switzerland. This article is an open access article distributed under the terms and conditions of the Creative Commons Attribution (CC BY) license (https://creativecommons.org/licenses/by/4.0/).

1. Introduction

Peripheral nerve injury is a challenging issue in neurological diseases, which can pose severe adverse effects on the health and well-being of patients [1–3]. Although autologous nerve grafts are considered the gold standard for repairing peripheral nerve injuries [4–6], their widespread use in clinical practice is limited due to the scarcity of available nerves, the need for secondary surgeries, and the permanent loss of function in the donor area [7]. In recent years, nerve grafts made from natural or synthetic biomaterials have been increasingly used for peripheral nerve regeneration [8,9]. However, their repairing effect is still insufficient to meet the clinical requirements. To realize better nerve regeneration, numerous scientists are focusing on the modification of the surface properties of biomaterials to facilitate nerve repair. Studies have shown that nerve conduits with electrical conductivity, porosity, appropriate mechanical properties, and appropriate topology can mimic the microenvironment of nerve cells and, thus, promote axonal growth [10]. Among these, the preparation of material surfaces with ordered topographical features (such as grooves, pillars, and pores) using photolithography or soft lithography is widely used due to its ease of operation and non-toxicity [11].

In addition to autologous nerve grafting, the artificial construction of nerve guidance conduits (NGCs) is also a common strategy for peripheral nerve repair. Materials used to construct NGCs need to have good biocompatibility, biodegradability and appropriate mechanical properties [12,13]. NGCs can be divided into natural materials and synthetic materials. Natural materials include collagen, fibrin, gelatin, chitosan, etc. Synthetic materials include poly (ε-caprolactone) (PCL), polylactic acid (PLA), polyglycolic acid (PGA), poly (lactic-co-glycolic acid) (PLGA), etc. However, these materials are biodegradable, where the surface structures gradually disappear with time, resulting in diminished cell–surface interaction. This paper used polystyrene (PS) as the film-forming material to prepare the submicron-grooved surface. PS has good cytocompatibility and adhesion as the raw material of cell tissue culture plates.

It is well known that Schwann cells are glial cells that envelop axons to form myelin sheaths, which are often used as model cells in the studies of peripheral nerve repair. Following peripheral nerve injury, Schwann cells can remove myelin debris and proliferate to form Büngner zone bands, promoting axon regeneration via distal nerve stumps [14]. Moreover, Schwann cells can secrete neurotrophic and growth factors that aid axonal regeneration [15–18]. Numerous studies have demonstrated that regularly oriented structures can be used to guide the directional alignment and growth of Schwann cells to promote axonal regeneration [19–24]. For instance, hierarchically arranged collagen micropatterns were prepared using micro-forming techniques and found to be effective in directing the growth of Schwann cells while upregulating specific genes related to axon regeneration and myelination [25]. Additionally, well-aligned electrospun fibers can promote Schwann cell migration and even neurite growth to repair the dorsal root ganglion [26]. Grooved patterns are another type of topographical cue that can be used to direct the alignment of Schwann cells, in which groove width [27,28], depth [29,30], and geometry [31,32] are key factors influencing cell behaviors.

While earlier studies have focused on the effect of grooves on the morphology of Schwann cells, there are few studies that have systematically examined the effect of grooves on the other behaviors of Schwann cells. Furthermore, to our knowledge, the grooves in most studies are in a micron range [33–35], and studies on how submicron grooves regulate the behaviors of Schwann cells have not been reported. More importantly, in these studies, bioactive molecules were used together to enhance the adhesion and bioactivity of Schwann cells on the grooved surface, making it difficult to determine which was the primary factor and impeding the investigation of related biological mechanisms. Therefore, a better understanding of how submicron grooves regulate Schwann cell behaviors and the underlying mechanisms is crucial at the present stage.

In this study, we prepared four diverse sizes of submicron-grooved PS films and systematically studied their effects on the properties of Schwann cells, including cell adhesion, morphology, proliferation, migration, transcriptome profile, and membrane potential. The relationship between the size of the submicron grooves and the behaviors of the Schwann cells was explored. Our findings provide valuable insights for understanding cell–pattern interactions and the preparation of advanced NGCs for peripheral nerve injury.

2. Materials and Methods

2.1. Fabrication of Submicron-Grooved Films

To produce topographical patterns on polystyrene (PS) films, we utilized four distinct sizes of grooved silicon substrates with the groove width/depth (in nanometers) of 800/400, 800/100, 400/400, and 400/100, according to our previous study [36]. Firstly, we mixed a 10:1 ratio of polydimethylsiloxane (PDMS) to curing agent thoroughly to create silicon rubber molds. The mixture was then poured onto the grooved silicon substrate and vacuumed to eliminate air bubbles. The molds were then heated and fixed in an oven at 80 °C for 3 h, after which they could be easily demolded from the grooved silicon substrates, obtaining PDMS molds with submicron-grooved topography. We also prepared PDMS molds on a flat silicon substrate as the control group.

A polyethylene terephthalate (PET) plate of 2 cm × 2 cm was utilized as the substrate to produce submicron-grooved PS films. Subsequently, a PS in toluene solution (5 wt%) was added dropwisely onto the PET plate. The patterned side of the PDMS mold was then inverted onto the PS solution, and pressure was added onto the back of the mold to ensure uniformity. The PS films with submicron grooves were dried naturally overnight and then cut into 1 cm × 1 cm squares for the subsequent experiments.

2.2. Characterization of Submicron-Grooved Films

The integrity of the submicron-grooved topography was assessed by observing the morphology of the PS submicron-grooved films using a scanning electron microscope (SEM, Carl Zeiss Supra 55, Jena, Germany). Additionally, as the hydrophilicity of the substrate surface is a critical factor for cell behaviors, we measured the water contact angle of the PS submicron-grooved films with or without air plasma treatment (0.4 mbar, 200 W and 3 min) using a contact angle measurement instrument (Theta Lite, Biolin, Espoo, Finland), with three parallel samples being used for the measurement of each group.

2.3. Cell Culture

The immortalized human-derived Schwann cells obtained from plexiform neurofibroma (ATCC, CRL-3390, Lot Number: 70024530) were purchased from the Cell Bank, Chinese Academy of Sciences (Shanghai, China). The Schwann cells were cultured in high-glucose Dulbecco's modified Eagle's medium (DMEM, Hyclone, Logan, UT, USA) and supplemented with 10% (*v*/*v*) fetal bovine serum (FBS, Gibco, NSW, Australia) and 1% antibiotics of penicillin/streptomycin (100 units/mL of penicillin and 100 mg/mL of streptomycin, Hyclone, Logan, UT, USA).

2.4. Cell Viability

The submicron-grooved films with various sizes were placed in 24-well plates, cleaned by air plasma for 3 min, sterilized with 75% ethanol, and then washed three times with PBS before cell seeding. For the cell proliferation assay, Schwann cells were seeded in 24-well plates at a density of 3×10^4 cells/mL. After 1, 3, and 5 days of cell culture, the medium was removed, and cells on the samples were washed thrice with PBS. Next, 10% Cell Counting Kit-8 reagent (CCK-8, Beyotime, Shanghai, China) was added and incubated at 37 °C for another 4 h, and 150 μL of supernatant was removed from each well and transferred to a 96-well plate. Finally, the absorbance value of the supernatant was measured at 450 nm using a SpectraMax M5 microplate spectrophotometer (Molecular Devices, San Jose, CA, USA). The absorbance value determined from the flat group was used as control to calculate the cell viability.

Live–dead staining was also performed to determine the viability of Schwann cells on submicron-grooved films. After 3 days of cell culture, the medium was removed, and the cells were stained with 2 μM Calcein-AM (Beyotime) and 4.5 μM propidium iodide (PI, Beyotime) for 15 min at 37 °C. The stained samples were then imaged using a fluorescence microscope (Olympus, Tokyo, Japan), while the live cells were identified by green fluorescence and the dead cells were identified by red fluorescence.

2.5. Cell Morphology

After 24 h of cell culture, the morphological features of Schwann cells cultured on submicron-grooved films were evaluated. The overall and individual morphological characteristics of Schwann cells on different samples were observed using optical and scanning electron microscopes (SEM), respectively. Thirty cells were randomly selected from different groups based on SEM images, and their cell area, and aspect ratio and orientation angle parameters were quantified using ImageJ software (v1.8.0).

2.6. Cell Adhesion

To evaluate the cell adhesion of Schwann cells on the submicron-grooved films, the Schwann cells cultured on different samples for 24 h were fixed using 4% paraformaldehyde for 20 min and permeabilized with 0.2% Triton X-100 in PBS for 10 min. After being washed thrice with PBS, the cells were blocked with 3% bovine serum albumin (BSA) at 37 °C for 1 h. Next, a primary antibody against vinculin (ab129002; Abcam, Cambridge, UK) was added at a dilution of 1:250 in 1% BSA and incubated at 4 °C overnight, followed by incubation with a secondary antibody of Alexa Fluor 488 for 3 h. In addition, Rhodamine 555 phalloidin (1:200) and DAPI (1:1000) were used to stain the cytoskeleton and nuclei of cells for 30 min. Finally, the stained samples were visualized and imaged using a Zeiss LSM-710 laser-scanning confocal microscope (LSM-710, Zeiss, Jena, Germany).

2.7. Cell Migration

The migration trajectory of Schwann cells on submicron-grooved films was recorded using the JuLI Stage Living cell monitoring system (Nano Entek, Seoul, South Korea) equipped with an incubator (37 °C, 5% CO_2). Cells were seeded onto the submicron grooves at a density of 3×10^4 cells/mL and allowed to adhere for 1 h before being transferred to the monitoring incubator. The movement of Schwann cells on the submicron grooves was recorded at 1 h intervals for 24 h.

The images of individual cells were manually tracked and analyzed using ImageJ software to obtain a series of X/Y coordinates corresponding to a time interval of 1 h. The original position of each cell was defined as (0, 0), and the trajectory of cells in the X/Y plane during the 24 h was reconstructed. Thirty cells were randomly selected from each group to record the trajectories.

2.8. Cell Cycle Analysis

Flow cytometry (BD Bioscience, San Jose, CA, USA) was utilized to evaluate the cell cycle progression of Schwann cells on submicron-grooved films. After 5 days of culturing on various samples, the Schwann cells were digested and collected in centrifuge tubes, which were then centrifuged at 1000× g for 5 min. Subsequently, the cells were fixed overnight by adding 70% ethanol pre-cooled in an ice bath. Next, the cells were rinsed with pre-cooled PBS and slowly resuspended in a prepared propidium iodide staining solution, which was then subjected to a 30 min incubation at 37 °C in darkness. Finally, the cell precipitate was filtered to obtain single cells, and the flow assay was completed within 1 h. ModFit LT5.0 analysis software was employed to analyze the cellular DNA content.

2.9. RNA Sequencing

The transcriptome sequencing was conducted according to the manufacturer's instructions (OE Biotech Co., Ltd., Shanghai, China). Total RNA from Schwann cells cultured on samples for 5 days was extracted using TRIzol reagent (Invitrogen, Carlsbad, NM, USA) and stored in liquid nitrogen. The RNA integrity was evaluated using an Agilent 2100 Bioanalyzer (Agilent Technologies, Santa Clara, CA, USA). The VAHTS Universal V5 RNA-seq Library Prep Kit (Vazyme, Nanjing, China) was used to construct transcriptome libraries per the manufacturer's instructions. The libraries were sequenced using the Illumina Novaseq 6000 sequencing platform to obtain read counts for each sample. Differentially expressed genes (DEGs, with a q value < 0.05 and fold change > 2) were analyzed using the DESeq2 software. Enrichment analysis of DEGs was performed using GO Reactome and KEGG based on the hypergeometric distribution.

2.10. Gene Expression

After culturing Schwann cells on various submicron-grooved films for 5 days, real-time quantitative polymerase chain reaction (RT-qPCR) was used to assess the expression levels of target genes. Total RNA was extracted from Schwann cells using TRIzol reagent (Invitrogen, Carlsbad, NM, USA), and 1 μg of total RNA was reverse transcribed to cDNA using Superscript III reverse transcriptase with random hexamer primers (RR037A, Takara, Beijing, China). RT-qPCR was performed on a real-time fluorescence quantitative PCR instrument (LightCycler 96, Roche, Basel, Switzerland) using the following amplification parameters: 95 °C for 2 min, 40 cycles at 95 °C for 15 s, 60 °C for 15 s, and 72 °C for 30 s. The primers of target genes (Table 1) were synthesized by Guangzhou IGE Biotechnology Ltd., and GAPDH was used as the housekeeping gene for normalization. The fold changes in the expression of each target gene were compared by calculating $2^{-\Delta\Delta Ct}$.

Table 1. The primer sequences used for RT-qPCR analysis.

Gene Name	Primer Sequences (5′→3′)
GAPDH	F: TCGGAGTCAACGGATTTGGT
	R: TTCCCGTTCTCAGCCTTGAC
Smad 6	F: GCTACCAACTCCCTCATCACT
	R: CGTACACCGCATAGAGGCG
MBP	F: CAGCACTAACCCTCCCTGAG
	R: GCTCAGCCCGTGTGTCTC
S100	F: CGGGGCACCCTTAAAACTTC
	R: ATCTGCATGGATGAGGAACGC
Sox10	F: CACAAGAAAGACCACCCGGA
	R: AAGTGGGCGCTCTTGTAGTG

2.11. Cell Membrane Potential

To record the membrane potential of cells on different samples, electrophysiological experiments were conducted using a HEKA EPC10 amplifier (HEKA Elektronik GmbH, Lambrecht, Goettingen, Germany) in the current clamp recording mode. After being cultured on submicron-grooved (800/400) and flat films for 3 days, Schwann cells were perfused with a standard extracellular solution containing 140 mM NaCl, 3 mM KCl, 1.5 mM $MgCl_2$, 2 mM $CaCl_2$, 10 mM HEPES, and 10 mM Glucose. The pH of the solution was adjusted to 7.4 with NaOH, and the osmolarity was between 300–320 mOsm. The internal solution contained 164 mM KCl, 2 mM $MgCl_2$, 1 mM $CaCl_2$, 10 mM HEPES, 11 mM EGTA, and 10 mM glucose, with a pH of 7.3 adjusted with KOH. All the electrophysiological experiments were performed at room temperature.

2.12. Statistical Analysis

The significant differences between groups were evaluated using ANOVA, with one-way ANOVA analysis performed using GraphPad Prism 8 software. The results are presented as the mean ± standard deviation (SD) of 3 independent experiments. A p value less than 0.05 was considered to be statistically significant.

3. Results

3.1. Characterization of Submicron-Grooved Films

Soft lithography was used to prepare grooved PS patterns. Four PDMS stamps with submicron groove structures were made from the master silicon wafers. A PS solution was dropped on a PET substrate and dried under the PDMS stamps (Scheme 1). The structural integrity of the grooves is crucial for the reliability of the cell culture platform. In this respect, SEM was utilized to assess the surface geometry of various submicron-grooved films. As show in Figure 1A, the submicron grooves possess a complete and uniform structure, with pronounced ridges and furrows typical of grating structures. The different light reflection from the submicron-grooved films leads to differences in brightness and darkness in the background, where the shallower grooves exhibit a lower contrast, whereas the deeper grooves show a higher contrast for the same width.

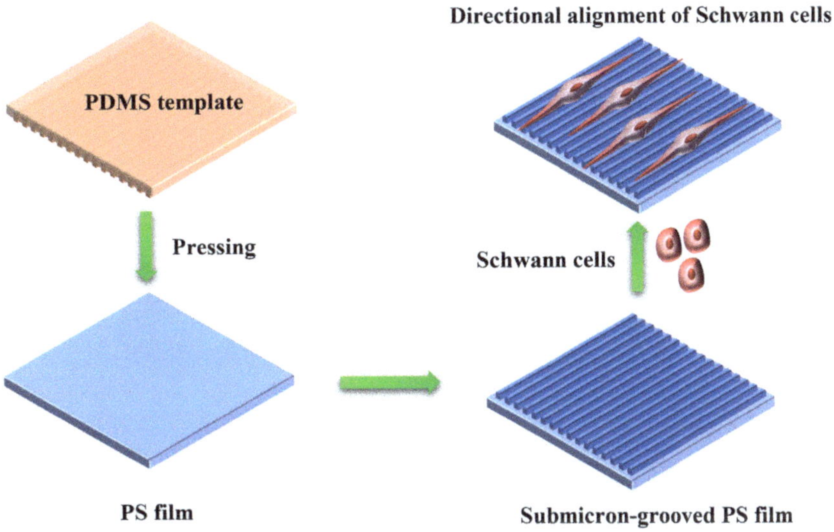

Scheme 1. A schematic illustration showing the fabrication process of submicron-grooved films, which can be utilized to direct the alignment of Schwann cells.

Surface wettability is another critical factor affecting cell adhesion. Therefore, water contact angles are measured on different submicron-grooved films with or without air plasma treatment using a contact angle measurement instrument. As depicted in Figure 1B,C, all the samples exhibit an explicit hydrophobicity with water contact angles greater than 90°. Notably, the water contact angle measured on the submicron-grooved films is considerably larger than that determined on the flat films. With increased groove depth, the water contact angle progressively rises to approximately 110° (800/100 and 400/100) and 120° (800/400 and 400/400). After plasma treatment, the grooved patterns became superhydrophilic due to a lateral capillary force inside the grooves, which is consistent with a previous study [37].

3.2. Cell Viability

Both live–dead staining and CCK-8 assays were performed to evaluate the viability of Schwann cells cultured on different samples. After 2 days of culturing on different submicron-grooved films, Schwann cells were stained with Calcein-AM and PI. Figure 2A shows the images of live–dead stained cells, and no significant difference can be found between the different groups. The abundant green fluorescence of viable cells demonstrates the excellent cytocompatibility of the groove samples. The CCK-8 assay evaluated the proliferation of the Schwann cells after 1, 3, and 5 days of culturing on the four samples. The proliferation analysis of the Schwann cells, shown in Figure 2B, indicates a minimal variation between the grooved and flat films at different cell culture times, and only the cell viability determined on the 800/400 grooved film is slightly lower than the other groups at day 5.

3.3. Cell Morphology

The impact of a submicron-grooved structure on cell morphology is primarily reflected in the directional arrangement of the Schwann cells. To evaluate the cell morphology, the cells were observed and analyzed after 24 h by optical microscopy and SEM. As shown in Figure 3A, the cells adhere randomly on the flat surfaces with circular or triangular cell shapes. However, on the grooved films, the cells show a directional arrangement, elongated in a linear pattern and growing in the same direction. In more detail, the degree of the directional alignment of the Schwann cells on different films follows the order of

800/400 > 400/400 > 400/100 > 800/100 > Flat. The SEM images shown in Figure 3B confirm that Schwann cells can span several ridges. The filopodia were located mainly in the two ends of the cells. Conversely, on the flat film, the cells had many filopodia and extensions surrounding the cells.

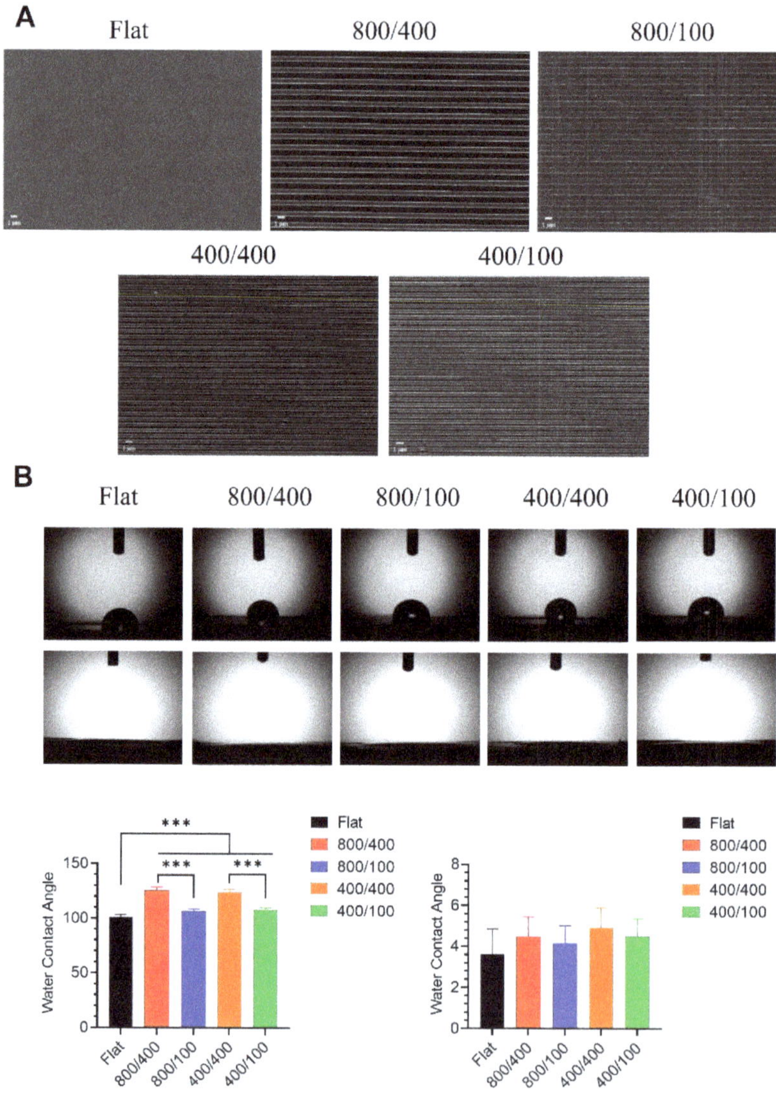

Figure 1. (**A**) Surface morphology of different submicron-grooved and flat films observed by SEM; scale bars are 1 μm. (**B**) Water contact angles detected on different samples without (above) or with (below) air plasma treatment and the corresponding statistical analysis. *** represent $p < 0.001$, respectively (n = 3).

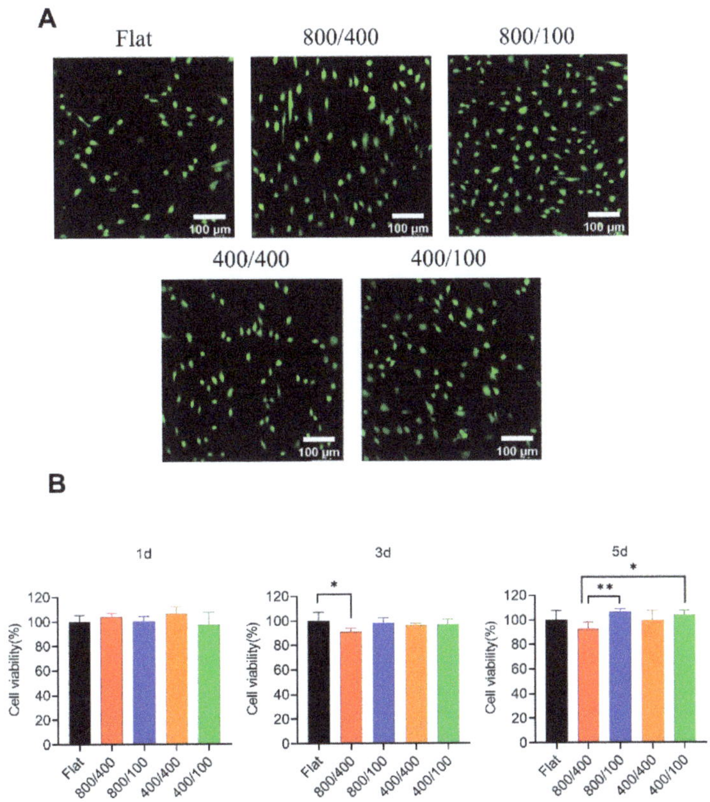

Figure 2. (**A**) Live–dead staining images of Schwann cells on different samples; scale bars are 100 µm. (**B**) Cell viability of Schwann cells cultured on different samples for 1, 3, and 5 days. * and ** represent $p < 0.05$, $p < 0.01$, respectively.

Other cell morphological indicators including area, aspect ratio or elongation (length/width, L/W), and orientation angle (OA) are also evaluated to assess the impact of different submicron grooves (Figure 4A). Figure 4B shows that the Schwann cells spread well on flat film, resulting in a significantly larger cell area than those cultured on the submicron-grooved samples. Among the samples with grooved structures, no significant difference in cell area can be observed. The cell elongation showed that the Schwann cells cultured on submicron grooves exhibit a significantly greater L/W value than those on flat film (Figure 4C). Moreover, the L/W values of cells on 800/400 grooves are significantly higher than those on 800/100 and 400/100 grooves. As direct evidence of cell arrangement, the orientation angles of all the cells on the different submicron grooves are below 20°, compared to the random orientation of cells on flat film (Figure 4D). The different grooves pose different effects on the orientation of Schwann cells, with the lowest orientation angle (the highest alignment of cells) being observed on the 800/400 grooves.

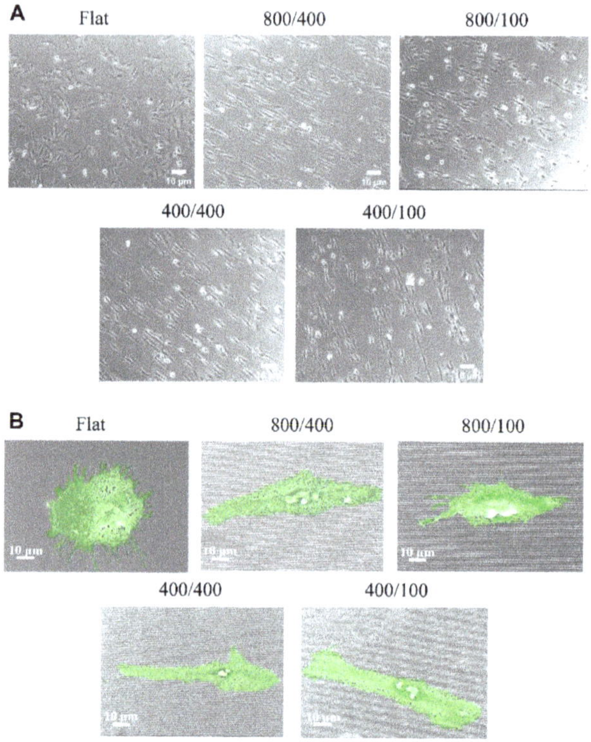

Figure 3. Cell morphology of Schwann cells cultured on different samples: (**A**) Optical microscope photographs and (**B**) SEM photographs. Scale bars are 10 μm.

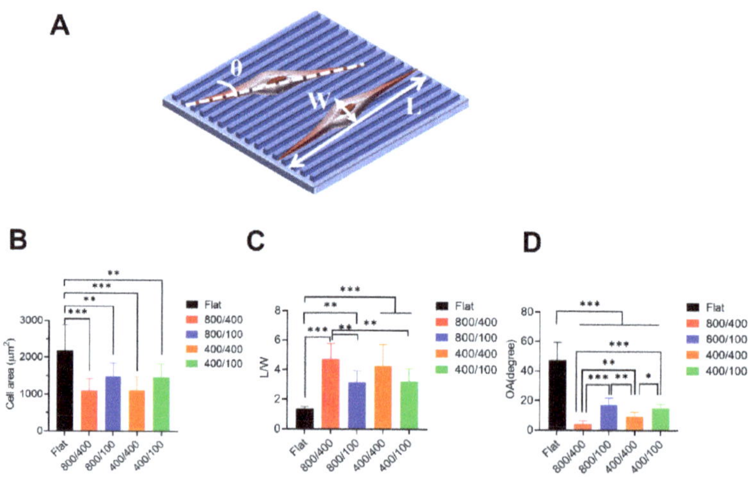

Figure 4. (**A**) A schematic illustration showing the calculation method of the length (L)-to-width (W) ratio and the orientation angle (θ) of Schwann cells; (**B**) cell area of Schwann cells cultured on different samples; (**C**) L/W ratio of Schwann cells cultured on different samples; (**D**) orientation angle (OA) of Schwann cells cultured on different samples. *, **, and *** represent $p < 0.05$, $p < 0.01$, and $p < 0.001$, respectively (n = 30).

3.4. Cell Adhesion and Cytoskeleton

In a next step, the effect of submicron grooves on the adhesion and skeleton of Schwann cells is evaluated by immunofluorescence staining. After the Schwann cells adhered to the submicron-grooved films for 24 h, the cells were immunofluorescence-stained with phalloidin and vinculin antibodies and observed under a laser confocal microscope. As shown in Figure 5, the F-actin microfilament structure of the Schwann cells forms bundles that elongate along the grooves, which is quite different from that observed on the flat film. Moreover, vinculin is evenly distributed at both ends of the elongated cells on the grooves, whereas it is primarily located around the nuclei of the Schwann cells cultured on the flat surface. These findings suggest that the groove structure can reshape the cytoskeletal structure and influence the distribution of adhesion proteins.

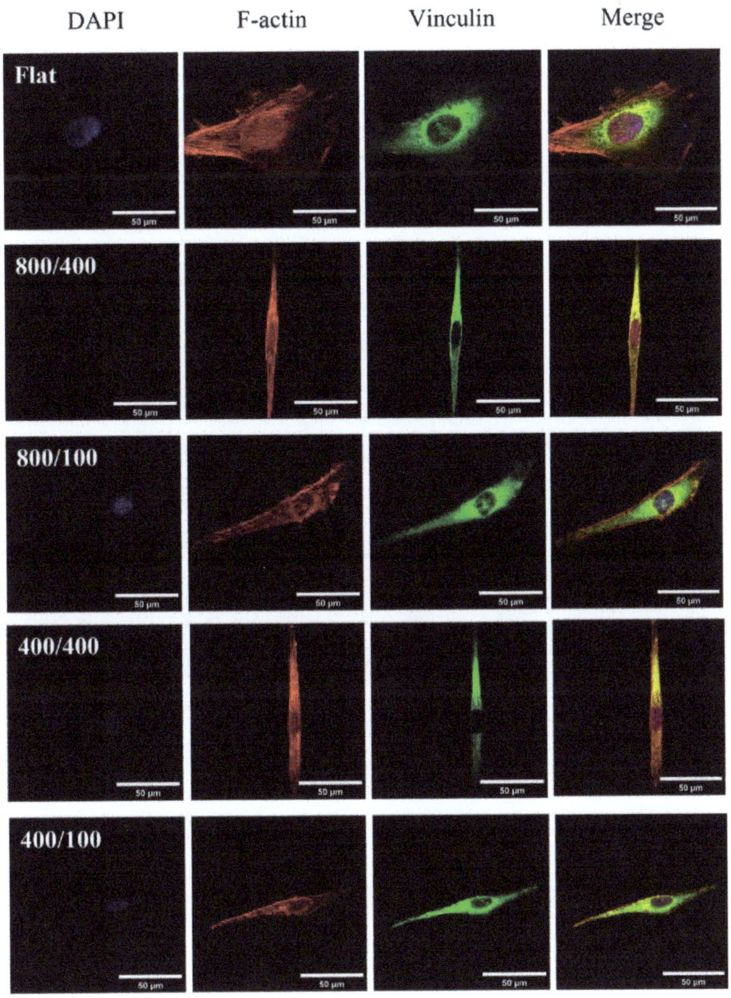

Figure 5. Immunofluorescent staining of Schwann cells cultured on different samples by laser confocal microscopy. The nuclei, F-actin, and vinculin were stained with DAPI (blue), phalloidin (red), and primary antibodies of anti-vinculin (green), respectively. Scale bars are 50 μm.

3.5. Cell Migration

To evaluate the cell migration, cell positions were captured every hour within 24 h after cells seeding on different samples. The cell movement trajectory is displayed in the X/Y axis coordinate plot (Figure 6), with the origin (0, 0) as the starting point. The results show that most of the Schwann cells migrate linearly along the Y-axis direction on the 800/400 and 400/400 grooves, and the movement of the Schwann cells is relatively less directional on the 400/100 and 800/100 grooves. In contrast, the movement of the Schwann cells on the flat surface was non-directional. These findings suggest that the depth of the grooves plays an essential role in directing the migration of Schwann cells.

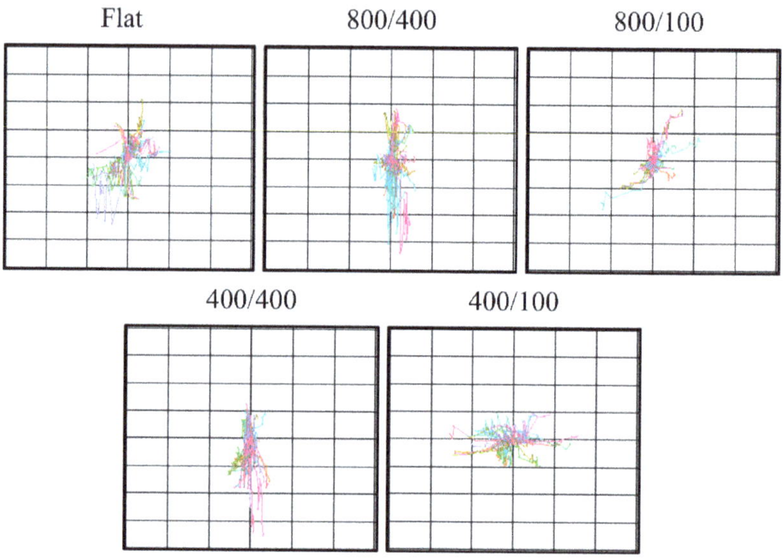

Figure 6. The migration traces of Schwann cells on different samples within 24 h.

3.6. Cell Cycle Analysis

After 5 days of culturing, the Schwann cells cultured on different samples were stained with PI. Then, the cell cycle was analyzed using flow cytometry. As depicted in Figure 7, the G1 phase is predominant for the Schwann cells cultured on both the grooved and flat samples, with no noticeable differences among them. The highest percentage of Schwann cells in the G2 phase can be observed on the flat sample (9.03%), followed by the 800/100 grooves (8.92%), while minimal differences can be observed among the 800/400, 400/400, and 400/100 groups (7.86–8.13%). The S phase is the DNA synthesis phase of cells. According to the results, the percentage of Schwann cells in the S phase is slightly higher in the flat (13.13%), 800/100 (13.03%), and 400/100 (12.96%) groups compared to the 800/400 (12.06%) and 400/400 (12.23%) groups. Our findings suggest that the depth relative to the width of the grooves has a slight effect on the cell cycle of Schwann cells.

3.7. Gene Expression

It is demonstrated by the aforementioned results that the Schwann cells exhibit the most conspicuous cell alignment behavior on the 800/400 groove among the various groups. Thus, RNA sequencing were performed on the 800/400 and flat films for 5 days. Trizol was used to collect total RNA from the cells cultured on different films after 5 days, and the gene expression was analyzed using an Illumina platform.

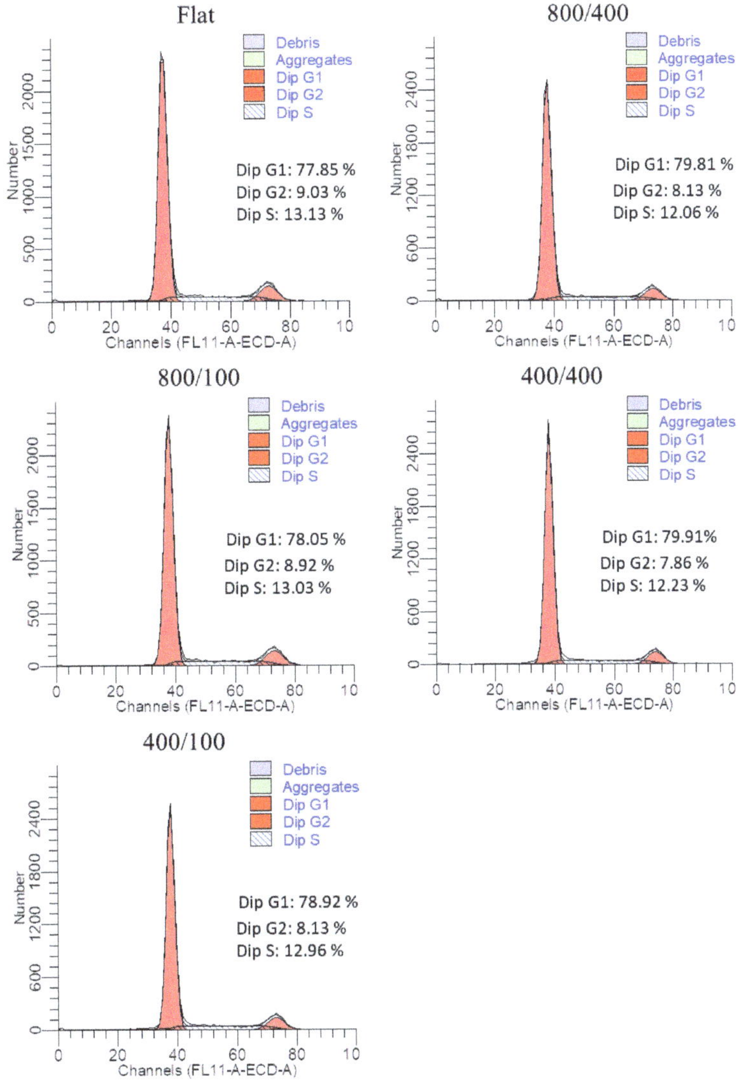

Figure 7. The cell cycle of Schwann cells after being cultured on different samples for 5 days.

The outcomes illustrated in Figure 8A,B reveal a total of 96 differently expressed genes (DEGs) between the 800/400 and the flat films, of which 45 genes are upregulated and 51 genes are downregulated. The upregulated DEGs are mainly associated with axon guidance, synaptic complex assembly, extracellular matrix secretion, cytoskeleton, and protein localization and transport, whereas the downregulated DEGs are primarily related to cell differentiation, cell membrane potential regulation, immune regulation, inflammatory response, and cell spreading. In addition, KEGG pathway analysis and GO enrichment were conducted to gain further insight into the functional annotations of the DEGs in both cases. Notably, the findings indicated that pathways influencing axon extension were predominantly enriched in the submicron-grooved group compared to the flat group, including the negative regulation of axon extension, neuroactive ligand–receptor interaction, calcium-ion-regulated exocytosis, etc.

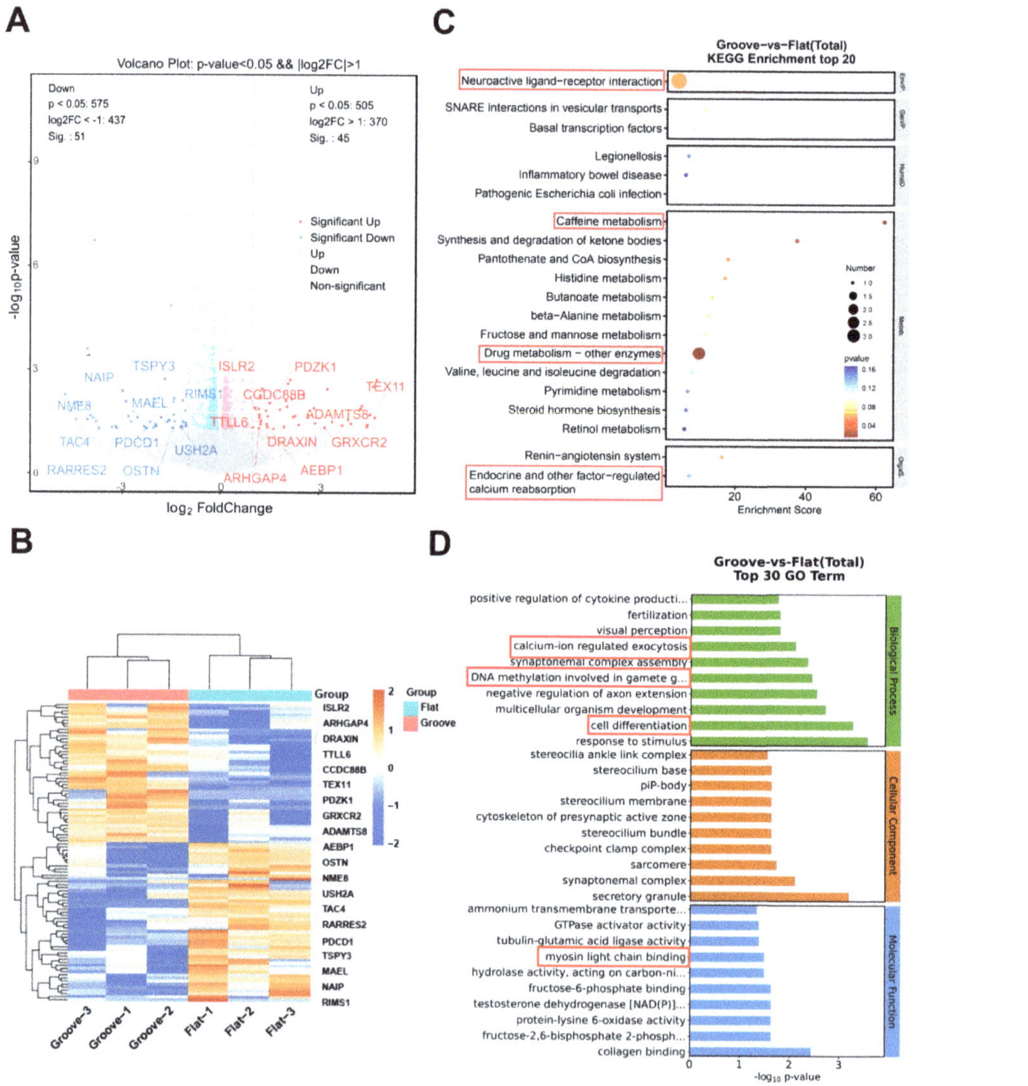

Figure 8. Transcriptome comparison of Schwann cells after being cultured on submicron-grooved (800/400) and flat films for 5 days (n = 3): (**A**) Volcano plots the total number of genes identified by RNA sequencing. (**B**) The heat map represents the differentially expressed genes. DEGs were defined as |log2(fold change)| > 1 and q value < 0.05. (**C**)The KEGG pathway enrichment of DEGs under submicron-grooved and flat films conditions. (**D**) Gene Ontology (GO) enrichment analysis of DEGs in both cases.

Afterwards, the expression of specific genes of the Schwann cells on all grooved patterns is determined by RT-qPCR. Figure 9 shows the gene expression of Schwann cells in different groups after 5 days of cultivation. The results demonstrate that the submicron grooves can upregulate the expression of MBP and Smad6 in Schwann cells, while their effect on the expression of Sox10 and S100 is negligible. Similarly to the previous findings, the MBP and Smad6 expressions are mainly affected by the depth of the grooves as a significantly higher expression of these two genes can be detected in the

800/400 and 400/400 groove groups, with the 800/400 groove group demonstrating the most pronounced effect.

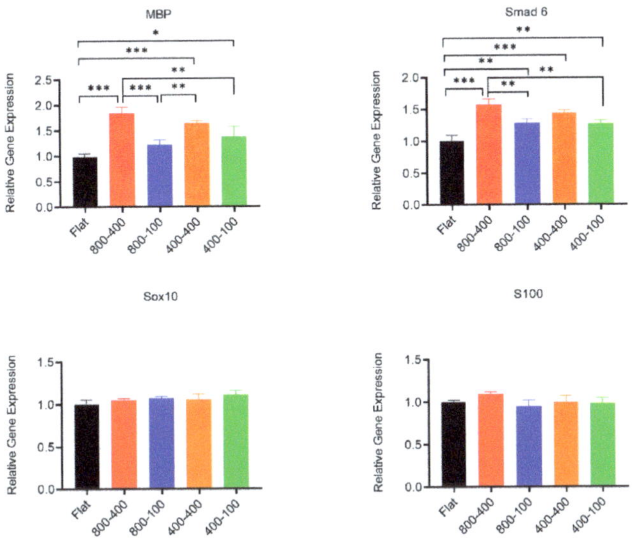

Figure 9. Relative gene expression associated with specific markers of Schwann cells after being cultured on different samples for 5 days. *, **, and *** represent $p < 0.05$, $p < 0.01$, and $p < 0.001$, respectively (n = 3).

3.8. Cell Membrane Potential

After 3 days of culturing, the patch clamp measurement is utilized to investigate the disparity in the membrane potential of the Schwann cells cultured on the 800/400 groove and flat films. As illustrated in Figure 10, the membrane potential of the Schwann cells cultured on the 800/400 groove is approximately −10 mV, while the value determined from the flat film group is approximately −40 mV. These findings imply that submicron grooves can considerably alter the membrane potential of cultured Schwann cells, which may be attributed to their effect on cell morphology and orientation.

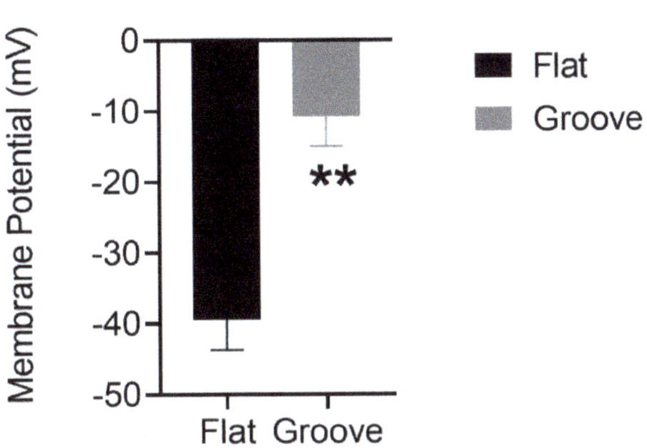

Figure 10. Membrane potential of Schwann cells after being cultured on submicron-grooved (800/400) and flat films for 3 days. ** represent $p < 0.01$ (n = 5).

4. Discussion

Peripheral nerve injury is a prevalent condition caused by accidents or natural disasters, which causes over one million people to suffer severe physical and psychological injuries annually [38–40]. Although peripheral nerves are self-repairable, such ability is limited for the repair of distant nerve damage (typically greater than 5 mm) [41]. Autologous nerve grafting, the gold standard for repairing peripheral nerve injury, is limited for clinical application due to the source limitations, secondary surgery, and permanent damage to the donor area [25,42]. Nerve conduits constructed by materials with surface topography have shown great potential in peripheral nerve repair as they can guide the growth of nerve protrusions and axonal extension [43,44]. Several studies have demonstrated that well-aligned groove structures can provide physical guidance and promote the alignment of Schwann cells [45]. However, most of the previous studies have focused on the effect of micron-sized grooves on the behaviors of Schwann cells [21,46]. In this study, we investigated the cellular behaviors, such as adhesion, proliferation, and migration, of Schwann cells on submicron grooves with different sizes.

Furthermore, the effects of different grooves on cell cycle, cell membrane potential, and gene expression profiles were investigated, aiming to better understand the biological functions and potential mechanisms of submicron grooves in regulating Schwann cells. Our experimental results show that Schwann cells exhibit significant directional growth and migration on the submicron-grooved films compared to the random orientation of the flat film and that this directional guidance effect was closely related to the depth of the grooves. Furthermore, the submicron groove structure elongates Schwann cells, resulting in restricted cell spreading. Thus, the area of the Schwann cells on the submicron grooves is much smaller than on the flat films. However, the cell viability and cell cycle of Schwann cells did not differ significantly between the grooved and flat surfaces. In addition, the submicron-grooved films significantly upregulated the expression of genes related to axonal regeneration and myelination relative to flat films. Finally, the submicron grooves could also substantially affect cell membrane potential compared to the flat surfaces.

In our study, submicron-grooved films were prepared using soft lithography, which offers advantages such as ease of preparation, high efficiency, and reusability [47,48]. Biocompatible and biodegradable polymers such as polylactide-poly (PCL) [46], polypropylene-glycolic acid copolymer (PLGA) [29], polylactic acid (PLA) [14], gelatin [49], and chitosan [33] are commonly used in peripheral nerve repair studies. Among them, PCL is one of the most used synthetic polymers to manufacture NGCs. PCL is an FDA-approved material with excellent mechanical properties and ease of processing [50]. Additionally, collagen, as a natural polymer, also has good biocompatibility and, thus, is widely used for axonal regeneration [51,52]. In this study, Polystyrene (PS) was used to prepare the submicron-grooved films because it provides good biocompatibility, excellent film-forming ability, and water stability, ensuring the formation of stable and precise groove structures. The SEM images reveal that the prepared submicron-grooved films have a well-defined ridge/furrow structure essential for the aligning and growth of Schwann cells. It is well known that the surface wettability of biomaterials is critical for the adhesion and proliferation of mammalian cells [53]. The results show that all the submicron-grooved films, as well as the flat film, are changed from hydrophobic to hydrophilic after plasma treatment, which is favorable for cell adhesion.

The morphology and orientation of cells are the most intuitive phenomena of cells cultured on biomaterials. Several previous studies have indicated that biomaterials with groove structures can direct the alignment and elongation of stem cells [49,54,55]. Nevertheless, most of the previous studies contribute to the optimization of groove dimensions (such as depth, width, and spacing) in the micron scale. To the best of our knowledge, this study is the first attempt to investigate the effect of submicron grooves on Schwann cell behaviors. Specifically, it is observed that submicron grooves can modulate the directional orientation of Schwann cells within 1 day, in which most of the cells are elongated and aligned in the same direction of the submicron grooves. In contrast, the cells cultured on the flat film are randomly distributed without orientation. A further SEM analysis reveals that Schwann cells, which have a size of approximately 10 μm, can span many submicron ridges and furrows. Interestingly, we found that the depth of the groove is a crucial factor influencing the orientation arrangement of Schwann cells on submicron grooves. Specifically, our results show that the degree of orientation of the Schwann cells increases with the depth of the groove, which is consistent with a previous study [29]. However, in contrast to our findings, some other studies have demonstrated groove width as the most critical parameter influencing the arrangement of Schwann cells [56]. This difference may be due to the varied sizes of grooves used in different studies. In our study, we selected submicron grooves over micron ones and found that the effect of groove depth on Schwann cell morphology is more pronounced than that of groove width. Compared to the flat film, submicron grooves can significantly reduce cell area, increase aspect ratio, and decrease the orientation angle of Schwann cells. These observations suggest that submicron groove structures may inhibit cell spreading.

These findings indicate that Schwann cells can rapidly respond to the structures of submicron grooves and undergo morphological changes. It is noteworthy that our findings are contradictory to the previous notion that only the groove structures with dimensions close to cells can be perceived by cells [57]. We propose that the effect of submicron grooves in guiding Schwann cell arrangement relies on the response of cytoskeleton and vinculin proteins. Immunofluorescence staining confirms that the groove structures can reshape the distribution of F-actin and vinculin along the groove direction, and the cytoskeleton and nucleus are significantly elongated by the guidance of the submicron grooves. Taken together, our results demonstrate that submicron grooves can regulate the morphology and arrangement of cultured Schwann cells, with groove depth being a more critical factor than groove width.

Generally, the alternation in cell morphology may impact the biological functions of cells. In our study, the outcomes of CCK8 indicate that the grooves do not compromise the proliferation of Schwann cells, which contrasts the results of previous studies that suggest that groove structures are unfavorable for cell proliferation [58,59]. This difference may be attributed to the relatively small size of grooves employed in our study, leading to stronger cell–groove contact than the boundary effects [30]. The cell cycle analysis can reflect intracellular DNA synthesis, which is closely related to cell proliferation [60]. Our results demonstrate that there is no significant difference of the cell cycle of Schwann cells among different groups, which is consistent with the results of cell proliferation.

The migration of Schwann cells also plays a crucial role in neural repair [61,62]. Consequently, we investigated the migration behavior of Schwann cells on submicron grooves and found that most Schwann cells maintain linear movement in the same direction of submicron grooves, whereas the movement of Schwann cells on flat film is random. This difference may be related to the traction force provided by F-actin in the cytoskeleton. The groove structures can reshape the cytoskeleton arrangement, and, thereby, enough traction force is contributed to the directional migration of the Schwann cells on the grooves [63]. Interestingly, the directional migration of the Schwann cells is closely associated with the depth of the grooves, with more Schwann cells migrating directionally along the grooves as the depth of the grooves increases. These results indicate that the groove depth can not only affect the morphology of Schwann cells but also provide a pronounced effect on cell migration.

To gain a better understanding of the regulatory role of groove structure on Schwann cell behaviors, we further investigated the effect of submicron grooves on the gene expression of Schwann cells using RNA sequencing and RT-qPCR. Ninety-six DEGs were identified by high-throughput screening. Subsequent GO enrichment analysis revealed that the genes associated with axon guidance, synaptonemal complex assembly, extracellular matrix secretion, cytoskeleton, and protein localization and transport are upregulated. In contrast, the genes related to cell differentiation, membrane potential regulation, immune response regulation, inflammatory response, and substrate adhesion-dependent cell spreading are downregulated. The gene expressions of the specific markers (MBP, Smad6, S100, and Sox10) in Schwann cells were also evaluated. Our results indicate that the submicron grooves can significantly increase the levels of Smad6 and MBP genes. Smad6 is a crucial axonal membrane protein that aids neuronal plasticity and regeneration [64]. The upregulation of Smad6 gene is beneficial for axonal growth. Similarly, MBP is essential for Schwann cells to form myelin around nerve axons [65,66]. An increase in the gene level of MBP suggests that Schwann cells are beginning to form myelin around nerve axons [67]. Taken together, our findings demonstrate that submicron grooves are favorable to axon regeneration and myelination, which is critical for peripheral nerve regeneration.

The alternation of cell morphology and cytoskeletons can possibly impact the membrane potential of cells. In this regard, we assessed the membrane potential of Schwann cells cultured on submicron-grooved and flat films. Our results demonstrate that the submicron groove structures can significantly alter the membrane potential of Schwann cells, indicating the remarkable change in the membrane proteins [68,69]. Additionally, micro-patterns have been found to interact with cell surface proteins, triggering a series of cascade reactions within cells [70]. This interaction could lead to changes in the expression of receptors related to membrane potential in Schwann cells, including sodium and potassium receptors, resulting in alterations in membrane potential. These findings verify that the submicron grooves substantially affect the behaviors of Schwann cells.

Normal nerve tissue is distributed in long strips with good directional growth characteristics, and simulating this structure will facilitate nerve regeneration. Previous reports have stated that ideal nerve guidance conduits must have the structural characteristics of longitudinally aligned regenerating axons [71,72]. Surfaces with nanoscale gratings, microgrooves, and channels are commonly used as substrates in neural differentiation studies because they mimic the natural neuronal and glial cell microenvironment in vivo [73,74], thus facilitating axon regeneration and development. In this paper, we use submicron groove structures to provide a linear 3D-like environment for Schwann cells. From a physiological perspective, submicron grooves can mimic and guide the directional growth and migration of Schwann cells in vivo, reducing axonal dispersion and mistargeted regeneration. The expression of key genes for axonal regeneration and myelination upregulated by submicron-grooved films could illustrate their potential application in peripheral nerve regeneration.

The in vitro construction of artificial nerve guidance conduits (NGCs) is currently the most common method for peripheral nerve repair. Many studies have reported that incorporating micro- and nano-scale filler materials into NGCs can promote axonal regeneration. These fillers include microlumen/microchannel structures, neatly aligned electrospun fibers, hydrogel matrices, and conduits rolled up by material sheets. In vivo, collagen–chitosan (CCH) scaffolds with longitudinal, parallel microchannels showed nerve regeneration and functional recovery similar to autografts when used to repair 15 mm sciatic nerve defects in rats. This micropatterning affects the migration and axonal regeneration of Schwann cells at the microscopic level of these cellular interactions [75]. Interestingly, many studies have used rolled up micro- or nanopatterned 2D surfaces, or patterned micro- or nanomaterials as filler materials for NGCs. For example, the surface of self-rolling silicon nitride microtubule arrays can provide growth channels and accelerate neuronal cell growth [76]. Another study also showed that micron-sized grooved surfaces can promote the development and directed axonal growth of adult neural stem cells and reduce axonal dispersion and mistargeted regeneration [77]. Therefore, we believe that these measures of nerve conduit grafts prepared from rolled-up 2D submicron-grooved films or replicating submicron groove structures on NGC fillers could provide practical applications for peripheral nerve repair.

5. Conclusions

In our study, soft lithography is employed to fabricate submicron-grooved films with four different sizes. These films are utilized as cell culture platforms to assess the behaviors of Schwann cells. Our results reveal that submicron groove structures possess an efficacy similar to micron groove structures reported previously in directing the alignment and migration of Schwann cells. However, this effect is contingent on the depth of the submicron grooves. While the submicron groove structures significantly transform the morphology and cytoskeleton of Schwann cells, their impact on cell proliferation and cell cycle is negligible. More importantly, the submicron grooves can upregulate the gene expressions of Schwann cells responsible for axonal regeneration and myelination, which is promising for peripheral nerve repair. In sum, this study lays the groundwork for developing advanced biomaterials used for peripheral nerve repair.

Author Contributions: Conceptualization, H.W., P.-Y.W. and Z.Z.; methodology, Z.Z., H.W. and P.-Y.W.; formal analysis, Y.L., J.H., J.S. and P.D.; data curation, J.S., P.O. and J.L.; writing—original draft preparation, Z.Z.; writing—review and editing, H.W. and P.-Y.W.; supervision, H.W. and P.-Y.W.; project administration, H.W. and P.-Y.W.; funding acquisition, H.W. and P.-Y.W. All authors have read and agreed to the published version of the manuscript.

Funding: This research was financially supported by the National Natural Science Foundation of China, grant number 82272157; Ministry of Science and Technology of China, grant number 2022YFA1105101; Chinese Academy of Sciences, grant number 172644KYSB20200002 and 172644KYSB20200048; Department of Science and Technology of Guangdong Province, grant number 2021A0505030055; Guangdong Basic and Applied Basic Research Foundation, grant number 2021A1515012163; Zhejiang Provincial Natural Science Foundation of China, grant number LZ23C070004; Science, Technology, and Innovation Commission of Shenzhen Municipality, grant number ZDSYS20190902093409851.

Data Availability Statement: All the RNAseq raw data has been deposited at NCBI Gene Expression Omnibus (GEO) with the GEO accession number GSE226857.

Conflicts of Interest: The authors declare no conflict of interest.

References

1. Jiang, X.; Lim, S.H.; Mao, H.Q.; Chew, S.Y. Current applications and future perspectives of artificial nerve conduits. *Exp. Neurol.* **2010**, *223*, 86–101. [CrossRef] [PubMed]
2. Lopez-Cebral, R.; Silva-Correia, J.; Reis, R.L.; Silva, T.H.; Oliveira, J.M. Peripheral Nerve Injury: Current Challenges, Conventional Treatment Approaches, and New Trends in Biomaterials-Based Regenerative Strategies. *ACS Biomater. Sci. Eng.* **2017**, *3*, 3098–3122. [CrossRef]
3. Hussain, G.; Wang, J.; Rasul, A.; Anwar, H.; Qasim, M.; Zafar, S.; Aziz, N.; Razzaq, A.; Hussain, R.; de Aguilar, J.L.G.; et al. Current Status of Therapeutic Approaches against Peripheral Nerve Injuries: A Detailed Story from Injury to Recovery. *Int. J. Biol. Sci.* **2020**, *16*, 116–134. [CrossRef]
4. Ashur, H.; Vilner, Y.; Finsterbush, A.; Rousso, M.; Weinberg, H.; Devor, M. Extent of Fiber Regeneration after Peripheral-Nerve Repair—silicon Splint vs. Suture, Gap Repair vs. Graft. *Exp. Neurol.* **1987**, *97*, 365–374. [CrossRef]
5. Tabakow, P.; Jarmundowicz, W.; Czapiga, B.; Fortuna, W.; Miedzybrodzki, R.; Czyz, M.; Huber, J.; Szarek, D.; Okurowski, S.; Szewczyk, P.; et al. Transplantation of Autologous Olfactory Ensheathing Cells in Complete Human Spinal Cord Injury. *Cell Transplant.* **2013**, *22*, 1591–1612. [CrossRef] [PubMed]
6. Wu, R.B.; Wang, L.; Chen, F.Y.; Huang, Y.M.; Shi, J.M.; Zhu, X.F.; Ding, Y.; Zhang, X.C. Evaluation of artificial nerve conduit and autografts in peripheral nerve repair in the rat model of sciatic nerve injury. *Neurol. Res.* **2016**, *38*, 461–466. [CrossRef]
7. May, F.; Buchner, A.; Matiasek, K.; Schlenker, B.; Stief, C.; Weidner, N. Recovery of erectile function comparing autologous nerve grafts, unseeded conduits, Schwann-cell-seeded guidance tubes and GDNF-overexpressing Schwann cell grafts. *Dis. Model. Mech.* **2016**, *9*, 1507–1511. [CrossRef]
8. Muir, D. The potentiation of peripheral nerve sheaths in regeneration and repair. *Exp. Neurol.* **2010**, *223*, 102–111. [CrossRef]
9. McCormick, A.M.; Maddipatla, M.; Shi, S.J.; Chamsaz, E.A.; Yokoyama, H.; Joy, A.; Leipzig, N.D. Micropatterned Coumarin Polyester Thin Films Direct Neurite Orientation. *ACS Appl. Mater. Interfaces* **2014**, *6*, 19655–19667. [CrossRef]
10. Vijayavenkataraman, S. Nerve guide conduits for peripheral nerve injury repair: A review on design, materials and fabrication methods. *Acta Biomater.* **2020**, *106*, 54–69. [CrossRef]
11. Wang, P.Y.; Bennetsen, D.T.; Foss, M.; Ameringer, T.; Thissen, H.; Kingshott, P. Modulation of Human Mesenchymal Stem Cell Behavior on Ordered Tantalum Nanotopographies Fabricated Using Colloidal Lithography and Glancing Angle Deposition. *ACS Appl. Mater. Interfaces* **2015**, *7*, 4979–4989. [CrossRef] [PubMed]
12. Zhang, S.; Vijayavenkataraman, S.; Chong, G.L.; Fuh, J.Y.H.; Lu, W.F. Computational Design and Optimization of Nerve Guidance Conduits for Improved Mechanical Properties and Permeability. *J. Biomech. Eng.—Trans. ASME* **2019**, *141*, 8. [CrossRef] [PubMed]

13. Amani, H.; Kazerooni, H.; Hassanpoor, H.; Akbarzadeh, A.; Pazoki-Toroudi, H. Tailoring synthetic polymeric biomaterials towards nerve tissue engineering: A review. *Artif. Cell. Nanomed. Biotechnol.* **2019**, *47*, 3524–3539. [CrossRef] [PubMed]
14. Sharma, A.D.; Zbarska, S.; Petersen, E.M.; Marti, M.E.; Mallapragada, S.K.; Sakaguchi, D.S. Oriented growth and transdifferentiation of mesenchymal stem cells towards a Schwann cell fate on micropatterned substrates. *J. Biosci. Bioeng.* **2016**, *121*, 325–335. [CrossRef]
15. Fu, S.Y.; Gordon, T. The cellular and molecular basis of peripheral nerve regeneration. *Mol. Neurobiol.* **1997**, *14*, 67–116. [CrossRef]
16. Krekoski, C.A.; Neubauer, D.; Zuo, J.; Muir, D. Axonal regeneration into acellular nerve grafts is enhanced by degradation of chondroitin sulfate proteoglycan. *J. Neurosci.* **2001**, *21*, 6206–6213. [CrossRef]
17. Gu, X.S.; Ding, F.; Yang, Y.M.; Liu, J. Construction of tissue engineered nerve grafts and their application in peripheral nerve regeneration. *Prog. Neurobiol.* **2011**, *93*, 204–230. [CrossRef]
18. Clements, M.P.; Byrne, E.; Guerrero, L.F.C.; Cattin, A.L.; Zakka, L.; Ashraf, A.; Burden, J.J.; Khadayate, S.; Lloyd, A.C.; Marguerat, S.; et al. The Wound Microenvironment Reprograms Schwann Cells to Invasive Mesenchymal-like Cells to Drive Peripheral Nerve Regeneration. *Neuron* **2017**, *96*, 98–114. [CrossRef]
19. Miller, C.; Jeftinija, S.; Mallapragada, S. Micropatterned Schwann cell-seeded biodegradable polymer substrates significantly enhance neurite alignment and outgrowth. *Tissue Eng.* **2001**, *7*, 705–715. [CrossRef]
20. Gumy, L.F.; Bampton, E.T.W.; Tolkovsky, A.M. Hyperglycaemia inhibits Schwann cell proliferation and migration and restricts regeneration of axons and Schwann cells from adult murine DRG. *Mol. Cell. Neurosci.* **2008**, *37*, 298–311. [CrossRef]
21. Li, G.C.; Zhao, X.Y.; Zhang, L.Z.; Wang, C.P.; Shi, Y.W.; Yang, Y.M. Regulating Schwann Cells Growth by Chitosan Micropatterning for Peripheral Nerve Regeneration In Vitro. *Macromol. Biosci.* **2014**, *14*, 1067–1075. [CrossRef] [PubMed]
22. Liu, C.; Kray, J.; Toomajian, V.; Chan, C. Schwann Cells Migration on Patterned Polydimethylsiloxane Microgrooved Surface. *Tissue Eng. Part C Methods* **2016**, *22*, 644–651. [CrossRef]
23. Li, G.C.; Zhao, X.Y.; Zhang, L.Z.; Yang, J.; Cui, W.G.; Yang, Y.M.; Zhang, H.B. Anisotropic ridge/groove microstructure for regulating morphology and biological function of Schwann cells. *Appl. Mater. Today* **2020**, *18*, 13. [CrossRef]
24. Li, G.C.; Xue, C.B.; Wang, H.K.; Yang, X.M.; Zhao, Y.X.; Zhang, L.Z.; Yang, Y.M. Spatially featured porous chitosan conduits with micropatterned inner wall and seamless sidewall for bridging peripheral nerve regeneration. *Carbohydr. Polym.* **2018**, *194*, 225–235. [CrossRef] [PubMed]
25. Li, G.C.; Chen, S.Y.; Zeng, M.; Yan, K.; Fei, Z.; Luzhong, Z.; Yumin, Y. Hierarchically aligned gradient collagen micropatterns for rapidly screening Schwann cells behavior. *Colloid Surf. B-Biointerfaces* **2019**, *176*, 341–351. [CrossRef]
26. Wang, H.B.; Mullins, M.E.; Cregg, J.M.; McCarthy, C.W.; Gilbert, R.J. Varying the diameter of aligned electrospun fibers alters neurite outgrowth and Schwann cell migration. *Acta Biomater.* **2010**, *6*, 2970–2978. [CrossRef]
27. Mitchel, J.A.; Hoffman-Kim, D. Cellular Scale Anisotropic Topography Guides Schwann Cell Motility. *PLoS ONE* **2011**, *6*, 13. [CrossRef] [PubMed]
28. Wu, T.; Xue, J.J.; Xia, Y.N. Engraving the Surface of Electrospun Microfibers with Nanoscale Grooves Promotes the Outgrowth of Neurites and the Migration of Schwann Cells. *Angew. Chem. Int. Edit.* **2020**, *59*, 15626–15632. [CrossRef]
29. Hsu, S.H.; Chen, C.Y.; Lu, P.S.; Lai, C.S.; Chen, C.J. Oriented Schwann cell growth on microgrooved surfaces. *Biotechnol. Bioeng.* **2005**, *92*, 579–588. [CrossRef]
30. Tonazzini, I.; Jacchetti, E.; Meucci, S.; Beltram, F.; Cecchini, M. Schwann Cell Contact Guidance versus Boundary Interaction in Functional Wound Healing along Nano and Microstructured Membranes. *Adv. Healthc. Mater.* **2015**, *4*, 1849–1860. [CrossRef]
31. Simitzi, C.; Efstathopoulos, P.; Kourgiantaki, A.; Ranella, A.; Charalampopoulos, I.; Fotakis, C.; Athanassakis, I.; Stratakis, E.; Gravanis, A. Laser fabricated discontinuous anisotropic microconical substrates as a new model scaffold to control the directionality of neuronal network outgrowth. *Biomaterials* **2015**, *67*, 115–128. [CrossRef]
32. Scaccini, L.; Mezzena, R.; De Masi, A.; Gagliardi, M.; Gambarotta, G.; Cecchini, M.; Tonazzini, I. Chitosan Micro-Grooved Membranes with Increased Asymmetry for the Improvement of the Schwann Cell Response in Nerve Regeneration. *Int. J. Mol. Sci.* **2021**, *22*, 7901. [CrossRef] [PubMed]
33. Li, G.C.; Zhao, X.Y.; Zhao, W.X.; Zhang, L.Z.; Wang, C.P.; Jiang, M.R.; Gu, X.S.; Yang, Y.M. Porous chitosan scaffolds with surface micropatterning and inner porosity and their effects on Schwann cells. *Biomaterials* **2014**, *35*, 8503–8513. [CrossRef] [PubMed]
34. Wang, Y.M.; Wang, W.J.; Wo, Y.; Gui, T.; Zhu, H.; Mo, X.M.; Chen, C.C.; Li, Q.F.; Ding, W.L. Orientated Guidance of Peripheral Nerve Regeneration Using Conduits with a Microtube Array Sheet (MTAS). *ACS Appl. Mater. Interfaces* **2015**, *7*, 8437–8450. [CrossRef] [PubMed]
35. Li, G.C.; Li, S.J.; Zhang, L.L.; Chen, S.Y.; Sun, Z.D.; Li, S.Q.; Zhang, L.Z.; Yang, Y.M. Construction of Biofunctionalized Anisotropic Hydrogel Micropatterns and Their Effect on Schwann Cell Behavior in Peripheral Nerve Regeneration. *ACS Appl. Mater. Interfaces* **2019**, *11*, 37397–37410. [CrossRef] [PubMed]
36. Wang, P.Y.; Yu, J.S.; Lin, J.H.; Tsai, W.B. Modulation of alignment, elongation and contraction of cardiomyocytes through a combination of nanotopography and rigidity of substrates. *Acta Biomater.* **2011**, *7*, 3285–3293. [CrossRef]
37. Wang, P.Y.; Yu, H.T.; Tsai, W.B. Modulation of Alignment and Differentiation of Skeletal Myoblasts by Submicron Ridges/Grooves Surface Structure. *Biotechnol. Bioeng.* **2010**, *106*, 285–294. [CrossRef]
38. Daly, W.; Yao, L.; Zeugolis, D.; Windebank, A.; Pandit, A. A biomaterials approach to peripheral nerve regeneration: Bridging the peripheral nerve gap and enhancing functional recovery. *J. R. Soc. Interface* **2012**, *9*, 202–221. [CrossRef]

39. Bell, J.H.A.; Haycock, J.W. Next Generation Nerve Guides: Materials, Fabrication, Growth Factors, and Cell Delivery. *Tissue Eng. Part B Rev.* **2012**, *18*, 116–128. [CrossRef]
40. Sullivan, R.; Dailey, T.; Duncan, K.; Abel, N.; Borlongan, C.V. Peripheral Nerve Injury: Stem Cell Therapy and Peripheral Nerve Transfer. *Int. J. Mol. Sci.* **2016**, *17*, 2101. [CrossRef]
41. Nectow, A.R.; Marra, K.G.; Kaplan, D.L. Biomaterials for the Development of Peripheral Nerve Guidance Conduits. *Tissue Eng. Part B Rev.* **2012**, *18*, 40–50. [CrossRef] [PubMed]
42. Ray, W.Z.; Mackinnon, S.E. Management of nerve gaps: Autografts, allografts, nerve transfers, and end-to-side neurorrhaphy. *Exp. Neurol.* **2010**, *223*, 77–85. [CrossRef] [PubMed]
43. Richardson, J.A.; Rementer, C.W.; Bruder, J.M.; Hoffman-Kim, D. Guidance of dorsal root ganglion neurites and Schwann cells by isolated Schwann cell topography on poly(dimethyl siloxane) conduits and films. *J. Neural Eng.* **2011**, *8*, 12. [CrossRef] [PubMed]
44. Oliveira, J.M.; Carvalho, C.R.; Costa, J.B.; Reis, R.L. Advanced Natural-based Biomaterials To Tackle The Current Challenges In Peripheral Nerve Regeneration. *Tissue Eng. Part A* **2016**, *22*, S106.
45. Georgiou, M.; Bunting, S.C.J.; Davies, H.A.; Loughlin, A.J.; Golding, J.P.; Phillips, J.B. Engineered neural tissue for peripheral nerve repair. *Biomaterials* **2013**, *34*, 7335–7343. [CrossRef]
46. Zhang, L.Z.; Chen, S.Y.; Liang, R.Y.; Chen, Y.; Li, S.J.; Li, S.Q.; Sun, Z.D.; Wang, Y.L.; Li, G.C.; Ming, A.J.; et al. Fabrication of alignment polycaprolactone scaffolds by combining use of electrospinning and micromolding for regulating Schwann cells behavior. *J. Biomed. Mater. Res. Part A* **2018**, *106*, 3123–3134. [CrossRef]
47. Heckele, M.; Schomburg, W.K. Review on micro molding of thermoplastic polymers. *J. Micromech. Microeng.* **2004**, *14*, R1–R14. [CrossRef]
48. Xie, R.X.; Zheng, W.C.; Guan, L.D.; Ai, Y.J.; Liang, Q.L. Engineering of Hydrogel Materials with Perfusable Microchannels for Building Vascularized Tissues. *Small* **2020**, *16*, 17. [CrossRef]
49. Tsai, C.Y.; Lin, C.L.; Cheng, N.C.; Yu, J.S. Effects of nano-grooved gelatin films on neural induction of human adipose-derived stem cells. *RSC Adv.* **2017**, *7*, 53537–53544. [CrossRef]
50. Woodruff, M.A.; Hutmacher, D.W. The return of a forgotten polymer-Polycaprolactone in the 21st century. *Prog. Polym. Sci.* **2010**, *35*, 1217–1256. [CrossRef]
51. Bozkurt, A.; Brook, G.A.; Moellers, S.; Lassner, F.; Sellhaus, B.; Weis, J.; Woeltje, M.; Tank, J.; Beckmann, C.; Fuchs, P.; et al. In vitro assessment of axonal growth using dorsal root ganglia explants in a novel three-dimensional collagen matrix. *Tissue Eng.* **2007**, *13*, 2971–2979. [CrossRef]
52. Gerardo-Nava, J.; Hodde, D.; Katona, I.; Bozkurt, A.; Grehl, T.; Steinbusch, H.W.M.; Weis, J.; Brook, G.A. Spinal cord organotypic slice cultures for the study of regenerating motor axon interactions with 3D scaffolds. *Biomaterials* **2014**, *35*, 4288–4296. [CrossRef]
53. Ahn, H.H.; Lee, I.W.; Lee, H.B.; Kim, M.S. Cellular Behavior of Human Adipose- Derived Stem Cells on Wettable Gradient Polyethylene Surfaces. *Int. J. Mol. Sci.* **2014**, *15*, 2075–2086. [CrossRef]
54. Wang, P.Y.; Li, W.T.; Yu, J.S.; Tsai, W.B. Modulation of osteogenic, adipogenic and myogenic differentiation of mesenchymal stem cells by submicron grooved topography. *J. Mater. Sci. Mater. Med.* **2012**, *23*, 3015–3028. [CrossRef] [PubMed]
55. Pan, F.; Zhang, M.; Wu, G.M.; Lai, Y.K.; Greber, B.; Scholer, H.R.; Chi, L.F. Topographic effect on human induced pluripotent stem cells differentiation towards neuronal lineage. *Biomaterials* **2013**, *34*, 8131–8139. [CrossRef]
56. Miller, C.; Shanks, H.; Witt, A.; Rutkowski, G.; Mallapragada, S. Oriented Schwann cell growth on micropatterned biodegradable polymer substrates. *Biomaterials* **2001**, *22*, 1263–1269. [CrossRef] [PubMed]
57. Li, J.G.; Zhang, K.; Yang, P.; Qin, W.; Li, G.C.; Zhao, A.S.; Huang, N. Human vascular endothelial cell morphology and functional cytokine secretion influenced by different size of HA micro-pattern on titanium substrate. *Colloid Surf. B Biointerfaces* **2013**, *110*, 199–207. [CrossRef] [PubMed]
58. Yim, E.K.F.; Pang, S.W.; Leong, K.W. Synthetic nanostructures inducing differentiation of human mesenchymal stem cells into neuronal lineage. *Exp. Cell Res.* **2007**, *313*, 1820–1829. [CrossRef]
59. Watari, S.; Hayashi, K.; Wood, J.A.; Russell, P.; Nealey, P.F.; Murphy, C.J.; Genetos, D.C. Modulation of osteogenic differentiation in hMSCs cells by submicron topographically-patterned ridges and grooves. *Biomaterials* **2012**, *33*, 128–136. [CrossRef]
60. Pardee, A.B.; Coppock, D.L.; Yang, H.C. Regulation of Cell-Proliferation at the Onset of DNA-Synthesis. *J. Cell Sci.* **1986**, *1986*, 171–180. [CrossRef]
61. Masciullo, C.; Dell'Anna, R.; Tonazzini, I.; Boettger, R.; Pepponi, G.; Cecchini, M. Hierarchical thermoplastic rippled nanostructures regulate Schwann cell adhesion, morphology and spatial organization. *Nanoscale* **2017**, *9*, 14861–14874. [CrossRef] [PubMed]
62. Motta, C.M.M.; Endres, K.J.; Wesdemiotis, C.; Willits, R.K.; Becker, M.L. Enhancing Schwann cell migration using concentration gradients of laminin-derived peptides. *Biomaterials* **2019**, *218*, 12. [CrossRef]
63. Ballestrem, C.; Wehrle-Haller, B.; Hinz, B.; Imhof, B.A. Actin-dependent lamellipodia formation and microtubule-dependent tail retraction control-directed cell migration. *Mol. Biol. Cell* **2000**, *11*, 2999–3012. [CrossRef]
64. Lei, L.; Tang, L. Schwann cells genetically modified to express S100A4 increases GAP43 expression in spiral ganglion neurons in vitro. *Bioengineered* **2017**, *8*, 404–410. [CrossRef] [PubMed]
65. Sternberger, N.H.; Itoyama, Y.; Kies, M.W.; Webster, H.D. Myelin Basic-Protein Demonstrated Immunocytochemically in Oligodendroglia Prior to Myelin Sheath Formation. *Proc. Natl. Acad. Sci. USA* **1978**, *75*, 2521–2524. [CrossRef]

66. Herbert, A.L.; Fu, M.M.; Drerup, C.M.; Gray, R.S.; Harty, B.L.; Ackerman, S.D.; O'Reilly-Pol, T.; Johnson, S.L.; Nechiporuk, A.V.; Barres, B.A.; et al. Dynein/dynactin is necessary for anterograde transport of Mbp mRNA in oligodendrocytes and for myelination in vivo. *Proc. Natl. Acad. Sci. USA* **2017**, *114*, E9153–E9162. [CrossRef]
67. Lemke, G.; Lamar, E.; Patterson, J. Isolation and Analysis of the Gene Encoding Peripheral Myelin Protein Zero. *Neuron* **1988**, *1*, 73–83. [CrossRef] [PubMed]
68. Van Simaeys, D.; Turek, D.; Champanhac, C.; Vaizer, J.; Sefah, K.; Zhen, J.; Sutphen, R.; Tan, W.H. Identification of Cell Membrane Protein Stress-Induced Phosphoprotein 1 as a Potential Ovarian Cancer Biomarker Using Aptamers Selected by Cell Systematic Evolution of Ligands by Exponential Enrichment. *Anal. Chem.* **2014**, *86*, 4521–4527. [CrossRef] [PubMed]
69. Yang, R.; Lirussi, D.; Thornton, T.M.; Jelley-Gibbs, D.M.; Diehl, S.A.; Case, L.K.; Madesh, M.; Taatjes, D.J.; Teuscher, C.; Haynes, L.; et al. Mitochondrial Ca^{2+} and membrane potential, an alternative pathway for Interleukin 6 to regulate CD4 cell effector function. *eLife* **2015**, *4*, 22. [CrossRef]
70. Sengupta, S.; Rothenberg, K.E.; Li, H.J.; Hoffman, B.D.; Bursac, N. Altering integrin engagement regulates membrane localization of K(ir)2.1 channels. *J. Cell Sci.* **2019**, *132*, 16. [CrossRef]
71. Sarker, M.; Naghieh, S.; McInnes, A.D.; Schreyer, D.J.; Chen, X.B. Strategic Design and Fabrication of Nerve Guidance Conduits for Peripheral Nerve Regeneration. *Biotechnol. J.* **2018**, *13*, 16. [CrossRef]
72. Liu, F.; Xu, J.W.; Wu, L.L.; Zheng, T.T.; Han, Q.; Liang, Y.Y.; Zhang, L.L.; Li, G.C.; Yang, Y.M. The Influence of the Surface Topographical Cues of Biomaterials on Nerve Cells in Peripheral Nerve Regeneration: A Review. *Stem Cells Int.* **2021**, *2021*, 13. [CrossRef]
73. Migliorini, E.; Grenci, G.; Ban, J.; Pozzato, A.; Tormen, M.; Lazzarino, M.; Torre, V.; Ruaro, M.E. Acceleration of Neuronal Precursors Differentiation Induced by Substrate Nanotopography. *Biotechnol. Bioeng.* **2011**, *108*, 2736–2746. [CrossRef]
74. Stukel, J.M.; Willits, R.K. Mechanotransduction of Neural Cells Through Cell-Substrate Interactions. *Tissue Eng. Part B Rev.* **2016**, *22*, 173–182. [CrossRef] [PubMed]
75. Hu, X.Y.; Huang, J.H.; Ye, Z.X.; Xia, L.; Li, M.; Lv, B.C.; Shen, X.F.; Luo, Z.J. A Novel Scaffold with Longitudinally Oriented Microchannels Promotes Peripheral Nerve Regeneration. *Tissue Eng. Part A* **2009**, *15*, 3297–3308. [CrossRef] [PubMed]
76. Froeter, P.; Huang, Y.; Cangellaris, O.V.; Huang, W.; Dent, E.W.; Gillette, M.U.; Williams, J.C.; Li, X.L. Toward Intelligent Synthetic Neural Circuits: Directing and Accelerating Neuron Cell Growth by Self-Rolled-Up Silicon Nitride Microtube Array. *ACS Nano* **2014**, *8*, 11108–11117. [CrossRef] [PubMed]
77. Beduer, A.; Vieu, C.; Arnauduc, F.; Sol, J.C.; Loubinoux, I.; Vaysse, L. Engineering of adult human neural stem cells differentiation through surface micropatterning. *Biomaterials* **2012**, *33*, 504–514. [CrossRef] [PubMed]

Disclaimer/Publisher's Note: The statements, opinions and data contained in all publications are solely those of the individual author(s) and contributor(s) and not of MDPI and/or the editor(s). MDPI and/or the editor(s) disclaim responsibility for any injury to people or property resulting from any ideas, methods, instructions or products referred to in the content.

Journal of
Functional
Biomaterials

MDPI

Review

Antibacterial-Based Hydrogel Coatings and Their Application in the Biomedical Field—A Review

Tai Peng [1,2,*], Qi Shi [1,2], Manlong Chen [1], Wenyi Yu [1,2] and Tingting Yang [1,2,*]

1 Key Lab of Oral Biomedical Materials and Clinical Application of Heilongjiang Province, Jiamusi University, Jiamusi 154007, China; shiqi@jmsu.edu.cn (Q.S.)
2 School of Materials Science and Engineering, Jiamusi University, Jiamusi 154007, China
* Correspondence: pt@jmsu.edu.cn (T.P.); yangtt@jmsu.edu.cn (T.Y.); Tel.: +86-454-8618-701 (T.P.)

Abstract: Hydrogels exhibit excellent moldability, biodegradability, biocompatibility, and extracellular matrix-like properties, which make them widely used in biomedical fields. Because of their unique three-dimensional crosslinked hydrophilic networks, hydrogels can encapsulate various materials, such as small molecules, polymers, and particles; this has become a hot research topic in the antibacterial field. The surface modification of biomaterials by using antibacterial hydrogels as coatings contributes to the biomaterial activity and offers wide prospects for development. A variety of surface chemical strategies have been developed to bind hydrogels to the substrate surface stably. We first introduce the preparation method for antibacterial coatings in this review, which includes surface-initiated graft crosslinking polymerization, anchoring the hydrogel coating to the substrate surface, and the LbL self-assembly technique to coat crosslinked hydrogels. Then, we summarize the applications of hydrogel coating in the biomedical antibacterial field. Hydrogel itself has certain antibacterial properties, but the antibacterial effect is not sufficient. In recent research, in order to optimize its antibacterial performance, the following three antibacterial strategies are mainly adopted: bacterial repellent and inhibition, contact surface killing of bacteria, and release of antibacterial agents. We systematically introduce the antibacterial mechanism of each strategy. The review aims to provide reference for the further development and application of hydrogel coatings.

Keywords: hydrogel coatings; antibacterial property; biomaterials

Citation: Peng, T.; Shi, Q.; Chen, M.; Yu, W.; Yang, T. Antibacterial-Based Hydrogel Coatings and Their Application in the Biomedical Field—A Review. *J. Funct. Biomater.* **2023**, *14*, 243. https://doi.org/10.3390/jfb14050243

Academic Editor: Shuilin Wu

Received: 28 March 2023
Revised: 15 April 2023
Accepted: 21 April 2023
Published: 25 April 2023

Copyright: © 2023 by the authors. Licensee MDPI, Basel, Switzerland. This article is an open access article distributed under the terms and conditions of the Creative Commons Attribution (CC BY) license (https://creativecommons.org/licenses/by/4.0/).

1. Introduction

In the current biomedical field, medical devices, such as catheters, hernia nets, implants, and wound dressings, are often adhered by bacteria, leading to varying degrees of infection and posing a threat to the health of patients [1]. Once bacteria adhere to the surface of the substrate, they will rapidly form a biofilm, which attracts more bacteria, affecting the antibacterial effect of the immune system [2]. Therefore, the prevention of bacterial infection in the process of biomaterial implantation has become the focus of researchers. Surface coating or modification, which preserves the original properties of the material and changes only the surface properties, has been recognized as a promising strategy for introducing antibacterial efficacy into biomaterials. Some bionic surface morphologies with high aspect ratios are effective against colonization by bacteria, although the mechanism of antibacterial activity is not clear [3–8]. For methods of tailoring surface chemistry, substrates can be chemically modified or physically coated with a variety of antibacterial substances, including polymers, functional groups, inorganic nanoparticles, hydrogels, and antibiotics [9]. Among these bactericidal materials, hydrogel coating has many advantages and has been widely studied.

Hydrogels represent a polymer network system [10], i.e., a polymer with a three-dimensional network structure that can crosslink water molecules in the system by combining hydrophilic residues with water molecules. They can absorb a large amount of water, are swellable but do not dissolve in water, can maintain a specific shape, and have good elasticity [11]. Hydrogels can be fabricated from natural or synthetic polymers [12–16]. The advantages and disadvantages of some hydrogel raw materials are shown in Table 1. Hydrogels are very similar to natural active materials and are an ideal material for the preparation of medical dressings [17], drug delivery systems [18–20], and tissue engineering [21]. However, most hydrogels exhibit poor mechanical properties due to their uneven network structure and lack of energy dissipation mechanism, which lead to their limited application in the biomedical field. For example, hydrogels are easily deformed after repair of bone defects and thus need to be combined with other biomaterials or chemical modification to improve their performance. At present, biomaterials with hydrogel coating combine the excellent mechanical properties of implant devices and the biological properties of hydrogels, exhibiting excellent development prospects in the biomedical field [22]. Currently, the listed medical hydrogel coating products are mainly used in the catheters, stent delivery systems, and guide wires, among which the catheter accounts for the largest proportion, reaching 70% of sales in 2022. Brands include the Futae hydrogel coated latex catheter bag, the BIP catheter, and the Weili catheter.

Table 1. Comparison of advantages and disadvantages of different hydrogels in the biomedical field.

Hydrogel Type	Raw Material	Advantages	Disadvantages
Natural hydrogel	Collagen (protein)	Low antigenicity, Low inflammatory response, Excellent biological properties	High cost, high possibility of thrombosis, low mechanical strength, and difficult modification
	Gelatin (protein)	Low cost, Low immunogenicity, Biodegradable and biocompatible,	Poor stability at high temperature
	Silk fibroin (protein)	Excellent mechanical properties, Low immunogenicity, Blood clots less likely to form	Difficult source and slow gelling
	Glycosaminoglycan—Hyaluronic acid and Chondroitin sulfate (polysaccharide)	Mimics extracellular matrix components, Biodegradable, Binding cytokines	Degrades rapidly in vivo and requires crosslinking to stabilize
	Chitosan (polysaccharide)	Antibacterial, Low cost, Biocompatibility and biodegradability	Poor mechanical performance
Synthetic hydrogel	Polyethylene glycol (PEG)	Biodegradable, non-immunogenic	Lack of adhesive support
	Polyglutamic acid (PGA)	Biodegradable by hydrolysis, Thermoplastic, Mechanical properties are adjustable	Physical crosslinking is weak, hydrolytic products can induce inflammatory reaction and degrade rapidly
	Polylactic acid (PLA) and copolymer	Biodegradable by hydrolysis, Good mechanical properties It is soluble in organic solvents	Hydrolyzed byproducts can cause an inflammatory response

The physical method of hydrogel coating to modify the surface of the biological substrate is simple, but it is not stable enough. The hydrogel coating is bonded to the surface of the substrate through a non-covalent bond, and the coating layer is easily delaminated from the surface [23,24]. Therefore, in order to improve the adhesion and stability of the hydrogel coating, it is necessary to use a chemical method (e.g., covalent bond) to attach the hydrogel coating [25,26]. Compared with other coatings (e.g., functional groups, self-assembled monolayers, and polymer brushes), hydrogel coatings exhibit many advantages. Firstly, hydrogel coatings have high grafting density and uniform coverage. Secondly, layer-by-layer (LbL) assembly coatings and polymer brush coatings are widely used, but the coating thickness has limitations. LbL hydrogel coatings are usually very thin, generally less than 100 nm, and they take much time to deposit layer by layer. The thickness of polymer brush coatings is related to the length of the polymer chain, generally less than several hundred nanometers [27]. In contrast, hydrogel coatings can be prepared by spin coating, electrochemical deposition, surface-initiated graft polymerization, and other methods, so the thickness of the hydrogel coating can be flexibly controlled from nanometer to micrometer, providing sufficient space to coat polymers, particles, small molecules, and other substances. Thirdly, unlike the non-crosslinked surface coating, which is attached with the substrate only by one covalent bond, the hydrogel coating is attached by multiple points. Therefore, hydrogel coatings are more stable for long-term modification.

We can classify antibacterial hydrogel coatings into two categories according to the type of active ingredients. The first is compositions containing antibacterial substances, such as natural or synthetic cationic polymers [28,29], amphoteric polymers [30], and antibacterial peptides [31]. The second is loaded antibacterial substances, such as antibiotics [32], antibacterial analogues [33], nano-silver [34], and zinc oxide, in the three-dimensional hydrogel network [35]. At present, researchers have conducted extensive research on the application of antibacterial hydrogels, but the research on hydrogels as a surface coating is not comprehensive. In this paper, the preparation methods for antibacterial hydrogel coating and its application in the biomedical antibacterial field are reviewed in detail.

2. Preparation Methods of Hydrogel Coatings

Hydrogels are prepared by chemical and physical crosslinking, but it is challenging to fix physically crosslinked hydrogels on the surface of materials due to the lack of binding sites for binding into the three-dimensional network [36,37]. In addition, the poor mechanical properties of physically crosslinked hydrogels compared to chemically crosslinked hydrogels limit the application of physically crosslinked hydrogels as durable coating materials [38]. The chemically crosslinked hydrogels can be gelatinized by monomer polymerization or conjugation reactions between polymer chains, whereby we can attach these hydrogels to the material surface in various ways to form stable hydrogel coatings [39]. The general strategies can be divided into three types: The first is surface-initiated graft crosslinking polymerization. The second is anchoring the hydrogel coating to the substrate surface. Third is the LbL self-assembly technique to coat crosslinked hydrogels [2].

2.1. Surface-Initiated Graft Crosslinking Polymerization

Free radical polymerization is a crucial way to prepare polymeric materials. The polymerization reaction is initiated by active radicals located on the surface of the substrate, so the adhesion of hydrogels can be promoted by generating active radicals on the substrate surface and introducing grafting sites on the surface [40].

2.1.1. Direct Generation of Reactive Radicals on the Substrate Surface

Polymer materials are widely used in biomedical fields (e.g., heart valves, catheters, contact lenses). Mature strategies have been established for surface modification of inorganic materials, but general methods for polymer surface chemical modification are still lacking due to the relatively low surface inertness and surface energy of polymers. The photografting reaction may introduce surface-active radicals to grow polymer brushes and hydrogel coatings on the surface [41]. In a typical photografting process, the Norrish II type photoinitiator, e.g., benzophenone (BP), is excited to a singlet state (BPS) under irradiation by UV light and is then transformed to a triplet state (BPT) by intersystem crossing. It is well recognized that BPT is capable of abstracting hydrogen from other molecules or even solid substrates (RH), which generates reactive radicals (R•) and relatively less reactive benzopinacol radicals (BP-OH•). In the presence of vinyl monomer, R• acts as a reactive species to initiate radical polymerization, while BP-OH• is stabilized by the conjugate effect from the benzene ring and is not able to attack monomers but will couple with reactive radicals. Since polymer materials have plenty of C–H bonds, it is convenient to conduct photografting reactions on them. There have been some reports on the preparation of thin hydrogel layers by photografting on polymer substrates, and hydrogel coatings with antibacterial properties have also been investigated. For example, Zhang Fanjun et al. prepared a hydrogel antibacterial coating that can be applied to various blood-contacting devices (PVC and polyurethane pipes) (Figure 1) [42]. Free radical polymerization on the material surface was initiated with the photoinitiator benzophenone using an ultraviolet lamp. The dehydrogenation of benzophenone forms reactive radicals on the PVC polymer backbone and initiates monomer polymerization, resulting in a hybrid acrylamide and acrylic hydrogel coating. The inhibition zone test and confocal laser scanning microscopy revealed that the hydrogel coating could maintain remarkable antimicrobial and antifouling properties for four weeks. Furthermore, the hydrogel coating decreased the platelet adhesion/activation without risk of hemolysis. The ex vivo blood circulation study confirmed the antithrombotic properties of the hydrogel coating. Using impregnation methods to coat antibacterial drugs, Ag^+, antibacterial peptides, and other substances may further improve the antibacterial properties of blood-contacting catheters. In another study, Fu Xiaoyi et al. prepared an ionic hydrogel coating doped with a Ni^{2+} trapping agent CS_2 on the surface of a Nitinol bone implant through radical polymerization technology initiated by photografting, which may effectively be antibacterial. CS_2 coated the released Ni^{2+} through chelating reaction to avoid cytotoxicity. This provides a research basis for embedding and detecting Ni^{2+} released from implantable biomaterials [43]. However, some polymers only have high-energy C(sp^2)-H bonds, (e.g., polytetrafluoroethylene (PTFE), polyimide (PI)), which cannot be surface-modified by photo-grafting. Because the irradiation energy of traditional ultraviolet light is relatively low, these inert surfaces can be surface-modified by photo-grafting with excimer lasers.

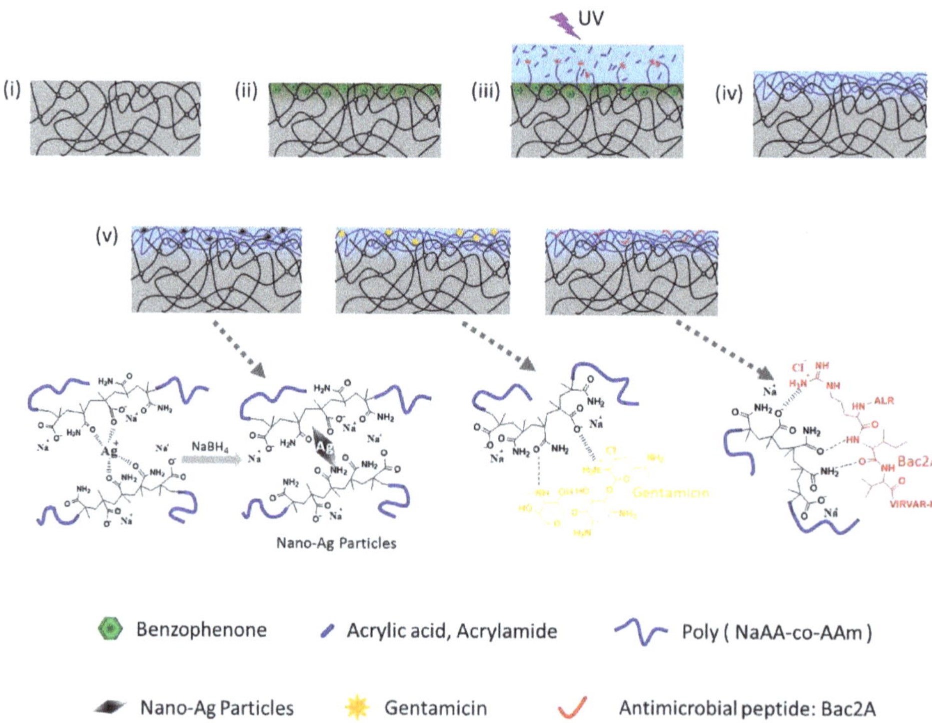

Figure 1. Preparation of the hydrogel coatings and antimicrobial agent loading [42]. (**i**) Pristine Polymer; (**ii**) Initiator Absorbing; (**iii**) In situ growth of hydrogel coating; (**iv**) As-formed hydrogel coating; (**v**) Antimicrobial agent loaded hydrogel coatings. Copyright 2021, The Royal Society of Chemistry.

2.1.2. Introduction of Peroxide Groups on the Substrate Surface

Redox reactions between peroxide initiators and reducing agents may generate free radicals and initiate monomer polymerization. Peroxide groups may be activated under the irradiation of ultraviolet lamps, which initiates free radical polymerization. Therefore, the graft-crosslinking polymerization of hydrogels can be easily carried out on the surface, functionalized with peroxide groups. For example, Ma et al. prepared hydrogel coatings through a redox diffusion method in which Fe^{2+} mingled with the substrate material before 3D printing; when Fe^{2+} leached from the 3D printed structures, a redox reaction occurred with the persulfate ions $S_2O_8{}^{2-}$ immersed in the solution, triggering the formation of crosslinked hydrogels [44,45]. However, Fe^{2+} leaching may have harmful effects on the properties of the substrate material; thus, Wancura Megan et al. developed a new method [46] in which the adsorption of Fe^{2+} onto the substrate would not have these effects. Iron (II) gluconate was adsorbed onto the substrate through a molecular adsorption method and immersed in a mixture of ammonium persulfate (APS) solution and polyethylene glycol diacrylate (PEGDA) to initiate a redox reaction between $S_2O_8{}^{2-}$ and Fe^{2+} to produce sulfate anions and free radicals. Then, the sulfate radicals triggered the vinyl groups of PEGDA, initiating free radical crosslinking from the surface to produce a hydrogel (Figure 2). The prepared hydrogel coating has a controllable thickness and is formed through a gentle method that has proven to be effective in improving biocompatibility and preventing thrombus formation in subsequent experiments.

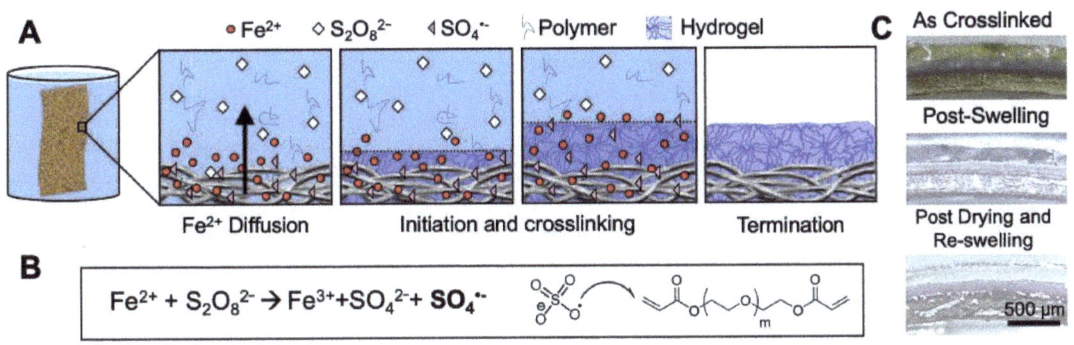

Figure 2. (**A**) Diffusion-mediated crosslinking, where Fe^{2+} diffuses away from the substrate into $S_2O_8{}^{2-}$ to generate radicals and crosslink PEGDA hydrogels; (**B**) reaction between iron gluconate and the persulfate anion to generate sulfate radicals that initiate poly(ethylene glycol) diacrylate (PEGDA) end-groups; (**C**) hydrogel coatings immediately after crosslinking, after swelling for 24 h, and after drying and re-swelling [46]. Copyright 2020, The Royal Society of Chemistry.

2.1.3. Introduction of Catechol Groups on the Substrate Surface

Catechol groups can become functional on the surface of most organic or inorganic substrates [47]. The commonly used surface initiators of catechol groups include 3,4-dihydroxyphenylalanine and dopamine, which may form coatings with adhesive functionality on the surface of substrates through autoxidative polymerization mechanism. The generated quinones, amino groups, and phenolic hydroxyl groups can be used as reaction sites to combine with free radical polymerization initiators and promote graft crosslinking of hydrogels on the surface [48]. For example, in one study, Zhou et al. first deposited a layer of polydopamine on the surface of polydimethylsiloxane (PDMS) (American Braintree Scientific, Braintree, MA, USA) and then bonded the atom transfer radical polymerization (ATRP) initiator (α-bromoisobutyl bromide, BiBB was purchased from Sigma-Aldrich, Darmstadt, Germany) with hydroxyl and amino groups and fixed it to the surface of the catheter; the cationic monomer or crosslinker (polyethylene glycol dimethacrylate, PEGDMA was purchased from Sigma-Aldrich) initiated ATRP polymerization on the surface and formed an antibacterial hydrogel coating. The in vivo antibacterial and antibiofilm effect of these non-leachable covalently linked coatings (using a mouse catheter model) can be tuned to achieve a 1.95 log (98.88%) reduction and 1.26 log (94.51%) reduction of clinically relevant pathogenic bacteria (specifically with Methicillin-resistant Staphylococcus aureus (MRSA, were kindly provided by Kimberly Kline's lab, which bought them from ATCC, Manassas, Virginia, USA) and Vancomycin-resistant Enterococcus faecalis (VRE)) [49]. This method of preparing antibacterial hydrogel coatings can also be used on small catheters with an inner diameter of 0.3 mm. In addition to radical polymerization, other methods can be used to prepare hydrogel coatings (e.g., Michael addition, Schiff-base addition, Diels-Alder addition). For example, Chen Yin et al. prepared an NOE hydrogel-coated stainless steel scaffold with nitric oxide elution [50]. The hydrogel precursor solution is directly coated on the surface of the scaffold. Due to the difference between melting point and body temperature, it is difficult to form a gel, so chemical crosslinking should be introduced. The study used the Michael addition reaction that promotes cytocompatibility. The previously used sulfhydryl-maleimide addition reaction was too fast and difficult to operate, so amine-maleimide addition was used instead of this reaction. Firstly, a P(DA-co-HDA) film layer was deposited on the stainless steel surface. The NOE hydrogel coating was prepared by coupling amine-maleimide and alginate, then crosslinking with gelatine and selenium species. The hydrogel coating can withstand balloon dilation, inhibiting excessive smooth muscle cell proliferation and preventing thrombus formation, and selenium species can also catalyze NO production and regulate cardiovascular homeostasis. Previous ISR prevention strategies rarely considered endothelial cell regeneration. This study increased endothelial cell adhesion and proliferation by grafting a hydrogel coating on the scaffold surface. The method has extensive research prospects.

Tugce et al. prepared a multifunctional "clickable" hydrogel coating through spin-coating [51], which enhanced the ability to move from protein fixation to cell attachment. Firstly, a dopamine methacrylamide layer was attached via catechol groups on the titanium surface. Then, furan-protected maleimide methacrylate (FuMaMA) hydrogels were prepared on the surface, and the furan groups were removed through a heating reaction, so that maleimide can react with sulfhydryl groups. Thus, the hydrogel can be functionalized by sulfhydryl maleimide nucleophilic addition and Diels-Alder cycloaddition reaction under mild conditions. The degree of functionalization of the hydrogel can be controlled by attachment of biotin-benzyltetrazine, followed by immobilization of TRITC-labelled ExtrAvidin, thus satisfying various biological properties. This method involves a simple synthetic principle and can couple various antibacterial substances, drugs, and biomolecules to achieve surface functionalization through various click reactions, allowing more diverse applications of hydrogel coatings. Liu prepared a polymer zwitterion coating based on poly(2-methacryloyloxyethylphosphorylcholine-co-dopamine methacrylate) (pMPCDA) copolymers with anti-inflammatory and antithrombotic properties [52], which are also used for blood-contacting catheters. In order to prepare such a uniform and stable coating, the PVC surface was amino functionalized by co-deposition of polydopamine (PDA) and polyvinyl imine. Then, based on various reaction mechanisms (e.g., Michael addition, Schiff-base addition) between catechol and amino groups, the zwitterionic pMPCDA copolymer was stably modified on the surface of PVC by using the mussel shell excitation chemical method. The hydrogel coating is not degradable, but degradable devices can reduce long-term complications in association with the residue of foreign materials in the body. Yang et al. prepared a membrane based on FDA-approved biodegradable Poly L-lactic acid (PLLA) material. First, dopamine (DA) groups are introduced on the surface of PLLA. Then, it is immersed in a synthetic Sulfobetaine methacrylate/Cerium oxide@Methacrylated gelatin dopamine conjugate (SBMA/CeO_2@GMDA) solution. After Michael addition and Schiff base reaction, the nanoceria-eluting degradable zwitterion hydrogel coating is fabricated (Figure 3) [53].

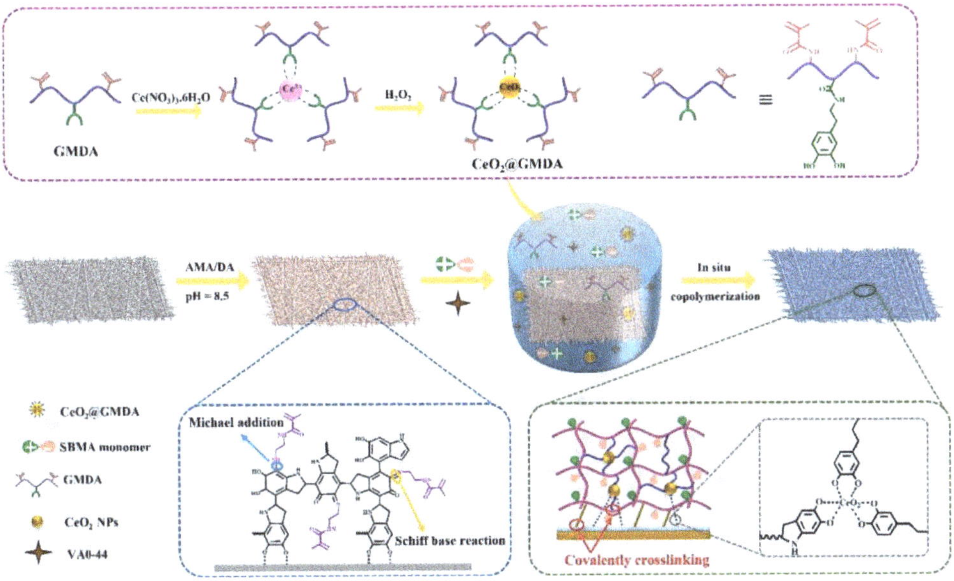

Figure 3. Schematic illustrations of preparation process of nanoceria-eluting degradable zwitterion hydrogel coating on PLLA membrane [53]. Copyright 2021, Chemical Engineering Journal.

2.1.4. Introduction of Silane Coupling Agents on the Substrate Surface

Silane coupling agents may modify the surface of most inorganic or organic materials by introducing various functional groups as reaction sites, such as amine, aldehyde, thiol, vinyl, and quaternary ammonium groups [54]. Among them, the quaternary ammonium salt group has been proven by many researchers to have a very effective antibacterial ability. These reactive sites contribute to the adhesion of hydrogel coatings [55]. For example, Wu Xiaofang et al. studied an antibacterial hydrogel-coated artificial joint prosthesis [56]. In this study, the silane-coupling agent was loaded on the surface of laser-treated titanium alloy with hydrogen bonds; then, chitosan gelatin mixed hydrogel was prepared, which was immersed in nano-silver solution to form CS-GT-nAg antibacterial hydrogel. Using the 3D printing method, the hydrogel was connected with silane coupling agents by covalent bond and hydrogen bond, and firmly attached to the titanium alloy. The hydrogel coating has a dual-scale porous network structure, and the composite coating constructed by applying this method is more uniform and less prone to peeling, improving biocompatibility and antibacterial properties compared with porous coatings prepared by sol-gel [57], microarc oxidation [58], and laser sintering methods [59]. Refer to Table 2 for the above information.

Table 2. Hydrogel coatings prepared by surface-initiated graft crosslinking polymerization.

First Author	Publication Year	Main Components of Hydrogel Coating	Loaded Substance	Coating Preparation Method	Coating Adhesion Principle	Crosslinking Agent	Function
Zhang [42]	2021	Acrylamide Acrylic acid	Ag nanoparticles/Antibiotics/Antimicrobial peptides	Photografted surfaces induce free radical polymerization	Dehydrogenation of benzophenone forms active free radicals on the main chain of the polyvinyl chloride polymer and induces monomer polymerization	UV crosslinking	Antithrombotic, antibacterial, effectively reduces platelet adhesion
Fu [43]	2021	HEMA, DMAPS, MMA	CS_2	Ultraviolet photografting	Free radical polymerization	PEGDMA	Antibacterial and captures Ni^+ through chelation of CS_2
Wancura [46]	2020	PEGDA, APS, Functional gelatin	Protein	Redox mediated crosslinking technique	$S_2O_8^{2-}$ and Fe^{2+} undergo redox reaction to form SO_4^{2-} and free radicals and trigger free radical cross-linking		Multi-layer structure with different functional characteristics can be generated and the thickness can be controlled to enhance the cell adhesion function
Zhou [49]	2017	Polyethylene glycol dimethacrylate (PEGDMA)		Atom transfer radical polymerization (ATRP)		BiBB	Has good anti-biofilm and antibacterial effect against methicillin-resistant Staphylococcus aureus (MRSA)
Chen [50]	2021	GelGA, GelMA	Organic selenium	Apply and then light cure			Withstands balloon dilation, inhibits smooth muscle cell hyperproliferation, prevents thrombosis, and promotes NO production
Tugce [51]	2020	FuMaMAPEGMEMA	Biotin-benzyltetrazine	Rotating coating method (involving click chemistry)	Dopamine methyl acrylamide is anchored to the surface of titanium by catechol group, and methacrylate group is bonded to it by covalent bond	DMPA	Multifunctional hydrogels promote cell adhesion and proliferation
Liu [52]	2021	Poly (2-methylacryloxyethyl phosphate choline—dopamine methacrylate) (pMPCDA) copolymer		Mussel shell excitation chemical method	Michael addition between catechol and amino group, Schiff base addition and other reaction mechanisms		Anti-inflammatory and antithrombotic
Wu [56]	2021	Chitosan, gelatin	Ag^+	3D printing technology	Silane coupling	Sodium Citrate	Promotes cell adhesion and bone growth

2.2. Anchoring the Hydrogel Coating to the Substrate Surface

The preparation of stable polymer brushes on the material surface is possible by initiating graft-crosslinking polymerization on the substrate surface, which introduces a high density of reaction sites. However, the reaction efficiency is low, and the preparation of hydrogel coatings on the material surface does not require too many reaction sites. In general terms, the hydrogel can be thought of as a large molecule; immobilization on the material surface improves the coating binding stability and reaction efficiency.

2.2.1. Click Chemistry for Anchoring Hydrogels to the Substrate Surface

Click chemistry has contributed significantly to the chemical synthesis field [60]. It has many applications, high yields, harmless byproducts, and simple reaction conditions. It is a widely used method in the field of biomedical research and includes the azide-alkyne cycloaddition reaction (AAC) [61], thiolene reaction [62], and Diels-Alder reaction [63]. The thiolene reaction is widely used for preparing hydrogel coatings due to its high reaction efficiency and fast gel formation. Magennis et al. functionalized polydimethylsiloxane (PDMS) with the silane coupling agent MTS [64]. Then, they used the introduced thiol groups as reaction sites to crosslink with multifunctional monomers to prepare hydrogel coatings immobilized on the substrate surface. The in vitro experiments showed effective bacterial growth inhibition compared to unmodified PDMS.

2.2.2. Dopamine Group Functionalized Hydrogels Anchored on the Substrate Surface

Introducing dopamine groups into the hydrogel structure and anchoring on the substrate surface may avoid complicated pretreatment procedures on the substrate surface and improve the reaction efficiency.

Leng Jin et al. constructed a ZnO layer with a nanoflower-like structure on a titanium surface by a hydrothermal method and then prepared a hybrid hydrogel of gelatine methacrylate (GelMA) and hyaluronic acid methacrylate (HAMA), which was firmly attached to the ZnO layer by grafting catechol motifs on the hydrogel and conducting photocrosslinking (Figure 4) [65]. The composite could self-adapt. Under normal conditions, the hydrogel coating is stable and may effectively reduce the toxicity of ZnO. However, when bacterial infection occurs, the hydrogel coating can be effectively degraded by enzymes to release ZnO for antibacterial purposes, which further regulates the biological behavior of fibroblasts and exhibits good soft tissue compatibility while effectively balancing biosafety and antibacterial activity.

He Ye et al. used the same method to prepare catechol motif-modified gelatine methacrylate containing photosensitizer Chlorin e6-loaded mesoporous polydopamine nanoparticles (GelMAc/MPDA@Ce6), which were firmly attached to the titanium after 365 nm UV for 10 min [66]. Then, the photobiomodulation (PBM) method and photodynamic therapy (PDT) was performed by laser irradiation at specific wavelengths, which enabled the Ce6 photosensitizer to be activated to produce reactive oxygen species (ROS) and combine with other biomolecules to promote tissue repair. The coating exhibited effective antibacterial properties and promoted cell adhesion and proliferation. Combining PBM methods with hydrogel coatings of biomaterials to achieve biological functions is relatively novel and promising. Both studies used catechol moieties as anchoring groups to graft hydrogel coatings. The anchoring principle of catechol is that large amine groups, phenolic hydroxyls, and quinone structures formed by the oxidation of the derivative deposition layer may function as reactive sites to couple to initiators of active/controllable radical polymerization, which can be helpful in grafting antibacterial hydrogel coatings onto substrates. Catechol is an important pharmaceutical intermediate that can firmly attach a hydrogel coating to the substrate and is now widely used in biomaterials.

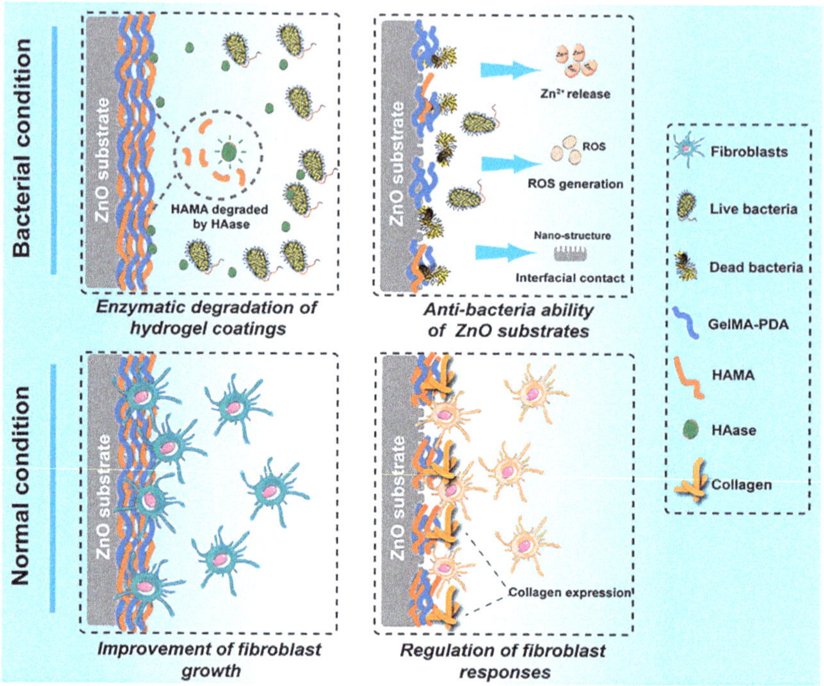

Figure 4. Schematic diagram of the self-adaptive strategy of [65]. Copyright 2021, Elsevier.

2.2.3. Anchoring Hydrogel Layers by Free Radical Polymerization

The attachment of the hydrogel layer involves first prefixing the initiator on the substrate surface; then, the hydrogel precursor solution is polymerized on the substrate surface. In addition, instead of initiating polymerization on the surface, the hydrogel layer may initiate polymerization in the precursor solution and then be chemically anchored to the substrate surface through grafting groups on the surface of the hydrogel layer.

Liu Chengde et al. prepared a bifunctional nanocomposite hydrogel coating on the surface of poly(aryl ether ketone) PAEK implants [67,68]. The raw materials for hydrogel preparation included type A gelatine, acrylic acid (AA), N-succinimidyl acrylate (AAc-NHS ester), nano-hydroxyapatite, and methacrylic anhydride (MA). As shown in Figure 5, bulk nanocomposite hydrogels were first prepared by dissolving AA, GelMA, AAc-NHS ester, and α-ketoglutaric acid (KGA) in deionized water, then mixed with nanohydroxyapatite in different proportions and cured in an ultraviolet light (UV) chamber for 15 min [69,70]. A novel poly (phthalazinone ether sulfone ketone) containing allyl groups (APPBAESK) was spin-coated on the surface of PPBESK. Then, the already prepared bulk hydrogel was spin-coated on the surface and crosslinked under UV irradiation to form the coating. NHS-ester activated groups and nano-HA were introduced to endow tissue adhesivity and osteogenic activity. Chemically inert PPBESK was successfully functionalized by spin-coating of APPBAESK containing allyl groups. The nanocomposite hydrogel coating containing nano-HA and NHS-ester activated groups was chemically anchored on the surface of modified PPBESK, which greatly improved its hydrophilicity. The resulting bifunctional PPBESK could adhere to tissues quickly and with high adhesion strength up to 300 KPa in pig skin. In addition, osteoblasts and fibroblasts could adhere and proliferate well on the surface of nanocomposite hydrogel coatings on PPBESK. Importantly, ALP expression and osteoblastic differentiation of preosteoblasts cultured with modified PPBESK were promoted by the nanocomposite hydrogel coating containing nano-HA. In conclusion, this research provides a new modification strategy of PAEK implants for integrating osteogenic activity and tissue adhesivity. Refer to Table 3 for the above information.

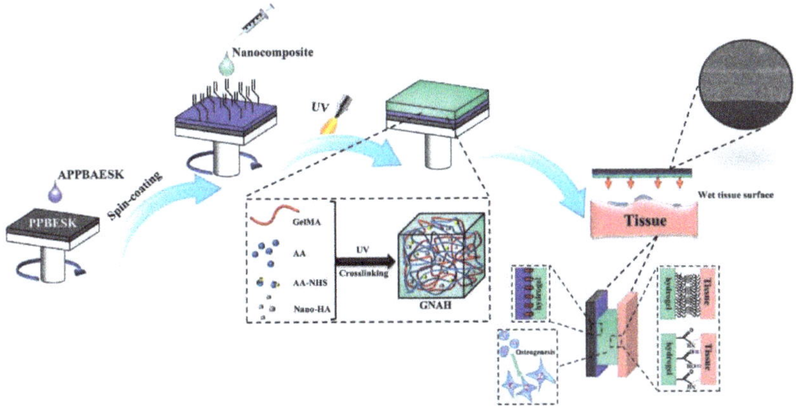

Figure 5. Preparation process of the chemically anchored bifunctional nanocomposite hydrogel coating. Reproduced with permission of [67]. Copyright 2021, John Wiley and Sons.

2.3. *LbL Self-Assembly Technique to Coat Crosslinked Hydrogels*

The layer-by-layer self-assembly technique may prepare coatings on substrate surfaces through alternating deposition methods [71], and the coatings can be attached mainly by hydrogen bonding, electrostatic interactions, and charge transfer interactions. Because of the low stability of such noncovalent bonding forces, each layer can be deposited conjugately by covalent bonding in the preparation of hydrogel coatings. The LbL self-assembly technique is more suitable for preparing ultrathin hydrogel coatings. The substrate surface needs pretreatment, and the covalent attachment of the layer to the substrate surface may be realized by grafting functional groups. For example, the surface of the silicon wafer undergoes functionalization through reactants containing dopamine groups [72]; an ultrathin hydrogel coating can be prepared by depositing alternating polyacrylic acid (PAA) and chitosan quaternary ammonium salts using the LbL self-assembly technique. The coating surface is positively charged and effectively inhibits bacterial adhesion.

Cai et al. prepared polyethylenimine (e-PEI) and alginate (Alg) conjugated with carboxyl-ebselen using the LbL self-assembly technique and then prepared multilayer films by crosslinking with the coupling agent EDC (Figure 6) [73]. Moreover, the functional membrane generated by the reaction of selenate with oxygen may produce superoxide, which inhibits bacterial adhesion. When (e-PEI/Alg)50 was placed in 105 CFU/mL *E. coli* at 37.5 °C, it was found that all the bacteria were killed after 24 h. The hydrogel coating may kill broad-spectrum bacteria.

Wang et al. synthesized Poly[oligo(ethylene glycol)fumarate]-co-poly[dodecyl bis(2-hydroxyethyl)methylammonium fumarate] (POEGDMAM) containing multi-enes and poly[oligo(ethylene glycol)mercaptosuccinate] (POEGMS) containing multi-thiols by polycondensation reaction, and the two synthesized functional polymers were deposited on thiosilicate substrates by the LbL self-assembly technique and "click" chemistry to prepare a crosslinked hydrogel coating [74]. The antibacterial activity of the ultrathin hydrogel films on silicon wafers was determined by the disc diffusion method against Gram-negative (Escherichia coli) and Gram-positive (*Staphylococcus aureus*) bacteria. The control groups without any hydrogel films generated no inhibition zones, while the others exhibited inhibition zones for both of these types of bacteria. The zones were quite obvious, considering the thickness of the films was just below 100 nm. Moreover, the inhibition zones increased significantly with the number of bilayers, which confirmed the LbL reaction on the substrates. Refer to Table 4 for the above information.

Table 3. Hydrogel coatings anchored to the surface of the substrate.

First Author	Publication Year	Main Components of Hydrogel Coating	Loaded Substance	Coating Preparation Method	Coating Adhesion Principle	Crosslinking Agent	Function
Leng Jin [65]	2021	GelMA, HAMA		UV crosslinking	ZnO layer anchored by catechol group on titanium surface	Photo-crosslinking	Reduce the toxicity of Zn^{2+}, improve soft tissue compatibility and antibacterial ability
He Ye [66]	2021	GelMA	Photosensitizer Ce6-loaded polydopamine nanoparticles	After application, UV lamp crosslinking is performed	Anchored to the titanium surface by catecholic groups	Photo-crosslinking	Antibacterial, promotes cell adhesion and proliferation
Liu [67]	2021	Type A gelatin, AA, AAc-NHS ester, methacrylate anhydride monomer	Nano hydroxyapatite	Spin-coating method	Chemical bond anchoring	Ultraviolet light crosslinking	Promotes osteoblast differentiation and cell adhesion, and promotes wound healing

Table 4. Hydrogel coatings prepared by LbL self-assembly technique.

First Author	Publication Year	Main Components of Hydrogel Coating	Coating Preparation Method	Coating Adhesion Principle	Crosslinking Agent	Function
Cai [73]	2012	e-PEI, Alg	Alternate deposition method	Two-step annealing stabilizes the film onto the surface of the material	EDC	Antithrombotic, antibacterial
Wang [74]	2014	POEGDMAM, POEGMS	LbL thiol–ene "click" reactions	POEGDMAM first reacted with the thiols on the surface to form a single layer of polymer. Subsequently, POEGMS reacted with the immobilized ene groups on the polymer surface to give the second layer. Repeated deposition of the polymers gave the corresponding multilayer films.		Antibacterial

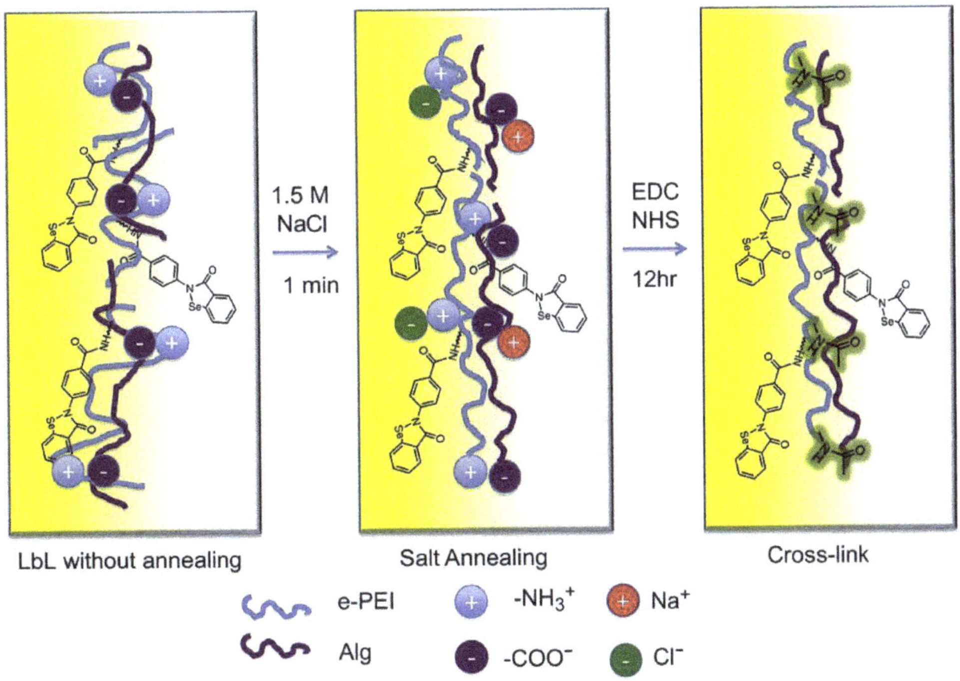

Figure 6. Schematic illustration of two-step annealing of LbL films, including salt annealing and cross-linking [73]. Copyright 2011, Elsevier.

In summary, the above three preparation strategies have their advantages. The tolerance of surface-initiated graft crosslinking polymerization to water and impurities as well as its compatibility with various functional monomers enable the hydrogel coating to be effectively loaded on the surface of the substrate; the photograft crosslinking method can achieve the regulation of the overall or local antibacterial properties of the material. Photo- driven polymerization can be carried out in vivo in a non-invasive way, which has a good biomedical application prospect. Anchoring the hydrogel coating to the substrate surface is simple and does not require too many reaction sites, and the entire composite has a "sandwich" structure, which has a positive impact on the subsequent antibacterial modification. The LbL self-assembly technique to coat crosslinked hydrogels has long-term stability and is suitable for use as a carrier for slow release of antibacterial drugs. The preparation process is simple, gentle, and fast, and the hydrogel coating can act on the human body for a long time. Researchers can design different grafts and functional hydrogel coatings according to their needs.

Thus far, this review has detailed methods for preparing hydrogel coatings on the substrate surface of biological materials. At present, it is necessary for biomedical materials to have effective antibacterial properties. Although hydrogel itself has certain antibacterial properties, the antibacterial effect of being implanted into human body as a coating is not sufficient; thus, it is necessary to combine antibacterial substances with hydrogel coatings through reasonable chemistry to enhance the antibacterial effect. Next, we will introduce the antibacterial mechanism of hydrogel coating by classification.

3. Hydrogel Coatings in Biomedical Antibacterial Applications

The following three antibacterial methods are the main methods for material surface modification by hydrogel coatings: The first is bacterial repellence and inhibition. The second is the contact surface killing of bacteria. The third is the release of antibacterial agents (Figure 7) [2].

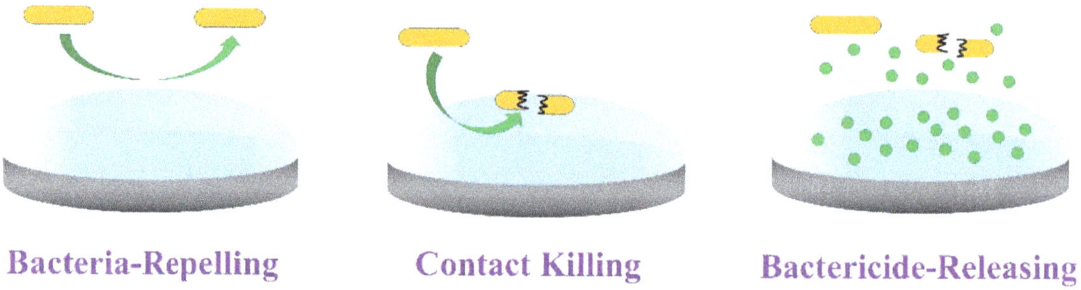

Figure 7. Strategies for antibacterial hydrogel coatings [2]. Copyright 2020, Elsevier.

3.1. Bacterial Repellence and Inhibition

In the earliest stage of bacterial biofilm formation, bacterial adhesion on the surface is reversible, so the introduction of hydrogel coatings that repell bacterial adhesion is the most direct antibacterial method. The hydrogel coating prepared by this method has better biocompatibility [75]. When biomaterials are implanted into the body, proteins tend to absorb nonspecifically on the surface, which promotes bacterial adhesion; thus, the repellent hydrogel coating should effectively inhibit the nonspecific absorption of proteins. Substances with hydrogen bonding acceptors and hydrophilic polar functional groups may inhibit the nonspecific absorption of proteins; they may form hydrogen bonds with water molecules in aqueous media and form a highly hydrated layer on the polymer surface to effectively achieve antibacterial properties. Polyethylene glycols (PEG) [76], polyvinyl alcohols (PVA) [77], polyacrylates [78], amphiphilic polymers [79], polysaccharides, and other hydrophilic substances are widely used as raw materials in the field of antibacterial hydrogels [80]. Next, the applications of hydrogel coatings containing the above substances in biomaterials are introduced.

PEG has excellent biocompatibility. The easiest way to prepare PEG hydrogels is to crosslink the PEG derivatives with two capped vinyl groups, such as polyethylene glycol diacrylate (PEGDA), or copolymerize the PEG macromonomer containing one vinyl group at the chain end, such as (PEGMA). Ekblad and his colleagues prepared PEGMA copolymer hydrogel coatings on the surface of biomaterials, reducing the sedimentation density of bacteria to less than 5% compared to that on the original surface [81]. Johnbosco Castro et al. used a spray coating technique to uniformly distribute a reactive hydrogel precursor on cobaltchromium (CoCr) vascular stents and left the solution polymerizing to form a hydrogel coating [82]. The hydrogel coating is a mixture of four-armed PEG and heparin with disilane and poly (ethylene-alt-maleic anhydride) (PEMA) as a bonding layer, ensuring that the hydrogel coating can provide covalent immobilization on the stent surface. Bioassay results showed that the PEG-based hydrogel coatings could effectively prevent the settlement and accumulation of bacteria.

PVA hydrogels can be prepared by physical and chemical crosslinking, have excellent biocompatibility and hydrophilicity, and are widely used in the field of biomaterials. Li et al. applied polyurethane prepolymer/polyvinyl alcohol (PPU/PVA) hydrogel coatings on the surfaces of polydimethylsiloxane (PDMS) nerve electrodes and compared these with the surfaces of nerve electrodes not coated with any substance and only coated with a polyurethane prepolymer (PPU). The surface nonspecific fibrinogen adsorption was reduced by 92%, proving that polyvinyl alcohols have effective antibacterial properties [83]. Yan et al. embedded quaternized chitosan-coated molybdenum disulfide (QCS-MoS_2) nanomaterials in PVA hydrogels. QCS-MoS_2 has photoresponsive properties that improve the mechanical strength of PVA hydrogels; it also has good photothermal conversion ability to generate reactive oxygen species under the irradiation of near-infrared light at 808 nm, which enables the hydrogels to exhibit excellent antibacterial activity while remaining non-cytotoxic to L929 cells [84].

Polyacrylate hydrogels may form free radical polymerizations initiated by carbon—carbon double bond functional groups. Polyester blocks may hydrolytically degrade these hydrogels after implantation as a surface coating for biomaterials. The degradation products eventually exit the body through internal circulation. Every year, millions of repair procedures occur in hospitals worldwide, and inguinal hernia repair is one of the most common procedures [85,86]. In the 1960s, polypropylene (PP) mesh emerged as a biomaterial for hernia repair. In addition to PP, polytetrafluoroethylene (PTFE), polyurethane (PU), and polyethylene (PE) can also be used for hernia repair [87–91]. Since a synthetic mesh implanted in the human body produces various adverse reactions in the physiological environment due to the presence of bacteria, leading to chronic pain and discomfort, a polyacrylate hydrogel coating was introduced to avoid bacterial adhesion on the implant surface while demonstrating antibacterial properties and promoting cell proliferation. For example, Andrada Serafim et al. prepared GelMA and MuMa hydrogel coatings on the surface of a PP mesh by EDC-NHS chemical grafting (Figure 8) [92]. The study showed that the GelMA hydrogel-coated scaffold interacted most strongly with fibroblasts, and this interaction was further enhanced when GelMA was combined with PRP, indicating that the coating could promote wound healing. The coating effectively prevents the adhesion of Gram-positive and Gram-negative bacteria, opening up a novel direction for the application of bioactive meshes for ventral hernia repair.

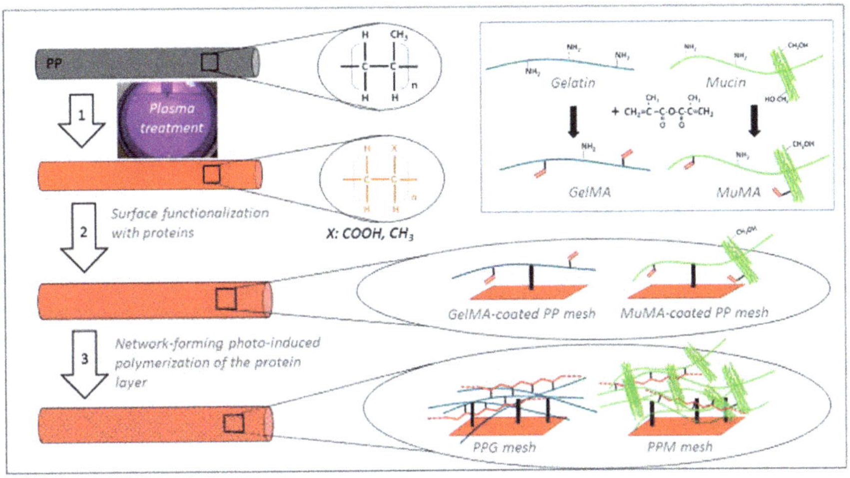

Figure 8. Coating characterization: Schematic representation of the three-step procedure used for the surface functionalization of PP meshes: 1—plasma treatment, 2—functionalization with protein analogues GelMA and MuMA (synthesis in inset), 3—generation of polymer-based hydrogels as coatings for the PP fibers. Reproduced with permission of [92]. Copyright 2020, MDPI polymers.

Polyacrylamide (PAM)-based hydrogel coatings have more antibacterial activity than polyacrylate coatings because of the higher surface hydration capacity of the amide groups. PAM-based hydrogels are polymers consisting of acrylamide (AAM) copolymerized with other monomers and are widely used in biomedical antibacterial materials. For example, Zhang J. et al. prepared a hydrogel by combining chitosan and poly(N-(2-hydroxyethyl)acrylamide) (PHEAA), which exhibited excellent mechanical properties with a fracture stress of 3.8 MPa and a strain of 640%. The antibacterial property of PHEAA may effectively inhibit the adsorption of nonspecific proteins, endowing the hydrogel excellent antibacterial properties [93].

We can combine the above substances in two or more reasonable combinations to obtain better antibacterial properties. Although the bacteria-repelling strategy has a good effect, it cannot kill pathogens in body fluids. Moreover, PEG-based coatings and other polymer coatings cannot completely prevent the adhesion of bacteria.

3.2. Contact Surface Killing of Bacteria

Researches show that the combination of bactericidal compounds and hydrogel coatings can effectively kill bacteria. Different from the passive antibacterial mechanism of bacterial-repelling hydrogel coatings, the bactericidal hydrogel coating can actively kill bacteria by destroying the cell membrane of bacteria, thus preventing the propagation of bacteria and achieving effective antibacterial activity. The bactericidal compounds commonly used generally contain cations and hydrophobic groups, and since bacteria have negative charges, they can be adsorbed by the cations of the bactericidal compounds. The hydrophobic groups of bactericides may also damage the lipid composition of the bacterial membranes. Some widely used bactericides are antimicrobial peptides (AMPs) and quaternary ammonium compounds (QACs) [94,95].

Antimicrobial peptides (AMPs), also known as host defense peptides, can be produced by plants, animals, humans, and bacteria. AMPs have effective antibacterial activity as the first line of defense against pathogenic invasion. The active bactericidal components of AMPs are arginine and lysine cations as well as a high proportion of hydrophobic amino acids. The antibacterial mechanism of AMPs is similar to that of quaternary ammonium compounds, which damage bacterial membranes through electrostatic interactions and hydrophobic groups. The AMPs may also self-assemble into physically crosslinked hydrogels with practical antibacterial ability. Tugce et al. anchored polymer films containing maleimide groups on the surface of titanium with catechol and then coupled them with AMPs by click chemistry. The antibacterial activity was enhanced, with an antibacterial rate of up to 80% (Figure 9) [96].

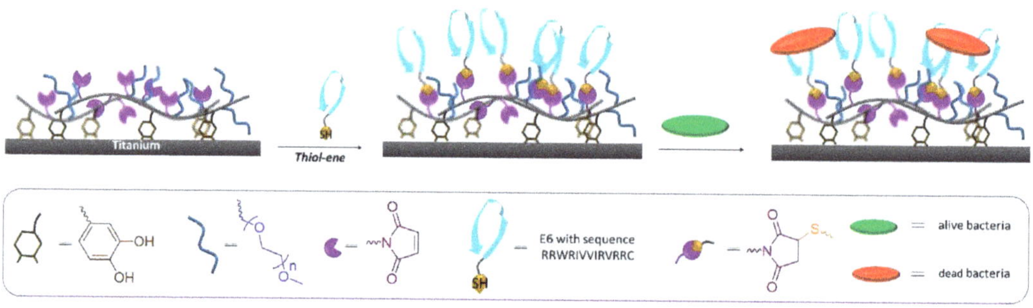

Figure 9. Schematic diagram of polymer hydrogel coating coupled with antimicrobial peptides to enhance the antibacterial performance. Reproduced with permission of [96]. Copyright 2019, American Chemical Society.

The bactericidal mechanism of quaternary ammonium compounds is the hydrophobic binding between cations and protein molecules, which adhere to bacteria, aggregate on the cell wall, and produce a chamber resistance effect to kill bacteria. In addition, the hydrophobic groups of QACs may interact with the hydrophilic groups of bacteria, changing the permeability of bacterial membranes and damaging the bacterial membrane structure. Combining QACs with hydrogel coatings may effectively confer antibacterial ability to the coatings and kill bacteria on contact, reaching nearly 100% antibacterial effect against Gram-positive and Gram-negative bacteria. Tao et al. prepared a hydrogel coating based on disulfide bonds by initiating a photopolymerization reaction of N-hydroxyethyl acrylamide (HEMAA), methylacryloxyethyl trimethyl ammonium chloride (DMC), and bis(2-methylpropene) ethoxydisulfide (DSDMA) on the surface. The coating combines the anti-adhesion property of poly(N-hydroxyethyl acrylamide) (PHEAA) with the bactericidal property of poly(quaternary ammonium salt), effectively inhibiting the adhesion and infection of E. coli and marine Vibrio. Ren et al. grafted QACs onto a polyurethane (PU) surface via surface-initiated atom transfer radical polymerization (SI-ATRP), and then grafted PVP hydrogel onto the surface by a Fenton-like reaction. QACs could improve the hydrophilicity of PU, and the surface water contact angle decreased from 93.6° to 60°, inhibiting bacterial adhesion and killing bacteria [97].

AMPs and QACs have excellent bactericidal properties but cannot distinguish between normal cells and bacteria, resulting in poor biocompatibility. A combination of repellent and bactericidal mechanisms have been used to overcome these drawbacks. For example, Yan et al. prepared a composite coating of poly (N, N-dimethyl aminoethyl methacrylate) block copolymer (PDMAEMA) and PSBMA using a surface-initiated photoiniferter-mediated polymerization (SI-PIMP) strategy. PSBMA formed a zwitterionic outer layer on the PDMAEMA layer. In the wet state, it can not only achieve the effect of repelling bacteria, but also reduces the erosion of normal cells on the cationic part. In the dry state, the amphoteric outer layer collapses to expose the cationic part, achieving a bactericidal effect. The method enhances the biocompatibility of the composite (Figure 10) [98].

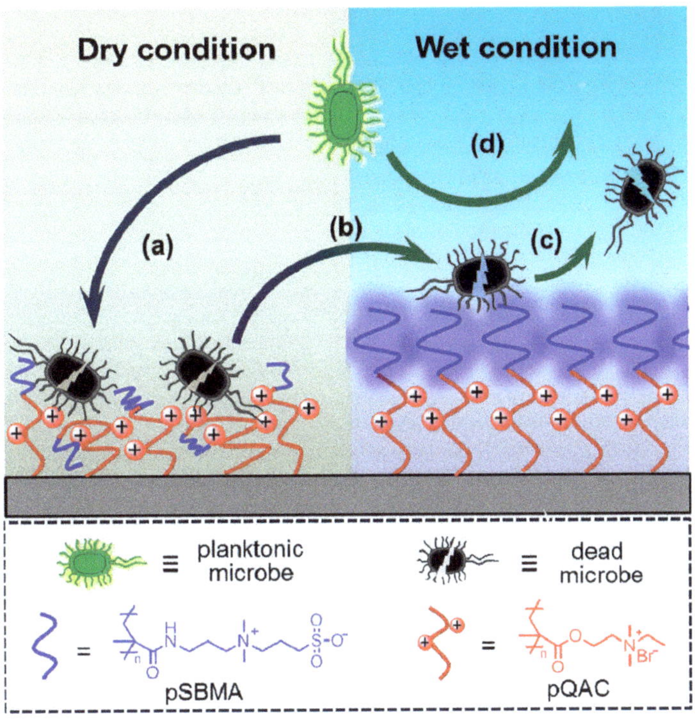

Figure 10. Antibacterial Surface in Dry and Wet Environment. In the dry state, the zwitterionic outer layer collapses, and the polycationic antibacterial layer kills bacteria (**a**); the collapsed zwitterionic outer layer swells (**b**) and allows the release of dead bacteria in the wet state (**c**). In a wet environment, the zwitterionic outer layer also prevents bacterial adhesion (**d**). Reproduced with permission of [98]. Copyright 2016, American Chemical Society.

Photocatalytic semiconductor composite hydrogels with reactive oxygen species (ROS)-generating capability have also attracted great attention for their contact bactericidal performance. Currently, such composite hydrogels are widely studied in the field of biological dressing coatings. Deng et al. prepared a hydrogel using oxidized sodium alginate and carbohydrazide-modified methacrylated gelatin as the matrix. Au@ZIF-8 semiconductor nanomaterials with MOF structure were embedded in the hydrogel. Au@ZIF-8 had the capability of photocatalysis, which can kill bacteria in contact with the wound surface by producing ROS under the irradiation of visible light [99]. Xing et al. prepared a chitosan sponge hydrogel coated with copper doped WO_{3-x} semiconductor material, which also had the same bactericidal mechanism (Figure 11) [100].

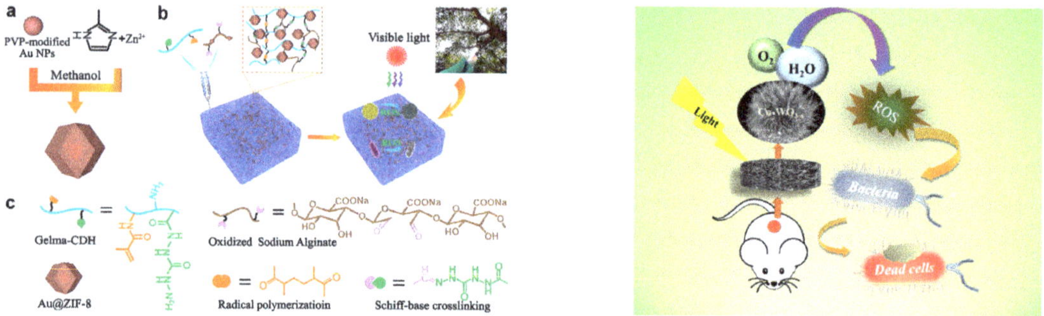

Figure 11. On the left is a diagram of reference [99]. (**a**) Schematic representation of Au@ZIF-8 synthesis; (**b**) Graphic illustration of Au@ZIF-8 imbedded into the hydrogels for light-driven treatment of the infection model; (**c**) Representaive chemical structures of individual components of the hydrogels. Copyright 2021, Elsevier. On the right is a diagram of reference [100]. Copyright 2023, American Chemical Society. The bactericidal mechanism of the two images is the same. Both produce reactive oxygen species (ROS) through photocatalytic semiconductor composite hydrogels to achieve bactericidal purposes.

3.3. Release of Antibacterial Agents

The three-dimensional network structure of hydrogels can be loaded with antibiotics, AMPs, cationic polymers, silver ions, copper ions, antibacterial drugs, and other bactericidal compounds to enhance the antibacterial properties of the biomaterials. Compared with the above two methods, this method is more flexible and controllable, repelling bacteria and killing bacteria via released antibacterial agents. However, the preparation process is relatively complex and requires consideration of the release rate and concentration of the antimicrobial agents. For example, Hoque et al. prepared a biocompatible hydrogel using dextran methacrylate (Dex-MA) as a monomer and encapsulated it with a small molecular cationic biocide by in situ loading during photopolymerization. The hydrogels showed a sustained release of biocide and displayed 100% activity against methicillin-resistant Staphylococcus aureus (MRSA) for an extended period of time (until day 5) [101].

Due to the limitation of the surface hydrogel coating function, the coated antibacterial substance will become depleted gradually and cannot maintain an excellent antibacterial effect for a long period. Therefore, the release rate of antibacterial substances should be reasonably controlled. Thus, an intelligently controlled release coating structure can be prepared to release antibacterial substances under specific conditions (pH, light reaction, temperature, and REDOX reaction) to kill bacteria and free bacteria attached to the material surface. The applications of these intelligent controlled-release hydrogel coatings will be described in the following sections.

Yan Kun et al. prepared composite hydrogel antibacterial coatings on stainless steel needle electrodes via a two-step electrochemical strategy [102]. First, electrochemical deposition was conducted to combine chitosan and other monomers on the surface of the stainless steel needle electrodes to form a hydrogel coating. Second, silver nanoparticles were synthesized in the hydrogel coating through an in situ electrochemical synthesis method. The hydrogel coating has a layered structure and releases silver nanoparticles in stages through the change in pH value to achieve controllable antibacterial properties.

This method is environmentally friendly and promotes the development of multifunctional nanomaterials in the biomedical field.

Xue et al. prepared a Dexp-loaded CuS nanoparticle cross-linked PEG hydrogel coating on the surface of 3D-printed polycaprolactone PCL scaffolds [103]. First, Dexp was loaded onto CuS NPs; then, the D-CuS-PEG hydrogel was prepared by crosslinking in PEG polymer, and the coating was formed on the surface of the PCL scaffold. CUS NPs had excellent photothermal conversion capability and high drug-loading capability. They could release Cu^{2+} and dexamethasone sodium phosphate under 1064 nm NIR irradiation and had excellent antibacterial activity.

Chandna et al. prepared a lignin-based hydrogel with both pH and light response properties; the hydrogels were doped with a photosensitizer (Rose Bengal, RB) and also with RB-conjugated lignin-derived silver nanocomplexes (RB@L-AgNCs). The RB@L-AgNCs were released under acidic conditions and irradiated with a green laser for 3 min. Reactive oxygen species (ROS) were produced, which greatly reduced the survival rate of bacteria. In subsequent studies, the hydrogel could be used in wound dressings and nanocoatings to achieve a stimulus-responsive antibacterial effect [104].

Li et al. loaded simvastatin into titanium dioxide nanotubes and prepared a thermosensitive chitosan-glycerin-hydroxypropyl methyl cellulose hydrogel (CGHH) coating on the surface of the nanotubes. At a normal body temperature of 37 °C, the CGHH coating was in the sol state, which resulted in the controlled release of simvastatin and promoted the differentiation of osteoblast *MC3T3-E1*. During bacterial infection, the CGHH coating transitioned into a gel state as the temperature rose to 40 °C, releasing glycerol and inducing macrophage polarization to the pro-inflammatory M1 phenotype to kill the bacteria [105].

Han et al. prepared a C-HA-Cys-allicin hydrogel coating with catechol-modified hyaluronic acid, cysteine, and allicin, and placed the hydrogel coating in an H_2O_2 solution. They found that the coating could be oxidized, which proved its REDOX response performance. Then, rhodamine was loaded into the hydrogel coating. It was found that the coating could release the drug in a REDOX environment for antibacterial purposes, and the best antibacterial effect occurred at a drug concentration of 5 μg/mL [106].

Each of the above three antibacterial methods has advantages and limitations. As mentioned above, dead microorganisms may accumulate after killing bacteria on the surface and affect the material properties. Releasing antibacterial agents also does not guarantee 100% elimination of bacteria; once bacteria are attached to the surface, they may multiply rapidly. To solve these limitations, we may combine two or more antibacterial mechanisms to prepare a hydrogel coating with excellent biocompatibility and antibacterial functions. For example, Jon et al. prepared HA-based hydrogel coatings on Ti6Al4V implants and used different crosslinking agents (1,4-butanediol diglycidyl ether, BDDE or divinyl sulfone, DVS) to modulate the physicochemical and nanomechanical properties of synthesized hydrogel coatings (Ti-HABDDE and Ti-HADVS) (Figure 12) [107]. Because the HA-based hydrogel coating has only excellent bacterial repellent function, in order to realize the bifunctional antibacterial mechanism (bacteria-repelling and bactericide-release), the researchers coated the antibacterial drug in the three-dimensional network structure of the hydrogel. Experiments showed that the coating had the ability to sustainedly release cefuroxime (CFX), tetracycline (TCN), amoxicillin (AMX, and acetylsalicylic acid (ASA). Relevant test data are as follows: HA-based hydrogel coatings demonstrated an outstanding multifunctional antibacterial activity: bacteria-repelling (51–55% of *S. aureus* and 27–40% of *E. coli*), bacteria-killing (82–119% of *S. aureus* and 83–87% of *E. coli*) and bactericide release killing (drug-loaded hydrogel coatings, $R > 2$).

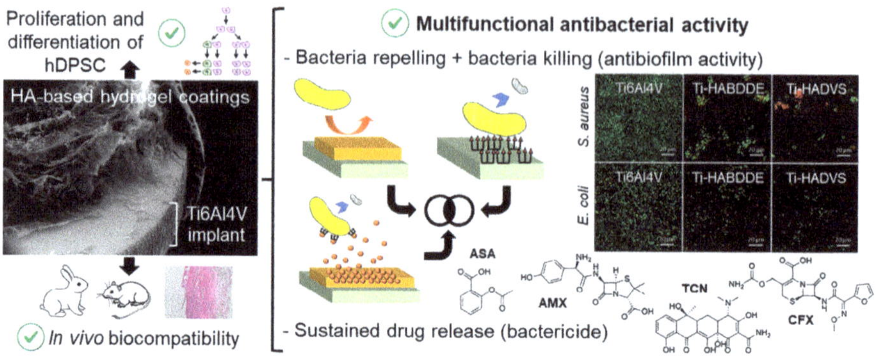

Figure 12. Graphical scheme and contact angle values. Reproduced with permission of [107]. Copyright 2023, Elsevier.

4. Conclusions and Prospects

This paper provides a detailed review of the application of hydrogel coatings in biomedical antibacterial applications and introduces the principles of adhesion on the surface of materials and antibacterial strategies. Hydrogels can be attached to the surface of biomaterials in three ways: The first is surface-initiated graft crosslinking polymerization. The second is anchoring the hydrogel coating to the substrate surface. The third is the LbL self-assembly technique to coat crosslinked hydrogels. Hydrogel coatings' antibacterial strategies are divided into three types: The first type is bacteria repellence and inhibition. The second type is the contact surface killing of bacteria. The third type is the release of antibacterial agents.

The antibacterial hydrogel coating interacts with organic and inorganic components as a biocompatible surface modifier, and the coating acts as a buffer between biomaterials and human tissues, making the biphasic interface of the material more stable and flexible and meeting the various needs of human tissue repair. The two key advantages of hydrogel coatings are as follows: Firstly, the coating can be firmly attached to the surface through chemical crosslinking and various anchoring reactions. Secondly, the coatings can attach to almost all kinds of materials, such as precious metals, oxides, polymers, and ceramics. Hydrogel coatings have excellent prospects for application, simple processing, stable performance, and wide application. Although significant progress has resulted from the research of antibacterial hydrogel coatings in biomedical applications, most of the research is only at the stage of cell and animal experiments, and further research on subsequent clinical applications needs to be conducted. The current research difficulties include the following: Firstly, the preparation method of antibacterial hydrogel coatings needs to be improved. Although the graft density of surface-initiated graft crosslinking polymerization is high, the initiator needs to be grafted to the surface, and the preparation process is relatively complex. The method of fixing the hydrogel coating to the substrate surface may result in uneven coverage of the hydrogel coating to the substrate surface due to the steric hinderance of the graft chain. Secondly, greater attention should be given to the study of the chemical stability of hydrogel coatings, including swelling, durability, degradability, mechanical properties, etc., which are important for the long-term effect of antibacterial hydrogel coatings on the human body. For example, swelling could be a problem for the coating of tubular medical devices, as the large swelling degree of hydrophilic hydrogels might block the tube. Finally, sterilization has been reported as an issue for most hydrogel coatings. In the complex environment of the body, the hydrogel coating needs to adapt to high temperature, high pressure, oxidation, and other conditions but also needs to maintain adhesion properties and bactericidal activity on the surface of the substrate. These problems will become the focus of future research on antibacterial hydrogel coatings. Future research directions may focus on the following aspects. First, there is a need to improve the adhesion of hydrogel coatings. The graft density should be large and uniform, so that the hydrogel coating uniformly covers the surface of the substrate, which is convenient for subsequent modification. Second, improvement of the mechanical properties of hydrogels and study of their long-term chemical stability are required, including improving the mechanical properties and breaking strength of hydrogels by crosslinking other chemicals. A third focus is the improvement of the swelling property of hydrogel, which is controlled by the change of pH and temperature. Improvement of this property is necessary for an intelligent and stable hydrogel coating. Fourth is the selection of materials that use monomers or segments with controllable degradation cycles and that have nontoxicity and harmless degradation products. This review provides a theoretical reference for follow-up research.

Author Contributions: Formal analysis, T.Y. and Q.S.; investigation, T.Y. and M.C.; writing—original draft preparation, Q.S. and W.Y.; review and editing, T.P. and T.Y. All authors have read and agreed to the published version of the manuscript.

Funding: This work was funded by the Talent Training Project of Reform and Development Funds of Central Government Supporting Local Universities, grant number 2021zyzcdf-01, the Fundamental Research Business Expenses Fundamental Research Project of Provincial Education Department Project of Heilongjiang Province of China, grant number 2021-KYYWF-0555, the Natural Science Foundation of Heilongjiang Province of China, grant number LH2021E116, the Doctoral Research Fund of Jiamusi University, grant number JMSUBZ2019-09, and the Open Project of Key Lab of Oral Biomedical Materials and Clinical Application of Heilongjiang Province, grant number KQSW2021-01.

Data Availability Statement: Not applicable.

Conflicts of Interest: All the authors declare that there is no conflict of interest.

References

1. Bao, Y.; Li, Z.; Li, Y.; Chen, T.; Cheng, Y.; Xu, M. Recent Advances of Biomedical Materials for Prevention of Post-ESD Esophageal Stricture. *Front. Bioeng. Biotechnol.* **2021**, *9*, 792929. [CrossRef] [PubMed]
2. Zhao, C.; Zhou, L.; Chiao, M.; Yang, W. Antibacterial hydrogel coating: Strategies in surface chemistry. *Adv. Colloid Interface Sci.* **2020**, *285*, 102280. [CrossRef] [PubMed]
3. Zhang, F.; Cheng, Z.; Ding, C.; Li, J. Functional biomedical materials derived from proteins in the acquired salivary pellicle. *J. Mater. Chem. B* **2021**, *9*, 6507–6520. [CrossRef] [PubMed]
4. Elbourne, A.; Crawford, R.J.; Ivanova, E.P. Nano-structured antimicrobial surfaces: From nature to synthetic analogues. *J. Colloid Interface Sci.* **2017**, *508*, 603–616. [CrossRef] [PubMed]
5. Modaresifar, K.; Azizian, S.; Ganjian, M.; Fratila-Apachitei, L.E.; Zadpoor, A.A. Bactericidal effects of nanopatterns: A systematic review. *Acta Biomater.* **2019**, *83*, 29–36. [CrossRef]
6. Jenkins, J.; Mantell, J.; Neal, C.; Gholinia, A.; Verkade, P.; Nobbs, A.H.; Su, B. Antibacterial effects of nanopillar surfaces are mediated by cell impedance, penetration and induction of oxidative stress. *Nat. Commun.* **2020**, *11*, 1626. [CrossRef]
7. Linklater, D.P.; Juodkazis, S.; Rubanov, S.; Ivanova, E.P. Comment on "Bactericidal Effects of Natural Nanotopography of Dragonfly Wing on *Escherichia coli*". *ACS Appl. Mater. Interfaces* **2017**, *9*, 29387–29393. [CrossRef]
8. Bandara, C.D.; Singh, S.; Afara, I.O.; Wolff, A.; Tesfamichael, T.; Ostrikov, K.; Oloyede, A. Bactericidal Effects of Natural Nanotopography of Dragonfly Wing on *Escherichia coli*. *ACS Appl. Mater. Interfaces* **2017**, *9*, 6746–6760. [CrossRef]
9. Hasan, J.; Crawford, R.J.; Ivanova, E.P. Antibacterial surfaces: The quest for a new generation of biomaterials. *Trends Biotechnol.* **2013**, *31*, 295–304. [CrossRef]
10. Takahashi, R.; Shimano, K.; Okazaki, H.; Kurokawa, T.; Nakajima, T.; Nonoyama, T.; King, D.R.; Gong, J.P. Tough Particle-Based Double Network Hydrogels for Functional Solid Surface Coatings. *Adv. Mater. Interfaces* **2018**, *5*, 1801018. [CrossRef]
11. Yao, H.; Wang, J.; Mi, S. Photo Processing for Biomedical Hydrogels Design and Functionality: A Review. *Polymers* **2017**, *10*, 11. [CrossRef]
12. Lee, S.C.; Kwon, I.K.; Park, K. Hydrogels for delivery of bioactive agents: A historical perspective. *Adv. Drug Deliv. Rev.* **2013**, *65*, 17–20. [CrossRef] [PubMed]
13. Peters, J.T.; Wechsler, M.E.; Peppas, N.A. Advanced biomedical hydrogels: Molecular architecture and its impact on medical applications. *Regen. Biomater.* **2021**, *8*, rbab060. [CrossRef]
14. Muir, V.G.; Burdick, J.A. Chemically Modified Biopolymers for the Formation of Biomedical Hydrogels. *Chem. Rev.* **2021**, *121*, 10908–10949. [CrossRef] [PubMed]
15. Oliva, N.; Shin, M.; Burdick, J.A. Editorial: Special Issue on Advanced Biomedical Hydrogels. *ACS Biomater. Sci. Eng.* **2021**, *7*, 3993–3996. [CrossRef] [PubMed]

16. Cadamuro, F.; Russo, L.; Nicotra, F. Biomedical Hydrogels Fabricated Using Diels–Alder Crosslinking. *Eur. J. Org. Chem.* **2021**, *2021*, 374–382. [CrossRef]
17. Chai, Q.; Jiao, Y.; Yu, X. Hydrogels for Biomedical Applications: Their Characteristics and the Mechanisms behind Them. *Gels* **2017**, *3*, 6. [CrossRef] [PubMed]
18. Francesko, A.; Petkova, P.; Tzanov, T. Hydrogel Dressings for Advanced Wound Management. *Curr. Med. Chem.* **2019**, *25*, 5782–5797. [CrossRef]
19. Jafari, H.; Ghaffari-Bohlouli, P.; Niknezhad, S.V.; Abedi, A.; Izadifar, Z.; Mohammadinejad, R.; Varma, R.S.; Shavandi, A. Tannic acid: A versatile polyphenol for design of biomedical hydrogels. *J. Mater. Chem. B* **2022**, *10*, 5873–5912. [CrossRef]
20. Taaca, K.L.M.; Prieto, E.I.; Vasquez, M.R. Current Trends in Biomedical Hydrogels: From Traditional Crosslinking to Plasma-Assisted Synthesis. *Polymers* **2022**, *14*, 2560. [CrossRef] [PubMed]
21. Guan, Z.; Yang, L.; Wang, W.; Zhang, J.; Liu, J.; Ren, C.; Wang, S.; Gao, Y.; Huang, P. Thermosensitive micellar hydrogel for enhanced anticancer therapy through redox modulation mediated combinational effects. *RSC Adv.* **2017**, *7*, 34755–34762. [CrossRef]
22. Fu, M.; Liang, Y.; Lv, X.; Li, C.; Yang, Y.Y.; Yuan, P.; Ding, X. Recent advances in hydrogel-based anti-infective coatings. *J. Mater. Sci. Technol.* **2021**, *85*, 169–183. [CrossRef]
23. Prucker, O.; Brandstetter, T.; Rühe, J. Surface-attached hydrogel coatings via C,H-insertion crosslinking for biomedical and bioanalytical applications (Review). *Biointerphases* **2018**, *13*, 010801. [CrossRef] [PubMed]
24. Riga, E.K.; Rühe, J.; Lienkamp, K. Non-Delaminating Polymer Hydrogel Coatings via C,H-Insertion Crosslinking (CHic)-A Case Study of Poly(oxanorbornenes). *Macromol. Chem. Phys.* **2018**, *219*, 1800397. [CrossRef] [PubMed]
25. Pan, J.; Jin, Y.; Lai, S.; Shi, L.; Fan, W.; Shen, Y. An antibacterial hydrogel with desirable mechanical, self-healing and recyclable properties based on triple-physical crosslinking. *Chem. Eng. J.* **2019**, *370*, 1228–1238. [CrossRef]
26. Islam, M.M.; AbuSamra, D.B.; Chivu, A.; Argüeso, P.; Dohlman, C.H.; Patra, H.K.; Chodosh, J.; González-Andrades, M. Optimization of Collagen Chemical Crosslinking to Restore Biocompatibility of Tissue-Engineered Scaffolds. *Pharmaceutics* **2021**, *13*, 832. [CrossRef] [PubMed]
27. Chollet, B.; Li, M.; Martwong, E.; Bresson, B.; Fretigny, C.; Tabeling, P.; Tran, Y. Multiscale Surface-Attached Hydrogel Thin Films with Tailored Architecture. *ACS Appl. Mater. Interfaces* **2016**, *8*, 11729–11738. [CrossRef]
28. Wang, H.; Chen, Z.; Cheng, S.; Li, R.; Pan, X.; Zhang, C.; Gu, H.; Xie, A.; Dong, W. Synthesis of cationic hydrogels with tunable physicochemical properties for antibacterial applications. *Eur. Polym. J.* **2022**, *173*, 111228. [CrossRef]
29. Demirkıran, E.; Başyiğit, B.; Altun, G.; Yücetepe, M.; Sağlam, H.; Karaaslan, M. Facile construction of fruit protein based natural hydrogel via intra/inter molecular cross-linking. *Food Hydrocoll.* **2022**, *133*, 107899. [CrossRef]
30. Kojima, C.; Koda, T.; Nariai, T.; Ichihara, J.; Sugiura, K.; Matsumoto, A. Application of Zwitterionic Polymer Hydrogels to Optical Tissue Clearing for 3D Fluorescence Imaging. *Macromol. Biosci.* **2021**, *21*, 2100170. [CrossRef]
31. Stor, D.S.; Andersen, O.S. Bilayer mediated regulation of pore formation by the antimicrobial peptide melittin. *Biophys. J.* **2022**, *121*, 217a–218a. [CrossRef]
32. Yu, H.; Zhu, Y.; Hui, A.; Yang, F.; Wang, A. Removal of antibiotics from aqueous solution by using porous adsorbent templated from eco-friendly Pickering aqueous foams. *J. Environ. Sci.* **2021**, *102*, 352–362. [CrossRef] [PubMed]
33. Varani, T.; Abdouss, M.; Azerang, P.; Tahghighi, A. Acetylenic Sulfones and Acetylenic Sulfonamide Analogs: A Novel and Preferable Antimicrobial Drugs Based on Computational Strategies. *J. Comput. Biophys. Chem.* **2022**, *21*, 115–122. [CrossRef]
34. Zhao, S.; Ke, H.; Yang, T.; Peng, Q.; Ge, J.; Yao, L.; Xu, S.; Zhirong, D.; Pan, G. Enhanced Thermal and Antibacterial Properties of Stereo-Complexed Polylactide Fibers Doped With Nano-Silver. *Front. Mater.* **2022**, *9*, 775333. [CrossRef]
35. Ahmad, N.J.; Ansari, M.J. The Use of Zinc Oxide in a Public Healthcare Organization in Riyadh Region. *J. Pharm. Res. Int.* **2021**, *33*, 8–12. [CrossRef]
36. Griffith, L.G.; Naughton, G. Tissue engineering—Current challenges and expanding opportunities. *Science* **2002**, *295*, 1009–1014. [CrossRef] [PubMed]
37. Spector, M. An interview with Jianwu Dai: Understanding the biological processes underlying regeneration to direct the implementation of biomedical materials. *Biomed. Mater.* **2020**, *15*, 030201. [CrossRef]
38. Beaman, H.T.; Howes, B.; Ganesh, P.; Monroe, M.B.B. Shape memory polymer hydrogels with cell-responsive degradation mechanisms for Crohn's fistula closure. *J. Biomed. Mater. Res. A* **2022**, *110*, 1329–1340. [CrossRef] [PubMed]
39. Bose, S.; Robertson, S.F.; Bandyopadhyay, A. Surface modification of biomaterials and biomedical devices using additive manufacturing. *Acta Biomater.* **2018**, *66*, 6–22. [CrossRef]
40. Zhang, Z.; Zhang, Z.; Hong, Z. Unscented Kalman Filter-Based Robust State and Parameter Estimation for Free Radical Polymerization of Styrene with Variable Parameters. *Polymers* **2022**, *14*, 973. [CrossRef] [PubMed]
41. Arango-Santander, S. Bioinspired Topographic Surface Modification of Biomaterials. *Materials* **2022**, *15*, 2383. [CrossRef] [PubMed]
42. Zhang, F.; Hu, C.; Yang, L.; Liu, K.; Ge, Y.; Wei, Y.; Wang, J.; Luo, R.; Wang, Y. A conformally adapted all-in-one hydrogel coating: Towards robust hemocompatibility and bactericidal activity. *J. Mater. Chem. B* **2021**, *9*, 2697–2708. [CrossRef]
43. Fu, X.; Liu, X.; Hao, D.; Xiao, W.; Nie, Q.; Meng, J. Nickel-Catcher-Doped Zwitterionic Hydrogel Coating on Nickel–Titanium Alloy Toward Capture and Detection of Nickel Ions. *Front. Bioeng. Biotechnol.* **2021**, *9*, 698745. [CrossRef] [PubMed]
44. Ma, S.; Yan, C.; Cai, M.; Yang, J.; Wang, X.; Zhou, F.; Liu, W. Continuous Surface Polymerization via Fe(II)-Mediated Redox Reaction for Thick Hydrogel Coatings on Versatile Substrates. *Adv. Mater.* **2018**, *30*, 1803371. [CrossRef] [PubMed]

45. Ma, S.; Rong, M.; Lin, P.; Bao, M.; Xie, J.; Wang, X.; Huck, W.T.S.; Zhou, F.; Liu, W. Fabrication of 3D Tubular Hydrogel Materials through On-Site Surface Free Radical Polymerization. *Chem. Mater.* **2018**, *30*, 6756–6768. [CrossRef]
46. Wancura, M.; Talanker, M.; Toubbeh, S.; Bryan, A.; Cosgriff-Hernandez, E. Bioactive hydrogel coatings of complex substrates using diffusion-mediated redox initiation. *J. Mater. Chem. B* **2020**, *8*, 4289–4298. [CrossRef] [PubMed]
47. Zhou, H.R.; Huang, J.; Chen, M.; Li, Y.; Yuan, M.; Yang, H. Effect of metal ions with reducing properties on hydrogels containing catechol groups. *Colloids Surf. Physicochem. Eng. Asp.* **2021**, *631*, 127657. [CrossRef]
48. Yao, D.; Feng, J.; Wang, J.; Deng, Y.; Wang, C. Synthesis of silicon anode binders with ultra-high content of catechol groups and the effect of molecular weight on battery performance. *J. Power Sources* **2020**, *463*, 228188. [CrossRef]
49. Zhou, C.; Wu, Y.; Thappeta, K.R.V.; Subramanian, J.T.L.; Pranantyo, D.; Kang, E.-T.; Duan, H.; Kline, K.; Chan-Park, M.B. In Vivo Anti-Biofilm and Anti-Bacterial Non-Leachable Coating Thermally Polymerized on Cylindrical Catheter. *ACS Appl. Mater. Interfaces* **2017**, *9*, 36269–36280. [CrossRef]
50. Chen, Y.; Gao, P.; Huang, L.; Tan, X.; Zhou, N.; Yang, T.; Qiu, H.; Dai, X.; Michael, S.; Tu, Q.; et al. A tough nitric oxide-eluting hydrogel coating suppresses neointimal hyperplasia on vascular stent. *Nat. Commun.* **2021**, *12*, 7079. [CrossRef]
51. Gevrek, T.N.; Degirmenci, A.; Sanyal, R.; Sanyal, A. Multifunctional and Transformable 'Clickable' Hydrogel Coatings on Titanium Surfaces: From Protein Immobilization to Cellular Attachment. *Polymers* **2020**, *12*, 1211. [CrossRef] [PubMed]
52. Liu, Y.; Zhang, F.; Lang, S.; Yang, L.; Gao, S.; Wu, D.; Liu, G.; Wang, Y. A Uniform and Robust Bioinspired Zwitterion Coating for Use in Blood-Contacting Catheters with Improved Anti-Inflammatory and Antithrombotic Properties. *Macromol. Biosci.* **2021**, *21*, 2100341. [CrossRef] [PubMed]
53. Yang, F.; Guo, G.; Wang, Y. Inflammation-directed nanozyme-eluting hydrogel coating promotes vascular tissue repair by restoring reactive oxygen species homeostasis. *Chem. Eng. J.* **2023**, *454*, 140556. [CrossRef]
54. Ran, W.; Zhu, H.; Shen, X.; Zhang, Y. Rheological properties of asphalt mortar with silane coupling agent modified oil sludge pyrolysis residue. *Constr. Build. Mater.* **2022**, *329*, 127057. [CrossRef]
55. Zhou, J.; Xu, W. Toward interface optimization of transparent wood with wood color and texture by silane coupling agent. *J. Mater. Sci.* **2022**, *57*, 5825–5838. [CrossRef]
56. Wu, X.; Liu, S.; Chen, K.; Wang, F.; Feng, C.; Xu, L.; Zhang, D. 3D printed chitosan-gelatine hydrogel coating on titanium alloy surface as biological fixation interface of artificial joint prosthesis. *Int. J. Biol. Macromol.* **2021**, *182*, 669–679. [CrossRef] [PubMed]
57. Pawar, K.N.; Nawpute, A.A.; Tambe, S.; Patil, P.; Ubale, Y.; Patil, A. Dextrose Assisted Sol-Gel Synthesis and Evaluation of Structural Parameters of $Li_{0.5}Fe_{2.5}O_4$ Nanoparticles for Microwave Device Application. *Adv. Mater. Res.* **2022**, *1169*, 27–33. [CrossRef]
58. Li, J.-W. Biosafety of a 3D-printed intraocular lens made of a poly(acrylamide-co-sodium acrylate) hydrogel in vitro and in vivo. *Int. J. Ophthalmol.* **2020**, *13*, 1521–1530. [CrossRef] [PubMed]
59. Efremov, D.V.; Gerasimova, A.A. Production of Fe–Cr–Co-Based Magnets by Selective Laser Sintering. *Steel Transl.* **2021**, *51*, 688–692. [CrossRef]
60. Deng, Y.; Shavandi, A.; Okoro, O.V.; Nie, L. Alginate modification via click chemistry for biomedical applications. *Carbohydr. Polym.* **2021**, *270*, 118360. [CrossRef]
61. Librando, I.L.; Mahmoud, A.G.; Carabineiro, S.A.C.; Guedes da Silva, M.F.C.; Maldonado-Hódar, F.J.; Geraldes, C.F.G.C.; Pombeiro, A.J.L. Heterogeneous Gold Nanoparticle-Based Catalysts for the Synthesis of Click-Derived Triazoles via the Azide-Alkyne Cycloaddition Reaction. *Catalysts* **2021**, *12*, 45. [CrossRef]
62. Jaisingh, A.; Kapur, G.S.; Nebhani, L. Tuning Melt Strength and Processability of Polyolefins by Addition of a Functionalized Additive Designed via the TEMPO-Driven Thiol-ene Reaction. *Ind. Eng. Chem. Res.* **2021**, *60*, 10155–10166. [CrossRef]
63. Soares, M.I.L.; Cardoso, A.L.; Pinho e Melo, T.M.V.D. Diels–Alder Cycloaddition Reactions in Sustainable Media. *Molecules* **2022**, *27*, 1304. [CrossRef] [PubMed]
64. Magennis, E.P.; Hook, A.L.; Williams, P.; Alexander, M.R. Making Silicone Rubber Highly Resistant to Bacterial Attachment Using Thiol-ene Grafting. *ACS Appl. Mater. Interfaces* **2016**, *8*, 30780–30787. [CrossRef]
65. Leng, J.; He, Y.; Yuan, Z.; Tao, B.; Li, K.; Lin, C.; Xu, K.; Chen, M.; Dai, L.; Li, X.; et al. Enzymatically-degradable hydrogel coatings on titanium for bacterial infection inhibition and enhanced soft tissue compatibility via a self-adaptive strategy. *Bioact. Mater.* **2021**, *6*, 4670–4685. [CrossRef]
66. He, Y.; Leng, J.; Li, K.; Xu, K.; Lin, C.; Yuan, Z.; Zhang, R.; Wang, D.; Tao, B.; Huang, T.J.; et al. A multifunctional hydrogel coating to direct fibroblast activation and infected wound healing via simultaneously controllable photobiomodulation and photodynamic therapies. *Biomaterials* **2021**, *278*, 121164. [CrossRef] [PubMed]
67. Liu, C.; Pan, L.; Liu, C.; Liu, W.; Li, Y.; Cheng, X.; Jian, X. Enhancing Tissue Adhesion and Osteoblastic Differentiation of MC3T3-E1 Cells on Poly(aryl ether ketone) by Chemically Anchored Hydroxyapatite Nanocomposite Hydrogel Coating. *Macromol. Biosci.* **2021**, *21*, 2100078. [CrossRef] [PubMed]
68. Liu, W.; Liu, C.; Liu, C.; Li, Y.; Pan, L.; Wang, J.; Jian, X. Surface chemical modification of poly(phthalazinone ether nitrile ketone) through rhBMP-2 and antimicrobial peptide conjugation for enhanced osteogenic and antibacterial activities in vitro and in vivo. *Chem. Eng. J.* **2021**, *424*, 130321. [CrossRef]
69. Khoffi, F.; Khalsi, Y.; Chevrier, J.; Kerdjoudj, H.; Tazibt, A.; Heim, F. Surface modification of polymer textile biomaterials by N2 supercritical jet: Preliminary mechanical and biological performance assessment. *J. Mech. Behav. Biomed. Mater.* **2020**, *107*, 103772. [CrossRef]

70. Wang, X.; Zhao, X.; Zhang, L.; Wang, W.; Zhang, J.; He, F.; Yang, J. Design and fabrication of carbon fibers with needle-like nano-HA coating to reinforce granular nano-HA composites. *Mater. Sci. Eng. C* **2017**, *77*, 765–771. [CrossRef]
71. Liu, Q.; Gao, S.; Zhao, Y.; Tao, W.; Yu, X.; Zhi, M. Review of layer-by-layer self-assembly technology for fire protection of flexible polyurethane foam. *J. Mater. Sci.* **2021**, *56*, 9605–9643. [CrossRef]
72. Zhu, D.; Guo, D.; Zhang, L.; Tan, L.; Pang, H.; Ma, H.; Zhai, M. Non-enzymatic xanthine sensor of heteropolyacids doped ferrocene and reduced graphene oxide via one-step electrodeposition combined with layer-by-layer self-assembly technology. *Sens. Actuators B Chem.* **2019**, *281*, 893–904. [CrossRef]
73. Cai, W.; Wu, J.; Xi, C.; Ashe, A.J.; Meyerhoff, M.E. Carboxyl-ebselen-based layer-by-layer films as potential antithrombotic and antimicrobial coatings. *Biomaterials* **2011**, *32*, 7774–7784. [CrossRef] [PubMed]
74. Wang, H.; Zha, G.; Du, H.; Gao, L.; Li, X.; Shen, Z.; Zhu, W. Facile fabrication of ultrathin antibacterial hydrogel films via layer-by-layer "click" chemistry. *Polym. Chem.* **2014**, *5*, 6489–6494. [CrossRef]
75. Hu, Q.; Liu, Y.; Pan, Y.; Wang, Y.; Jiang, L.; Lin, H.; Cheng, Y.; Xu, C.; Lin, D.; Cheng, H. Assessments of ionic release and biocompatibility of Co-Cr and CP-Ti produced by three different manufacturing techniques. *Mater. Today Commun.* **2022**, *30*, 103100. [CrossRef]
76. Hu, F.; Lu, H.; Xu, G.; Lv, L.; Chen, L.; Shao, Z. Carbon quantum dots improve the mechanical behavior of polyvinyl alcohol/polyethylene glycol hydrogel. *J. Appl. Polym. Sci.* **2022**, *139*, e52805. [CrossRef]
77. Zhang, M.-K.; Ling, X.-H.; Zhang, X.-H.; Han, G.-Z. A novel alginate/PVA hydrogel -supported Fe3O4 particles for efficient heterogeneous Fenton degradation of organic dyes. *Colloids Surf. Physicochem. Eng. Asp.* **2022**, *652*, 129830. [CrossRef]
78. Patrick, D.K.; Karasawa, A.; Sonoyama, N. Sodium Polyacrylate Hydrogel Electrolyte Hybridized with Layered Double Hydroxide for Solid-State NiCo/Zinc Battery. *J. Electrochem. Soc.* **2022**, *169*, 040559. [CrossRef]
79. Nakano, H.; Kakinoki, S.; Iwasaki, Y. Long-lasting hydrophilic surface generated on poly(dimethyl siloxane) with photoreactive zwitterionic polymers. *Colloids Surf. B Biointerfaces* **2021**, *205*, 111900. [CrossRef] [PubMed]
80. Mondal, P.; Chatterjee, K. Injectable and self-healing double network polysaccharide hydrogel as a minimally-invasive delivery platform. *Carbohydr. Polym.* **2022**, *291*, 119585. [CrossRef]
81. Ekblad, T.; Bergström, G.; Ederth, T.; Conlan, S.L.; Mutton, R.; Clare, A.S.; Wang, S.; Liu, Y.; Zhao, Q.; D'Souza, F.; et al. Poly(ethylene glycol)-Containing Hydrogel Surfaces for Antifouling Applications in Marine and Freshwater Environments. *Biomacromolecules* **2008**, *9*, 2775–2783. [CrossRef] [PubMed]
82. Johnbosco, C.; Zschoche, S.; Nitschke, M.; Hahn, D.; Werner, C.; Maitz, M.F. Bioresponsive starPEG-heparin hydrogel coatings on vascular stents for enhanced hemocompatibility. *Mater. Sci. Eng. C* **2021**, *128*, 112268. [CrossRef] [PubMed]
83. Li, Y.; Han, M.; Cai, Y.; Jiang, B.; Zhang, Y.; Yuan, B.; Zhou, F.; Cao, C. Muscle-inspired MXene/PVA hydrogel with high toughness and photothermal therapy for promoting bacteria-infected wound healing. *Biomater. Sci.* **2022**, *10*, 1068–1082. [CrossRef] [PubMed]
84. Yan, P.; Li, M.; Liu, J.; Song, L.; Tang, K. Near-infrared responsive quaternized chitosan-coated MoS2/poly(vinyl alcohol) hydrogel with improved mechanical and rapid antibacterial properties. *Eur. Polym. J.* **2022**, *180*, 111593. [CrossRef]
85. Vinshtok, Y.; Cassuto, D.; Belenky, I. Pneumatic Delivery of Hyaluronan for Skin Remodeling: A Comparative Review. *J. Drugs Dermatol.* **2020**, *19*, 170–175. [CrossRef]
86. Herrero, A.; Gonot Gaschard, M.; Bouyabrine, H.; Perrey, J.; Picot, M.-C.; Guillon, F.; Fabre, J.-M.; Souche, R.; Navarro, F. Étude comparative des prothèses biologiques versus synthétiques dans le traitement des éventrations abdominales classées grade II/III du Ventral Hernia Working Group. *J. Chir. Viscérale* **2022**, *159*, 102–111. [CrossRef]
87. Kalaba, S.; Gerhard, E.; Winder, J.S.; Pauli, E.M.; Haluck, R.S.; Yang, J. Design strategies and applications of biomaterials and devices for Hernia repair. *Bioact. Mater.* **2016**, *1*, 2–17. [CrossRef]
88. Guillaume, O.; Pérez-Tanoira, R.; Fortelny, R.; Redl, H.; Moriarty, T.F.; Richards, R.G.; Eglin, D.; Petter Puchner, A. Infections associated with mesh repairs of abdominal wall hernias: Are antimicrobial biomaterials the longed-for solution? *Biomaterials* **2018**, *167*, 15–31. [CrossRef]
89. Bilsel, Y.; Abci, I. The search for ideal hernia repair; mesh materials and types. *Int. J. Surg.* **2012**, *10*, 317–321. [CrossRef]
90. Schreinemacher, M.H.F.; van Barneveld, K.W.Y.; Dikmans, R.E.G.; Gijbels, M.J.J.; Greve, J.-W.M.; Bouvy, N.D. Coated meshes for hernia repair provide comparable intraperitoneal adhesion prevention. *Surg. Endosc.* **2013**, *27*, 4202–4209. [CrossRef]
91. Wolf, M.T.; Carruthers, C.A.; Dearth, C.L.; Crapo, P.M.; Huber, A.; Burnsed, O.A.; Londono, R.; Johnson, S.A.; Daly, K.A.; Stahl, E.C.; et al. Polypropylene surgical mesh coated with extracellular matrix mitigates the host foreign body response: Polypropylene Surgical Mesh Coated With ECM. *J. Biomed. Mater. Res. A* **2014**, *102*, 234–246. [CrossRef]
92. Serafim, A.; Cecoltan, S.; Olăreț, E.; Dragusin, D.-M.; Vasile, E.; Popescu, V.; Manolescu Mastalier, B.S.; Iovu, H.; Stancu, I.-C. Bioinspired Hydrogel Coating Based on Methacryloyl Gelatin Bioactivates Polypropylene Meshes for Abdominal Wall Repair. *Polymers* **2020**, *12*, 1677. [CrossRef]
93. Zhang, J.; Chen, L.; Shen, B.; Chen, L.; Feng, J. Ultra-high strength poly(N-(2-hydroxyethyl)acrylamide)/chitosan hydrogel with "repelling and killing" bacteria property. *Carbohydr. Polym.* **2019**, *225*, 115160. [CrossRef]
94. Li, R.; He, S.; Yin, K.; Zhang, B.; Yi, Y.; Zhang, M.; Pei, N.; Huang, L. Effects of N-terminal modifications on the stability of antimicrobial peptide SAMP-A4 analogues against protease degradation. *J. Pept. Sci.* **2021**, *27*, e3352. [CrossRef] [PubMed]

95. Cao, J.; He, G.; Ning, X.; Chen, X.; Fan, L.; Yang, M.; Yin, Y.; Cai, W. Preparation and properties of O-chitosan quaternary ammonium salt/polyvinyl alcohol/graphene oxide dual self-healing hydrogel. *Carbohydr. Polym.* **2022**, *287*, 119318. [CrossRef] [PubMed]
96. Gevrek, T.N.; Yu, K.; Kizhakkedathu, J.N.; Sanyal, A. Thiol-Reactive Polymers for Titanium Interfaces: Fabrication of Antimicrobial Coatings. *ACS Appl. Polym. Mater.* **2019**, *1*, 1308–1316. [CrossRef]
97. Ren, Y.; Zhou, H.; Lu, J.; Huang, S.; Zhu, H.; Li, L. Theoretical and Experimental Optimization of the Graft Density of Functionalized Anti-Biofouling Surfaces by Cationic Brushes. *Membranes* **2020**, *10*, 431. [CrossRef] [PubMed]
98. Yan, S.; Luan, S.; Shi, H.; Xu, X.; Zhang, J.; Yuan, S.; Yang, Y.; Yin, J. Hierarchical Polymer Brushes with Dominant Antibacterial Mechanisms Switching from Bactericidal to Bacteria Repellent. *Biomacromolecules* **2016**, *17*, 1696–1704. [CrossRef]
99. Deng, Z.; Li, M.; Hu, Y.; He, Y.; Tao, B.; Yuan, Z.; Wang, R.; Chen, M.; Luo, Z.; Cai, K. Injectable biomimetic hydrogels encapsulating Gold/metal–organic frameworks nanocomposites for enhanced antibacterial and wound healing activity under visible light actuation. *Chem. Eng. J.* **2021**, *420*, 129668. [CrossRef]
100. Qi, X.; Gan, J.; Zhao, Z.; Li, N.; Chen, Y.; Jin, T. Chitosan Sponge/Cu–WO_{3-x} Composite for Photodynamic Therapy of Wound Infection. *Langmuir* **2023**, *39*, 2631–2640. [CrossRef]
101. Hoque, J.; Haldar, J. Direct Synthesis of Dextran-based Antibacterial Hydrogels for Extended Release of Biocides and Eradication of Topical Biofilms. *ACS Appl. Mater. Interfaces* **2021**, *9*, 15975–15985. [CrossRef] [PubMed]
102. Yan, K.; Xu, F.; Wei, W.; Yang, C.; Wang, D.; Shi, X. Electrochemical synthesis of chitosan/silver nanoparticles multilayer hydrogel coating with pH-dependent controlled release capability and antibacterial property. *Colloids Surf. B Biointerfaces* **2021**, *202*, 111711. [CrossRef] [PubMed]
103. Xue, X.; Zhang, H.; Liu, H.; Wang, S.; Li, J.; Zhou, Q.; Chen, X.; Ren, X.; Jing, Y.; Deng, Y.; et al. Rational Design of Multifunctional CuS Nanoparticle-PEG Composite Soft Hydrogel-Coated 3D Hard Polycaprolactone Scaffolds for Efficient Bone Regeneration. *Adv. Funct. Mater.* **2022**, *32*, 2202470. [CrossRef]
104. Chandna, S.; Paul, S.; Kaur, R.; Gogde, K.; Bhaumik, J. Photodynamic Lignin Hydrogels: A Versatile Self-Healing Platform for Sustained Release of Photosensitizer Nanoconjugates. *ACS Appl. Polym. Mater.* **2022**, *4*, 8962–8976. [CrossRef]
105. Li, B.; Zhang, L.; Wang, D.; Peng, F.; Zhao, X.; Liang, C.; Li, H.; Wang, H. Thermosensitive -hydrogel-coated titania nanotubes with controlled drug release and immunoregulatory characteristics for orthopedic applications. *Mater. Sci. Eng. C* **2021**, *122*, 111878. [CrossRef]
106. Xiao, H. *Preparation and Biocompatibility of REDOX Responsive C-HA-Cys-Allicin Hydrogel Coatings*; Southwest Jiaotong University: Chengdu, China, 2021.
107. Andrade del Olmo, J.; Alonso, J.M.; Sáez-Martínez, V.; Benito-Cid, S.; Pérez-González, R.; Vilas-Vilela, J.L.; Pérez-Álvarez, L. Hyaluronic acid-based hydrogel coatings on Ti6Al4V implantable biomaterial with multifunctional antibacterial activity. *Carbohydr. Polym.* **2023**, *301*, 120366. [CrossRef] [PubMed]

Disclaimer/Publisher's Note: The statements, opinions and data contained in all publications are solely those of the individual author(s) and contributor(s) and not of MDPI and/or the editor(s). MDPI and/or the editor(s) disclaim responsibility for any injury to people or property resulting from any ideas, methods, instructions or products referred to in the content.

Journal of
Functional
Biomaterials

MDPI

Article

Cell Responses to Calcium- and Protein-Conditioned Titanium: An In Vitro Study

Qiang Zhi [1,2,3,†], Yuehua Zhang [2,3,4,†], Jianxu Wei [1,2,3], Xiaolei Lv [1,2,3], Shichong Qiao [1,2,3,*] and Hongchang Lai [1,2,3,*]

1 Department of Implant Dentistry, Shanghai Ninth People's Hospital, College of Stomatology, Shanghai Jiao Tong University School of Medicine, Shanghai 200011, China
2 National Clinical Research Center for Oral Diseases, Shanghai 200011, China
3 Shanghai Key Laboratory of Stomatology, Shanghai Research Institute of Stomatology, Shanghai 200125, China
4 Department of Orthodontics, Shanghai Ninth People's Hospital, College of Stomatology, Shanghai Jiao Tong University School of Medicine, Shanghai 200011, China
* Correspondence: shichong_qiao@139.com (S.Q.); lhc9@hotmail.com (H.L.)
† These authors contributed equally to this work.

Abstract: Dental implants have become the leading choice for patients who lose teeth; however, dental implantation is challenged by peri-implant infections. Here, calcium-doped titanium was fabricated by the combinational use of thermal evaporation and electron beam evaporation in a vacuum; then, the material was immersed in a calcium-free phosphate-buffered saline solution containing human plasma fibrinogen and incubated at 37 °C for 1 h, creating calcium- and protein-conditioned titanium. The titanium contained 12.8 ± 1.8 at.% of calcium, which made the material more hydrophilic. Calcium release by the material during protein conditioning was able to change the conformation of the adsorbed fibrinogen, which acted against the colonization of peri-implantitis-associated pathogens (*Streptococcus mutans*, UA 159, and *Porphyromonas gingivalis*, ATCC 33277), while supporting the adhesion and growth of human gingival fibroblasts (*h*GFs). The present study confirms that the combination of calcium-doping and fibrinogen-conditioning is a promising pathway to meeting the clinical demand for suppressing peri-implantitis.

Keywords: titanium; dental implant; bacterial infection; protein adsorption; peri-implantitis

Citation: Zhi, Q.; Zhang, Y.; Wei, J.; Lv, X.; Qiao, S.; Lai, H. Cell Responses to Calcium- and Protein-Conditioned Titanium: An In Vitro Study. *J. Funct. Biomater.* **2023**, *14*, 253. https://doi.org/10.3390/jfb14050253

Academic Editors: Josette Camilleri and Masaaki Nakai

Received: 17 March 2023
Revised: 20 April 2023
Accepted: 28 April 2023
Published: 1 May 2023

Copyright: © 2023 by the authors. Licensee MDPI, Basel, Switzerland. This article is an open access article distributed under the terms and conditions of the Creative Commons Attribution (CC BY) license (https://creativecommons.org/licenses/by/4.0/).

1. Introduction

Tooth loss has become an urgent problem for the aged population of China and the rest world [1], and dental implantation is now the leading choice for partial and full edentulism restoration [2]. However, the practice of dental implants is seriously affected by peri-implant infections, which are characterized by peri-implant mucositis and alveolar bone resorption, leading to implant loosening and shedding after osseointegration [2]. The prevalence of peri-implantitis for dental implants serving over 5 and 10 years is 9.6% and 26%, respectively [3]. In the Swedish population, the incidence of peri-implantitis can be even high up to 45% nine years after prosthodontics [4]. Bacterial colonization is a principal causal factor in the development of peri-implant infections, and the microbiota In peri-implantitis appears to be different from periodontal diseases. Higher counts of *Streptococcus mutans* and *Porphyromonas gingivalis* have been reported in peri-implantitis compared with periodontitis [5–7]. It is believed that the lack of a "biological seal" around the implants' neck allows greater potential for bacterial colonization and peri-implant infections [8,9]. Therefore, improvement of the implant–fibroblast interactions in the transgingival portion of dental implants is also needed to provide effective soft tissue sealing against peri-implantitis.

Various disinfection strategies have been developed in recent decades to deal with peri-implantitis. These include near-infrared light triggered nitric oxide release by N,N′-Di-sec-butyl-N,N′-dinitroso-1,4-phenylene diamine (BNN6)-loaded molybdenum disulfide nanoflowers [10]; the combinational effect of cathodic proton consumption and anodic calcium production aroused by silver/calcium (Ag/Ca) galvanics on titanium surfaces [11]; the rapid release of silver nanoparticles promoted by the double Schiff base bonds in mussel-inspired hydrogels [12]; the galvanic effect of silver nanoparticles embedded in titanium on the proton motive force in bacterial membranes [13]; the prolonged antibacterial efficacy via slow hydrolysis of the hemiaminal ether linkage in polymers and sustained release of antibacterial agents [14]; the lysis action of the Schottky barriers between silver nanoparticles and a titanium oxide support mediated by storage of the bacteria-extruded electrons during bacterial adhesion in the dark [15]; and immune-instructive pathways to restore the host's capability in bacterial clearance [16]. However, most of these designs are using toxic materials (e.g., DNA damage was found after exposure of human mesenchymal stem cells to 0.1 μg/mL of silver nanoparticles for 1, 3, and 24 h [17]; the molybdenum released by molybdenum disulfide can cause damage to the plasma membrane and inhibit efflux pump activities in mammalian cells [18]; and nitric oxide can react with DNA via multiple pathways and cause considerable damage to human cells and tissues [19]), targeting the pathogenic bacteria directly, which probably impact the implants' tissue integration because it induced undesired host responses.

The placement of a dental implant initiates numerous biochemical interactions between the implant surface and the peri-implant fluid or tissues. The first process is the wetting of blood, which serves as a source of active ingredients, such as growth factors, cytokines, and chemokines that guarantee proper tissue integration of the implant [20,21]. Therefore, the wetting of blood or its components on implant surfaces and the subsequent biological effects have become a major concern in the field of surgical procedures and implantology [22,23]. Very recently, Cao et al. demonstrated that non-bactericidal calcium could unlock the antimicrobial sequence in human fibrinogen adsorbing to titanium, leading to high inhibition efficacy against *Pseudomonas aeruginosa* [24]. Previously, calcium-doping was found to be superior in promoting bone tissue integration [25] and gingival sealing around titanium implants [26], both of which are important to the long-term stability of dental implants. More importantly, fibrinogen adsorption is an important process which mediates the adhesion and growth of cells to implant materials and is the major focus of the biomaterial communities [27,28]. However, previous studies demonstrated that fibrinogen adsorption enhanced the adhesion of *Pseudomonas aeruginosa* and *Staphylococcus epidermidis* on various biomaterials, including polyurethane, polyvinylchloride, glass, and titanium [29,30], which likely undermines the prevention of peri-implant infections. Since calcium is an essential metal to humans [31,32], and protein adsorption is an immediate phase during medical device implantations, the use of calcium for endowing implantable devices of antibacterial activity is promising for clinical applications. Nonetheless, clinical translation of such a design needs further efforts that include testing its effect on bacterial strains on mammalian cells specific to an "*intended use*".

Accordingly, the objective of present study is to test the effect of calcium doping and protein conditioning on the adhesion of pathogens and human gingival fibroblasts contributing to peri-implantitis and biological sealing, respectively. The results demonstrated that calcium- and fibrinogen-conditioned titanium has apparent activity against *Streptococcus mutans* (*Sm*, UA 159) and *Porphyromonas gingivalis* (*Pg*, ATCC 33277) colonization, but possess no side effect on human gingival fibroblasts growth.

2. Materials and Methods

2.1. Material Preparation

The pure titanium (Ti) and calcium-doped titanium (Ti-Ca) sample groups were fabricated by using a Leybold physical vapor deposition system (Univex 350, Köln, Germany), which is equipped with appendixes of thermal evaporation and an electron beam evaporation in a single vacuum chamber. As shown in Figure 1, the Ti samples were prepared by merely taking advantage of the electron beam evaporation appendix to source titanium (which has a purity of 99.99%, MaTeck Material Technologie and Kristalle GmbH, Germany), while the Ti-Ca samples were prepared by simultaneously using the thermal evaporation appendix to source calcium (which has a purity of 99.5%, MaTeck Material Technologie anda Kristalle GmbH, Jülich, Germany) and the electron beam evaporation system to deposit titanium on the substrates. The evaporation rate and the film thickness were kept at 0.3 nm/s and 200 nm, respectively. BOROFLOAT® B33 glass discs (Jena 4 H Engineering GmbH, Jena, Germany) with a diameter of 15 mm were used as substrates to support the deposited materials, i.e., titanium and calcium-doped titanium (Figure 1).

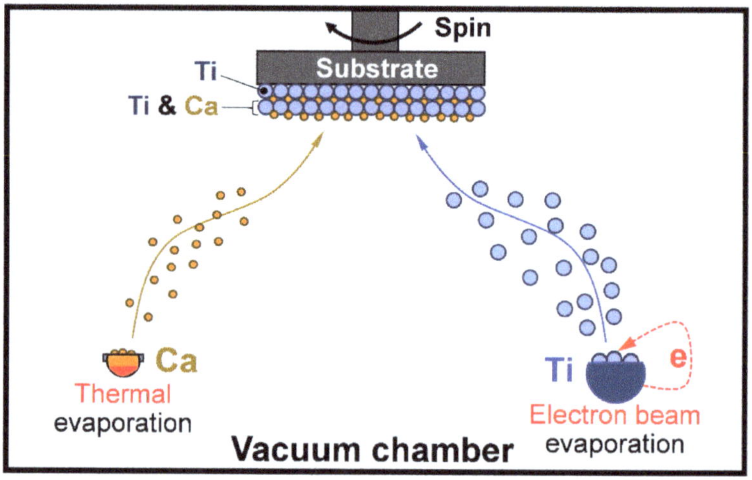

Figure 1. Illustrate the process for fabricating calcium-doped titanium by combining thermal evaporation and electron beam evaporation.

2.2. Material Characterization

Atomic Force Microscopy (AFM): A Dimension 3100 AFM system (Digital Instruments, Santa Barbara, CA, USA) was employed to examine the surface morphology of the Ti and Ti-Ca samples. The AFM system is equipped with a standard Si_3N_4 tip on a cantilever beam.

X-ray photoelectron spectroscopy (XPS): A Quantum 2000 XPS system (PHI Co., Chanhassen, MN, USA) was used to determine the chemical states of associated constituents in the material surface. The XPS system was excited with a monochromatic Al Kα source (1486.6 eV). The Multipak software supplied by the manufacturer was used for data analysis.

Contact angle: The Ti and Ti-Ca samples' wettability was measured using an SL200B system (Solon, Shanghai, China). A suspended water droplet (1 μL) was advanced toward the sample surfaces by using a microliter syringe; then, the water-material contact images were recorded by a camera and analyzed with the manufacturer-supplied software. Every group was tested three times. Statistical analysis was conducted by using the one-way analysis of variance, and all the results were presented as the mean ± standard deviation.

Inductively-coupled plasma optical emission spectrometry (ICP-OES): The samples were immersed in 10 mL calcium-free phosphate-buffered saline solution (PBS) and incubated at 37 °C for 4 h, 1 day, 3 days, 7 days, 14 days, and 28 days. The PBS was refreshed at each time point and the obtained solutions were analyzed by inductively-coupled plasma optical emission spectrometry (ICP-OES) to determine the calcium released to the solutions by the samples (in mg/L/cm^2). Every group was tested three times. Statistical analysis was conducted by using the one-way analysis of variance, and all the data were presented as the mean ± standard deviation.

Fourier-transform infrared spectroscopy (FTIR): The protein-conditioned samples (Ti-F and Ti-Ca-F) were further examined by an ALPHA-P FTIR system (Bruker, Ettlingen, Germany). The FTIR spectra were recorded in a range of 4000–400 cm^{-1} with a resolution of 8 cm^{-1}. The amide I and II regions of the spectra acquired from the Ti-Ca-F group were further analyzed by using the Fityk 1.3.1 software with a fixed half-maximum of 15.5 cm^{-1}.

2.3. Protein Conditioning

The pure titanium and calcium-doped titanium were placed in 24-well plates and rinsed with calcium-free phosphate-buffered saline solution (PBS, preheated at 37 °C) twice; then, 2.5 mL of human plasma fibrinogen (HPF, Calbiochem, EMD Chemicals, Boston, MA, USA) contained phosphate-buffered saline solution (calcium-free, HPF concentration of 1 mg/L) was added to each well. The samples together with the HPF solution were incubated at 37 °C for 1 h. After that, the samples were rinsed with calcium-free PBS (preheated at 37 °C) once and pure water (preheated at 37 °C) twice, then dried at room temperature. As a result, fibrinogen-conditioned pure titanium (Ti-F) or calcium- and fibrinogen-conditioned titanium (Ti-Ca-F) were prepared for further studies.

2.4. Responses of Mammalian Cells

Human gingival fibroblasts (*h*GFs) were obtained from gingival tissue during the dental implantation from patients aged 18 to 40 years at Shanghai Ninth People's Hospital, affiliated with Shanghai Jiaotong University School of Medicine, and all patients provided written informed consent (approved by the Medical Ethics Committee of the Ninth People's Hospital, affiliated with Shanghai Jiao Tong University School of Medicine).

The dermal tissue without epithelial layer was cut into tiny pieces and stored in Dulbecco's Modified Eagle's Medium (Thermo Fisher Scientific, Waltham, MA, USA) with 20% fetal bovine serum (Hyclone, Logan, UT, USA) and 1% penicillin/streptomycin (Gibco, Waltham, MA, USA) with 5% CO_2 and 95% air at 37 °C. Cells were passaged by trypsinization and *h*GFs from the fourth passage were used in this study.

The concentration of the suspension of the *h*GFs was adjusted to 1×10^4 cells/mL and then 1.0 mL cell suspension was seeded on the materials. After incubating at 37 °C for various durations (three samples per each group for every time point), i.e., 24 h, 72 h, and 168 h, the samples together with the adherent *h*GFs were rinsed with PBS three times and fixed with 4% paraformaldehyde for 10 min. Then the adherent cells on those samples were permeabilized with 0.5% Triton X-100 for 5 min. After that, the samples were stained with 200 μL (100 nM) phalloidin conjugated fluorescein isothiocyanate (FITC-Phalloidin, green color, Sigma, Livonia, MI, USA) for 1 h and 200 μL (100 nM) 4′,6-diamidino-2-phenylindole (DAPI, blue color, Sigma, Livonia, MI, USA) for 30 s in the dark. Then, the morphologies of the *h*GFs cells were examined under a confocal laser scanning microscope (Zeiss LSM 800, Jena, Germany) to evaluate their responses to the calcium-doped and fibrinogen-conditioned materials.

2.5. Responses of Bacterial Cells

To prepare proper bacterial suspensions, a colony of *Streptococcus mutans* (*Sm*, UA159) was cultivated at 37 °C in the brain–heart infusion broth medium with 5% CO_2; while a colony of *Porphyromonas gingivalis* (*Pg*, ATCC 33277) was grown in the brain–heart infusion broth medium with hemin (0.5 mg/mL) and menadione (10 mg/mL) in an atmosphere of 10% CO_2, 10% H_2, and 80% N_2.

Then, the responses of the bacterial cells to the materials were evaluated by seeding 10^7 CFU/mL bacterial cells on the materials and incubating them at 37 °C for 5 h (three samples per each group). After that, the samples were rinsed three times using PBS and stained with 500 μL of the combination dye of SYTO 9 and propidium iodide (PI, LIVE/DEAD BacLight bacteria viability kits; Thermo Fisher Scientific, Waltham, MA, USA) in the dark for 30 min. All the sample groups were examined with a confocal laser scanning microscopy (Zeiss LSM 800, Jena, Germany).

3. Results

3.1. Features of the Materials

The surface morphology of the materials was investigated by atomic force microscopy. As shown in Figure 2, both the pure titanium (Ti) and calcium-doped titanium (Ti-Ca) were composed of very small grains, whose size was in a range of 31–62 nm, forming a relatively smooth surface. The mean roughness (Sa) for the Ti and Ti-Ca groups was 2.1 $\pm$ 0.5 nm and 1.6 $\pm$ 0.1 nm, respectively. In addition, the maximum height (Sz) for the Ti and Ti-Ca groups was 22.8 $\pm$ 7.3 nm and 16.7 $\pm$ 3.6 nm, respectively. Although there is no significant difference in the surface roughness between the Ti and Ti-Ca groups, their wettability was significantly different. As shown by the inserts in Figure 2, the water contact angle for the Ti-Ca group was 52.5 $\pm$ 10.3 degrees, which was significantly lower than that of the Ti group (the record was 94.4 $\pm$ 1.9 degrees).

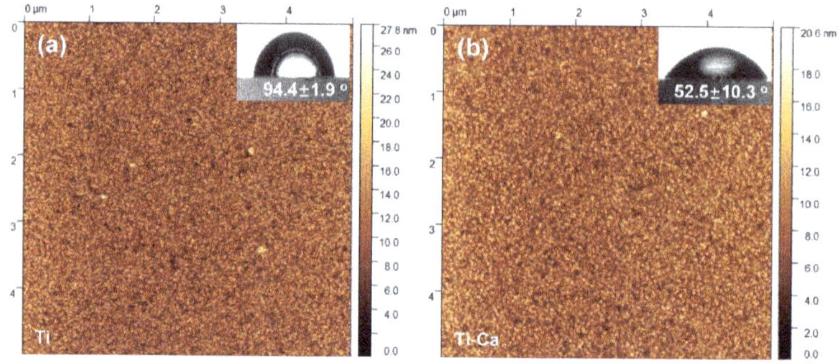

Figure 2. Atomic force microscopy (AFM) images of the pure titanium (**a**) and calcium-doped titanium (**b**) surfaces. The inserts are the corresponding water contact angle images.

The surface chemistry of the materials was further checked by using X-ray photoelectron spectroscopy (XPS). As shown by the XPS survey spectra (Figure 3a), both the Ti and Ti-Ca groups were contaminated by carbon (C), oxygen (O), and nitrogen (N); while the calcium peaks were only sharply recorded in the Ti-Ca group. The atomic concentration of the doped calcium in the Ti-Ca group determined by XPS was 12.8 $\pm$ 1.8 at.%. A high-resolution Ca2p doublet at 349.9eV and 346.3 eV was identified as the Ti-Ca group (Figure 3b), which indicates the doped calcium was oxide [33]. The cumulative calcium concentration released by the Ti-Ca group was determined by inductively-coupled plasma optical emission spectrometry (ICP-OES). As shown in Table 1, the calcium-doped titanium was able to release 3.8 $\pm$ 0.1 mg/L/cm^2 calcium, and this concentration increased to 9.7 $\pm$ 0.6 mg/L/cm^2 as the incubation duration was prolonged to 28 days.

Table 1. The cumulative release of calcium by the Ti-Ca group.

Time Point	Calcium Concentration (mg/L/cm^2)
4 h	3.8 $\pm$ 0.1
1 day	4.2 $\pm$ 0.1
3 days	4.6 $\pm$ 0.1
7 days	5.6 $\pm$ 0.2
14 days	7.1 $\pm$ 0.5
28 days	9.7 $\pm$ 0.6

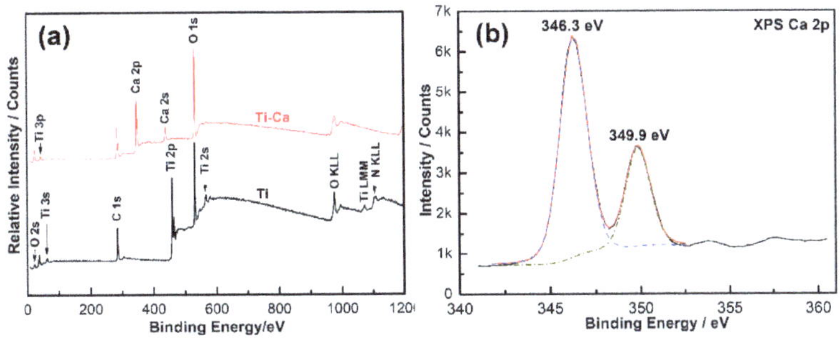

Figure 3. X-ray photoelectron spectroscopy (XPS) of the pure titanium and calcium-doped titanium surfaces: (**a**) XPS survey spectra and (**b**) XPS Ca 2p spectrum acquired from the calcium-doped titanium.

The materials fabricated by PVD were further conditioned by immersing in a calcium-free PBS containing 1mg/L of human plasma fibrinogen and incubated at 37 °C for 1 h. Then, both the protein-conditioned pure titanium (Ti-F) and calcium-doped titanium group (Ti-Ca-F) were examined by Fourier-transform infrared spectroscopy (FTIR). As shown by Figure 4a, the FTIR spectra of the fibrinogen's amide I and II bands were sharply different. The absorbance peak of the amide I band for the Ti-F group was 1677.1 cm^{-1}, which was over 36 wavenumbers higher than that of the Ti-Ca-F group (1640.3 cm^{-1}). In addition, the absorbance peak of the amide II band for the Ti-F group was 1527.0 cm^{-1}, which was over 25 wavenumbers lower than that of the Ti-Ca-F group (1652.5 cm^{-1}). Moreover, the amide II and amide I bands (which correspond to the vibrations of in-plane N-H bending and C-N stretching [34]) for the Ti-Ca-F group overlaid each other, while those for the Ti-F group did not. In addition, a band at 1594 cm^{-1}, arising from the stretching vibration of the -COO^- group [34], was identified in the deconvolution of the FTIR spectrum acquired from the Ti-Ca-F group (Figure 4b). These data indicated that the calcium released by the Ti-Ca tended to react with carboxyl groups and chelate with nitrogen in the adsorbed fibrinogen.

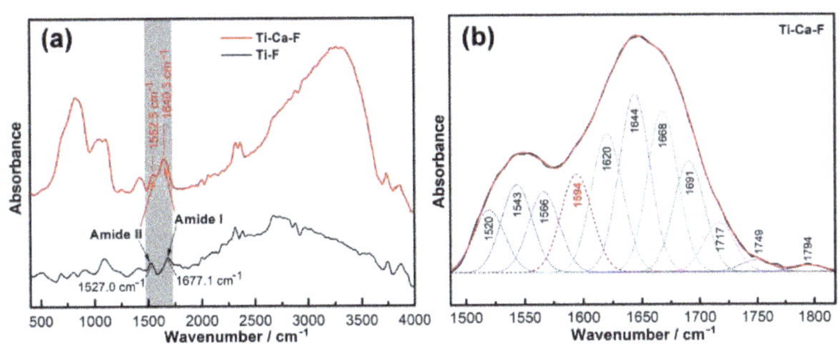

Figure 4. Fourier-transform infrared spectroscopy (FTIR) of the pure titanium and calcium-doped titanium surfaces pre-incubated in calcium-free PBS with 1 mg/L fibrinogen (**a**) and deconvolution of the amide band I and II regions of the FTIR spectrum acquired from Ti-Ca-F (**b**).

3.2. Bacterial Responses

The responses of pathogenic microbes associated with peri-implantitis were tested by seeding the *Streptococcus mutans* (UA 159) and *Porphyromonas gingivalis* (ATCC 33277) on the materials and incubated at 37 °C for 5 h. The bacterial colonization on the materials was evaluated by combinational staining with dye SYTO 9 (green, which indicates the total number of the adherent microbes) and propidium iodide (red, which indicates those bacteria of disintegrative membranes) and examined under a confocal laser scanning microscope. As to the *Streptococcus mutans* strain (Figure 5, fluorescent measurement was shown in Figure S1, Supplementary Materials), the intensity of green spots on Ti-Ca

(Figure 5c-2) was stronger than that of the Ti (Figure 5a-2), Ti-F (Figure 5b-2), and Ti-Ca-F (Figure 5d-2) groups, which is consistent with previous reports that calcium doping itself facilitates bacterial colonization and growth [11,35]. Although the intensity of green spots on the Ti, Ti-F, and Ti-Ca-F groups was comparable, the intensity of the red spots on the Ti-Ca-F (Figure 5d-1) group is remarkably stronger than that of the Ti (Figure 5a-1), Ti-F (Figure 5b-1), and Ti-Ca (Figure 5c-1) groups, indicating that the Ti-Ca-F group has significant activity against the colonization of *Streptococcus mutans* strain. As to *Porphyromonas gingivalis* species (Figure 6, Fluorescent measurement was shown in Figure S2, Supplementary Materials), the intensity of green fluorescence in all the concerned groups is comparable (Figure 6a-2,b-2,c-2,d-2); whereas the intensity of red fluorescence on the Ti (Figure 6a-1), Ti-F (Figure 6b-1), and Ti-Ca (Figure 6c-1) groups was prominently weaker than that of the Ti-Ca-F (Figure 6d-1) group. These results demonstrate that the Ti-Ca-F group has stronger antibacterial activity compared with the pure titanium groups (both the Ti and Ti-F groups) and the calcium-doped titanium group without the condition of fibrinogen (Ti-Ca).

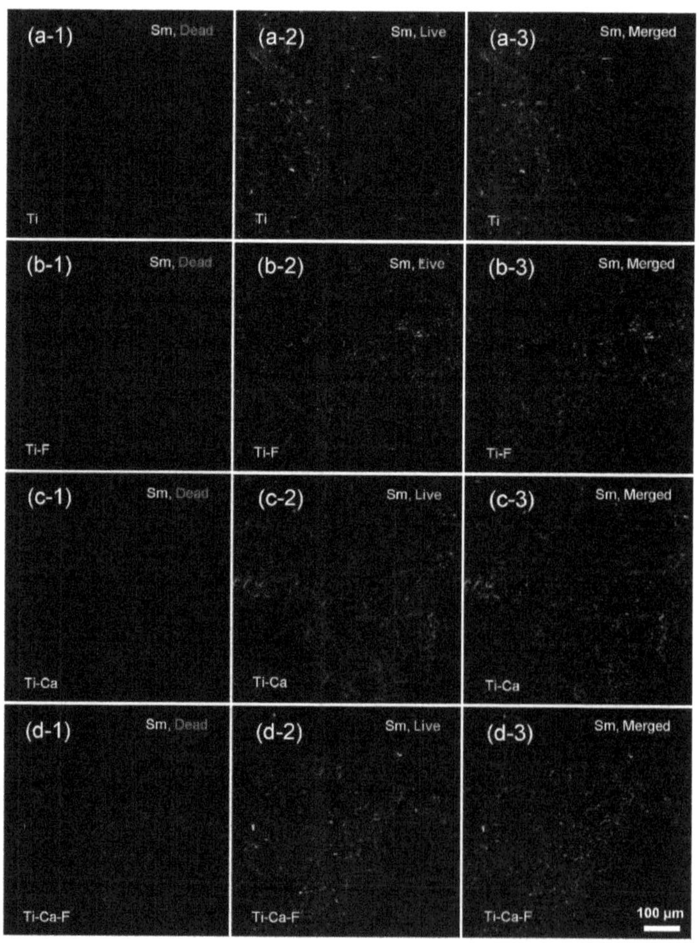

Figure 5. Fluorescent images of *Streptococcus mutans* (*Sm*, UA 159) cultured on pure titanium without (**a**-i) or with (**b**-i) protein conditioning; calcium-doped titanium without (**c**-i) or with (**d**-i) protein conditioning. (*i = 1, 2, 3 represent staining with dye of propidium iodide (PI, Red), SYTO 9 (Green), and the merged images.*)

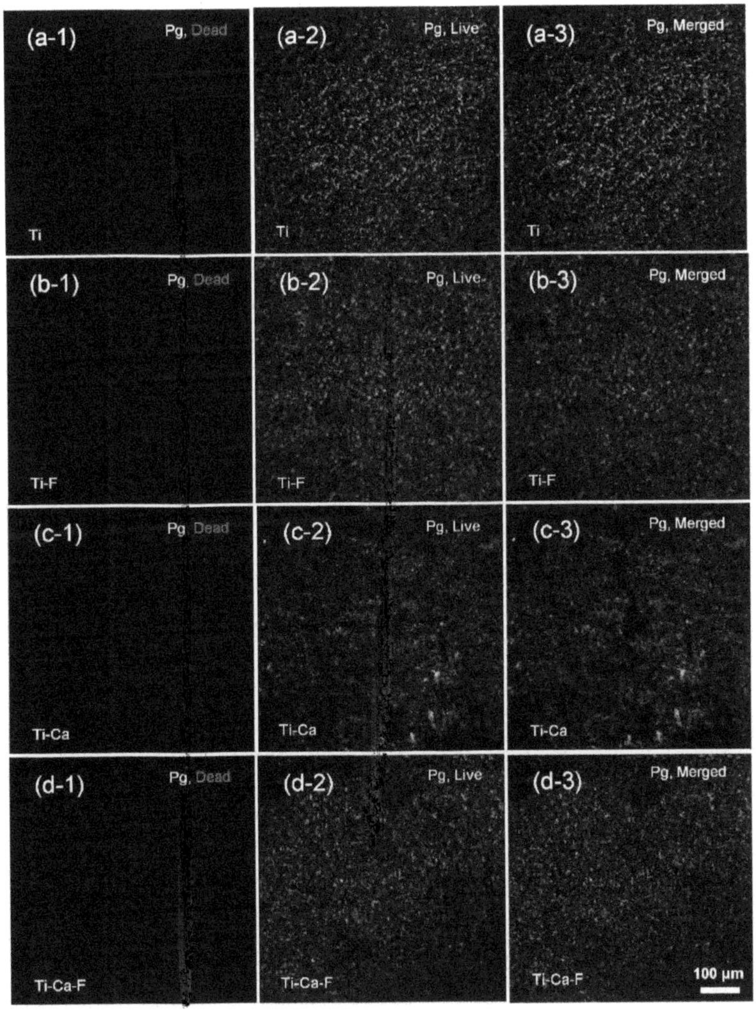

Figure 6. Fluorescent images of *Porphyromonas gingivalis* (*Pg*, ATCC 33277) cultured on pure titanium without (**a**-i) or with (**b**-i) protein conditioning; calcium-doped titanium without (**c**-i) or with (**d**-i) protein conditioning. (*i = 1, 2, 3 represent staining with dye of propidium iodide (PI, Red), SYTO 9 (Green), and the merged images.*)

3.3. Response of the hGFs

The responses of mammalian cells were tested by seeding human gingival fibroblasts on the materials and combinational staining with FITC (green, which shows the adhesion morphology of the cells) and DAPI (blue, indicates the corresponding cell nucleus); then, they were examined under a confocal laser scanning microscope after incubating at 37 °C for 24 h, 72 h, and 168 h. As shown in Figure 7 (cell number measurement was shown in Figure S3, Supplementary Materials), the fibroblasts on the Ti-F (Figure 7b-1), Ti-Ca (Figure 7c-1), and Ti-Ca-F (Figure 7d-1) groups were larger and spreading more rapidly than those on the Ti (Figure 7a-1) group after culturing the cells on the materials for 24 h. However, after culturing for 72 h, this difference in the growth of hGFs between Ti (Figure 7a-2), Ti-F (Figure 7b-2), Ti-Ca (Figure 7c-2), and Ti-Ca-F (Figure 7d-2) groups became weaker, all the sample groups demonstrated good support on the growth of hGFs

(Figure 7a-3,b-3,c-3,d-3). These results demonstrated that the Ti-Ca-F group has good compatibility with the adhesion and growth of human gingival fibroblasts.

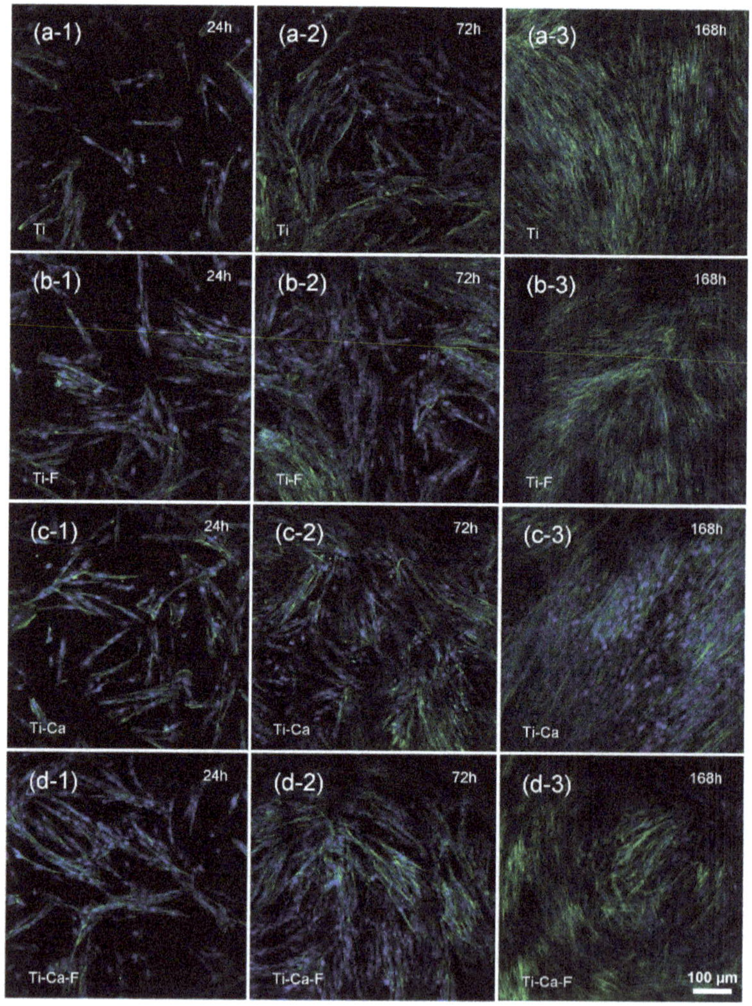

Figure 7. Fluorescent images (F-actin stained with FITC/green and nuclei stained with DAPI/blue) of human gingival fibroblasts cultured on pure titanium without (**a**-i) or with (**b**-i) protein conditioning; calcium-doped titanium without (**c**-i) or with(**d**-i) protein conditioning. (*i = 1, 2, 3 represent the culture duration of 24 h, 72 h, and 168 h, respectively*).

4. Discussion

Bacterial infection is a major complication associated with the practice of dental implants, and the development of safe and effective prevention strategies for peri-implantitis is a long-standing clinical demand in the community. Unlike most previous material designs targeting directly bacterial cells, calcium-doped titanium was very recently found capable of unlocking the activity of adsorbent fibrinogen against *Pseudomonas aeruginosa* colonization [24]. Since fibrinogen is a major player during the placement of implantable medical devices, this finding indicates a new paradigm for the development of implantable antibacterial surfaces. To verify the validity of this paradigm in dental implants, the present follow-up study tested the responses of peri-implantitis-associated pathogens, i.e., *Strepto-*

coccus mutans (*Sm*, UA 159) and *Porphyromonas gingivalis* (*Pg*, ATCC 33277), and biological sealing-related human gingival fibroblasts to calcium- and fibrinogen-conditioned titanium.

The present study successfully doped 12.8 ± 1.8 at.% of calcium into titanium (Figure 3a) by combining thermal evaporation and the electron beam evaporation systems (Figure 1). The doped calcium was relatively easy to react with water, which makes the titanium surface more hydrophilic (the insert in Figure 2) and facilitates the release of calcium (Table 1). The released calcium tends to bind to fibrinogen, which was evidenced by a new FTIR band at 1594 cm^{-1} (Figure 4b). Moreover, the calcium in titanium was oxide because the metal is highly reactive and ready to react with the oxygen in the air (Figure 3b). This chemical feature of the doped calcium likely creates a basic local pH adjacent to the titanium surface and further facilitates the deprotonation of the amino acid in fibrinogen [36,37]. As a result, remarkable shifts of the amide II and amide I were detected by using FTIR (Figure 4a). These data demonstrated that the doped calcium changed the conformation of the adsorbed fibrinogen on titanium. This conformation change in fibrinogen endows the calcium- and protein-conditioned titanium (Ti-Ca-F) with activity against *Streptococcus mutans* and *Porphyromonas gingivalis* (Figures 5 and 6), the pathogens associated with peri-implantitis, and good compatibility to the adhesion and growth of human gingival fibroblasts (Figure 7), which contribute to biological sealing.

The aforementioned antibacterial property and cytocompatibility of the calcium-doped and fibrinogen-conditioned titanium indicated a safe pathway toward solving the problem of peri-implant infections. A dental implant system has interfaces with both hard and soft tissue [38], which requires the dental implant surface to be compatible with or more soundly to promote the functions of both bone and gingival cells. Calcium has been proven effective in guaranteeing the integration of bone tissue and sealing of gum to titanium in vivo [25,26]; whereas previous studies found that calcium-doped titanium was insufficient to protect titanium from bacterial colonization in vitro [35,39], or that it even increased the adhesion of bacterial cells [40], which compromises the strategy of using calcium to build effective antibacterial titanium. The data in the present study also show that calcium doping itself has a minimal effect against the adhesion of both *Streptococcus mutans* and *Porphyromonas gingivalis* (Figures 5c-1 and 6c-1), which is consistent with those reports. However, calcium doping in combination with fibrinogen adsorption was found to be effective against the colonization of the bacterial species highly associated with peri-implantitis (Figure 5d-1 and Figure 6d-1), which was likely because of the calcium-dependent conformation changes in the protein (Figure 4). The effect of calcium on the control of the structure of fibrinogen was well-known. It was demonstrated that the binding of calcium is usually mediated by the carboxyl residues in the protein [41]. The two αC regions interacted intramolecularly with each other and with the N-termini of Bβ chains of fibrinogen so that fibrinopeptide B could be removed upon switching these intramolecular interactions of αC regions to intermolecular manners [42]. Previous studies demonstrated that calcium at a higher concentration (about 0.05 M) could turn the carboxyl terminal regions of the Aα chains in fibrinogen from an inward position to a more solvent-exposed orientation, facilitating intermolecular interactions, and the calcium uptake by human fibrinogen was parallelly correlated with the release of the fibrinopeptide B [43,44]. In addition, the effect of calcium doping on changing the conformation of adsorbed fibrinogen was evidenced by the Fourier-transform infrared spectroscopy data (Figure 4) in our study. The removal of fibrinopeptide B likely unlocked the peptide Gly-His-Arg-Pro (Gly, glycine; His, histidine; Pro, proline; Arg, arginine) at the N-terminal end of the β chain, where the antibacterial peptide Bβ15-42 was located [45], further contributing to the prevention of the bacterial colonization as shown in Figures 5 and 6.

Fibrinogen is a key player in reinforcing hemostasis and acts as the first line of host defense against bacterial infections. Virtually all tissue damages trigger local activation of the coagulation cascades, during which fibrinogen is transformed into fibrin, which accumulates outside the blood vessels within and around the impaired region of the tissue and physically prevents the invasion of pathogenic bacteria [46]. In addition, the efficacy of bacterial clearance within infected organs by the host innate immunity cells, i.e., neutrophils and macrophages, is also associated with fibrin depositions [47]. It was also well-established that fibrinogen adsorbs immediately to the surfaces of biomedical devices (or biomaterials) which come in contact with biological fluids during surgeries, and this determines the devices' further interactions with immune effector cells and associated foreign body responses [27]. Unfortunately, uncontrolled adsorption of fibrinogen to biomaterials likely induces undesirable cell responses. Many previous studies have evidenced fibrinogen denaturation while adsorbing to titanium surfaces [48–50], and taking advantage of the semiconductive nature of previously non-stoichiometrical titanium oxide films was proposed to reduce such denaturation on titanium and subsequently improved compatibility [51,52]. In the present study, the activity against bacterial adhesion for the fibrinogen adsorbed on pure titanium (the Ti-F group, Figure 5b-1 and Figure 6b-1) was inferior to that adsorbed on calcium-doped titanium (the Ti-Ca-F group, Figures 5c-1 and 6c-1). The inferiority of the Ti-F group against peri-implants-associated *Streptococcus mutans* and *Porphyromonas gingivalis* was probably the result of fibrinogen denaturation. As shown in Figure 4a, the conformation of the fibrinogen adsorbed on calcium-doped titanium (Ti-Ca-F) was different from that adsorbed on titanium.

Taken together, the data solidly confirm that the indirect antibacterial strategy originally proposed by Cao et al. [24], i.e., the combination of calcium-doping and fibrinogen-conditioning, is a promising strategy to prevent the colonization of pathogenic microbes and meet the clinical demand for suppressing peri-implantitis. The results obtained in our study demonstrated that antibacterial titanium implants do not need to target the bacterial cells directly, which may be a good pathway to designing and developing advanced surface modification processes that can properly balance the tissue integration and antimicrobial adhesion in implantable medical devices, especially dental implant systems. However, there are major limitations in the present study. The bacterial and mammalian cells' adhesion to the materials is only considered in a short time. Since adsorbed fibrin(ogen) is likely degraded in normal healing processes, the effect of the adsorbed fibrinogen on cell adhesion should be checked for longer culture durations in future studies. Moreover, to promote the clinical translation of such a strategy, more efforts should be made. These include detailed examinations on the safety and efficacy of the disinfection of dental implants by calcium doping and protein conditioning *in vivo*.

5. Conclusions

In this study, calcium- and protein-conditioned titanium was fabricated first by a duplex deposition process composed of thermal evaporation deposition and electron beam evaporation deposition in a vacuum, and then by fibrinogen adsorption in a phosphate-buffered saline solution free of calcium. The responses of peri-implantitis and biological sealing associated cells were evaluated by seeding the cells on the materials and examining their morphology and growth behaviors. The results obtained in this study confirmed that calcium can regulate the conformation of the adsorbed fibrinogen, act against bacterial colonization (*Streptococcus mutans*, UA 159 and *Porphyromonas gingivalis*, ATCC 33277), and facilitate the adhesion and growth of human gingival fibroblasts, demonstrating that calcium- and protein-conditioning is a promising strategy for tackle peri-implant infections.

Supplementary Materials: The following supporting information can be downloaded at: https://www.mdpi.com/article/10.3390/jfb14050253/s1, Figure S1: Fluorescent measurement (Integrated intensity) of the images was obtained by culturing Streptococcus mutans (UA 159) on the materials for 5 h (Figure 5). The RED BARS represent the results from images of the propidium iodide (PI, Red) channel, while the GREEN BARS represent the results from images of the SYTO 9 (Green) channel. The measurements were made by using ImageJ bundled with 64-bit Java 8. * $p < 0.05$; Figure S2. Fluorescent measurement (Integrated intensity) of the images obtained by culturing Porphyromonas gingivalis (Pg, ATCC 33277) on the materials for 5 h (Figure 6). The RED BARS represent the results from images of the propidium iodide (PI, Red) channel, while the GREEN BARS represent the results from images of the SYTO 9 (Green) channel. The measurements were made by using ImageJ bundled with 64-bit Java 8. * $p < 0.05$; Figure S3. The number of adherent fibroblasts on the materials (in an area of about 4.2 × 105 square microns). Cell counts were made according to fluorescent images (Figure 7), on which the cell nuclei were stained with DAPI/blue. * $p < 0.05$.

Author Contributions: Investigation, data curation, writing—original draft preparation, Q.Z. and Y.Z.; Conceptualization, methodology, resources, funding acquisition, supervision, writing—review and editing, S.Q., J.W., X.L. and H.L. All authors have read and agreed to the published version of the manuscript.

Funding: This research was funded by Science and Technology Commission of Shanghai Municipality (20S31906100, 16441903600).

Data Availability Statement: The data presented in this study are available on request from the corresponding authors.

Conflicts of Interest: The authors declare no conflict of interest.

References

1. Guo, J.; Ban, J.H.; Li, G.; Wang, X.; Feng, X.P.; Tai, B.J.; Hu, D.Y.; Lin, H.C.; Wang, B.; Si, Y.; et al. Status of Tooth Loss and Denture Restoration in Chinese Adult Population: Findings from the 4th National Oral Health Survey. *Chin. J. Dent. Res.* **2018**, *21*, 249–257.
2. Wu, X.; Qiao, S.; Wang, W.; Zhang, Y.; Shi, J.; Zhang, X.; Gu, W.; Zhang, X.; Li, Y.; Ding, X.; et al. Melatonin prevents peri-implantitis via suppression of TLR4/NF-κB. *Acta Biomater.* **2021**, *134*, 325–336. [CrossRef] [PubMed]
3. Cao, H.; Qiao, S.; Qin, H.; Jandt, K.D. Antibacterial designs for implantable medical devices: Evolutions and challenges. *J. Funct. Biomater.* **2022**, *13*, 86. [CrossRef]
4. Derks, J.; Schaller, D.; Hakansson, J.; Wennstrom, J.L.; Tomasi, C.; Berglundh, T. Effectiveness of implant therapy analyzed in a Swedish population: Prevalence of peri-implantitis. *J. Dent. Res.* **2016**, *95*, 43–49. [CrossRef] [PubMed]
5. Persson, G.R.; Renvert, S. Cluster of bacteria associated with peri-implantitis. *Clin. Implant. Dent. Relat. Res.* **2014**, *16*, 783–793. [CrossRef]
6. Lafaurie, G.I.; Sabogal, M.A.; Castillo, D.M.; Rincón, M.V.; Gómez, L.A.; Lesmes, Y.A.; Chambrone, L. Microbiome and Microbial Biofilm Profiles of Peri-Implantitis: A Systematic Review. *J. Periodontol.* **2017**, *88*, 1066–1089. [CrossRef] [PubMed]
7. Belibasakis, G.N.; Manoil, D. Microbial Community-Driven Etiopathogenesis of Peri-Implantitis. *J. Dent. Res.* **2021**, *100*, 21–28. [CrossRef] [PubMed]
8. Alani, A.; Bishop, K. Peri-implantitis. Part 2: Prevention and maintenance of peri-implant health. *Br. Dent. J.* **2014**, *217*, 289–297. [CrossRef]
9. Guy, S.C.; McQuade, M.J.; Scheidt, M.J.; McPherson, J.C.; Rossmann, J.A.; Van Dyke, T.E. In vitro attachment of human gingival fibroblasts to endosseous implant materials. *J. Periodontol.* **1993**, *64*, 542–546. [CrossRef]
10. Liu, L.; Pan, X.; Liu, S.; Hu, Y.; Ma, D. Near-infrared light-triggered nitric oxide release combined with low-temperature photothermal therapy for synergetic antibacterial and antifungal. *Smart Mater. Med.* **2021**, *2*, 302–313. [CrossRef]
11. Cao, H.; Tang, K.; Liu, X. Bifunctional galvanics mediated selective toxicity on titanium. *Mater. Horiz.* **2018**, *5*, 264–267. [CrossRef]
12. Hu, C.; Long, L.; Cao, J.; Zhang, S.; Wang, Y. Dual-crosslinked mussel-inspired smart hydrogels with enhanced antibacterial and angiogenic properties for chronic infected diabetic wound treatment via pH-responsive quick cargo release. *Chem. Eng.* **2021**, *411*, 128564. [CrossRef]
13. Cao, H.; Liu, X.; Meng, F.; Chu, P.K. Biological actions of silver nanoparticles embedded in titanium controlled by micro-galvanic effects. *Biomaterials* **2011**, *32*, 693–705. [CrossRef]
14. Phoungtawee, P.; Seidi, F.; Treetong, A.; Warin, C.; Klamchuen, A.; Crespy, D. Polymers with hemiaminal ether linkages for pH-responsive antibacterial materials. *ACS Macro Lett.* **2021**, *10*, 365–369. [CrossRef] [PubMed]
15. Cao, H.; Qiao, Y.; Liu, X.; Lu, T.; Cui, T.; Meng, F.; Chu, P.K. Electron storage mediated dark antibacterial action of bound silver nanoparticles: Smaller is not always better. *Acta Biomater.* **2013**, *9*, 5100–5110. [CrossRef] [PubMed]
16. Cao, H.; Qin, H.; Li, Y.; Jandt, K.D. The Action-Networks of Nanosilver: Bridging the Gap between Material and Biology. *Adv. Healthc. Mater.* **2021**, *10*, e2100619. [CrossRef]
17. Hackenberg, S.; Scherzed, A.; Kessler, M.; Hummel, S.; Technau, A.; Froelich, K.; Ginzkey, C.; Koehler, C.; Hagen, R.; Kleinsasser, N. Silver nanoparticles: Evaluation of DNA damage, toxicity and functional impairment in human mesenchymal stem cells. *Toxicol. Lett.* **2011**, *201*, 27–33. [CrossRef]
18. Liu, S.; Shen, Z.; Wu, B.; Yu, Y.; Hou, H.; Zhang, X.; Ren, H. Cytotoxicity and efflux pump inhibition induced by molybdenum disulfide and boron nitride nanomaterials with sheetlike structure. *Environ. Sci. Technol.* **2017**, *51*, 10834–10842. [CrossRef]
19. Burney, S.; Caulfield, J.L.; Niles, J.C.; Wishnok, J.S.; Tannenbaum, S.R. The chemistry of DNA damage from nitric oxide and peroxynitrite. *Mutat. Res.* **1999**, *424*, 37–49. [CrossRef]
20. Lackington, W.A.; Fleyshman, L.; Schweizer, P.; Elbs-Glatz, Y.; Guimond, S.; Rottmar, M. The response of soft tissue cells to Ti implants is modulated by blood-implant interactions. *Mater. Today Bio* **2022**, *15*, 100303. [CrossRef]
21. Xiao, L.; Ma, Y.; Crawford, R.; Mendhi, J.; Zhang, Y.; Lu, H.; Zhao, Q.; Cao, J.; Wu, C.; Wang, X.; et al. The interplay between hemostasis and immune response in biomaterial development for osteogenesis. *Mater. Today* **2022**, *54*, 202–224. [CrossRef]

22. Temelci, A.; Yılmaz, H.G.; Ünsal, G.; Uyanik, L.O.; Yazman, D.; Ayali, A.; Minervini, G. Investigation of the wetting properties of thalassemia patients' blood samples on grade 5 titanium implant surfaces: A pilot study. *Biomimetics* **2023**, *8*, 25. [CrossRef] [PubMed]
23. Scarano, A.; Rexhep, S.T.; Leo, L.; Lorusso, F. Wettability of implant surfaces: Blood vs autologous platelet liquid (APL). *J. Mech. Behav. Biomed. Mater.* **2022**, *126*, 104773. [CrossRef]
24. Cao, H.; Dauben, T.J.; Helbing, C.; Jia, Z.; Zhang, Y.; Huang, M.; Müller, L.; Gu, S.; Zhang, X.; Qin, H.; et al. The antimicrobial effect of calcium-doped titanium is activated by fibrinogen adsorption. *Mater. Horiz.* **2022**, *9*, 1962–1968. [CrossRef] [PubMed]
25. Hanawa, T.; Kamiura, Y.; Yamamoto, S.; Kohgo, T.; Amemiya, A.; Ukai, H.; Murakami, K.; Asaoka, K. Early bone formation around calcium-ion-implanted titanium inserted into rat tibia. *J. Biomed. Mater. Res.* **1997**, *36*, 131–136. [CrossRef]
26. Ayukawa, Y.; Oshiro, W.; Atsuta, I.; Furuhashi, A.; Kondo, R.; Jinno, Y.; Koyano, K. Long term retention of gingival sealing around titanium implants with $CaCl_2$ hydrothermal treatment: A rodent study. *J. Clin. Med.* **2019**, *8*, 1560. [CrossRef]
27. Horbett, T.A. Fibrinogen adsorption to biomaterials. *J. Biomed. Mater. Res. Part A* **2018**, *106*, 2777–2788. [CrossRef]
28. Szott, L.M.; Horbett, T.A. Protein interactions with surfaces: Cellular responses, complement activation, and newer methods. *Curr. Opin. Chem. Biol.* **2011**, *15*, 677–682. [CrossRef]
29. Mohammad, S.F.; Topham, N.S.; Burns, G.L.; Olsen, D.B. Enhanced bacterial adhesion on surfaces pretreated with fibrinogen and fibronectin. *ASAIO J.* **1988**, *34*, 573.
30. Vasconcelos, D.M.; Falentin-Daudre, C.; Blanquaert, D.; Thomas, D.; Granja, P.L.; Migonney, V. Role of protein environment and bioactive polymer grafting in the S. epidermidis response to titanium alloy for biomedical applications. *Mater. Sci. Eng. C* **2014**, *45*, 176–183. [CrossRef]
31. Zoroddu, M.A.; Aaseth, J.; Crisponi, G.; Medici, S.; Peana, M.; Nurchi, V.M. The essential metals for humans: A brief overview. *J. Inorg. Biochem.* **2019**, *195*, 120–129. [CrossRef]
32. Sigel, H.; Sigel, A. The bio-relevant metals of the periodic table of the elements. *Z. Nat.* **2019**, *74*, 461–471. [CrossRef]
33. Wangner, C.D.; Riggs, W.M.; Davis, L.E.; Moulder, J.F.; Muilenberg, G.E. *Handbook of X ray Photoelectron Spectroscopy: A Reference Book of Standard Spectra for Identification and Interpretation of Xps Data*; Perkin-Elmer Corporation, Physical Electronics Division: Eden Prairie, MN, USA, 1992; pp. 64–65.
34. SYu, V.; Kalnin, N.N. Quantitative IR spectrophotometry of peptide compounds in water (H_2O) solutions. I. Spectral parameters of amino acid residue absorption bands. *Biopolymers* **1990**, *30*, 1243.
35. Yoshinari, M.; Oda, Y.; Kato, T.; Okuda, K. Influence of surface modifications to titanium on antibacterial activity in vitro. *Biomaterials* **2001**, *22*, 2043–2048. [CrossRef] [PubMed]
36. Miñana, M.; Carnes, D.L., Jr.; Walker, W.A., 3rd. PH changes at the surface of root dentin after intracanal dressing with calcium oxide and calcium hydroxide. *J. Endod.* **2001**, *27*, 43–45. [CrossRef]
37. Jung, S.; Lim, S.; Albertorio, F.; Kim, G.; Gurau, M.C.; Yang, R.D.; Holden, M.A.; Cremer, P.S. The Vroman effect: A molecular level description of fibrinogen displacement. *J. Am. Chem. Soc.* **2003**, *125*, 12782–12786. [CrossRef]
38. Kim, J.C.; Lee, M.; Yeo, I.L. Three interfaces of the dental implant system and their clinical effects on hard and soft tissues. *Mater. Horiz.* **2022**, *9*, 1387–1411. [CrossRef]
39. Yoshinari, M.; Oda, Y.; Kato, T.; Okuda, K.; Hirayama, A. Influence of surface modifications to titanium on oral bacterial adhesion in vitro. *J. Biomed. Mater. Res.* **2000**, *52*, 388–394. [CrossRef]
40. Kang, S.N.; Jeong, C.M.; Jeon, Y.C.; Byon, E.; Jeong, Y.; Cho, L. Effects of Mg-ion and Ca-ion implantations on *P. gingivalis* and *F. nucleatum* adhesion. *Tissue Eng. Regen. Med.* **2014**, *11*, 39–46. [CrossRef]
41. Marguerie, G.; Chagniel, G.; Suscillon, M. The binding of calcium to bovine fibrinogen. *Biochim. Biophys. Acta* **1977**, *490*, 94–103. [CrossRef]
42. Litvinov, R.I.; Yakovlev, S.; Tsurupa, G.; Gorkun, O.V.; Medved, L.; Weisel, J.W. Direct evidence for specific interactions of the fibrinogen αC-domains with the central e region and with each other. *Biochemistry* **2007**, *46*, 9133–9142. [CrossRef] [PubMed]
43. Apap-Bologna, A.; Webster, A.; Raitt, F.; Kemp, G. The influence of calcium ions on fibrinogen conformation. *Biochim. Biophys. Acta* **1989**, *995*, 70–74. [CrossRef] [PubMed]
44. Mihalyi, E. Review of some unusual effects of calcium binding to fibrinogen. *Biophys. Chem.* **2004**, *112*, 131–140. [CrossRef] [PubMed]
45. Jennewein, C.; Tran, N.; Paulus, P.; Ellinghaus, P.; Eble, J.A.; Zacharowski, K. Novel aspects of fibrin(ogen) fragments during inflammation. *Mol. Med.* **2011**, *17*, 568–573. [CrossRef] [PubMed]
46. Luyendyk, J.P.; Schoenecker, J.G.; Flick, M.J. The multifaceted role of fibrinogen in tissue injury and inflammation. *Blood* **2019**, *133*, 511–520. [CrossRef] [PubMed]
47. Prasad, J.M.; Gorkun, O.V.; Raghu, H.; Thornton, S.; Mullins, E.S.; Palumbo, J.S.; Ko, Y.P.; Höök, M.; David, T.; Coughlin, S.R.; et al. Mice expressing a mutant form of fibrinogen that cannot support fibrin formation exhibit compromised antimicrobial host defense. *Blood* **2015**, *126*, 2047–2058. [CrossRef] [PubMed]
48. Chen, H.; Li, X.; Zhao, Y.; Li, J.; Chen, J.; Yang, P.; Maitza, M.F.; Huang, N. Construction of a multifunctional coating consisting of phospholipidsand endothelial progenitor cell-specific peptides on titaniumsubstrates. *Appl. Surf. Sci.* **2015**, *347*, 169–177. [CrossRef]
49. Chen, J.; Cao, J.; Wang, J.; Maitz, M.F.; Guo, L.; Zhao, Y.; Li, Q.; Xiong, K.; Huang, N. Biofunctionalization of titanium with PEG and anti-CD34 for hemocompatibility and stimulated endothelialization. *J. Colloid Interface Sci.* **2012**, *368*, 636–647. [CrossRef]

50. Sundgren, J.E.; Bodö, P.; Ivarsson, B.; Lundström, I. Adsorption of Fibrinogen on Titanium and Gold Surfaces Studied by ESCA and Ellipsometry. *J. Colloid Interface Sci.* **1986**, *113*, 530–543. [CrossRef]
51. Huang, N.; Yang, P.; Cheng, X.; Leng, Y.; Zheng, X.; Cai, G.; Zhen, Z.; Zhang, F.; Chen, Y.; Liu, X.; et al. Blood compatibility of amorphous titanium oxide films synthesized by ion beam enhanced deposition. *Biomaterials* **1998**, *19*, 771–776.
52. Huang, N.; Yang, P.; Leng, Y.X.; Chen, J.Y.; Sun, H.; Wang, J.; Wang, G.J.; Ding, P.D.; Xi, T.F.; Leng, Y. Hemocompatibility of titanium oxide films. *Biomaterials* **2003**, *24*, 2177–2187. [CrossRef] [PubMed]

Disclaimer/Publisher's Note: The statements, opinions and data contained in all publications are solely those of the individual author(s) and contributor(s) and not of MDPI and/or the editor(s). MDPI and/or the editor(s) disclaim responsibility for any injury to people or property resulting from any ideas, methods, instructions or products referred to in the content.

MDPI AG
Grosspeteranlage 5
4052 Basel
Switzerland
Tel.: +41 61 683 77 34

Journal of Functional Biomaterials Editorial Office
E-mail: jfb@mdpi.com
www.mdpi.com/journal/jfb

Disclaimer/Publisher's Note: The statements, opinions and data contained in all publications are solely those of the individual author(s) and contributor(s) and not of MDPI and/or the editor(s). MDPI and/or the editor(s) disclaim responsibility for any injury to people or property resulting from any ideas, methods, instructions or products referred to in the content.

www.ingramcontent.com/pod-product-compliance
Lightning Source LLC
Chambersburg PA
CBHW061826210926
51850CB00006B/815
* 9 7 8 3 7 2 5 8 2 0 1 3 9 *